# Computational Biomechanics of the Musculoskeletal System

# Computational Biomechanics of the Musculoskeletal System

Edited by
Ming Zhang and Yubo Fan

CRC Press is an imprint of the
Taylor & Francis Group, an **informa** business

First published in paperback 2024

First published 2015
by CRC Press
2385 NW Executive Center Drive, Suite 320, Boca Raton FL 33431

and by CRC Press
4 Park Square, Milton Park, Abingdon, Oxon, OX14 4RN

First issued in hardback 2019

*CRC Press is an imprint of Taylor & Francis Group, LLC*



**Library of Congress Cataloging-in-Publication Data**

Computational biomechanics of the musculoskeletal system / edited by Ming Zhang and Yubo Fan.
p.; cm.
Includes bibliographical references and index.
ISBN 978-1-4665-8803-5 (hardback : alk. paper)
I. Zhang, Ming (Professor of biomedical engineering), editor. II. Fan, Yubo, editor.
[DNLM: 1. Musculoskeletal Diseases--physiopathology. 2. Biomechanical Phenomena. 3.

Computational Biology--methods. 4. Models, Biological. 5. Musculoskeletal System--physiopathology. WE 140]

RC925.5
616.7--dc23 2014020109

ISBN: 978-1-466-58803-5 (hbk)
ISBN: 978-1-032-92047-4 (pbk)
ISBN: 978-0-429-16832-1 (ebk)

DOI: 10.1201/b17439

**Visit the Taylor & Francis Web site at**
**http://www.taylorandfrancis.com**

**and the CRC Press Web site at**
**http://www.crcpress.com**

# Contents



## SECTION I Foot and Ankle Joint



## SECTION II Knee Joint

## SECTION III Hip and Pelvis



## SECTION IV Lower Limb for Rehabilitation



## SECTION V Spine

## SECTION VI Head and Hand



## SECTION VII Bone

# Editors

**Ming Zhang** is a professor of biomedical engineering and the director of the Research Center for Musculoskeletal Bioengineering, Interdisciplinary Division of Biomedical Engineering, The Hong Kong Polytechnic University. He obtained a BSc degree in automation control engineering and an MSc degree in mechanical engineering from Beijing Institute of Technology and a PhD degree in medical engineering and physics from King's College, University of London. He is the secretary of the World Association for Chinese Biomedical Engineers (WACBE), council member of the World Council of Biomechanics, and a standing council member for the Chinese Society of Biomedical Engineering (CSBME) and the Chinese Rehabilitation Devices Association (CRDA). He has published about 200 journal papers and book chapters. His current research interests include computational biomechanics, bone biomechanics, foot biomechanics and footwear design, body support biomechanics, prosthetic and orthotic bioengineering, human motion, and body vibration analysis.

**Yubo Fan** is the dean of the School of Biological Science and Medical Engineering at Beihang University and director of the Key Laboratory for Biomechanics and Mechanobiology of the Chinese Education Ministry. He is also currently the president of the Chinese Biomedical Engineering Society (2008–), the vice president of the Chinese Strategic Alliance of Medical Device Innovation (2010–), an American Institute for Medical and Biological Engineering (AIMBE) fellow (2014), and council member of the World Council of Biomechanics. Professor Fan specializes in biomechanics, with particular interest in computational biomechanics, the biomechanical design of implanting, musculoskeletal and dental mechanics, cardiovascular fluid mechanics, biomaterials, and rehabilitation engineering. He has successfully bid for research grants for over 30 million Chinese Yuan in the past 5 years. He has published over 100 international journal papers and has supervised more than 50 graduate students including MSc, PhD, and postdoctoral fellows in the area of biomechanical engineering.

# Contributors

**Jason Tak-Man Cheung**
Li Ning Sports Science Research Center
Beijing, China

**Yan Cong**
Interdisciplinary Division of Biomedical Engineering
The Hong Kong Polytechnic University
Hong Kong SAR, China

**Cheng-fei Du**
Key Laboratory for Biomechanics and Mechanobiology of the Ministry of Education
School of Biological Science and Medical Engineering
Beihang University
Beijing, China

**Yubo Fan**
Key Laboratory for Biomechanics and Mechanobiology of the Ministry of Education
School of Biological Science and Medical Engineering
Beihang University
Beijing, China

**He Gong**
School of Biological Science and Medical Engineering
Beihang University
Beijing, China

**Lixin Guo**
School of Mechanical Engineering and Automation
Northeastern University
Shenyang, China

**Ying He**
Department of Modern Mechanics
University of Science and Technology
Hefei, China

**Winson C.C. Lee**
Interdisciplinary Division of Biomedical Engineering
The Hong Kong Polytechnic University
Hong Kong SAR, China

**Aaron Kam-Lun Leung**
Interdisciplinary Division of Biomedical Engineering
The Hong Kong Polytechnic University
Hong Kong SAR, China

**Deyu Li**
Key Laboratory for Biomechanics and Mechanobiology of the Ministry of Education
School of Biological Science and Medical Engineering
Beihang University
Beijing, China

**Qi Li**
Key Laboratory for Biomechanics and Mechanobiology of the Ministry of Education
School of Biological Science and Medical Engineering
Beihang University
Beijing, China

**Xiaoyu Liu**
Key Laboratory for Biomechanics and Mechanobiology of the Ministry of Education
School of Biological Science and Medical Engineering
Beihang University
Beijing, China

**Xuan Liu**
Shenzhen Research Institute
The Hong Kong Polytechnic University
Shenzhen, China

and

Interdisciplinary Division of Biodmedical Engineering
The Hong Kong Polytechnic University
Hong Kong SAR, China

**Zhan Liu**
Provincial Key Laboratory of Biomechanical Engineering
Sichuan University
Chengdu, Sichuan, China

**Jiong Mei**
Tongji Hospital
Tongji University School of Medicine
Shanghai, China

**Irina Mizeva**
Institute of Continuous Media Mechanics
Academy Koroleva
Perm, Russia

**Zhongjun Mo**
Key Laboratory for Biomechanics and Mechanobiology of the Ministry of Education
School of Biological Science and Medical Engineering
Beihang University
Beijing, China

and

Interdisciplinary Division of Biomedical Engineering
The Hong Kong Polytechnic University
Hong Kong SAR, China

**Ming Ni**
Tongji Hospital
Tongji University School of Medicine
and
Department of Orthopedics
Pudong New Area People's Hospital
Shanghai, China

**Wen-Xin Niu**
Tongji Hospital
Tongji University School of Medicine
and
Shanghai Key Laboratory of Orthopaedic Implants
Shanghai, China

and

Key Laboratory for Biomechanics and Mechanobiology of the Ministry of Education
School of Biological Science and Medical Engineering
Beihang University
Beijing, China

and

Interdisciplinary Division of Biomedical Engineering
The Hong Kong Polytechnic University
Hong Kong SAR, China

**Zhihui Pang**
First Affiliated Hospital of Guangzhou University of Chinese Medicine
Guangzhou, China

**Ying-li Qian**
AVIC 611 Institute
Chengdu, Sichuan, China

**Ling Qin**
Department of Orthopaedics and Traumatology
Chinese University of Hong Kong
Hong Kong SAR, China

**Hongwei Shao**
Department of Modern Mechanics
University of Science and Technology
Hefei, China

**Ting-Ting Tang**
Shanghai Key Laboratory of Orthopaedic Implants
Shanghai, China

**Yuanliang Tang**
Department of Modern Mechanics
University of Science and Technology
Hefei, China

**Ee-Chon Teo**
School of Mechanical and Aerospace Engineering
Nanyang Technological University
Singapore

**Chao Wang**
Key Laboratory for Biomechanics and Mechanobiology of the Ministry of Education
School of Biological Science and Medical Engineering
Beihang University
Beijing, China

**Lizhen Wang**
Key Laboratory for Biomechanics and Mechanobiology of the Ministry of Education
School of Biological Science and Medical Engineering
Beihang University
Beijing, China

**Yan Wang**
Interdisciplinary Division of Biomedical Engineering
The Hong Kong Polytechnic University
Hong Kong SAR, China

**Ya-wei Wang**
Key Laboratory for Biomechanics and Mechanobiology of the Ministry of Education
School of Biological Science and Medical Engineering
Beihang University
Beijing, China

**Yuxing Wang**
National Key Lab of Virtual Reality Technology
Beihang University
and
Key Laboratory for Biomechanics and Mechanobiology of the Ministry of Education
School of Biological Science and Medical Engineering
Beijing, China

**Duo Wai-Chi Wong**
Interdisciplinary Division of Biomedical Engineering
The Hong Kong Polytechnic University
Hong Kong SAR, China

**Peng Xu**
Key Laboratory for Biomechanics and Mechanobiology of the Ministry of Education
School of Biological Science and Medical Engineering
Beihang University
Beijing, China

**Jie Yao**
Key Laboratory for Biomechanics and Mechanobiology of the Ministry of Education
School of Biological Science and Medical Engineering
Beihang University
Beijing, China

**Wei Yao**
Department of Biomedical Engineering
University of Strathclyde
Glasgow, Scotland, United Kingdom

**Jia Yu**
Interdisciplinary Division of Biomedical Engineering
The Hong Kong Polytechnic University
Hong Kong SAR, China

**Hengdi Zhang**
Institute of Continuous Media Mechanics
Academy Koroleva
Perm, Russia

**Ming Zhang**
Interdisciplinary Division of Biomedical Engineering
The Hong Kong Polytechnic University
Shenzhen Research Institute
Hong Kong SAR, China

**Yi Zhang**
Chongqing Research Center for Oral Diseases and Biomedical Science
College of Stomatology
Chongqing Medical University
Chongqing, China

**Yuan-li Zhang**
Provincial Key Laboratory of Biomechanical Engineering
Sichuan University
Chengdu, Sichuan, China

# MATLAB Statement

MATLAB® is a registered trademark of The MathWorks, Inc. For product information, please contact:

The MathWorks, Inc.
3 Apple Hill Drive
Natick, MA 01760-2098 USA
Tel: 508-647-7000
Fax: 508-647-7001
Email: info@mathworks.com
Web: www.mathworks.com

# *Section I*

## *Foot and Ankle Joint*

# 1 Foot Model for Investigating Foot Biomechanics and Footwear Design

*Ming Zhang, Jia Yu, Yan Cong, Yan Wang, and Jason Tak-Man Cheung*

## CONTENTS



## SUMMARY

Information on load transfer of foot internal structures as well as the foot-support interface during various activities is useful in enhancing biomechanical knowledge for foot support design. Modeling of the human foot and ankle is challenging because of the very complex structures. Computational models can be used to understand joint biomechanics and design proper foot supports. Three-dimensional geometrically accurate finite element (FE) models of the human foot-ankle structures were developed from reconstructed magnetic resonance (MR) images. The foot FE model consists of 28 separate bones, 72 ligaments, and the plantar fascia, embedded in a volume of encapsulated soft tissue. The main bone interactions were simulated as contacting deformable bodies. The analyses took into consideration the nonlinearities of material properties, large deformations, and interfacial slip/friction conditions. A series of experiments on human subjects and cadavers were conducted to validate the model measurements. These experiments recorded plantar pressure distribution, foot arch and joint motion, and plantar fascia strain under different simulated weight-bearing and orthotic conditions of the foot. The validated models can be used for parametric studies to investigate the biomechanical effects of tissue stiffness, and orthotic performances on the foot-ankle complex.

## 1.1 INTRODUCTION

The human foot is a complicated structure comprising 28 bones (Figure 1.1) and numerous muscles, ligaments, and other connective tissues. To facilitate the foot in weight bearing, locomotion, and support, man has clad the foot in a variety of footwear. Proper footwear should provide a comfortable environment for the foot, protect the foot from injury, and enable the foot to function in support and locomotion over various surfaces.

Progressing from a simple external support, attention has shifted to using footwear to alleviate foot problems such as hypermobility, limited movement, rigidous, degeneration, deformities, ulceration, soft tissue injuries, and bone fractures. With an aging population and high rates of obesity, proper fitting of footwear is also becoming increasingly important.

A clear picture of the biomechanical behavior of the foot and ankle structures with supports under loading can help to demonstrate the rationale behind foot support design. Proper fitting of footwear is influenced by many factors relating to the foot itself (shape, structure, static and dynamic performance), footwear (size, shape, materials, insole/outsole design), and environmental parameters. The biomechanical fundamentals of footwear fit and comfort are not fully understood.

Experimental techniques, such as in vivo motion analysis, force platforms, in-shoe pressure measurements, and cadaveric experiments were developed for the quantification of foot biomechanics and

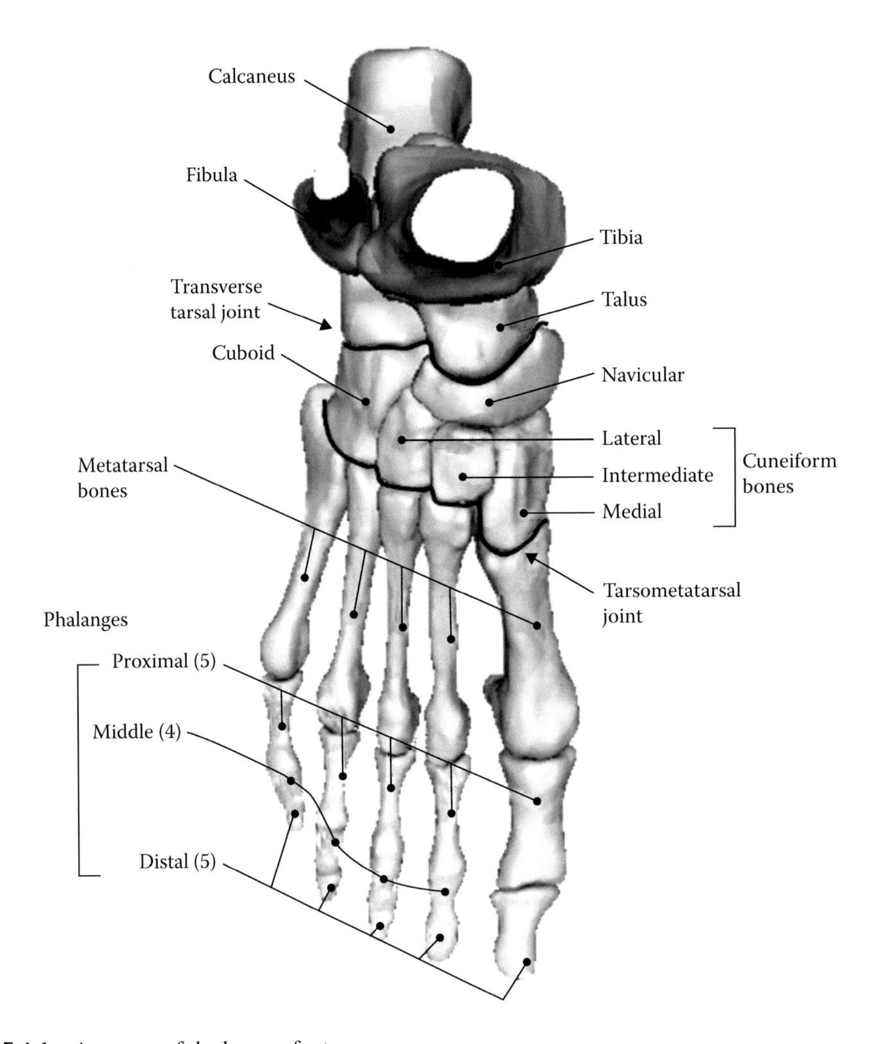


**FIGURE 1.1** Anatomy of the human foot.

are commonly used in predicting joint motion and quantifying plantar pressure distributions. However, the load transfer mechanism and the states of internal stress within the soft tissues and the bony structures remain unaddressed due to the difficulties and limitations of the experimental approach. The functional role of different anatomical components on load distribution and stabilizing ability depends merely on gross pressure distribution, as recorded from experimental measurements. Experimental measurements are time-consuming and would need to be conducted on a significant number of patients or specimens with different characteristics to yield generalized and promising results.

Computational models based on FE methods have been used increasingly in many biomechanical investigations with great success due to the capability of modeling structures with irregular geometry and complex material properties, and varying boundary and loading conditions. Computational modeling can be easily used to vary factors, including geometrical features such as the shapes of articular surfaces and the arrangement of ligaments, mechanical properties of the bones and soft tissues, muscle forces, external loads, and different supporting conditions.

A number of FE models were developed to investigate foot biomechanics and a few studies provided biomechanical information for footwear design (Lemmon et al. 1997; Chen, Ju, and Tang 2003; Verdejo and Mills 2004; Cheung and Zhang 2008). The simplified models available for stress-strain analyses were either two-dimensional (Nakamura, Crowninshield, and Cooper 1981; Lewis 2003; Lemmon et al. 1997; Erdemir, Saucerman, and Lemmon 2005; Verdejo and Mills 2004; Goske et al. 2006; Spears et al. 2007) or partial three-dimensional foot skeleton or connected bony structure (Chu, Reddy, and Padovan 1995; Chen, Ju, and Tang 2003). More geometrically accurate three-dimensional FE models were developed later (Gefen et al. 2000; Cheung and Zhang 2005). Successful FE analyses (Chu, Reddy, and Padovan 1995; Chen, Ju, and Tang 2003; Cheung et al. 2005) have been carried out on insoles and ankle-foot orthoses. A comprehensive review of the development of foot FE models was undertaken in 2009 (Cheung et al. 2009). In the last five years, the number of FE studies on the foot and its support has increased remarkably, as summarized in Table 1.1. FE modeling, if conducted properly, could potentially make significant contributions to the understanding of foot biomechanics and improvements in footwear design.

**TABLE 1.1**
**Finite Element Models and Applications in the Literature since 2009**

| Reference | Geometrical Properties | Parameters of Interest |
|---|---|---|
| Garcia-Aznar et al. (2009) | Three-dimensional (3D), computerized tomography (CT) | Load transfer mechanism of different metatarsal geometries |
| Garcia-Gonzalez et al. (2009) | images (foot bones) | The effect of tendon transfer on dorsal displacement to claw toe deformity |
| Bayod et al. (2010) | | Comparison of proximal interphalangeal joint fusion and flexor tendon transfer |
| Bayod et al. (2012) | | Effect of bone graft harvesting on calcaneus |
| Halloran, Erdemir, and van den Bogert (2009) | Two-dimensional (2D; bone, encapsulated soft tissue) | Demonstration of surrogate modeling system in comparison of finite element analysis |
| Halloran et al. (2010) | | Alteration of movement to soft tissue loading |
| Chen et al. (2010) | 3D, CT images (foot bones, encapsulated soft tissue) | Effect of internal stress concentration to plantar soft tissue |
| Chen et al. (2012) | | Effect of muscle forces on forefoot pressure, metatarsophalangeal, and ankle joint motion |
| Gu et al. (2010a) | 3D, CT, and magnet resonance (MR) | Different skin stiffness at heel-strike |
| Gu et al. (2010b) | images (hindfoot bones, encapsulated soft tissue, fat | Different inversion landing angles on forefoot pressure and metatarsal stress |
| Gu et al. (2011) | pad, skin) | Different foot support to metatarsal loading |

*Continued*

**TABLE 1.1** ***(Continued)***
**Finite Element Models and Applications in the Literature since 2009**

| Reference | Geometrical Properties | Parameters of Interest |
|---|---|---|
| Jamshidi, Hanife, and Rostami (2010) | 3D, MR images (bones, encapsulated soft tissue) | Different thickness and materials of ankle-foot orthosis on tension in bone and muscle |
| Matzaroglou et al. (2010) | 2D (first metatarsal) | Comparison of compressive stress of chevron osteotomy |
| Tao et al. (2010) | 3D, MR images (bones, encapsulated soft tissue) | Arch height, bone shift, and rotation after simulated plantar ligament release |
| Liang et al. (2011) | 3D, CT images (foot bones and encapsulated soft tissue) | Planar pressure and stress of bones after plantar fascia release |
| Luo et al. (2011) | 3D, MR images (calcaneus and heel pad tissue) | Effect of insole design and material to heel pad |
| Qiu et al. (2011) | 3D, CT images (foot bone, encapsulated soft tissue) | Demonstration of foot-boot model platform |
| Sopher et al. (2011) | 3D, data from Visible Human Project (calcaneus, heel fat pad, and skin) | Influence of foot posture, support stiffness, heel pad loading, and tissue properties to ulceration |
| Spyrou and Aravas (2012) | 3D, CT images (foot bones, encapsulated soft tissue) | Muscle-driven simulation |
| Sun et al. (2012) | 3D, CT images (foot bones, encapsulated soft tissue) | Three arch heights<br>Different plantar fascia strain on von Mises stress of bones |
| Brilakis et al. (2012) | 3D, CT images (bone, encapsulated tissue) | Principal strain and von Mises stress on the fifth metatarsal after fracture in different loading conditions |
| Chokhandre et al. (2012) | 3D, CT images of cadaver (foot bones, soft tissue) | Heel pad material property by inverse finite element analysis |
| Fontanella et al. (2012)<br>Forestiero et al. (2012) | 3D, MR images (fat pad, calcaneus soft tissue on heel region, ankle joint, and ligaments) | Displacement and compressive stress in compression test<br>Displacement and compressive stress with barefoot and different midsole/outsoles<br>Comparison of load-displacement results with cadaveric experiments, maximum principal stress, displacement |
| Isvilanonda et al. (2012) | 3D, CT images of cadaver (foot bones, encapsulated soft tissue) | Postulating claw toe deformity by applying different muscle forces.<br>Simulated modified Jones procedure and flexor hallucis longus transfer, and assessing metatarsal head pressure |
| Kim et al. (2012) | 3D, MR image (foot bones, cartilage, ligament, shoe) | Ground reaction force, acceleration transfer, and frequency response on different sports ground |
| Shin, Yue, and Untaroiu (2012) | 3D, CT, and MR images (foot bones, encapsulated tissue) | Force-displacement and torque-angle response until failure |
| Xu et al. (2012) | 3D, MR images (rear-foot bones) | Various tenodesis procedures on ankle flexibility and ligaments stress |
| Luboz et al. (2012) | 3D, CT images (foot bones, soft tissue) | Plantar pressure with orthoses |
| Liu and Zhang (2013) | 3D, MR images (knee-ankle-foot complex) | Laterally wedged insole support to knee joint |

The development of comprehensive computational models of the human foot was suggested to be one of the most important directions for future research in podiatric biomechanics (Kirby 2001). In reviewing models developed so far for the foot and ankle, improvements may be made in the following aspects:

- Accurate representation of the complicated three-dimensional geometries of the human foot and ankle, incorporating complete ligamentous and articulating features, as well as support.
- Bone interactions and their relative movement.
- Proper mechanical properties for various bodies.
- Accurate boundary and loading conditions, including joint and muscle forces under various activities.
- Model validation with help of experimental studies.

## 1.2 DEVELOPMENT OF A COMPREHENSIVE FOOT MODEL

A flowchart of FE model development, validation, and applications is shown in Figure 1.2. The normal procedures for FE model development and validation involve data acquisition, parameter extraction, model processing and post-processing, and analysis.

### 1.2.1 Geometric Model of Human Foot and Support

The geometry of the human foot and ankle for building the FE model was obtained from three-dimensional reconstruction of coronal MR images taken at 2 mm intervals from the right foot of a normal male subject of age 26, height 174 cm, and weight 70 kg in the neutral foot position. A custom ankle-foot orthosis fabricated during upright sitting was used to maintain the neutral foot position of the supine subject during MR scanning.

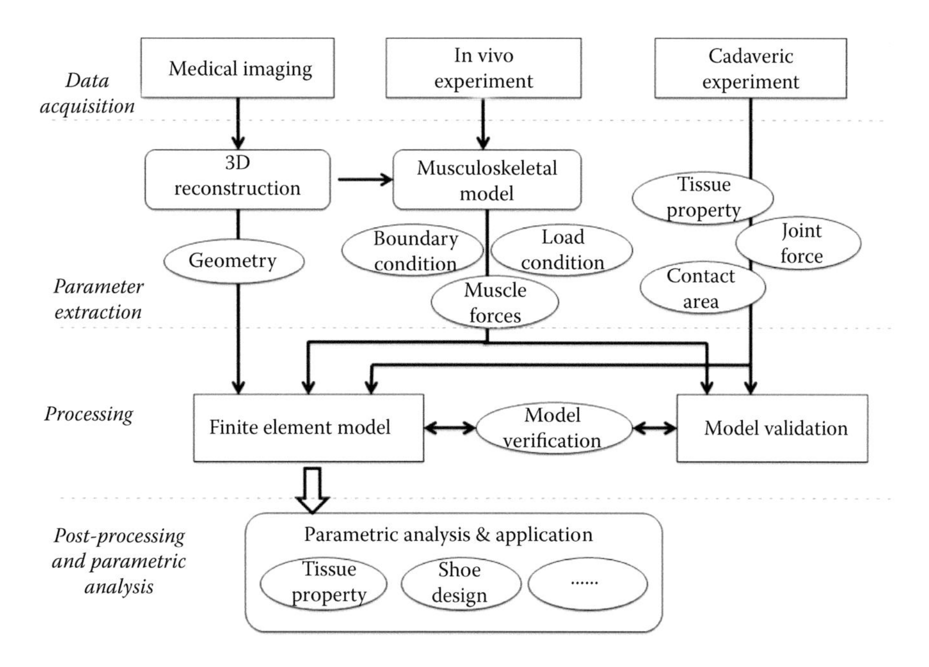


**FIGURE 1.2** Flowchart of the finite element model and validation.

The MR images were segmented using MIMICS (Materialise, Leuven, Belgium) to obtain the tissue boundaries. For the sake of simplification, the articular cartilages of the bones are fused with their corresponding bone surfaces in the segmentation process. The boundary surfaces of the skeletal and skin components were processed using SolidWorks (SolidWorks Corporation, Massachusetts) to form solid models for each bone and the whole foot surface. The solid model was then imported and assembled in the FE package, ABAQUS (Simulia, United States).

The FE model of the human foot and ankle consisted of 28 bony segments, including the distal segments of the tibia and fibula and 26 foot bones, namely, talus, calcaneus, cuboid, navicular, 3 cuneiforms, 5 metatarsals, and 14 components of the phalanges embedded in a volume of encapsulated bulk soft tissue (Figure 1.3). The phalangeal bones were connected together and spaced by 2 mm using solid elastic elements, which represented the thickness of the articulating cartilage layers and simulated the connection of the cartilage and other connective tissues. The interaction among the metatarsals, cuneiforms, cuboid, navicular, talus, calcaneus, tibia, and fibula were defined as contacting elastic bodies to allow the simulation of relative bone movement.

A total of 72 ligaments and the plantar fascia were included and defined by connecting the corresponding attachment points on the bones. Information on the attachment regions of the ligamentous structures was obtained from the Interactive Foot and Ankle (Primal Picture Ltd., London, 1999). The attachment points of the major plantar ligamentous structures such as the plantar fascia, long plantar ligament, short plantar ligament, and spring ligament are depicted in Figure 1.3. The number of attachment points defined for individual ligaments depended on the width of the ligamentous structures. For instance, the plantar fascia was divided into five rays of separate sections, linking the insertions between the calcaneus and the metatarsophalangeal joints (Figure 1.3). The plantar ligaments and spring ligament were defined by three and two rays of separate sections, while only a single ray was defined for small ligaments. The attachment points were defined close to the geometrical center of the insertion regions of the ligamentous structures. All the bony and ligamentous structures were embedded in a volume of soft tissue.

A variety of solids elements in the ABAQUS package can be used to model the foot and ankle structures. The bony and encapsulated soft tissue structures were meshed into four-noded

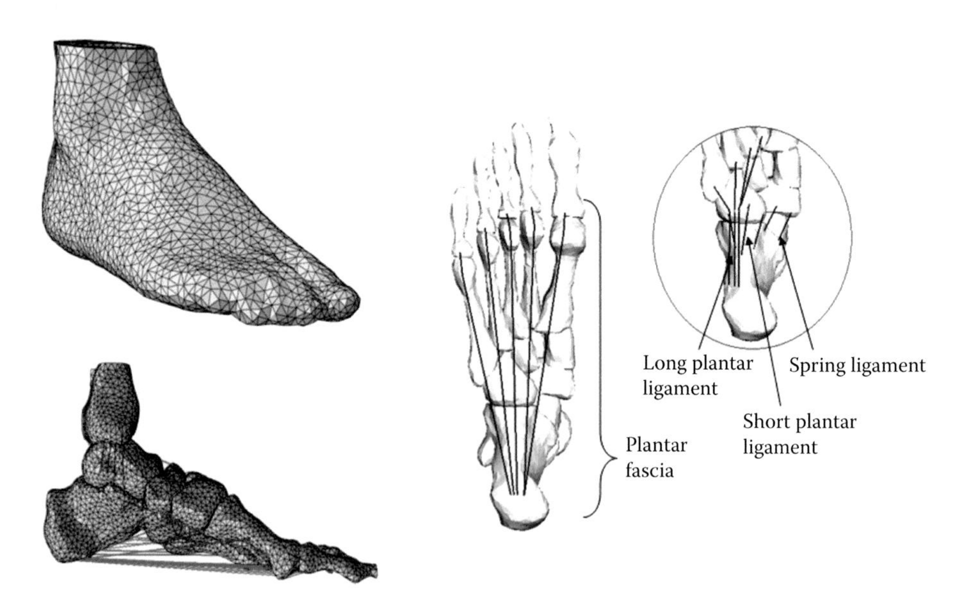


**FIGURE 1.3** The finite element (FE) meshes of the encapsulated soft tissue, bony structures in the lateral view, and the attachment points of the plantar fascia, spring ligaments, and long and short plantar ligaments of the FE model.

tetrahedral elements. For the sake of geometrical simplification, truss elements were chosen to model the ligaments. Truss elements are typically designated to model slender, line-like structures that can only transmit force along the axis or the center line of the element. Truss elements cannot resist loading perpendicular to their axis. The distance between the two connecting nodes defines the length of each truss element and the cross-sectional area is specified by the user. As the ligaments were assumed to sustain tensile force only, the No-Compression option in ABAQUS was used to modify the elastic behavior of the material so that compressive stress could not be generated. In the current FE model, a total number of 98 tension-only truss elements were used to represent the ligaments and the plantar fascia.

To simulate the surface interactions among the bony structures, ABAQUS's automated surface-to-surface contact algorithm was used. A pair of contacting surfaces consisted of a master and a slave surface. Because of the lubricating nature of the articulating surfaces, the contact behavior between the articulating surfaces can be considered as frictionless. The overall joint stiffness against shear loading was assumed to be governed by the surrounding ligamentous and encapsulated soft tissue structures together with the contacting stiffness between the adjacent contoured articulating surfaces. Frictionless surface-to-surface contact behavior was defined between the contacting bony structures. Contact stiffness resembling the softened contact behavior of the cartilaginous layers (Athanasiou et al. 1998) was prescribed between each pair of contact surfaces to simulate the covering layers of articular cartilage.

To simulate a barefoot stance, a horizontal plate consisting of an upper concrete layer and a rigid bottom layer was used to establish the foot-ground interface. The horizontal ground support was meshed with hexahedral elements. The same contact modeling algorithm was used to establish the contact simulation of the foot-ground interface with an additional frictional property assigned to model the frictional contact behavior at the foot-support interface. During the contact phase, sliding was allowed only when the shear stress exceeded the critical shear stress value. During the sliding phase, if the shear stress was reduced and lower than the critical shear stress value, sliding stopped. It was assumed that the static and kinetic coefficients of friction were the same in this model. The coefficient of friction between the foot and ground was taken as 0.6 (Zhang and Mak 1999).

The geometry of the foot orthosis was obtained from the shape of the subject's bare foot. The three-dimensional foot shape of the subject was obtained from surface digitization via a three-dimensional laser scanner (INFOOT Laser Scanner, I-Ware Laboratory Co. Ltd.). The foot shape was obtained under three different weight-bearing conditions: single-limb standing (full-weight-bearing), double-limb standing (semi-weight-bearing), and upright sitting (non-weight-bearing). Algorithms were established in MATLAB® software (Mathworks, Inc.) to create surface models for the insole and midsole from the digitized foot surface for each scanning position. The surface models were transferred to SolidWorks software (SolidWorks Corporation, Massachusetts) for creation of solid models of variable thicknesses.

The solid model of the foot orthosis was then imported into ABAQUS for meshing. In order to enhance the accuracy of the FE analysis, the foot orthosis was properly partitioned for meshing with hexahedral elements. The FE mesh of the foot orthosis (Figure 1.4) is composed of an insole layer, a midsole layer, and an outsole layer. The same frictional contact modeling approach was used to establish the contact simulation of the foot-insole interface.

### 1.2.2 Material Properties

A number of material property models can be used, from the simplest linear elasticity to nonlinear elasticity and even viscoelasticity. The material properties used in this model are listed in Table 1.2. To reduce the complexity and the size of the problem, except for the encapsulated soft tissue, all tissues, including bony, ligamentous, and cartilaginous structures and ground supports, were idealized as homogeneous, isotropic, and linearly elastic. The linearly elastic properties can be defined by providing any two constants: Young's modulus $E$, shear modulus $G$, and Poisson's ratio $\nu$.

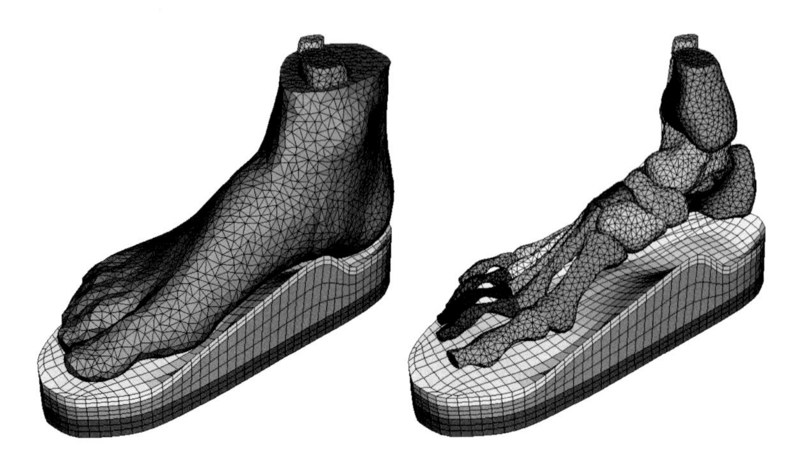

**FIGURE 1.4** **(See color insert.)** Foot and foot orthosis model.

**TABLE 1.2**
**Material Properties and Element Types Defined in the Finite Element Model**

| Component | Element Type | Young's Modulus *E* (MPa) | Poisson's Ratio *v* | Cross-sectional Area (mm²) |
|---|---|---|---|---|
| Bony structures | 3D-Tetrahedra | 7,300 | 0.3 | – |
| Encapsulated soft tissue | 3D-Tetrahedra | Hyperelastic | – | – |
| Cartilage | 3D-Tetrahedra | 1 | 0.4 | – |
| Ligaments | Tension-only Truss | 260 | – | 18.4 |
| Fascia | Tension-only Truss | 350 | – | 58.6 |
| Ground Support | 3D-Brick | 17,000 upper layer 1,000,000 lower layer | 0.1 | – |

*Sources:* Bones (Nakamura, Crowninshield, and Cooper, *Bulletin Prosthetics Research*, 18, 27–34, 1981); cartilage (Athanasiou et al., *Clinical Orthopaedics and Related Research*, 348, 269–81, 1998); ligaments (Siegler, Block, and Schneck, *Foot & Ankle*, 8, 234–42, 1988); plantar fascia (Wright and Rennels, *Journal of Bone and Joint Surgery, American Volume*, 46, 482–92, 1964).

The Young's modulus and Poisson's ratio for the bony structures were assigned as 7300 MPa and 0.3, respectively (Nakamura, Crowninshield, and Cooper 1981). The Young's modulus and Poisson's ratio for foot bony structures were obtained by averaging the elasticity values of cortical and trabecular bones in terms of their volumetric contribution. The Young's modulus of the cartilage (Athanasiou et al. 1998), ligaments (Siegler, Block, and Schneck 1988), and the plantar fascia (Wright and Rennels 1964) were selected from the literature. The cartilage was assigned a Poisson's ratio of 0.4 for its nearly incompressible nature. The ligaments and the plantar fascia were assumed to be incompressible. The encapsulated soft tissue of the FE model was defined as nonlinearly elastic. The stress-strain data on the plantar heel pad were adopted from in vivo ultrasonic measurements (Lemmon et al. 1997) to represent the stiffness of the encapsulated soft tissue.

ABAQUS offers a hyperelastic material model to simulate highly incompressible, elastic materials. The hyperelastic material model defined in ABAQUS is isotropic and nonlinear, which is especially useful in representing materials that exhibit instantaneous elastic response up to large strains, such as rubber and bulk soft tissue. Given isotropy and additive decomposition of the deviatoric and volumetric strain energy contributions in the presence of incompressible or almost incompressible behavior, the strain energy potential can be represented with a polynomial expression. A second-order polynomial strain energy potential (ABAQUS) was adopted with the form

$$U = \sum_{i+j=1}^{2} C_{ij}(\bar{I}_1 - 3)^i(\bar{I}_2 - 3)^j + \sum_{i=1}^{2} \frac{1}{D_i}(J_{el} - 1)^{2i} \tag{1.1}$$

where U is the strain energy per unit of reference volume; $C_{ij}$ and $D_i$ are material parameters; $\bar{I}_1$ and $\bar{I}_2$ are the first and second deviatoric strain invariants, defined as

$$\bar{I}_1 = \bar{\lambda}_1^2 + \bar{\lambda}_2^2 + \bar{\lambda}_3^2 \tag{1.2}$$

$$\bar{I}_2 = \bar{\lambda}_1^{(-2)} + \bar{\lambda}_2^{(-2)} + \bar{\lambda}_3^{(-2)} \tag{1.3}$$

with the deviatoric stretches $\bar{\lambda}_i = J_{el} - 1/3\ \lambda_i$ and $J_{el}$ and $\lambda_i$ are the elastic volume ratio and the principal stretches, respectively.

The coefficients of the hyperelastic model extracted from the soft tissue experimental data are 0.08556, −0.05841, 0.03900, −0.02319, 0.00851, 3.65273, and 0.0 for C10, C01, C20, C11, C02, D1, and D2, respectively.

Foot orthoses or insoles made of rubber-like elastomeric foam are elastic but compressible and can be modeled as hyperelastic. Common examples of elastomeric foam materials are cellular polymers for cushions, padding, and packaging materials that utilize the excellent energy absorption properties of foams. Three distinct stages can be distinguished during compression of elastomeric foam. Within small strains of about 5%, the foam deforms in a linear elastic manner due to cell wall bending. The next stage is a plateau of deformation at almost constant or slowly increasing stress caused by the elastic buckling of cell edges or walls. Finally, a region of densification occurs, where the cell walls crush together, resulting in a rapid increase of compressive stress. The ultimate compressive nominal strain usually ranges from 0.7 to 0.9.

### 1.2.3 Loading and Boundary Conditions

In order to simulate the physiological loading, accurate ground reaction and muscle forces should be applied. The major active muscles can be applied by adding force vectors with respect to their lines of action. The muscle and joint forces can be determined using inverse dynamics based on the kinematic and kinetic data obtained by any motion analysis system and force platform. Forces from the major muscles can be obtained from either musculoskeletal models (Anybody, OpenSim) or with the help of electromyography (EMG) and optimization (Rohrle et al. 1984; Glitsch and Baumann 1997).

Simulating a subject with a body mass of 70 kg, a vertical force of approximately 350 N was applied on each foot during balanced standing. The standing line of gravity was about 6 cm in front of the ankle (Opila et al. 1988). Therefore, the plantar flexors act to balance the forward moment of the body about the ankle in order to achieve an equilibrium balanced standing position. The triceps surae provides the major stabilization of the foot during balanced standing, with minimal reaction from all other intrinsic and extrinsic muscles. Therefore, only the Achilles tendon loading was considered during simulated balanced standing, while other intrinsic and extrinsic muscle forces were neglected.

For simulated balanced standing, force vectors, corresponding to half of the body weight, and the reaction of the Achilles tendon were applied. Five equivalent force vectors representing the Achilles tendon tension were applied at the points of insertion by defining contraction forces via five axial connector elements. The ground reaction force was applied as a concentrated force underneath the ground support.

While simulating the stance phase of gait, ground reaction and the active extrinsic muscle forces were applied. Geometrical information on the muscular insertion points was obtained from the Interactive Foot and Ankle. The extrinsic muscle forces during midstance were estimated from the physiological cross-sectional area (PCSA) of the muscles (Dul 1983) and normalized EMG data during normal walking (Perry 1992), assuming a linear EMG-force curve with a muscle gain of 25 N/cm$^2$ (Kim, Kitaoka, and Luo 2001). Fine adjustments of the applied muscle forces were made to match the measured center of pressure of the subject. Musculotendon forces were applied at their corresponding points of insertion by defining contraction forces via axial connector elements. Again, the ground reaction force was applied as a concentrated force underneath the ground support, and the superior surface of the soft tissue, distal tibia, and fibula was fixed throughout the analysis.

### 1.2.4 Model Validation

Model validation is extremely challenging, and is normally carried out through comparison of FE prediction results with experimental measurements. Experimental information from cadaveric studies is typically limited to either in vivo plantar pressure distribution or joint contact forces. Figure 1.5 shows the plantar pressure distributions obtained from FE prediction and Tekscan measurement.

The model predicted peak pressures of about 0.23 MPa at the heel region, while the corresponding actual peak pressures were about 0.18 MPa. The predicted plantar pressure distribution pattern was in general comparable to the experimental measurements. Relatively large pressure was measured beneath the fifth metatarsal head, with peak values lower than those predicted. Figure 1.6 shows the von Mises stress distribution on bone.

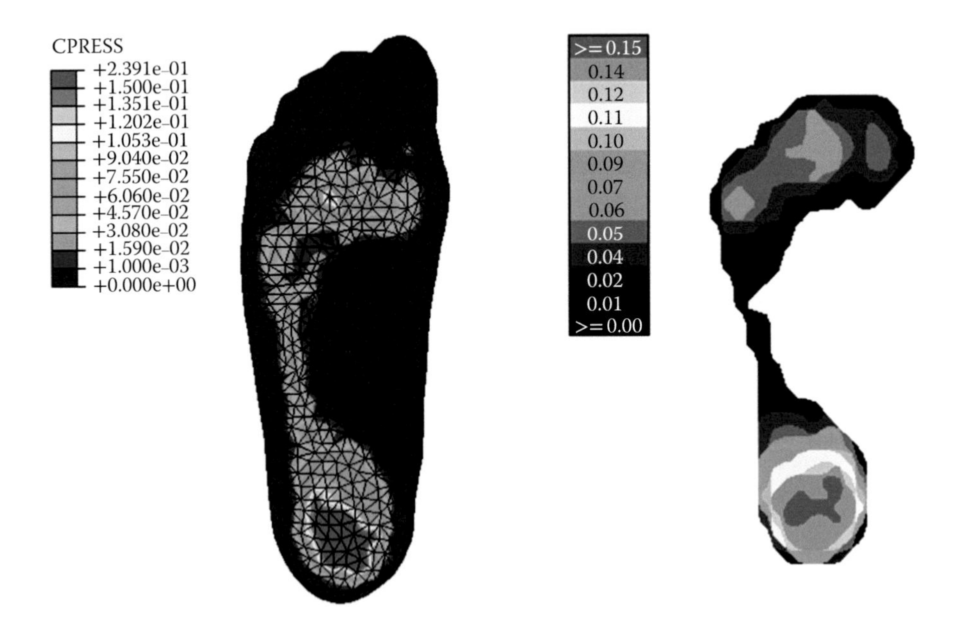


**FIGURE 1.5** **(See color insert.)** Plantar pressure distributions predicted by finite element modeling and measured by Tekscan.

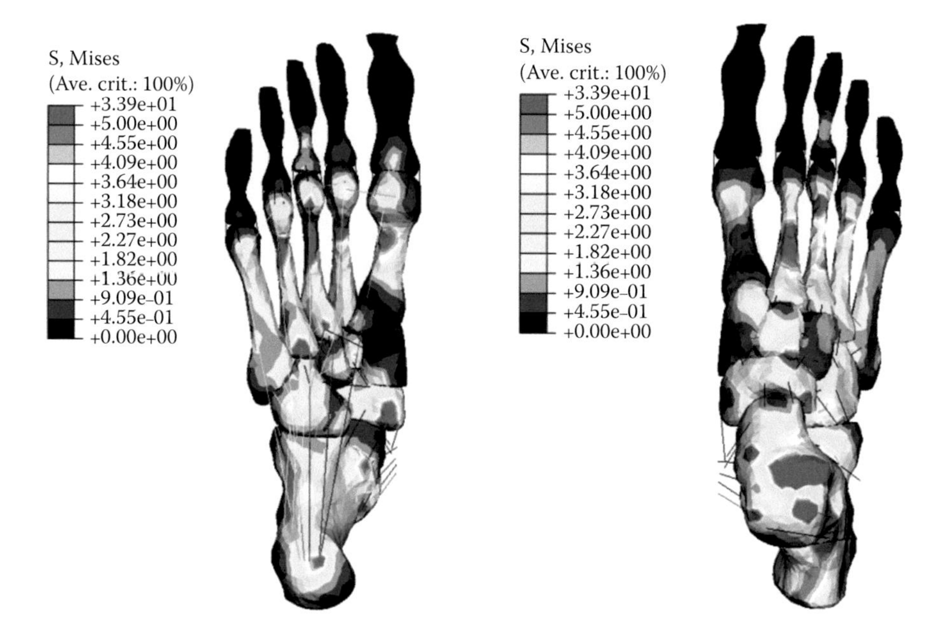


**FIGURE 1.6** von Mises stress distributions of foot bony structures (bottom view and top view).

## 1.3 APPLICATIONS OF THE FINITE ELEMENT MODEL

The FE models developed and validated have many potential applications. Two examples, foot biomechanics and foot support design, are discussed in the following section.

### 1.3.1 Effect of Soft Tissue Stiffening

Heel pain and ulceration of the diabetic foot are some of the most common complaints among patients with foot and ankle problems (Selth and Francis 2000; Holewski et al. 1989; Lavery et al. 2003; Pham et al. 2000; Boyko et al. 1999). One of the major causes of diabetic ulceration and painful heel syndrome is thought to be the presence of abnormally high plantar pressure (Mueller et al. 1994; Onwuanyi 2000; Reiber, Smith, and Wallace 2002; Sage, Webster, and Fisher 2001; Lavery et al. 2003; Pham et al. 2000; Veves et al. 1992; Boulton et al. 1987), which can be attributed to bony prominences, calluses, structural deformities, or poorly fitting footwear. Knowledge of the effect of soft tissue compliance or other structural characteristics on the stress distribution of the plantar foot surface and bony structures is essential for devising an appropriate individualized treatment strategy.

In this sensitivity study, the developed FE model was able to document the effect of varying bulk soft tissue stiffness on the plantar pressure distributions and stress of the bony structures during weight bearing of the foot. The nominal stress values at corresponding nominal strains adopted from the in vivo measurements (Lemmon et al. 1997) were multiplied by factors of two, three, and five to simulate stiffening of soft tissue (Figure 1.7).

The predicted plantar pressure distributions during balanced standing are shown in Figure 1.8. With increased plantar soft tissue stiffness, the pressure tended to be concentrated beneath the heel and the medial metatarsal heads, especially for the second and third metatarsals. In all calculated cases, the peak plantar pressure was located at the center of the heel and beneath the second and third metatarsal heads. This complies with the frequent observation of plantar foot ulcers at the medial forefoot and heel regions of diabetic patients (Mueller et al. 1994; Raspovic et al. 2000). The FE model suggested that stiff plantar soft tissue will decrease the ability of the foot to accommodate

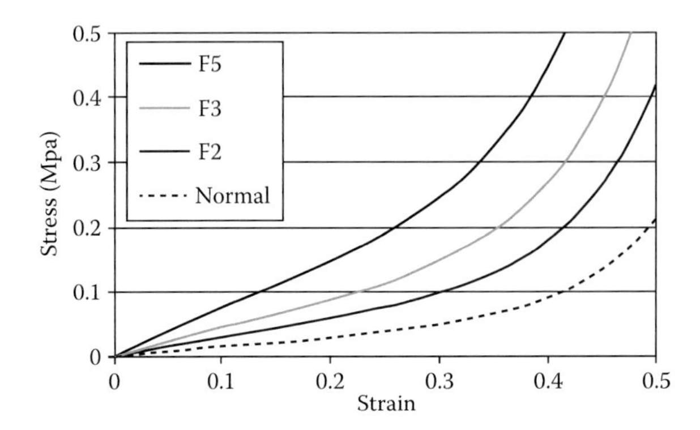


**FIGURE 1.7** Nonlinear stress-strain response of soft tissue assigned for the finite element models. The nominal stress values at corresponding nominal strains adopted from the in vivo measurements (Lemmon et al., 1997) were multiplied by factors of 2, 3, and 5, as F2, F3, and F5, to simulate stiffening of soft tissue.

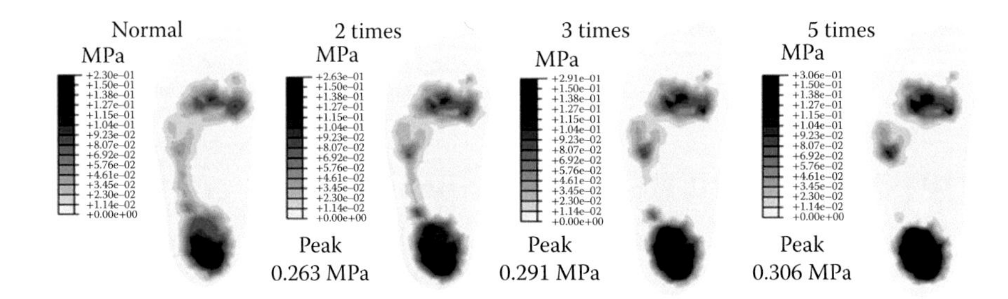


**FIGURE 1.8** Plantar pressure distributions with stiffening soft tissue.

and assimilate the plantar pressures, which can be a possible factor for igniting plantar foot pain and further tissue breakdown and ulcer development. The predicted shear stress also increased with tissue stiffening, which can also be a direct contribution to tissue breakdown in diabetic feet (Lord and Hosein 2000). In fact, the predicted percentage increase in plantar shear stress was more pronounced than the plantar pressure with stiffening of plantar soft tissue. Further investigations should be done to correlate the incidence of diabetic ulceration in terms of plantar shear stress.

From the FE model, the rate of increase in peak plantar pressure was found to be lower than the corresponding increase of soft tissue stiffness. A fivefold increase in soft tissue stiffness resulted in only about 1.33 and 1.35 times increase in peak plantar pressure at the heel and forefoot regions, respectively. From the FE predictions, it is clear that the increase in soft tissue stiffness will lead to an intensified peak plantar pressure. However, the rate of increase in peak plantar pressure was found to be much lower than the corresponding increase of soft tissue stiffness. The results suggest that screening of soft tissue stiffness might be an important procedure in addition to plantar pressure screening for early detection of susceptible ulceration sites.

### 1.3.2 Effect of Insole Design

Footwear and orthotics play an important role in reducing peak plantar pressure–related foot ulceration in people with diabetes, especially for those with neuropathy or lack of sensation. Foot orthoses for patients with diabetes should have increased cushioning for shock absorption. Such orthoses should also redistribute and reduce the excessive plantar pressure or shear from the vulnerable areas to more tolerable areas by enabling total contact with the plantar foot of the patient. In addition, foot orthoses should support and align the joints of the foot in a proper position for weight bearing and propulsion and accommodate or correct foot deformities.

Using our developed FE model, flat and custom-molded insole supports were simulated. The custom-molded insole was made from the unloaded shape of the subject's bare foot. The bare-foot shape was obtained by an impression cast with the subject sitting in the neutral position. The positive cast was digitized and imported to SolidWorks to form the solid models of the insole. A 5-mm-thick insole was meshed into three-dimensional brick elements with a Poisson ratio of 0.4 and a varied Young's modulus of 0.3 (soft), 1 (firmer), and 1000 MPa (rigid) for simulation of (1) open-cell polyurethane foams, such as Professional Protective Technology's PPT material; (2) high-density ethylene vinyl acetate; and (3) polypropylene materials, respectively. A very rigid, 1-mm-thick bottom layer was used to simulate the ground support and to facilitate the application of concentrated ground reaction forces. Figure 1.9 shows the peak plantar pressures over three regions of the foot during balanced standing using six different insoles.

From the FE parametric analysis of the foot orthosis, the use of a custom arch support was found to be an important design factor for reducing peak plantar pressure. Another important factor for reducing pressure is the use of a soft cushioning insole material. The FE model predicted a pronounced inverse relationship between insole thickness and peak plantar pressure, consistent with the experimental findings in the literature (Linge 1996). To find the optimal insole thickness one needs to consider the issue of reduced proprioception and propulsion efficiency, which deserves further investigation.

By and large, a custom pressure-relieving foot orthosis should provide the best total contact fit of the plantar foot during weight bearing. The cushioning insole layer should contribute the majority of the thickness of the foot orthosis to reduce the peak pressure. The use of extra-depth footwear for patients with diabetes is highly recommended to comfortably accommodate a thick, cushioning insole layer. The FE analysis suggested that the custom-molded total-contact foot orthosis was effective in reducing peak plantar pressures and would play an important role in the prevention of foot ulceration in diabetic patients.

It should be noted that the design guidelines for pressure-relieving foot orthoses were established from FE simulations and validation of a single subject, representing individual geometrical properties. In addition, only a single instance of the stance phase of gait was simulated in a static condition and the influence of the shoe-orthosis-foot interface was not considered. Therefore, the orthotic performance during different load-bearing and dynamic conditions and the generalization of the current FE simulations for the response of the general population deserves further computational and experimental investigation.

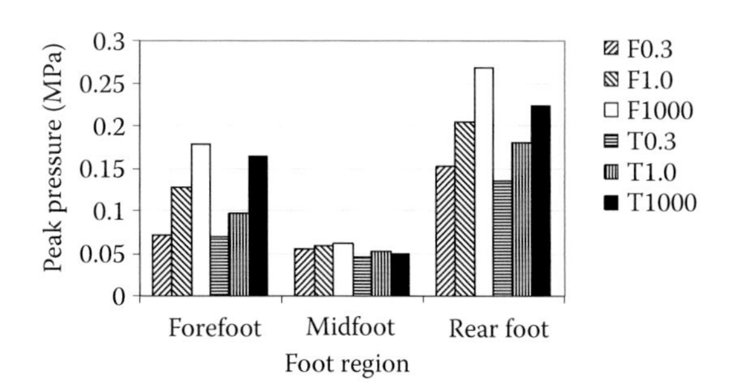


**FIGURE 1.9** Effects of stiffness of flat and custom-molded insoles on peak plantar pressure. The key indicates insole type (F, flat; M, custom molded) and rigidity (0.3, soft; 1.0, firmer; 1000, rigid).

## ACKNOWLEDGMENTS

This work was supported by the Research Grant Council of Hong Kong (GRF Project nos. PolyU5352/08E, PolyU5326/11E) and NSFC (11272273, 11120101001).

## REFERENCES

Athanasiou, K. A., G. T. Liu, L. A. Lavery, D. R. Lanctot, and R. C. Schenck. 1998. Biomechanical topography of human articular cartilage in the first metatarsophalangeal joint. *Clinical Orthopaedics and Related Research* 348:269–81.

Bayod, J., R. Becerro-de-Bengoa-Vallejo, M. E. Losa-Iglesias, and M. Doblaré. 2012. Mechanical stress redistribution in the calcaneus after autologous bone harvesting. *Journal of Biomechanics* 45:1219–26.

Bayod, J., M. Losa-Iglesias, R. Becerro de Bengoa-Vallejo, J. C. Prados-Frutos, K. T. Jules, and M. Doblaré. 2010. Advantages and drawbacks of proximal interphalangeal joint fusion versus flexor tendon transfer in the correction of hammer and claw toe deformity: A finite-element study. *Journal of Biomechanical Engineering* 132:051002.

Boulton, A. J., R. P. Betts, C. I. Franks, J. D. Ward, and T. Duckworth. 1987. The natural history of foot pressure abnormalities in neuropathic diabetic subjects. *Diabetes Research* 5:73–7.

Boyko, E. J., J. H. Ahroni, V. Stensel, R. C. Forsberg, D. R. Davignon, and D. G. Smith. 1999. A prospective study of risk factors for diabetic foot ulcer. The Seattle Diabetic Foot Study. *Diabetes Care* 22:1036–42.

Brilakis, E., E. Kaselouris, F. Xypnitos, C. G. Provatidis, and N. Efstathopoulos. 2012. Effects of foot posture on fifth metatarsal fracture healing: A finite element study. *Journal of Foot and Ankle Surgery* 51:720–8.

Chen, W. M., T. Lee, P. V. Lee, J. W. Lee, and S. J. Lee. 2010. Effects of internal stress concentrations in plantar soft-tissue: A preliminary three-dimensional finite element analysis. *Medical Engineering & Physics* 32:324–31.

Chen, W. M., J. Park, S. B. Park, V. P. Shim, and T. Lee. 2012. Role of gastrocnemius-soleus muscle in forefoot force transmission at heel rise: A 3D finite element analysis. *Journal of Biomechanics* 45:1783–9.

Chen, W. P., C. W. Ju, and F. T. Tang. 2003. Effects of total contact insoles on the plantar stress redistribution: A finite element analysis. *Clinical Biomechanics* 18:S17–S24.

Cheung, J. T., J. Yu, D. W. Wong, and M. Zhang. 2009. Current methods in computer-aided engineering for footwear design. *Footwear Science* 1:31–46.

Cheung, J. T., and M. Zhang. 2005. A 3-dimensional finite element model of the human foot and ankle for insole design. *Archives of Physical Medicine and Rehabilitation* 86:353–8.

Cheung, J. T., and M. Zhang. 2008. Parametric design of pressure-relieving foot orthosis using statistics-based finite element method. *Medical Engineering and Physics* 30:269–77.

Cheung, J. T., M. Zhang, A. K. Leung, and Y. B. Fan. 2005. Three-dimensional finite element analysis of the foot during standing a material sensitivity study. *Journal of Biomechanics* 38:1045–54.

Chokhandre, S., J. P. Halloran, A. J. van den Bogert, and A. Erdemir. 2012. A three-dimensional inverse finite element analysis of the heel pad. *Journal of Biomechanical Engineering* 134:031002.

Chu, T. M., N. P. Reddy, and J. Padovan. 1995. Three-dimensional finite element stress analysis of the polypropylene ankle-foot orthosis: Static analysis. *Medical Engineering & Physics* 17: 372–9.

Dul, J. 1983. Development of a minimum-fatigue optimization technique for predicting individual muscle forces during human posture and movement with application to the ankle musculature during standing and walking. PhD dissertation. Vanderbilt University.

Erdemir, A., J. J. Saucerman, D. Lemmon et al. 2005. Local plantar pressure relief in therapeutic footwear: Design guidelines from finite element models. *Journal of Biomechanics* 38:1798–1806.

Fontanella, C. G., S. Matteoli, E. L. Carniel et al. 2012. Investigation on the load-displacement curves of a human healthy heel pad: In vivo compression data compared to numerical results. *Medical Engineering & Physics* 34:1253–9.

Forestiero, A., E. L. Carniel, and A. N. Natali. 2012. Biomechanical behaviour of ankle ligaments: Constitutive formulation and numerical modelling. *Computer Methods in Biomechanics and Biomedical Engineering* 17:395–404.

García-Aznar, J. M., J. Bayod, A. Rosas et al. 2009. Load transfer mechanism for different metatarsal geometries: A finite element study. *Journal of Biomechanical Engineering* 131:021011.

García-González, A., J. Bayod, J. C. Prados-Frutos et al. 2009. Finite-element simulation of flexor digitorum longus or flexor digitorum brevis tendon transfer for the treatment of claw toe deformity. *Journal of Biomechanics* 42:1697–1704.

Gefen, A., M. Megido-Ravid, Y. Itzchak, and M. Arcan. 2000. Biomechanical analysis of the three-dimensional foot structure during gait: A basic tool for clinical applications. *Journal of Biomechanical Engineering* 122:630–9.

Glitsch, U., and W. Baumann. 1997. The three-dimensional determination of internal loads in the lower extremity. *Journal of Biomechanics* 30:1123–31.

Goske, S., A. Erdemir, M. Petre, S. Budhabhatti, and P. R. Cavanagh. 2006. Reduction of plantar heel pressures: Insole design using finite element analysis. *Journal of Biomechanics* 39:2363–70.

Gu, Y., J. Li, X. Ren, M. J. Lake, and Y. Zeng. 2010. Heel skin stiffness effect on the hind foot biomechanics during heel strike. *Skin Research and Technology* 16:291–6.

Gu, Y. D., X. J. Ren, J. S. Li, M. J. Lake, Q. Y. Zhang, and Y. J. Zeng. 2010. Computer simulation of stress distribution in the metatarsals at different inversion landing angles using the finite element method. *International Orthopaedics* 34:669–76.

Gu, Y. D., X. J. Ren, G. Q. Ruan, Y. J. Zeng, and J. S. Li. 2011. Foot contact surface effect to the metatarsals loading character. *International Journal of Numerical Methods in Biomedical Engineering* 27:476–84.

Halloran, J. P., M. Ackermann, A. Erdemir, and A. J. van den Bogert. 2010. Concurrent musculoskeletal dynamics and finite element analysis predicts altered gait patterns to reduce foot tissue loading. *Journal of Biomechanics* 43:2810–5.

Halloran, J. P., A. Erdemir, and A. J. van den Bogert. 2009. Adaptive surrogate modeling for efficient coupling of musculoskeletal control and tissue deformation models. *Journal of Biomechanical Engineering* 131:011014.

Holewski, J. J., K. M. Moss, R. M. Stess, P. M. Graf, and C. Grunfeld. 1989. Prevalence of foot pathology and lower extremity complications in a diabetic outpatient clinic. *Journal of Rehabilitation Research Development* 26:35–44.

Isvilanonda, V., E. Dengler, J. M. Iaquinto, B. J. Sangeorzan, and W. R. Ledoux. 2012. Finite element analysis of the foot: Model validation and comparison between two common treatments of the clawed hallux deformity. *Clinical Biomechanics* 27:837–44.

Jamshidi, N., H. Hanife, M. Rostami et al. 2010. Modelling the interaction of ankle-foot orthosis and foot by finite element methods to design an optimized sole in steppage gait. *Journal of Medical Engineering & Technology* 34:116–23.

Kim, K. J., H. B. Kitaoka, Z. P. Luo et al. 2001. An in vitro simulation of the stance phase in human gait. *Journal of Musculoskeletal Research* 5:113–21.

Kim, S. H., J. R. Cho, J. H. Choi, S. H. Ryu, and W. B. Jeong. 2012. Coupled foot-shoe-ground interaction model to assess landing impact transfer characteristics to ground condition. *Interaction and Multiscale Mechanics* 5:75–90.

Kirby, K. A. 2001. What future direction should podiatric biomechanics take? *Clinics in Podiatric Medicine and Surgery* 18:719–23.

Lavery, L. A., D. G. Armstrong, R. P. Wunderlich, J. Tredwell, and A. J. Boulton. 2003. Predictive value of foot pressure assessment as part of a population-based diabetes disease management program. *Diabetes Care* 26:1069–73.

Lemmon, D., T. Y. Shiang, A. Hashmi, J. S. Ulbrecht, and P. R. Cavanagh. 1997. The effect of insoles in therapeutic footwear: A finite-element approach. *Journal of Biomechanics* 30:615–20.

Lewis, G. 2003. Finite element analysis of a model of a therapeutic shoe: Effect of material selection for the outsole. *Bio-medical Materials and Engineering* 13:75–81.

Liang, J., Y. Yang, G. Yu, W. Niu, and Y. Wang. 2011. Deformation and stress distribution of the human foot after plantar ligaments release: A cadaveric study and finite element analysis. *Science China. Life Science* 54:267–71.

Linge, K. 1996. A preliminary objective evaluation of leprosy footwear using in-shoe pressure measurement. *Acta Orthopaedica Belgica* 62:S18–S22.

Liu, X., and M. Zhang. 2013. Redistribution of knee stress using laterally wedged insole intervention: Finite element analysis of knee-ankle-foot complex. *Clinical Biomechanics* 28:61–7.

Lord, M., and R. Hosein. 2000. A study of in-shoe plantar shear in patients with diabetic neuropathy. *Clinical Biomechanics* 15:278–83.

Luboz, V., A. Perrier, N. Vuillerme, M. Bucki, B. Diot, F. Cannard, and Y. Payan. 2012. Foot biomechanical modelling to study orthoses influence. *Computer Methods in Biomechanics and Biomedical Engineering* 15(Suppl. 1):360–2.

Luo, G., V. L. Houston, M. A. Garbarini, A. C. Beattie, and C. Thongpop. 2011. Finite element analysis of heel pad with insoles. *Journal of Biomechanics* 44:1559–65.

Matzaroglou, C., P. Bougas, E. Panagiotopoulos, A. Saridis, M. Karanikolas, and D. Kouzoudis. 2010. Ninety-degree chevron osteotomy for correction of hallux valgus deformity: Clinical data and finite element analysis. *Open Orthopaedics Journal* 4:152–6.
Mueller, M. J., D. R. Sinacore, S. Hoogstrate, and L. Daly. 1994. Hip and ankle walking strategies: Effect on peak plantar pressures and implications for neuropathic ulceration. *Archives of Physical Medicine and Rehabilitation* 75:1196–1200.
Nakamura, S., R. D. Crowninshield, and R. R. Cooper. 1981. An analysis of soft tissue loading in the foot: A preliminary report. *Bulletin Prosthetics Research* 18:27–34.
Onwuanyi, O. N. 2000. Calcaneal spurs and plantar heel pad pain. *The Foot* 10:182–5.
Opila, K. A., S. S. Wagner, S. Schiowitz, and J. Chen. 1988. Postural alignment in barefoot and high-heeled stance. *Spine* 13:542–7.
Perry, J. 1992. *Gait Analysis: Normal and Pathological Function*. Thorofare, NJ: SLACK.
Pham, H., D. G. Armstrong, C. Harvey, L. B. Harkless, J. M. Giurini, and A. Veves. 2000. Screening techniques to identify people at high risk for diabetic foot ulceration: A prospective multicenter trial. *Diabetes Care* 23:606–11.
Qiu, T. X., E. C. Teo, Y. B. Yan, and W. Lei. 2011. Finite element modeling of a 3D coupled foot-boot model. *Medical Engineering & Physics* 33:1228–33.
Raspovic, A., L. Newcombe, J. Lloyd, and E. Dalton. 2000. Effect of customized insoles on vertical plantar pressures in sites of previous neuropathic ulceration in the diabetic foot. *Foot* 10:133–8.
Reiber, G. E., D. G. Smith, C. Wallace et al. 2002. Effect of therapeutic footwear on foot reulceration in patients with diabetes: a randomized controlled trial. *Journal of American Medical Association* 287:2552–8.
Rohrle, H., R. Scholten, C. Sigolotto, W. Sollbach, and H. Kellner. 1984. Joint forces in the human pelvis–leg skeleton during walking. *Journal of Biomechanics* 17:409–24.
Sage, R. A., J. K.Webster, and S. G. Fisher. 2001. Outpatient care and morbidity reduction in diabetic foot ulcers associated with chronic pressure callus. *Journal of the American Podiatric Medical Association* 91:275–9.
Selth, C. A., and B. E. Francis. 2000. Review of non-functional plantar heel pain. *Foot* 10:97–104.
Shin, J., N. Yue, and C. D. Untaroiu. 2012. A finite element model of the foot and ankle for automotive impact applications. *Annals of Biomedical Engineering* 40:2519–31.
Siegler, S., J. Block, and C. D. Schneck. 1988. The mechanical characteristics of the collateral ligaments of the human ankle joint. *Foot & Ankle* 8:234–42.
Sopher, R., J. Nixon, E. McGinnis, and A. Gefen. 2011. The influence of foot posture, support stiffness, heel pad loading and tissue mechanical properties on biomechanical factors associated with a risk of heel ulceration. *Journal of the Mechanical Behavior of Biomedical Materials* 4:572–82.
Spears, I. R., J. E. Miller-Young, J. Sharma, R. F. Ker, and F. W. Smith. 2007. The potential influence of the heel counter on internal stress during static standing: A combined finite element and positional MRI investigation. *Journal of Biomechanics* 40:2774–80.
Spyrou, L. A., and N. Aravas. 2012. Muscle-driven finite element simulation of human foot movements. *Computer Methods in Biomechanics and Biomedical Engineering* 15:925–34.
Sun, P. C., S. L. Shih, Y. L. Chen, Y. C. Hsu, R. C. Yang, and C. S. Chen. 2012. Biomechanical analysis of foot with different foot arch heights: A finite element analysis. *Computer Methods in Biomechanics and Biomedical Engineering* 15:563–9.
Tao, K., W. T. Ji, D. M. Wang, C. T. Wang, and X. Wang. 2010. Relative contributions of plantar fascia and ligaments on arch static stability: A finite element study. *Biomedizinische Technik* 55:265–71.
Verdejo, R., and N. J. Mills. 2004. Heel-shoe interactions and the durability of EVA foam running-shoe midsoles. *Journal of Biomechanics* 37:1379–86.
Veves, A., H. J. Murray, M. J. Young, and A. J. Boulton. 1992. The risk of foot ulceration in diabetic patients with high foot pressure: A prospective study. *Diabetologia* 35:660–3.
Wright, D., and D. Rennels. 1964. A study of the elastic properties of plantar fascia. *Journal of Bone and Joint Surgery, American Volume* 46:482–92.
Xu, C., M. Y. Zhang, G. H. Lei et al. 2012. Biomechanical evaluation of tenodesis reconstruction in ankle with deltoid ligament deficiency: A finite element analysis. *Knee Surgery, Sports Traumatology Arthroscopy: Official Journal of the ESSKA* 20:1854–62.
Zhang, M., and A. F. Mak. 1999. In vivo frictional properties of human skin. *Prosthetics and Orthotics International* 23:135–41.

# 2 Female Foot Model for High-Heeled Shoe Design

*Jia Yu, Yubo Fan, and Ming Zhang*

## CONTENTS



## SUMMARY

Despite frequent complaints of shoe discomfort and foot problems, women continue to wear high-heeled shoes for cosmetic purposes and high-heeled shoes are expected to become increasingly popular with global urbanization. Due to limitations of the experimental approach, direct measurements of internal stress/strain of the foot are impossible or invasive, and developing a comprehensive computational foot-shoe model to increase the biomechanical knowledge of high-heeled shoes is very important for addressing foot problems related to their wear and optimization of their design. Computational modeling based on finite element analysis is a useful tool for understanding foot and footwear biomechanics (Cheung et al. 2009; Kirby 2001). The knock-on effect of heel height and high-heeled shoe donning and walking on the foot are investigated in this chapter.

## 2.1 INTRODUCTION

There are many reasons ladies want to wear high heels. The primary reason for increasing the heel height of the shoe is to add an extra degree of attractiveness. Whether women are standing or walking, wearing high-heeled shoes creates an optical illusion of a smaller foot, shapes the contour of the ankle and leg, contributes to a long-legged look, thrusts the buttocks backwards, and increases height to generate the sensation and appearance of power and status.

The anthropologists Smith and Helms (1999) comprehensively analyzed how high heels provided a good example of an evolved cultural display directed by sexual selection. Conforming to culturally prescribed patterns of dress must have some positive fitness-enhancing aspects. A woman wearing high heels sends a variety of messages that reveal her receptivity, sexuality, confidence, and power.

### 2.1.1 Adverse and Beneficial Effects of High-Heeled Shoes

The Gallup organization reported that 59% of women wore high-heeled shoes for one to eight hours per day (Gallup Organization Inc. 1986). Medical scientists have documented the adverse effects of high-heeled shoes, which can be traced back over four centuries (Linder and Saltzman 1998).

Many researchers have reported that overworked or injured leg muscles, lower back pain, and knee osteoarthritis (OA) may be linked to the abnormal posture that high heels induce (Kerrigan et al. 2005). It is important to note that high heels do cause cumulative damage to the feet. Common foot problems associated with high heels include corn, callus, hammer, mallet and claw toes, hallux valgus, Morton's neuroma, metatarsalgia, pump bump, and tight heel cords. If women frequently wear high-heeled shoes that are too narrow or too short for their feet, they could be setting themselves up for one or more of these conditions. Based on a survey by the American Orthopaedic Foot and Ankle Society on women's shoes in 1993, 88% wore shoes that were smaller than their feet (average 12 mm), 80% had foot pain or deformity, and 76% had one or more forefoot deformities.

Apart from the aforementioned foot problems, high-heeled shoes can also be used as an alternative therapeutic measure in the treatment of tendinitis and partial ruptures of the Achilles tendon (Kogler et al. 2001). This conservative treatment strategy for plantar fasciitis can reduce the strain in the plantar fascia (Gordon 1984; Cole, Seto, and Gazewood 2005; Marshall 1988).

#### 2.1.1.1 Review of Finite Element Analysis Research on the Foot and Footwear

Due to the lack of technology and the invasive nature of experimental measurements, experimental studies are often restricted to studying the plantar pressure distribution, and gross motion of the foot, while the evaluation of internal joint movements and load distributions are usually unaddressed. As an alternative, researchers propose modeling approaches. Computer simulations, such as the finite element (FE) method, are versatile tools and have the potential to provide detailed biomechanical information. Such simulations are capable of modeling structures with irregular geometry and complex material properties, and can accurately simulate complicated boundary and loading conditions. In terms of footwear design, the FE method allows prediction of plantar pressure and joint movement, as well as contact stress and internal stress/strain of the foot under various loading and supporting conditions. These models can isolate the variable of interest, which is not always possible using conventional experimental methods.

For human musculoskeletal joint modeling, the challenge remains to produce geometrically, kinematically, and mechanically accurate models that can then be used in fundamental investigations, as well as injury simulation and prediction (Penrose et al. 2002). FE models may be useful for deciding footwear design parameters such as material, heel height, and sole shape. The development of comprehensive computational models of the human foot was suggested to be one of the most important directions for future research in podiatric biomechanics (Kirby 2001). Reviewing FE models related to the foot and footwear developed so far in the literature, modeling accuracy could be improved in the following ways:

- Limited finite element analysis (FEA) is available for interaction of assembly footwear and the foot under different loading conditions due to complications with FEA simulation. A comprehensive three-dimensional foot-footwear model could offer unique insight into footwear design by sensitivity analysis.
- If the bones are modeled as separate components, the ligamentous connections will play an important role in realistic joint movement simulation in multiple directions. It is necessary to apply more realistic loading boundary conditions, including muscle forces, for the gait simulation, because muscle activity plays an important role in load balance.
- FEA is a powerful tool that has been extensively used in biomechanics, but it is "easy to do poorly and very hard to do well" (Viceconti et al. 2005). Few models were carefully assigned with physiological based boundary conditions (Speirs et al. 2007) and well validated. Therefore, only validated FE models can be a platform for parametric studies.
- Several male foot models have been developed (Jacob and Patil 1999; Gefen et al. 2000; Chen, Ju, and Tang 2003; Cheung et al. 2005), whereas few FE models of a female foot have been reported. According to morphological studies, a female foot is not merely a scaled-down version of a male foot. A female foot has its own shape characteristics (Manna et al. 2001; Wunderlich and Cavanagh 2001). An FE model that takes into account the morphological features of a female foot is a prerequisite for studying the biomechanical behavior of a female foot and the evaluation of high-heeled shoes.

The FE model may help to detail the effects of high-heeled shoe design parameters such as heel height and toe box. Few FE analyses are available for investigating the interaction of assembly footwear and the foot under different loading conditions. Thus, developing an FE model of a female foot is essential for investigating the biomechanical response of the female foot in response to footwear design, especially for high-heeled shoes.

#### 2.1.1.2 Finite Element Modeling

To develop an anatomically detailed FE foot model, coronal magnetic resonance (MR) images of the right foot in a neutral, non-weight-bearing condition were obtained at 1-mm intervals using a 3.0-T MR scanner (Siemens, Erlangen, Germany). The subject was a healthy female adult of age 28, height 165 cm, and mass 54 kg. A custom-made ankle-foot orthosis was used to fix the ankle in a neutral unloaded position during the MR scanning procedure (Figure 2.1).

The MR images were segmented using MIMICS v9.10 (Materialise, Leuven, Belgium) to obtain the boundaries of each bone and skin surface. Afterward a three-dimensional surface model of bones and skin was generated from the stacked outlines. The surface model of the foot was imported into SolidWorks 2001 (SolidWorks Corporation, Massachusetts) to create the solid model. The encapsulated soft tissue was subtracted from the whole foot volume by the bony structures. Thereafter, the FE package ABAQUS v6.7 was used for creating the FE mesh and subsequent analysis. Except for the collateral ligaments of the four lateral phalanges and other connective tissue, a total number of 78 ligaments and the plantar fascia were included and defined by connecting the corresponding attachment points on the bones. All the ligamentous and bony structures were embedded in the bulk volume of soft tissue. The FE model developed consisted of 28 distinct bony segments, including the distal tibia, distal fibula, talus, calcaneus, cuboid, navicular, three cuneiforms, five metatarsals, and five phalanges embedded in the soft tissue. The plantar fascia and foot ligaments, excluding those ligaments at fused interphalangeal joints, were modeled as tension-only truss elements, connecting their corresponding attachment points

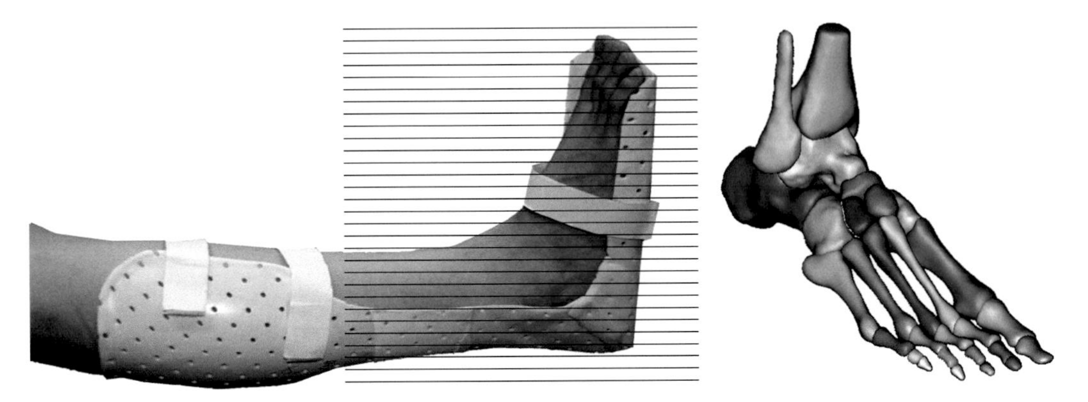

**FIGURE 2.1** **(See color insert.)** Fixing foot and ankle in a neutral position by foot orthosis (left), and foot bones surface model (right).

on the bony surfaces. The tension-only truss element was used to reflect the tensile-resistive but noncompression-resistive mechanical characteristics of ligaments. Musculotendon forces were applied at their corresponding sites of insertion by defining contraction forces via axial connector elements. To simulate the surface interactions among joint contact pairs, an automated surface-to-surface contact algorithm in ABAQUS was used. Because of the lubricating nature of articulating surfaces, the contact behavior between the contacting bony segments was idealized as frictionless.

The same contact modeling algorithm was used to simulate the contact between the foot and supporting interface. The surface interaction between the plantar foot and external foot supporting surface was assigned with a coefficient of friction of 0.6 (Zhang and Mak 1999). The surface interaction between the high-heeled shoe and ground support was assigned a coefficient of friction of 0.5 (Hanson, Redfern, and Mazumdar 1999).

### 2.1.2 Material Properties

In order to reduce the complexity of the FE model, except for the bulk soft tissue, the foot bones, cartilages, and ligaments were idealized as homogeneous, isotropic, and linearly elastic. The Young's modulus ($E$) and the Poisson's ratio ($\nu$) determine the linearly elastic properties. The mechanical properties of bone, cartilage, plantar fascia, and ligaments were selected from the literature. The foot support was made of Pedilen® rigid foam 300 with material properties adopted from Shiina, Hamamoto, and Okumura (2006). High-density polyethylene is a common material for the outsole (Lewis 2003). A long, thin steel shankpiece is typically embedded within the middle of the outsole to reinforce the shank of high-heeled shoes. The shankpiece was assigned the material properties of steel. A flat support, with an upper layer assigned the properties of rigid foam (Shiina, Hamamoto, and Okumura 2006) and a lower layer assigned the properties of a rigid body, was used to simulate the ground support.

## 2.2 APPLICATIONS OF THE FINITE ELEMENT MODEL FOR EXAMINING EFFECTS OF HEEL HEIGHT

This study aimed at evaluating the biomechanical effects of high-heeled support on the ankle-foot complex using an FE model. In this study, balanced standing on different heel-elevated foot supports was simulated. Physiological muscle forces on the foot and location of ground reaction force (GRF) of each condition were derived and prescribed. Plantar pressure measurements and motion analysis were conducted to obtain the loading and boundary conditions and validate the FE model response.

### 2.2.1 Experimental Validation: In Vivo and In Vitro Approach

#### 2.2.1.1 In Vivo

Foot supports with 1-, 2-, and 3-inch heels were used. The metatarsophalangeal (MTP) joint of the subject's feet were aligned and positioned according to the profile of the foot supports. As shown in Figure 2.2, the F-scan sensors for plantar pressure measurements were placed underneath a sheet of plain paper. A pair of F-scan sensors was fixed on the top of each foot support. In order to standardize foot alignment and foot shape measurements, the foot supports were put on top of a 45-cm platform. During balanced standing, real-time plantar contact pressure data was recorded at a sampling frequency of 50 Hz for 10 seconds. The medial longitudinal arch height was determined as the height of the navicular tuberosity from a line joining the posterior point of the plantar heel pad and the plantar first metatarsal head (Shimizu and Andrew 1999). These three points were marked by the digitizer.

#### 2.2.1.2 In Vitro

In the cadaveric experiments, the contact pressure of the first MTP joint and the foot/ankle kinematics of normal ankles with high-heeled shoes were obtained. Two right female ankle-foot specimens amputated at the tibial plateau level, with the same foot size (Continental size 38), were used in the study. The specimens were evaluated by both clinical examination and radiography and were free from any observable pathology and deformity. A custom-made multi-axis testing device (Mayo Clinic, USA), was used to generate planar motions of the foot and ankle. This device utilized one motorized rotatory stage (Newport Corporation, Irvine, CA) and two custom-made motorized tilting stages that result in a three-axis gimbal. Unconstrained linear sliding in the vertical, anterior-posterior, and medial-lateral axes provided a pure moment configuration of the rotations at the ankle and subtalar joint by eliminating the shear forces. The integration of the three rotatory stages resulted in a tilting platform that represented the floor on which the specimen rested. The tibia was rigidly fixed to the testing frame and an axial load was applied to the tibia. The platform under the foot could be programmed to induce motion in the sagittal plane (plantarflexion-dorsiflexion), coronal plane (inversion-eversion), or transverse plane (internal-external rotation). Above the fixed tibia, a platform was incorporated with a muscle actuator.

Joint contact area and pressure were obtained with the K-Scan System (sensor model 6900; Tekscan, Inc., South Boston, MA). The K-Scan sensor has four independent sensing regions. Four infrared sensors embedded in rigid bodies were inserted into the tibia, calcaneus, the first metatarsal, and the first proximal phalanx of the specimen to measure their motion (Figure 2.3). An

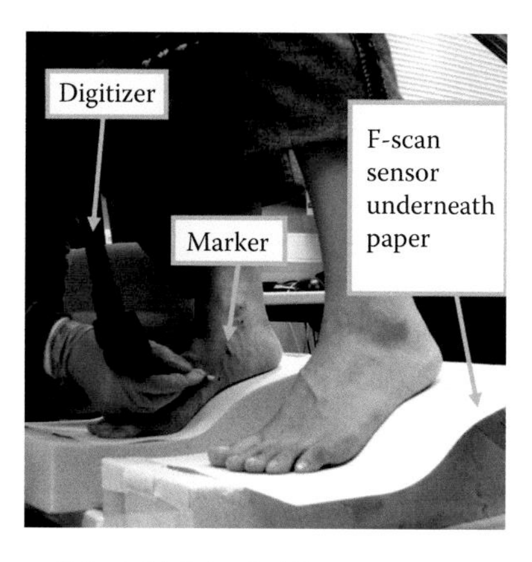


**FIGURE 2.2** Balanced standing on 2-inch high-heeled foot support.

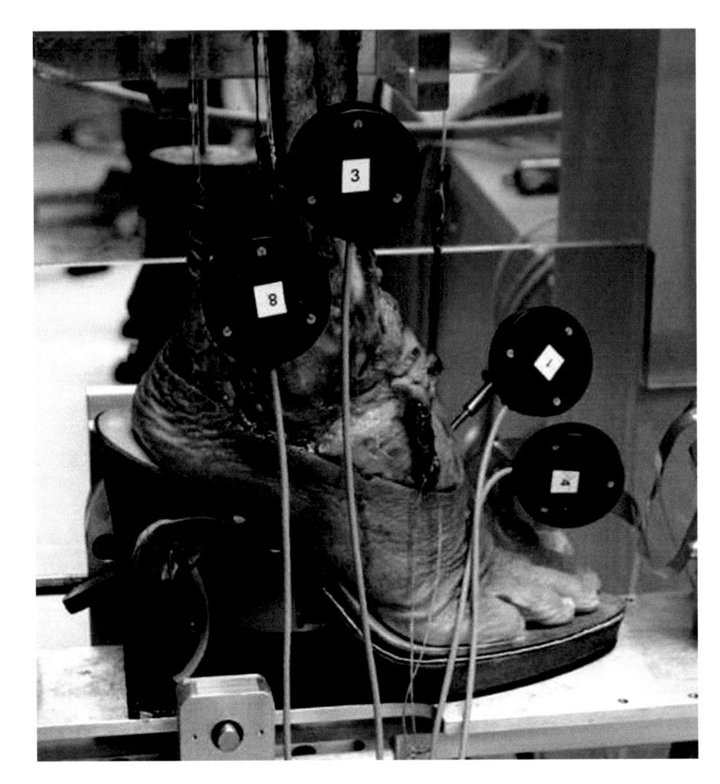

FIGURE 2.3 Mounting four infrared sensors in the specimen.

optoelectric tracking device (Optotrak Certus, Northern Digital Inc., Ontario, Canada) was used to measure MTP joint kinematics. Relative angular motion between the bones was calculated with the MotionMonitor software (Innovative Sports Training, Inc., Chicago, IL), and expressed using Euler angles.

An axial load (212 N) was applied to the tibia. At first, static loads were applied to the Achilles tendon, peroneus tertius, peroneus longus, flexor hallucis longus, and extensor hallucis longus. The applied forces were controlled with a programmable servo-pneumatic regulator (Proportion-Air, McCordsville, IN), and tendon excursions were measured with a potentiometer incorporated in the pulley of the actuator unit. An initial preconditioning cyclically loaded each specimen at 10 mm/s for 10 cycles under axial loading (212 N) to establish a mechanical stabilized state just prior to testing. Measurements were made with each specimen mounted in the testing machine at the end of preconditioning.

Each specimen was subjected to the following four tests, which were a flat support (control) and three different high-heeled shoes of the same size (Continental size 38). To follow the high-heeled shoe used in the FE model and for easy mounting of specimens onto the high-heeled shoe, the upper material body of the shoe was carefully removed. For each testing condition, five test runs were performed so that an average value could be obtained for data analysis. The protocol for all tests was kept the same for each specimen. Specimens were loaded for 10 seconds for each condition and then data were collected in real time for an additional 3 seconds.

### 2.2.2 Finite Element Simulation

In this study, shoes with 1-, 2-, and 3-inch heels were used. The loading and boundary conditions for balanced standing on 2-inch high-heeled shoes is shown in Figure 2.4. To simulate balanced

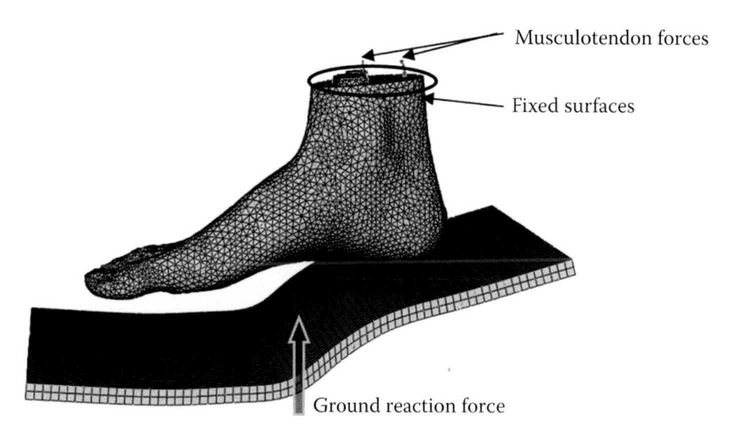


**FIGURE 2.4** Loading and boundary conditions for standing on high-heeled support.

standing on a flat support and high-heeled foot supports with different heel heights, a vertical force corresponding to half body weight (BW) was applied to the bottom of the supports, which were only allowed to move vertically. The upper extremity of soft tissue, distal tibia, and distal fibula were fixed throughout the analysis to serve as the boundary conditions in all standing simulations. For a subject with body mass of 54 kg, a vertical force of approximately 270 N was applied to each foot during balanced standing. Due to the fact that the line of gravity was in front of the ankle joint during both barefoot and high-heeled standing (Opila et al. 1988), the plantar flexors must act to balance the forward moment of the body on the ankle in order to remain balanced. It was found that the triceps surae played the major stabilization role of the foot during balanced standing on flat support and the reactions of all other intrinsic and extrinsic muscles were minimal (Basmajian and Stecko 1963). Therefore, only the tension in the Achilles tendon was considered during simulated balanced standing on flat support while all intrinsic and remaining extrinsic muscle forces were neglected. The Achilles tendon force was estimated by matching the FE predictions with the measured plantar pressure distribution and location of the center of pressure (COP) of the subject. The foot and ankle are multiplanar. With only the Achilles tendon force without peroneus brevis and peroneus longus muscle forces, the foot has a plantarflexion movement combining with inversion and adduction movement. Thereafter, for balanced standing on high-heeled foot supports, additional muscle forces of peroneus brevis and peroneus longus were applied to match the foot position according to experimental measurements.

As carried out with the flat supports, a sensitivity analysis of the Achilles tendon loading was conducted for all balanced tests using high-heeled foot supports. From the sensitivity analysis, the Achilles tendon forces required for simulating the upright, balanced standing posture were estimated by matching the FE predictions with the location of COP and the measured plantar pressure distribution. For the high-heeled testing, small muscle forces for extensor hallucis longus (10 N) and extensor digitorum longus (5 N) were applied to better accommodate MTP joints upon toe spring. Since the subject stood along the midline of the foot support, muscle forces for peroneus brevis (20 N) and peroneus longus (25 N) were estimated and applied to match foot orientation.

When simulating balanced standing on each high-heeled support, the Achilles tendon force was applied with magnitudes ranging from 50% to 200% of half BW, at intervals of 5%. The Achilles tendon forces were estimated by matching the FE predictions with the measured plantar pressure distributions and locations of COP. It was found that Achilles tendon forces of 65%, 80%, and 160% of half BW was proper in terms of the closest COP, for 1-, 2-, and 3-inch high-heeled foot supports, respectively. During balanced standing on 1-, 2-, and 3-inch heels, the COP was 59, 41, and 92 mm anterior to the lateral malleolus, respectively. The COP of each condition was about 12 mm lateral to the lateral malleolus. The maximum deviations in all calculated cases between FE predictions and experimental measurements were less than 3 mm in the anterior-posterior direction and 2 mm in the medial-lateral direction.

Foot deformations during balanced standing on four different heel heights are demonstrated in Figure 2.5. In all cases, the foot was aligned with the curve of the foot support. The maximum foot adduction angle among all conditions was less than 2 degrees. The plantar pressure distributions during balanced standing on three different high-heeled foot supports from FE predictions and F-scan measurements are compared in Figure 2.6. These measurements were, in general, comparable, but the FE model predicted a slightly larger magnitude. The peak plantar pressure region shifted from the central heel region to the central forefoot region with a 3-inch heel for both FE and F-scan measurements.

There was a general increase in maximum von Mises stress in foot bones with increasing heel height. The peak von Mises stresses in major bones with different high-heeled foot supports are compared in Figure 2.7. Peak von Mises stress appeared at the plantar junction of the calcaneal-cuboid joint. In the forefoot region, relatively high von Mises stresses concentrated at the second to the fourth metatarsal shafts as well. From the FE predictions, an increase in heel height from 0 to 3 inches resulted in a decrease in arch deformation from 8.8 mm to 1.1 mm, which was consistent with the measured trend of arch deformation (Figure 2.8). The FE-predicted strain and tension of the plantar fascia with different heel heights during balanced standing is shown in Figure 2.9. With a 2-inch heel, the strain and total tension force of plantar fascia was minimum in all calculated cases.

It was found that wearing high-heeled shoes may help to reduce arch deformation of the weight-bearing feet. There was a general increase in predicted maximum von Mises stress of foot bones with increasing heel height from 0 to 3 inches. In the forefoot region, relatively high von Mises stresses concentrated at the second to the fourth metatarsals. With 2-inch heels, the strain and total

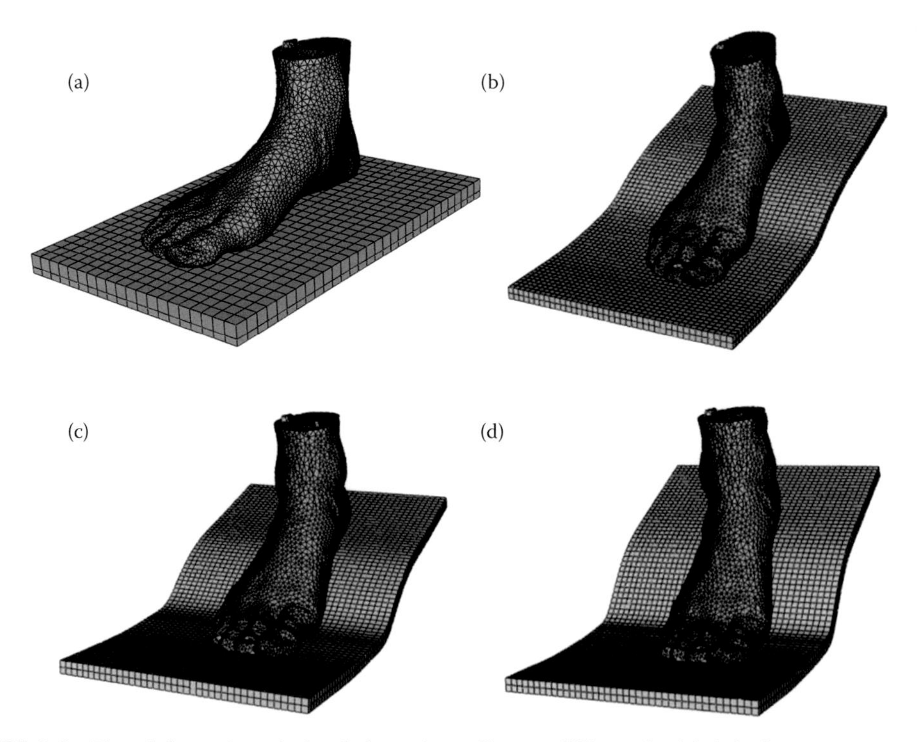


**FIGURE 2.5** Foot deformations during balanced standing on different heel-height foot supports: (a) flat, (b) 1-inch, (c) 2-inch, and (d) 3-inch.

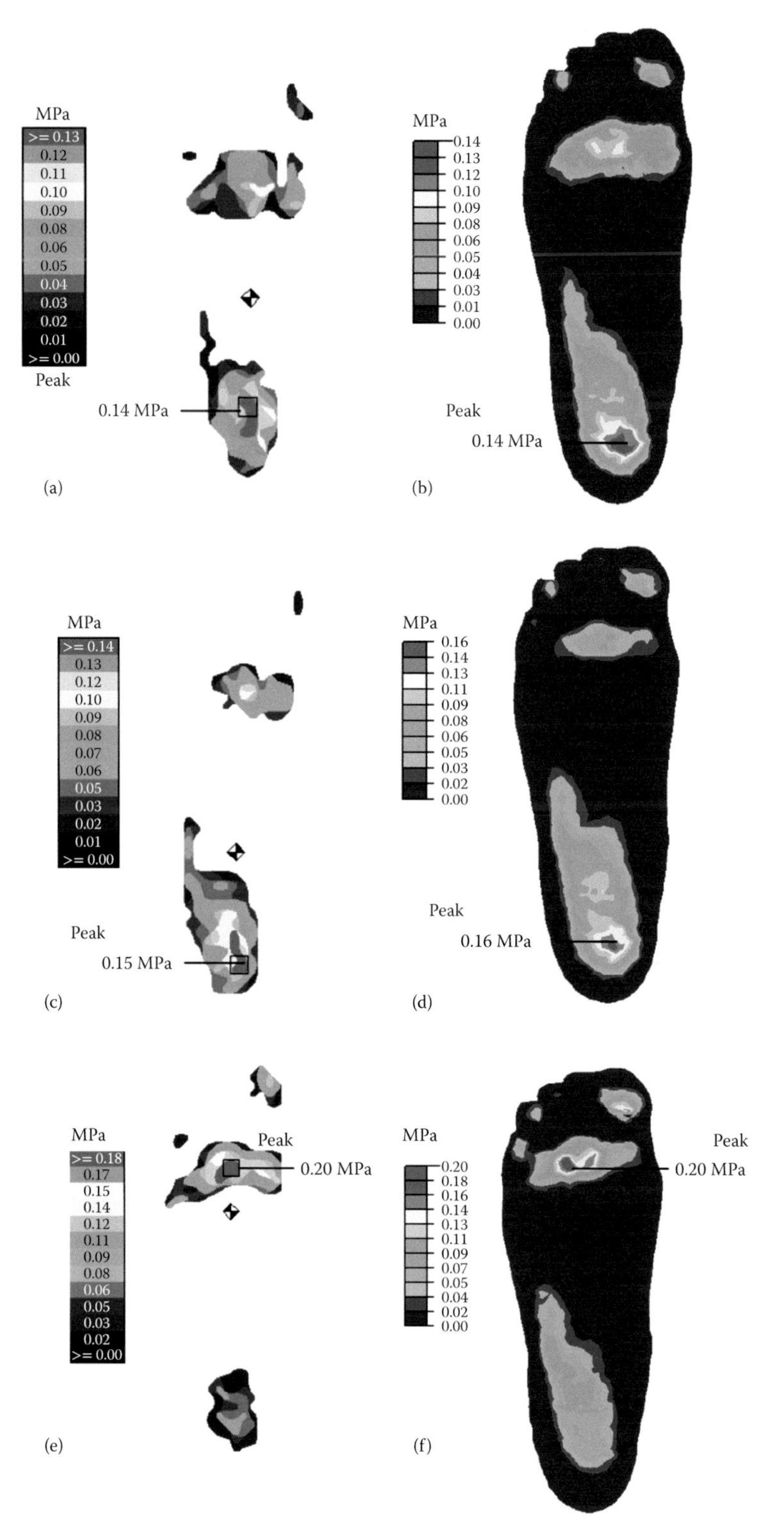


**FIGURE 2.6** **(See color insert.)** Plantar pressure distributions: (a) 1-inch from F-scan measurement, (b) 1-inch from finite element (FE) prediction, (c) 2-inch from F-scan measurement, (d) 2-inch from FE prediction, (e) 3-inch from F-scan measurement, and (f) 3-inch from FE prediction.

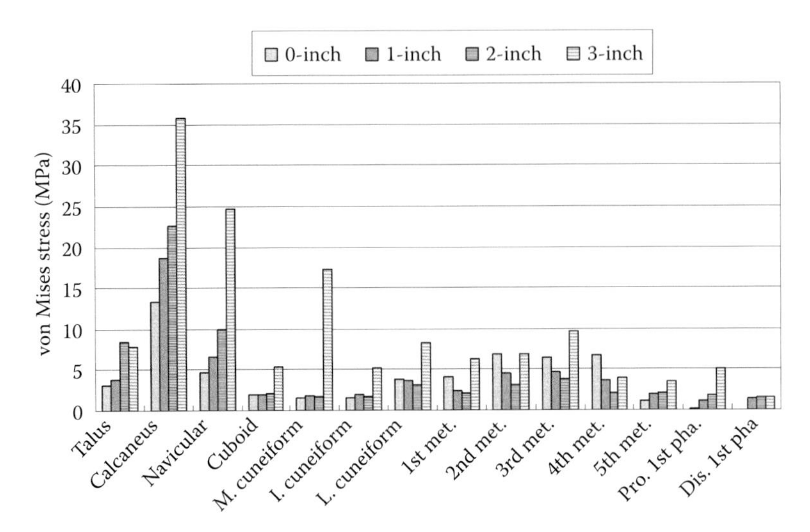


**FIGURE 2.7** Finite element–predicted peak von Mises stress of the foot bones during balanced standing on different high-heeled foot supports.

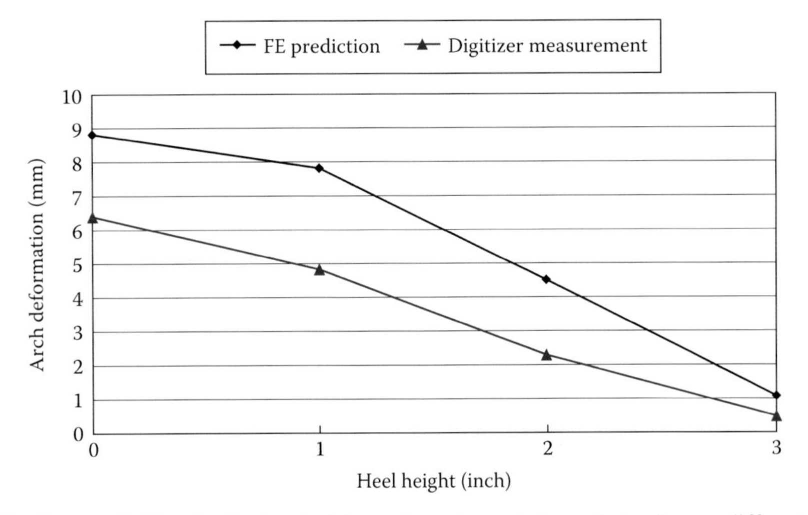


**FIGURE 2.8** Foot medial longitudinal arch deformations during balanced standing on different heel height foot supports: (a) flat, (b) 1-inch, (c) 2-inch, and (d) 3-inch.

tension force in the plantar fascia was minimum in all calculated cases. Moderate heel elevation may help to reduce the strain in the plantar fascia. This finding corresponds with existing conservative treatment strategies for plantar fasciitis. Comparing the FE predictions of static standing on flat support and high-heeled shoes, no noticeable rotation movement in the transverse plane of the first MTP segment was found, which was consistent with the cadaveric experiment. A pronounced increase in peak von Mises stress in the first MTP joint was predicted with high-heeled shoes when compared to flat support. Therefore, heel elevation was not found to be a direct biomechanical risk factor for hallux valgus deformity. However, combined effects with a tight toe box may impose risks of hallux valgus deformity. Heel elevation could be a triggering factor and should be confirmed in further study.

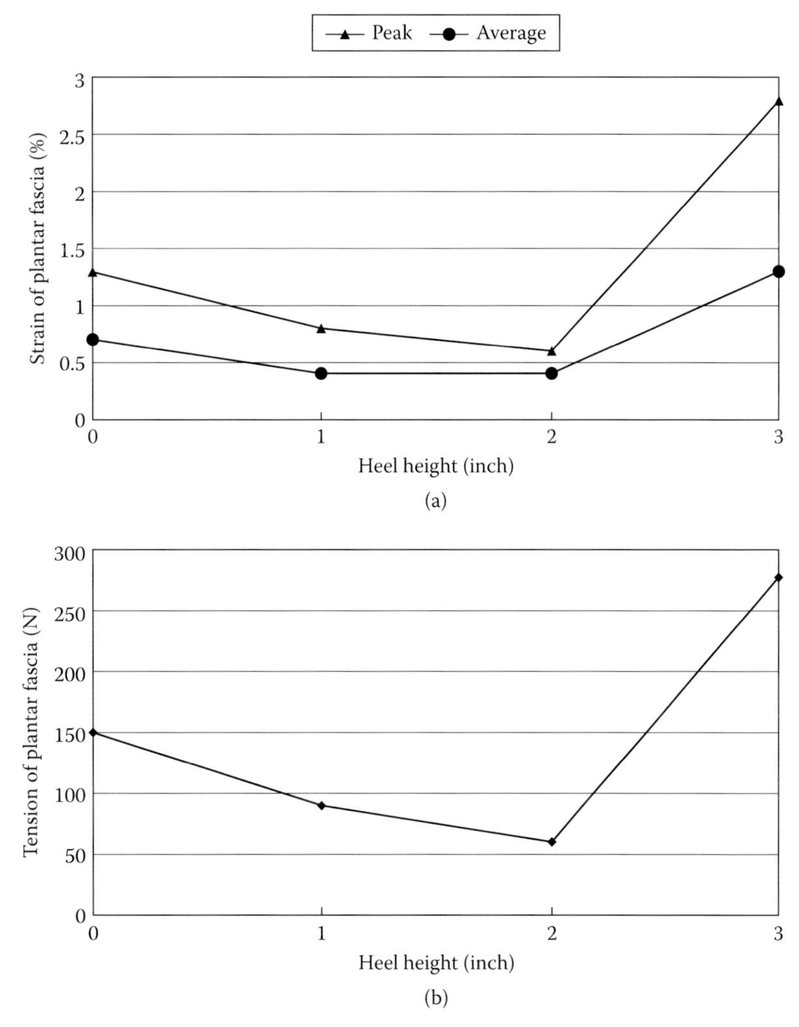


**FIGURE 2.9** Effects of heel height on plantar fascia: (a) finite element (FE)–predicted peak and average strain; and (b) FE-predicted total tension force.

## 2.3 APPLICATION OF THE FINITE ELEMENT MODEL FOR HIGH-HEELED SHOE DONNING AND WALKING

Validated FE foot and footwear models have been used for improving therapeutic and functional footwear designs (Cheung et al. 2009). Foot-sole FE models were developed to examine the effects of sole design on plantar pressure and bone stress (Chen, Ju, and Tang 2003; Cheung and Zhang 2008). Foot-shoe-ground models were also developed to study the effect of different midsole plug designs on plantar pressure (Gu et al. 2011) and the influence of varying sporting ground materials on impact force during landing (Kim et al. 2012). Recent studies showed the potentials and versatility of the FE method to simulate comprehensive foot-shoe interface (Ruperez, Monserrat, and Alcaniz 2008; Cheung et al. 2009; Qiu et al. 2011). However, these existing models usually simplified the footwear conditions, and the donning procedure with a complete shoe was not considered in the simulation. In some models, the footwear model was often simulated based on initial foot shape at a neutral position without consideration of the shoe's upper construction (Gu et al. 2011; Kim et al. 2012). The foot shape will deform after a shoe is donned, especially tight or high-heeled shoes, and the mechanical interaction between the foot and shoe will surely influence the loading characteristics. Therefore, incorporation of realistic donning is an important step for realistic biomechanical simulation of the foot and footwear.

This study incorporated a complete high-heeled shoe into a previously developed female FE foot model (Yu et al. 2008; Luximon et al. 2012) with an aim of extending the accuracy of shoe donning and shod walking simulations using the FE method. The objective of this study was thus to evaluate the validity and versatility of the proposed FE simulation, which could potentially contribute to the establishment of standard simulation techniques for foot-shoe interaction from donning to locomotion.

### 2.3.1 Simulation of High-Heeled Shoe Donning

Prior to establishing the donned condition, the high-heeled shoe model was first visually positioned and aligned in parallel and below the initially unloaded neutral foot model according to its central heel location and longitudinal foot axis (Figure 2.10a). Right at the start and throughout the whole donning simulation, the proximal end of the ankle (soft tissue, distal tibia, and fibula) was encrusted and gravity was applied. A four-step donning procedure was then initiated by first manipulating the

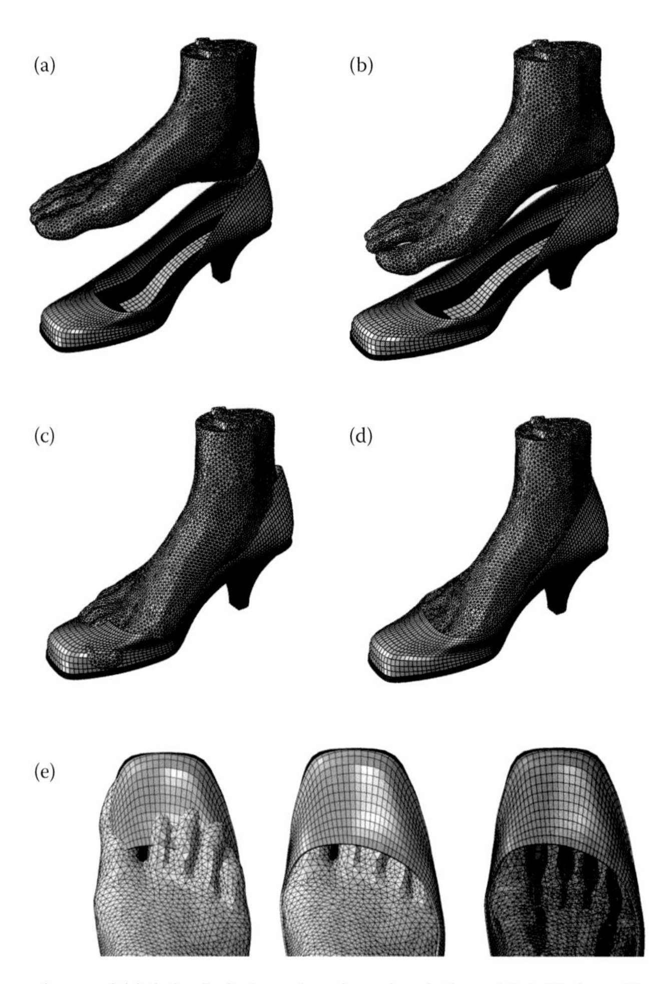


**FIGURE 2.10** Procedures of high-heeled shoe donning simulation: (a) initial position; (b) muscle forces applied to manipulate foot position; (c) contact established between foot and shoe sole; (d) contact established between foot and shoe upper; (e) localized view of shoe upper contact establishment.

foot into a plantarflexed position to match the shank profile of the high-heeled shoe. The Achilles tendon force (216 N, 40% of body weight) was first determined for matching the sagittal plane inclination between the high-heeled shoe support and the plantar foot (Figure 2.10b). This was then followed by defining the magnitude of other extrinsic muscle forces, including extensor hallucis longus (30 N) and extensor digitorum longus (20 N) for toe dorsiflexion, and peroneus brevis (50 N) and peroneus longus (60 N) for foot eversion in which their relative contributions to the ankle plantarflexion motion were based on estimations from EMG data and respective cross-sectional areas (Stefanyshyn et al. 2000; Perry, 1992). The tendon forces were properly adjusted according to the above criteria and sequenced until the foot was properly aligned relative to the shoe. In the second step, the high-heeled shoe was vertically displaced to initiate the surface contact between the shoe sole and the plantar foot. Fine adjustment of Achilles tendon force was also done at the same time to minimize the surface overlapping between the dorsal foot and upper shoe layer (Figure 2.10c). During the third step, the contact between the dorsal foot and upper shoe was established iteratively by reducing the area of overlapping surfaces using the interference fit contact simulation algorithm of ABAQUS (ABAQUS, 2011). A 50-mm initial allowable interference which ramped down to zero linearly over the step was given at the beginning of this iterative process for gradually resolving the surface overclosure (Figure 2.10d). The relaxed donned position of the foot with the high-heeled shoe (Figure 2.10e) was achieved in the final step by releasing all musculotendon forces and displacement constraints of the high-heeled shoe. At this donned position, there were about five degrees of ankle plantarflexion and three degrees of heel varus for the ankle-foot complex.

### 2.3.2 Simulation of High-Heeled Shod Walking

Following the shoe donning simulation and prior to the shod walking simulation with high-heeled shoes, the ground support was vertically displaced to establish the contact with the shoe heel. Loading and boundary conditions for the three simulated stance instances, heel strike, midstance, and push off, were then applied accordingly for the subsequent steps. The vertical GRF, location of center of pressure, and tibial inclination relative to the ground in the sagittal plane were collected during high-heeled shod walking for the same subject. For the sake of simplification, other GRF and tibial angle components in the transverse and coronal planes were not defined. The measured vertical GRFs (114%, 95%, and 120% of body weight, 540 N) and tibial inclination angles (-4, 0, and 19 degrees relative to the vertical ground axis) were prescribed via the rigid bottom layer of the ground support for heel strike, midstance, and push off phases, respectively, while the Achilles tendon forces were adjusted until the predicted center of pressure matched the experimental results. An increasing Achilles tendon load using an interval of 5% of body weight was applied and Achilles tendon forces of 75%, 110%, and 140% of the subject's body weight enabled the closest match with the corresponding measured centers of pressure. The remaining extrinsic muscle forces were estimated from the physiological cross-sectional area of the muscles (Dul 1983) and normalized normal walking EMG data (Perry 1992) assuming a linear relationship (Kim et al. 2001). Figure 2.11 shows the deformed plot of the ankle-foot and high-heeled shoe complex at different stages of the donning procedures and at the three simulated stance instances.

The predicted plantar and dorsal foot pressure, major joint contact pressures, and metatarsal bone stresses for the donned and shod walking conditions were compared. The FE predicted plantar pressure and its distributions were compared to the measured in-shoe plantar pressure using a 0.2-mm-thick thin-film insole pressure sensor (Tekscan, Boston, MA), which wore the same designed 2-inch high-heeled shoes.

### 2.3.3 Prediction Results

A coupled three-dimensional FE ankle-foot and high-heeled shoe model complex was established and used to simulate the biomechanical response of the foot-shoe interface during donning and three major stance instances, from heel strike to midstance to push off. The predicted interfacial

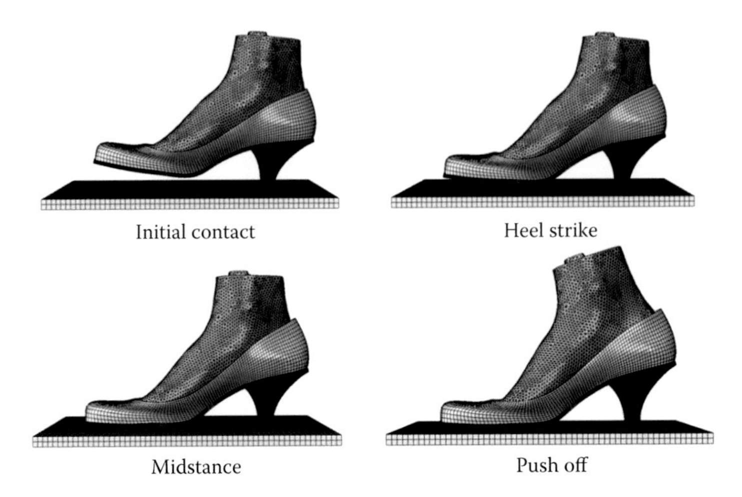


**FIGURE 2.11** Finite element–predicted walking shod in high-heeled shoes.

pressure, internal stress, and movement of different foot regions and bony structures in corresponding loading conditions are compared in the following subsections.

#### 2.3.3.1 Interfacial Foot Pressure during Donned and Shod Walking Conditions

The FE-predicted plantar pressure distribution patterns in general showed good agreement with the experimental measurements, while the predicted peak pressures at heel strike (0.47 MPa) and push off (0.63 MPa) phases were 14% and 26% larger than the measured values. During heel strike, peak pressure was located at the center of the heel. The pressurized area was concentrated at the lateral midfoot, ball of the foot, and toes during midstance. Localized peak pressure was found underneath the second and third metatarsal heads and at the center of the heel but there was a large area having no contact at the forefoot region during midstance. At push off, the heel and lateral midfoot were unloaded and the peak pressure at the forefoot intensified. The FE-predicted interfacial pressures, including plantar and dorsal pressure, during donned and shod walking conditions are tabulated in Table 2.1. After donning, slight contact pressure was found at the heel and toe regions, while noticeable contact pressure concentrated at the dorsum of the first, fourth, and fifth toes. During

**TABLE 2.1**
**Finite Element–Predicted Plantar and Dorsal Foot Contact Pressure (MPa) for Simulated High-Heeled Shoe Donning and Walking Conditions**

| | | Donned | | Heel strike | | Midstance | | Push off | |
|---|---|---|---|---|---|---|---|---|---|
| **Foot Regions** | | **Plantar** | **Dorsal** | **Plantar** | **Dorsal** | **Plantar** | **Dorsal** | **Plantar** | **Dorsal** |
| Toes | 1st | 0.03 | 0.07(M)<br>0.10(S) | 0.07 | 0.10(M)<br>0.10(S) | 0.15 | 0.18(M)<br>0.10(S) | 0.36 | 0.42(M)<br>0.11(S) |
| | 4th | - | 0.10 | 0.02 | 0.07 | 0.06 | 0.14 | 0.17 | 0.12 |
| | 5th | - | 0.13 | - | 0.11 | 0.10 | 0.31 | 0.36 | 0.32 |
| Forefoot | | - | 0.04 | - | 0.07 | 0.23 | 0.06 | 0.63 | 0.11 |
| Midfoot | | 0.04 | - | 0.10 | 0.05 | 0.08 | 0.06 | - | - |
| Rear foot | | - | 0.03 | 0.47 | 0.06 | 0.22 | 0.03 | - | - |

*Note:* M means medial side surface of the dorsum of the first toe as shown in Figure 2.12; S means superior side surface of the dorsum of the first toe as shown in Figure 2.12.

shod walking, dorsal contact pressure at the forefoot, midfoot, and rear foot had lower values than pressure at the dorsal toe regions. At the toe region, the first and fifth toes (0.42 and 0.32 MPa) had considerably large dorsal pressure at push off. Specifically, the dorsal contact pressure at the medial side of the first toe increased over four times (0.10 to 0.42 MPa) while the dorsal contact pressure at the superior side remained almost unchanged (0.10 MPa). Meanwhile, the dorsal contact pressure at the fifth toe increased about three times (0.11 to 0.32 MPa) from heel strike to push off.

#### 2.3.3.2 Joint Contact Pressure, Movement, and Bone Stress during Donned and Shod Walking

The predicted peak joint contact pressure at the MTP joints, tarsometatarsal joint (TMJ), midtarsal joint (MTJ), subtalar joint (STJ), and talocrural joint (TCJ) during donned and shod walking conditions are shown in Figure 2.12. The contact pressure at all MTP joints intensified and reached their maximum at push off. The maximum contact pressure at MTP joints during push off increased by at least four times (third MTP) and up to 11 times (first MTP) in comparison to the heel strike phase, and increased by about three to five times in comparison to the midstance phase. For midfoot and rearfoot joints, TMJ (22.1 MPa) contact pressure reached its maximum with a similar trend to the MTP joints for donned and shod conditions. However, the maximum contact pressure of TCJ (34.6 MPa) and STJ (25.9 MPa) occurred earlier during midstance and decreased at push off. The peak contact pressure of MTJ from heel strike to push off remained almost the same at around 15 MPa.

Regarding movements of the MTP joints, the first and fifth MTP joints showed the maximum changes among all MTP joints. The first MTP joint had 3.1 degrees of valgus while the fifth MTP joint had 1.5 degrees of varus in the transverse plane at donned position compared to the initial position (Figure 2.10e). During shod walking, the maximum deformation of MTP joints occurred at push off, of which the first MTP joint had 4.9 degrees of valgus and the fifth MTP joint had 1.7 degrees of varus in the transverse plane. In the sagittal plane, the first MTP joint had an increased dorsiflexion of 3.7 and 7.1 degrees at midstance and push off compared to the heel strike phase, while the fifth MTP joint increased by 3.2 and 6.1 degrees accordingly. The predicted joint movement characteristics of the remaining three MTP joints were similar to the first and fifth MTP joints during shod walking but with less movement, especially in the transverse plane. Meanwhile, the strain in the

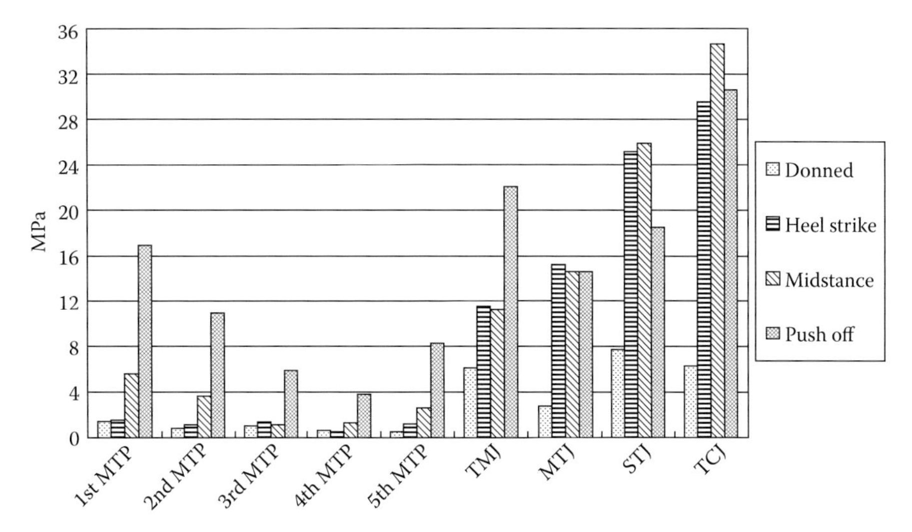


**FIGURE 2.12** Finite element–predicted peak joint contact pressure during donned and shod walking conditions (MTP: metatarsophalangeal joint; TMJ: tarsometatarsal joint; MTJ: midtarsal joint; STJ: subtalar joint; TCJ: talocrural joint).

medial collateral ligament of the first MTP joint at push off increased to 106% compared to the heel strike phase, while the lateral collateral ligament of the fifth MTP joint had only increased by 20% from the FE prediction.

### 2.3.4 Interpretation and Discussion

#### 2.3.4.1 Footwear Fitting and Donning

Using an FE foot-shoe model, interfacial foot pressure, bone and joint movement, and internal stress during the donning process could be visualized and quantitatively predicted. After the donned high-heeled shoe simulation, noticeable contact pressures tended to concentrate at the dorsum of the first, fourth, and fifth toes. It was predicted that the hallux deviated laterally and the fifth toe had a varus movement. The results showed that foot pressure and joint movements could be significantly altered by physical interaction with footwear, such as toe box constraint. The described shoe donning simulation approach allowed realistic simulation of complex foot shapes and shoe structures, which could be extensively used to investigate footwear design for neutral or deformed foot shapes.

#### 2.3.4.2 Interfacial Foot Pressure during Shod Walking

Despite high peak pressure values being predicted over some small regions at heel strike and push off due to the large deformation of the plantar foot tissue, the FE model showed close agreement in plantar pressure distribution patterns with the corresponding in-shoe plantar pressure measurements. Localized peak plantar pressure at the forefoot was found underneath the first, second, and third metatarsal heads, which was consistent with a previous study of high-heeled shoes by Mandato and Nester (1999), who reported that peak pressures increased by 30% to 40% at the forefoot and pressure shifted to the medial forefoot, especially the first metatarsal head and the hallux, with high-heeled shoes. Peak dorsal contact pressure mainly concentrated at the toe regions during shod walking. The dorsal medial contact pressure on the first toe increased significantly while the superior side remained almost unchanged. This could be one of the mechanical contributing factors to hallux valgus deformity. The in-shoe contact pressure on the plantar and dorsal sides of the toes could contribute to different kinds of soft tissue–related diseases or discomfort, such as skin irritation, callus, bunions, and metatarsalgia.

#### 2.3.4.3 Joint Contact Pressure, Movement, and Bone Stress during Shod Walking

During shod walking, the maximum contact pressures on MTP joints were predicted at the push-off phase as the forefoot sustained high loading with increased muscle forces in order to propel the body forward. McBride et al. (1991) reported from their locomotion analysis that the MTP joint reaction force was twice as large with high-heeled shoes as with barefoot walking. Wearing high-heeled shoes increases the load on the forefoot and reduces the load on the rear foot when compared to flat shoes (Broch, Wyller, and Steen 2004). During shod walking, the predicted maximum valgus angle on the first MTP joint increased by 51% (1.8 degrees) while the maximum varus of the fifth MTP joint only increased by 13% (0.2 degree), which correlated with the predicted ligament strain. The predicted strain of the medial collateral ligament of the first MTP joint was over five times greater than the strain in the lateral collateral ligament of the fifth MTP joint at push off. Meanwhile, the first MTP joint had a 16% larger increase (1.0 degree) in dorsiflexion at push off in comparison to the fifth MTP joint. These results indicated that the first MTP joint had a larger movement in both the transverse and sagittal planes than the fifth MTP joint at push off during high-heeled walking. These findings coincided with the clinical observations on the etiology of hallux valgus deformity after adoption of westernized shoe styles, especially high-heeled shoes, over traditional sandals in Japan (Kato and Watanabe 1981).

During high-heeled walking, major joints experienced different loading patterns. From the FE analysis, TMJ had the largest contact pressure at push off, which could be the result of a larger bending moment than MTJ at the midfoot, which had similar values of contact pressure throughout the stance phases. The relative high von Mises stress predicted at the second and third metatarsal shafts during push off could be a result of their confined positions with reduced joint mobility as well as longer shaft lengths of respective metatarsal bones. The mid-shafts of the second and the third metatarsals were found to be the most vulnerable locations to stress fracture or fatigue failure, consistent with clinical imaging observations in women (Goud et al. 2011) and previous FE-predicted bone stress distribution of a weight-bearing male foot (Cheung et al. 2005).

## 2.4 CONCLUSIONS

Systematic simulation procedures for shoe donning from the neutral foot position to the high-heeled shod position were established. Shod walking simulation considering a three-dimensional foot-shoe interface for the three major stance instances, from heel strike to midstance to push off, was done subsequently to the high-heeled shoe donning simulation. The described FE approach could enhance the accuracy and validity of FE predictions involving complex interfaces between the foot and footwear and provide a quantitative prediction of interfacial pressure, internal joint contact pressure, and movements during donning and shod walking conditions. The established high-heeled shoe donning and walking simulation in this study proved the versatility and promising potential of computational approaches for realistic biomechanical evaluation and optimization of footwear design in a virtual environment.

## ACKNOWLEDGMENTS

This work was supported by the Research Grant Council of Hong Kong (GRF Project nos. PolyU5352/08E, PolyU5326/11E), Hong Kong Polytechnic University PhD scholarship, and NSFC (11272273, 11120101001).

## REFERENCES

Basmajian, J.V., and G. Stecko. 1963. The role of muscles in arch support of the foot. *Journal of Bone and Joint Surgery (American)*. 45:1184–1190.

Broch, N.L., T. Wyller, and H. Steen. 2004. Effects of heel height and shoe shape on the compressive load between foot and base: A graphic analysis of principle. *Journal of the American Podiatric Medical Association*. 94:61–469.

Chen, W.P., C.W. Ju, and F.T. Tang. 2003. Effects of total contact insoles on the plantar stress redistribution: A finite element analysis. *Clinical Biomech.* 18, S17–24.

Cheung, J.T., J. Yu, D.W. Wong, and M. Zhang. 2009. Current methods in computer-aided engineering for footwear design. *Footwear Sci.* 1:31–46.

Cheung, J.T., and M. Zhang. 2008. Parametric design of pressure-relieving foot orthosis using statistics-based finite element method. *Med. Eng. Phys.* 30:269–277.

Cheung, J.T., M. Zhang, A.K. Leung, and Y.B. Fan. 2005. Three-dimensional finite element analysis of the foot during standing: A material sensitivity study. *J. Biomech.* 38:1045–1054.

Cole, C., C. Seto and J. Gazewood. 2005. Plantar fasciitis: Evidence-based review of diagnosis and therapy. *Am. Fam. Physician* 72:2237–2242.

Dul, J. 1983. Development of a minimum-fatigue optimization technique for predicting individual muscle forces during human posture and movement with application to the ankle musculature during standing and walking. PhD thesis, Vanderbilt University.

Gallup Organization Inc., 1986. Women's attitudes and usage of high-heeled shoes. The Gallup Organization Inc. Surrey, England.

Gefen, A., M. Megido-Ravid, Y. Itzchak, and M. Arcan. 2002. Analysis of muscular fatigue and foot stability during high-heeled gait. *Gait Posture* 15:56–63.

Gordon, G.M. 1984. Evaluation and prevention of injuries. *Clinical Podiatr.* 1:401–416.

Goud, A., B. Khurana, C. Chiodo, and B.N. Weissman. 2011. Women's musculoskeletal foot conditions exacerbated by shoe wear: An imaging perspective. *Am. J. Orthop.* 40:183–191.

Gu, Y.D., J.S. Li, M.J. Lake, Y.J. Zeng, X.J. Ren, and Z.Y. Li. 2011. Image-based midsole insert design and the material effects on heel plantar pressure distribution during simulated walking loads. *Comp. Meth. Biomech. Biomed. Eng.* 14:747–753.

Hanson, J.P., M.S. Redfern, and M. Mazumdar. 1999. Predicting slips and falls considering required and available friction. *Ergonomics* 42:1619–1633.

Jacob, S., and M.K. Patil. 1999. Stress analysis in three-dimensional foot models of normal and diabetic neuropathy. *Front. Med. Biol. Eng.* 9:211–227.

Kato, T., and S. Watanabe. 1981. The etiology of hallux valgus in Japan. *Clin. Orthop.* 157:78–81.

Kerrigan, C.D., J.L. Johansson, M.G. Bryant, and J.A. Boxer. 2005. Moderate-heeled shoes and knee joint torques relevant to the development and progression of knee osteoarthritis. *Arch. Phys. Med. Rehabil.* 86:871–875.

Kim, S.H., J.R. Cho, J.H. Choi, S.H. Ryu, and W.B. Jeong. 2012. Coupled foot-shoe-ground interaction model to assess landing impact transfer characteristics to ground condition. *Interaction and Multiscale Mechanics* 5:75–90.

Kirby, K.A. 2001. What future direction should podiatric biomechanics take? *Clin. Podiatr. Med. Surg.* 18:719–723.

Kogler, G.F., F.B. Veer, S.J. Verhulst, S.E. Solomonidis, and J.P. Paul. 2001. The effect of heel elevation on strain within the plantar aponeurosis: In vitro study. *Foot Ankle International* 22:433–439.

Lewis, G. 2003. Finite element analysis of a model of a therapeutic shoe: Effect of material selection for the outsole. *Biomed. Mater. Eng.* 13:75–81.

Linder, M., and C.L. Saltzman. 1998. A history of medical scientists on high heels. *Int. J. of Health Services.* 28:201–225.

Luximon, Y., A. Luximon, J. Yu, and M. Zhang. 2012. Biomechanical evaluation of heel elevation on load transfer: Experimental measurement and finite element analysis. *Acta Mech. Sinica* 28:232–240.

Mandato, M.G., and E. Nester. 1999. The effects of increasing heel height on forefoot peak pressure. *J. Am. Podiatr. Med. Assoc.* 89:75–80.

Manna, I., D. Pradhan, S. Ghosh, S.K. Kar, and P. Dhara. 2001. A comparative study of foot dimension between adult male and female and evaluation of foot hazards due to using of footwear. *J. Physiol. Anthropol. Appl. Human Sci.* 20:241–246.

Marshall, P. 1988. The rehabilitation of overuse foot injuries in athletes and dancers. *Clin. Sports Med.* 7:175–191.

McBride, I.D., U.P. Wyss, T.V. Cooke, B. Chir, L. Murphy, J. Phillips, and S. Olney. 1991. First metatarsophangeal joint reaction forces during high-heel gait. *Foot Ankle* 11:282–288.

Opila, K.A., S.S. Wagner, S. Schiowitz, and J. Chen. 1988. Postural alignment in barefoot and high-heeled stance. Spine 13:542–547.

Penrose, J.M., G.M. Holt, M. Beaugonin, and D.R. Hose. 2002. Development of an accurate three-dimensional finite element knee model. *Comp. Meth. Biomech. Biomed. Eng.* 5:291–300.

Perry, J. 1992. *Gait Analysis: Normal and Pathological Function.* Thorofare, NJ: SLACK Inc.

Qiu, T.X., E.C. Teo, Y.B. Yan, and W. Lei. 2011. Finite element modeling of a 3D coupled foot-boot model. *Med. Eng. Phys.* 33:1228–1233.

Ruperez, M.J., C. Monserrat, and M. Alcaniz. 2008. Simulation of the deformation of materials in shoe uppers in gait: Force distribution using finite elements. *Int. J. Interactive Design Manufacturing* 2:59–68.

Shiina, Y., Y. Hamamoto, and K. Okumura. 2006. Fracture of soft cellular solids: Case of non-crosslinked polyethylene foam. *Europhys. Lett.* 76:588–594.

Shimizu, M., and P.D. Andrew. 1999. Effect of heel height on the foot in unilateral standing. *J. Phys. Ther. Sci.* 11:95–100.

Smith, E.O., and W.S. Helms. 1999. Natural selection and high heels. *Foot Ankle Int.* 20:55–57.

Speirs, A.D., M.O. Heller, G.N. Duda, and W.R. Taylor. 2007. Physiologically based boundary conditions in finite element modelling. *J. Biomech.* 40:2318–2323.

Stefanyshyn, D.J., B.M. Nigg, V. Fisher, B. O'Flynn, and W. Liu. 2000. The influence of high heeled shoes on kinematics, kinetics and muscle EMG of normal female gait. *J. Appl. Biomech.* 16:309–319.

Viceconti, M., S. Olsen, L.P. Nolte, and K. Burton. 2005. Extracting clinically relevant data from finite element simulations. *Clin. Biomech.* 20:451–454.

Wunderlich, R.E., and P.R. Cavanagh. 2001. Gender differences in adult foot shape: Implications for shoe design. *Med. Sci. Sports Exerc.* 33:605–611.

Yu. J., J.T. Cheung, Y. Fan, Y. Zhang, A.K. Leung, and M. Zhang. 2008. Development of a finite element model of female foot for high-heeled shoe design. *Clin. Biomech.* 23 (Suppl. 1):31–38.

Zhang, M., and A.F. Mak. 1999. In vivo skin frictional properties. *Prosthetics Orthotics Int.* 23:135–141.

# 3 Foot and Ankle Model for Surgical Treatment

*Yan Wang and Ming Zhang*

## CONTENTS



## SUMMARY

People commonly succumb to many foot-related problems or diseases, such as plantar foot pain, diabetic foot, arthritic foot, flexible flatfoot, joint sprain, bone fracture, and other sports-related injuries. Surgical treatments such as joint fusion (arthrodesis) and joint replacement (arthroplasty) are often used to promptly alleviate the symptoms and restore normal functions. Concerns often arise as to the effectiveness and biomechanical consequences of surgical treatments and how the treatments may influence the biomechanical structure of the entire foot. A clear picture of the internal biomechanics, such as stress/strain distribution, contact pressure, and deformation, can help us to better understand the situation. Because it is difficult, costly, and invasive to conduct experimental measurements on the human body, we developed finite element (FE) methods to address these questions. As an example, fusion of the tarsometatarsal joints is analyzed and reported on in this chapter.

## 3.1 INTRODUCTION

Midfoot injury occurs around the five tarsal bones and the attached ligaments, covering a wide spectrum of soft tissues and bony damage. The tarsometatarsal (TMT) joints comprise the bases of the five metatarsal bones and their articulation with the three cuneiforms and the cuboid bone. It was reported that fractures and dislocations of the TMT joints accounted for 0.2% of foot injuries (Ghate et al. 2012; Myerson 1989; Panchbhavi et al. 2009). The actual incidence may be higher than reported (Ghate et al. 2012; Myerson et al. 1986; Philbin, Rosenberg, and Sferra 2003; Rammelt et al. 2008), because such injuries often go unreported.

Foot-related problems can be solved or alleviated by either nonsurgical methods, including orthosis, physical therapy, medicine, and injections, or surgical methods, such as bone fixation and joint replacement. Normally, all kinds of injuries should ideally be treated by nonsurgical means, with surgery being used if an expected functional level is not achieved. Subtle ligamentous damage to TMT joints is generally treated using nonsurgical methods, while bone fractures, dislocations, and severe ligamentous injuries should be treated by closed or open reduction with bone fixation.

Surgical treatments with open reduction and internal fixation are reliable means of securing and maintaining reduction of these injuries (Arntz, Veith, and Hansen 1988; Myerson 1989; Kuo et al. 2000). Delayed treatment or misdiagnosis of these injuries, even minor injuries with subtle subluxation and diastasis of the articulation, can lead to significant complications, such as long-term pain, degenerative arthritis, chronic instability, and other disabilities (Eleftheriou, Rosenfeld, and Calder 2013). These eventually must be treated with bone fixation or arthrodesis (Aronow 2006).

Surgical treatment is expected to provide patients with plantigrade foot with normal function. However, the outcome of the operation may not always be positive. In some cases patients may be inflicted with new disabilities or require further surgery.

With surgical treatment of TMT injuries, short-term complications such as compartment syndrome, infection, neurovascular injury, and wound-associated problems have been reported. In the long term, the majority of complaints were post-traumatic arthritis, flat foot deformity, and instability (vanRijn et al. 2012; Myerson 1999). It has been debated that stressful interaction between fixed segments could cause bone irritation, loosening, and sometimes breakage (Cottom, Hyer, and Berlet 2008). A study investigating mid-term outcomes of TMT joint fixation found that 13.8% of patients sustained post-traumatic arthritis, 3.45% developed subluxation, 6.9% developed flatfoot, and another 6.9% developed severe symptoms with limited functionality which led to further arthrodesis (Ghate et al. 2012).

Clinically, the recurrence of fractures at new sites is also quite common. For female track and field athletes, half of those who reported a history of stress fracture had experienced a stress fracture on more than one previous occasion (Bennell et al. 1995). Similarly, a study conducted on male military personnel presented comparable results (Milgrom et al. 1985).

Those findings indicate that surgical interventions may alter foot biomechanics, such as internal load transfer and pressure distribution, and modification of any part may affect other segments, especially adjacent regions. Biomechanical disorders resulting from surgical interventions are the most common problem after operation. It is thus very important to estimate the mechanical effects of surgeries on the functionality of the foot and ankle to assist in surgical planning for an optimal outcome.

The foot is an intricate and synergetic system. It is necessary to consider the possible consequences of surgical treatment not only around the surgical regions, but also over the entire foot and ankle. Changes in stress/strain distribution on the foot structures are important contributors for understanding the consequences of treatment. Unfortunately, biomechanical information cannot be easily gathered through experimental means.

## 3.2 CURRENT METHODS IN BIOMECHANICAL STUDIES

### 3.2.1 Experimental Studies

The methods currently used to estimate the effect of surgery on the foot and ankle are mainly focused on follow-up investigations, human motion analysis, and cadaveric experiments. Follow-up investigations involve radiography estimation and score systems like the American Orthopedic Foot and Ankle Society score and Maryland foot score. They are explored to find functional outcomes of surgeries and the radiography method is mainly used to assess alignment after operations. In most cases, postoperative radiograph combined with a score system are used to evaluate corrections, failures, and even fractures (Weatherall, Chapman, and Shapiro 2013; Adam et al. 2011; Coetzee and Wickum 2004). Gait analysis is the most commonly adopted method of examining human motion used to estimate surgical outcome and rehabilitation. It provides kinematic information on the foot and ankle. It could be employed to identify interaction between muscles and bones (Burg et al. 2013), to investigate outcomes of treatments (Goetz et al. 2013; Lins et al. 2013; Graf et al. 2012; Hetsroni et al. 2011; Garrido et al. 2010), and to study characteristics of the pathological foot (Watanabe et al. 2012; Baan et al. 2012; Sawacha et al. 2009). In order to obtain accurate kinematic information, even invasive in vivo research on the human foot has been reported (Nester 2009). Cadaveric experiments are employed as a testing or training ground before surgery to predict possible outcomes and complications (Lui, Chan, and Chan 2013; Kamiya et al. 2012; Cook et al. 2009). These methods are widely adopted in clinical practice.

The score system can only be used to provide an overview of outcomes after surgery, rather than the biomechanical changes within the foot, and thus will not provide information on complications and passive outcomes. Radiograph images show the position of different segments and assist in alignment, but more detailed information on the inner foot and ankle cannot be visualized pre- and postoperatively. Cadaveric experiments can detail contact pressure on the interfaces of some of the articulations, but the measurements are quite limited by the sensor technique and the complex contours at the articular interfaces. Gait analysis reveals the effect of clinical interventions on the whole locomotion system rather than the influence on the inner foot. Despite the readily available data on alterations to the gait pattern, the most common complications may be localized to the modified foot and ankle complex, such as recurrence of stress fracture and arthritis. Gait analysis offers kinematic parameters of the marked segments in detail, instead of the kinetic parameters of individual segments of the foot and ankle that may be linked to surgical complications. Cadaveric experiments, radiographs, and gait analysis can estimate the overall status of the outcome of surgeries, but none of them can reveal the initiating factors of the complications and passive outcomes.

### 3.2.2 Computational Analysis

Since direct measurement of the internal biomechanical parameters is difficult to achieve, computational approaches such as FE analysis are widely used for such purposes. FE methods were developed for analyzing structural mechanical problems. With the advancements in material properties and computational modeling techniques, the FE method became widely used for biomechanical applications. This method offers a feasible alternative to experimental methods and presents a realistic simulation of in vivo conditions. Computational analysis offers insight into the internal stress distribution, contact pressure, and deformation of each individual subject under load, despite the complex geometries and material properties included in the model. Input parameters such as loading conditions, material properties, and geometries may be rapidly modified within the simulation and the model will update the results accordingly. The flexibility in simulating complex loading and pathological conditions of the foot and ankle complex offers great advantages in terms of speed, cost effectiveness, and detailed results. As such, the FE analysis could be used to assist in surgical decision making for various injuries to the foot and ankle.

FE models of the foot and leg have been developed to offer a better understanding of the mechanisms of injury to the ankle and subtalar joints under impact loading conditions during vehicle collisions (Shin, Yue, and Untaroiu 2012). Their model was developed based on a geometric model of the bones and skins constructed by the Center for Injury Biomechanics, Virginia Tech-Wake Forest University (Gayzik et al. 2011). In this model, only the bones of the hind foot and those surrounding the ankle were modeled as deformable bodies with elastoplastic material properties. All other bones were defined as rigid body models with no deformation. A 12% variation of failure moment was observed in the range of axial foot rotations (±15 degrees). This simulation provided limited biomechanical information on the fore- and mid-foot, which is necessary for investigation of the load transfer path.

Liang et al. (2011) investigated the deformation and stress distribution of the human foot after plantar ligaments release using FE analysis and cadaveric experiments, aiming to explore the role of the plantar ligaments in foot arch biomechanics. The FE model was developed from computer tomography images and reconstructs the majority of the foot musculoskeletal structures, including bone segments, major ligaments, and plantar soft tissue. There are no soft tissues embedded in the bony structure, and thus the interaction between the tissue's inner surface and the bones was neglected, which may be an important factor in bone stress distribution. The plantar ligament was represented by a solid model and was assumed to be linearly elastic, which is much different from reality. The cadaveric experiment was carried out to validate the FE model. They found that the release of the plantar fascia decreased arch height without causing the total collapse of the foot arch and the longitudinal foot arch was destroyed after sectioning of the four major plantar ligaments. The FE model indicated that the release of the plantar fascia may relieve focal stress associated with heel pain.

An FE study was conducted to evaluate the effects of different foot postures on fifth metatarsal fracture healing (Brilakis et al. 2012). The model combined 53 plantar and dorsal ligaments, the plantar fascia, 28 bones, and surrounding soft tissue, all of which were set as homogeneous, isotropic, and linearly elastic. The muscle attachment points were not positioned exactly at their anatomical locations, and boundaries conditions, including the muscle forces, were simplified to be of the same value in three different foot postures. As stress in the bone is mainly caused by the transfer of ground reaction forces (GRFs) and muscle contraction, inaccurate application of GRFs and muscle forces may lead to erroneous results. Different postures were simulated by simply turning the model through different angles in relation to the ground, instead of simulating the actual interaction of different segments during different postures. The study concluded that different foot postures did not significantly influence the peak strains at the fracture site of the fifth metatarsal and eversion of the foot caused further torsional strain on the fracture site of the fifth metatarsal.

Although the aforementioned studies offer much insight into the biomechanical environment within the foot and corresponding forces between segments, there are inherent limitations to these models that may affect their reliability. It is necessary to develop a more accurate model of the foot and ankle for clinical application.

## 3.3 FINITE ELEMENT SIMULATION OF TARSOMETATARSAL JOINT FUSION

An FE model of the foot and ankle was modified from a previous study, the details of which are provided in Chapter 1. The FE model was based on the right foot of a normal female adult with body weight and mass of 164 cm and 54 kg, consisting of 28 bony segments, 72 ligaments, and the plantar fascia embedded in a volume of encapsulated soft tissue (Cheung and Zhang 2005; Cheung et al. 2005). For simplicity, it was assumed that the TMT joint fusion does not change the foot kinematics much during gait, based on the fact that the motion of TMT joints is quite limited in a normal foot.

### 3.3.1 Boundary and Loading Conditions during Walking

The kinematic and kinetic information during walking was obtained from human locomotion analysis using a motion monitoring system, force platform, and electromyography (EMG) measurements. The gait tracking system provides kinematic information on the lower limbs with markers positioned on regions of interest. The curve of the GRFs obtained from the force platform, with the first peak, midstance, and the second peak, are shown in Figure 3.1.

The instant of the first peak in terms of the vertical GRF is at about 25% and the second peak located around 70% of the stance phase. The midstance instant was chosen as the valley in the curve between the first and second peak. To simulate different gait instants, the active extrinsic muscle forces, in addition to the GRF, were applied. The muscle forces were estimated from physiological cross-sectional areas (Dul 1983) of muscles and EMG data with a linear EMG-force assumption (Kyu-Jung Kim et al. 2001). All of the boundary and loading conditions obtained from the four simulated instants are listed in Table 3.1.

The muscles in the FE model were represented by lines connecting the anatomical attachment points of muscles to bones. The Achilles tendon was represented by five axial connector elements, on which five equivalent force vectors were applied at the points of insertion. Other muscle forces were applied to the corresponding muscle structures represented by solid lines, as shown in Figure 3.2,

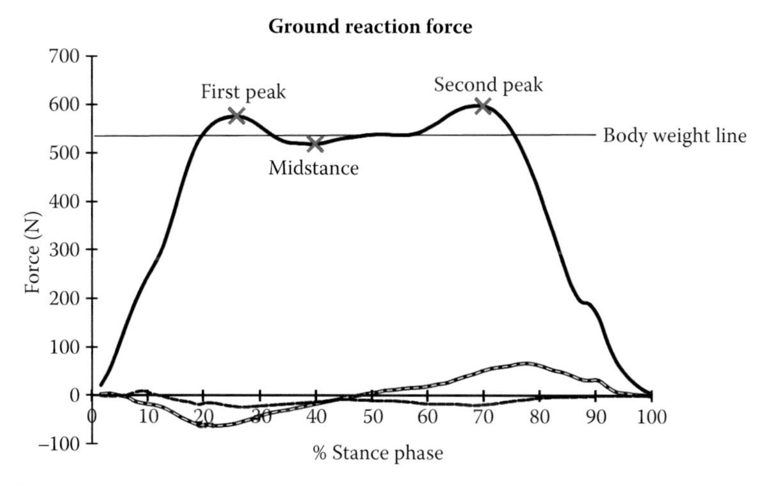


**FIGURE 3.1** The vertical ground reaction force and the three investigated gait instants.

**TABLE 3.1**
**Boundary Conditions of the Four Simulated Instants: Balanced Standing, First Peak, Midstance, and Second Peak**

| Loadings | Balanced Standing | First Peak | Midstance | Second Peak |
|---|---|---|---|---|
| GRF (N) | 270 | 578.6 | 519.3 | 600 |
| Ankle-shank angle (rad) | 0 | 0.113 | 0.216 | 0.485 |
| Tibialis anterior (N) | - | 0 | 0 | 0 |
| Tibialis posterior (N) | - | 34 | 42.5 | 0 |
| Achilles tendon (N) | 135 | 500 | 900 | 1100 |
| Extensor digitorum longus (N) | 0 | 0 | 0 | 0 |
| Flexor digitorum longus (N) | 0 | 40 | 20 | 96 |
| Flexor hallucis longus (N) | 0 | 30 | 0 | 284 |
| Peroneus longus (N) | 0 | 0 | 41.25 | 0 |
| Peroneus brevis (N) | 0 | 20 | 22 | 91.8 |

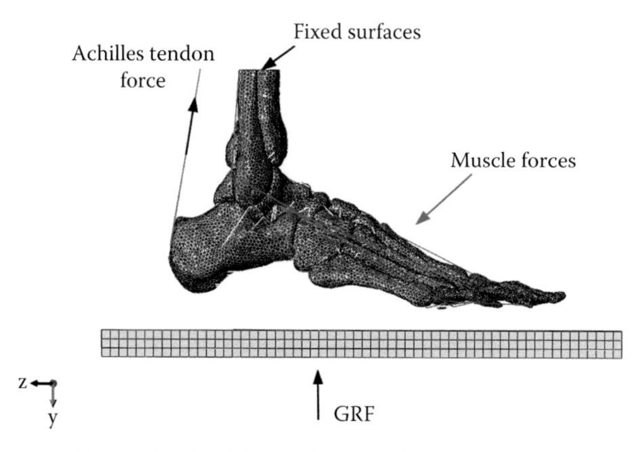


**FIGURE 3.2** Boundary conditions for simulation of the four instants.

by defining contraction forces via axial connector elements. The GRF was applied as a concentrated force to the rigid plate. The superior surfaces of the encapsulated soft tissue, distal tibia, and fibula were fixed throughout the simulation and the foot shank positions were represented by turning the rigid plate to the same angles. The boundary conditions applied are shown in Figure 3.2.

Balanced standing was also studied to investigate the arch deformation. Half of the body weight, which was approximately 270 N for a 54-kg subject, was applied vertically on the foot. During balanced standing, it was simplified such that only the Achilles tendon force was applied, neglecting the other muscle forces.

### 3.3.2 Simulation of First and Second Tarsometatarsal Joints Fusion

To simulate midfoot joint fusion, the first and second TMT joints (Figure 3.3) were fused. The tied articulating pairs were the medial cuneiform and first metatarsal, the medial cuneiform and intermediate cuneiform, the intermediate cuneiform and second metatarsal, and the second metatarsal and medial cuneiform (Figure 3.3). The three instants were simulated in the normal foot model and the joint fusion model. The stress distribution on the bones, with particular interest to the metatarsal bones, and the contact pressure on the adjacent joints of the fused region were reported.

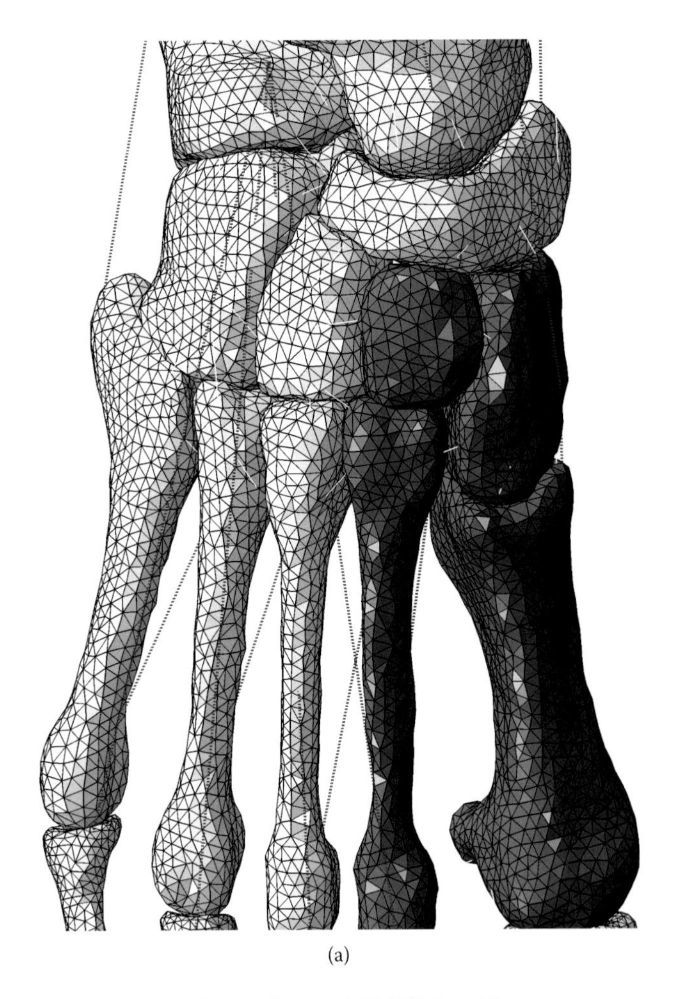

(a)

**FIGURE 3.3** The fixed bones of the first and second TMT joints (a).

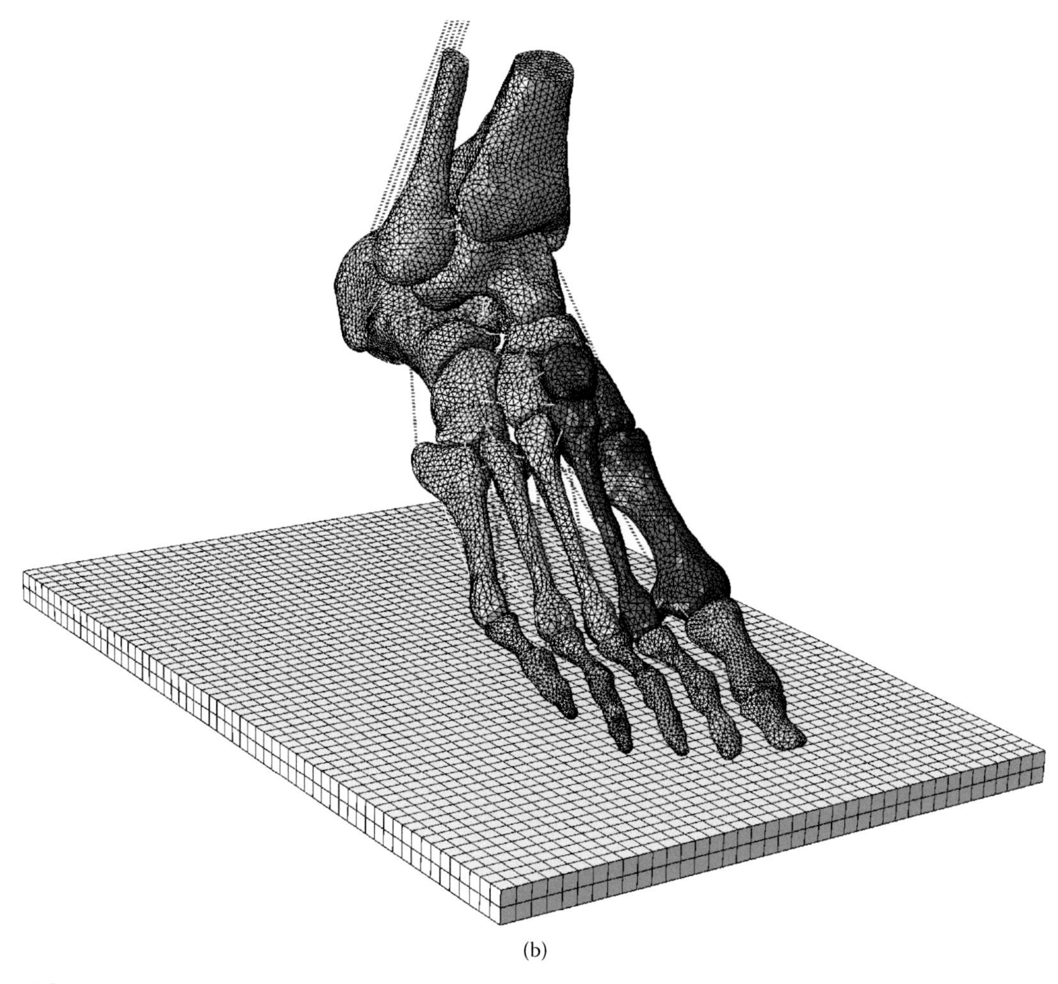

(b)

**FIGURE 3.3** (*Continued*) The model for simulation (b).

## 3.4 EFFECTS OF TARSOMETATARSAL JOINT FUSION

In order to reveal the effect of the fixation of the TMT joints, the results of the normal and fused joint models were compared. The contact pressures at the articulating interfaces in the midfoot and hindfoot, von Mises stress in the five metatarsal bones, and contact pressure distribution on the plantar foot were analyzed.

### 3.4.1 Contact Pressure of Plantar Foot and Joints

The maximum contact pressure between plantar foot and support increased by 0.42%, 19%, and 60%, corresponding to the first peak, midstance, and the second peak. A normal arch deforms to interact with the environment in a most effective way, protecting the segments from excessive loads. It was found that a foot with fused joint limited arch deformation capability and was subjected to higher plantar pressure, especially during push-off.

The first metatarsal bone was observed to be significantly dorsiflexed in flatfoot relative to the talus bone (Blackman et al. 2009; Kido et al. 2013). The fusion resulted in a limited range of movement of the first metatarsal bone and a stiffer arch. In order to investigate the effect of fusion on a flexible flatfoot, the two models in a balanced standing state were compared. The arch height was measured by the distance between the dorsal peak of the intermediate cuneiform and the plantar peak of the calcaneus bone in the superior-interior direction and was found to differ between the normal foot and the foot with fused joints. The fused foot had a 24% less variation in arch height compared to the normal foot. This could be attributed to the fact that the fused foot was more capable of resisting arch deformation because the relative motion among the four fused bones was totally limited. Fusion of the first TMT joint could be a way to correct flatfoot.

The contact pressures at the joints of the ankle, subtalar, talonavicular, calcaneocuboid, navico-mcuneiform, navico-icuneiform, navico-lcuneiform, lcunecuboid, and the third, fourth, and fifth TMT were investigated. It was found that joint fusion increased the contact pressure at the joints of the ankle, talonavicular, navico-icuneiform, navico-cuboid, and fifth meta-cuboid.

The maximum contact pressure on the ankle joint in the normal foot model increased as the gait cycle progressed, from 14 MPa in the first peak to 28 MPa in midstance and 48 MPa in the second peak. In the fusion model, these instants increased further by 12%, 14%, and 0.58%, respectively. The talonavicular joint was subjected to higher contact pressure by the fusion, increasing by 5.2%, 1.7%, and 11% of the normal foot model. The maximum contact pressure in the navico-icuneiform joint was 7.5, 10, and 16 MPa during the three gait instants and grew 8.4, 12, and 20 MPa after fusion.

Among all the fluctuations in contact pressure on the articulating interfaces in the hind- and midfoot, the navicular and cuboid contact pair showed the most considerable variation during midstance. The following two were the fifth meta-cuboid and navico-icuneiform joints. The ankle joint sustained the highest magnitudes among all joints, reaching a maximum of 48 MPa. The limited motion of the fixed bones induced higher contact pressure in mid- and hindfoot joints. These joints were subject to greater risk of cartilage damage and deformation from a normal anatomical position under a continual and long-term excessive loading condition. This could be regarded as a predictor of foot pain and malalignment. Malalignment of foot segments could further affect normal functioning of the upper parts of the foot and ankle, for example the knee joint. A disordered mechanical environment also contributes to disturbing the maintenance of the articular cartilage and underlying bones. Higher contact pressure on joints may leave them more susceptible to fatigue wear of the contact surfaces. Heightened pressure over a prolonged period often leads to the development of arthritis, which has been reported as a common postoperative complication (Myerson 1999; van Rijn et al. 2012; Ghate et al. 2012).

### 3.4.2 von Mises Stress in the Five Metatarsal Bones

The von Mises stress is often considered one predictor for bony stress fracture (Keyak and Rossi 2000). The five metatarsal bones are thought to be most susceptible for recurring stress fracture because of the long and thin shape and the function of loading transfer. Figure 3.4 shows the von Mises stress during midstance.

Based on the comparison between the two models, the most change in von Mises stress was observed in the second metatarsal bone during midstance, showing a 22% increase, from 26 MPa to 31 MPa, after fusion. The increase was 16% in the first peak and 14% in the second peak. The fifth metatarsal bone increased by 5.1% and 9.5% in the first peak and midstance after fusion. The stress in the first and fourth metatarsal bones did not change substantially after fusion.

Metatarsal stress fractures are most commonly seen in the second and the third metatarsals and the fracture of the second metatarsal is reported to be one of the most common problem after surgeries in foot and ankle (Weatherall, Chapman, and Shapiro 2013). FE analysis of fusion of the first and second TMT joints shows that the second metatarsal bone is more likely to sustain a fracture, considering the 22% increase in von Mises stress during midstance.

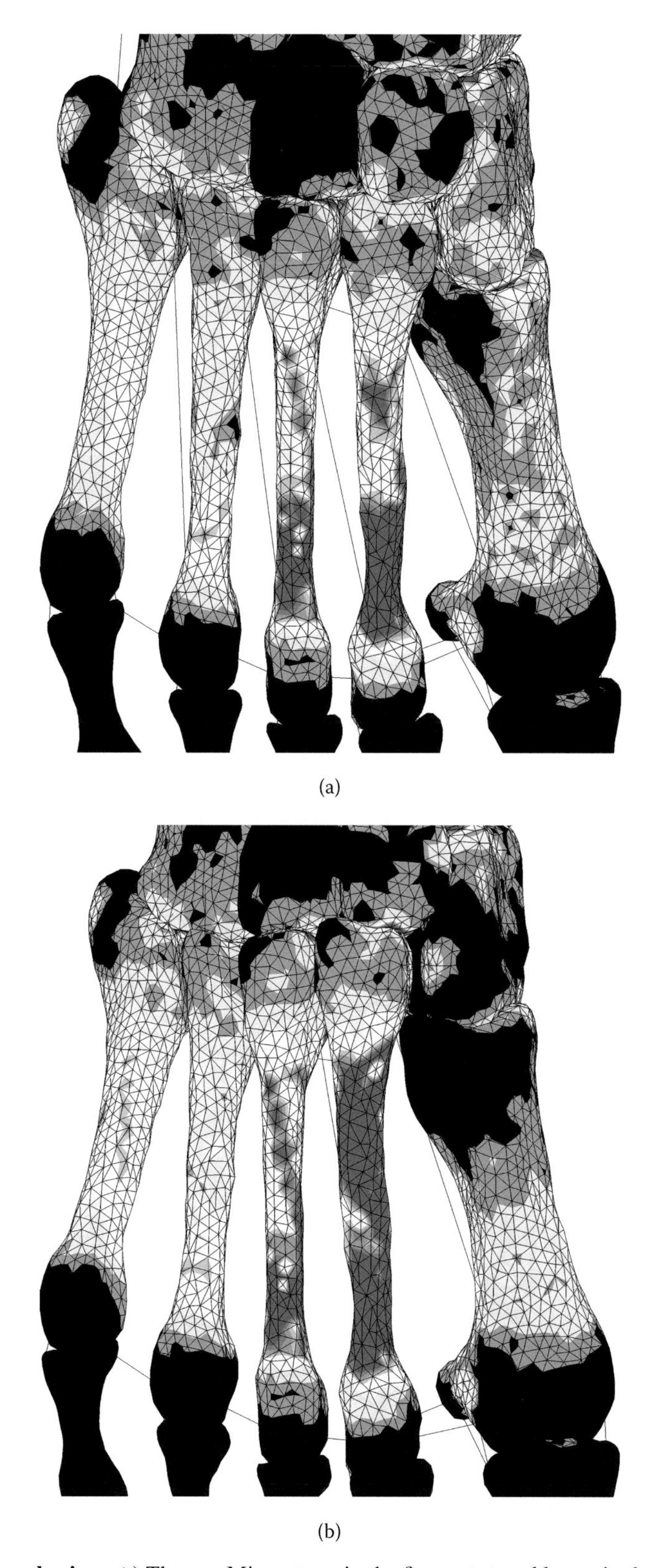


**FIGURE 3.4 (See color insert.)** The von Mises stress in the five metatarsal bones in the normal foot model (a) and model with the first and second TMT joints fused (b) in midstance.

## 3.5 CONCLUSIONS

This study reported on the possible outcomes and complications associated with surgical intervention on the first and second TMT joints. A number of biomechanical parameters, including plantar contact pressure, von Mises stress in the five metatarsal bones, and the contact pressure of joints, were analyzed during walking instants to identify the biomechanical consequences of joint fusion. The analysis suggested that fixation of the first two metatarsals and the first two cuneiforms would influence the stress and pressure distribution of other segments of the foot. The plantar contact pressure increased in all three instants due to fusion, with the maximum variation occurring during push off. The fusion model is stiffer than the normal foot. Fusion of the first TMT joint could be a way to correct flatfoot.

In the analysis of the contact pressure in joints of the hind- and midfoot, the navicular and cuboid contact pair showed the most considerable variation in midstance. The ankle joint, talonavicular, navico-icuneiform, and the fifth meta-cuboid joints also sustained increased contact pressure. Depending on the results, it is possible that the navico-cuboid and the ankle joints have the greatest potential to succumb to arthritis. Among the five metatarsal bones, the von Mises stress in the second metatarsal varied the most in midstance. Since the second and third metatarsal bones are reported to be the segments most susceptible to stress fracture, the increased von Mises stress in the second metatarsal resulting from the fusion of the two joints could be an indication of an increased risk of stress fracture.

Optimal surgeries are expected to decrease the complications and negative long-term outcomes, permitting effective surgical intervention to address foot problems associated with pain, decreased ambulation, and decreased quality of life. Sufficient understanding of biomechanics could provide surgeons with more low-risk, sophisticated treatment options that are currently not well known or considered too risky to undertake. FE analysis could be an effective method to explore the rationale of biomechanical changes undergone after surgery and provide direct guidelines for surgery planning.

## ACKNOWLEDGMENTS

This work was supported by the Research Grant Council of Hong Kong (GRF Project no. PolyU5326/11E), and NSFC (11272273, 11120101001).

## REFERENCES

Adam, S. P., S. C. Choung, Y. Gu, and M. J. O'Malley. 2011. Outcomes after scarf osteotomy for treatment of adult hallux valgus deformity. *Clin Orthop Relat Res* 469 (3):854–9.

Arntz, C. T., R. G. Veith, and S. T. Hansen, Jr. 1988. Fractures and fracture-dislocations of the tarsometatarsal joint. *J Bone Joint Surg Am* 70 (2):173–81.

Aronow, M. S. 2006. Treatment of the missed Lisfranc injury. *Foot Ankle Clin* 11 (1):127–42, ix.

Baan, H., R. Dubbeldam, A. V. Nene, and M. A. van de Laar. 2012. Gait analysis of the lower limb in patients with rheumatoid arthritis: a systematic review. *Semin Arthritis Rheum* 41 (6):768–788 e8.

Bennell, K. L., S. A. Malcolm, S. A. Thomas, P. R. Ebeling, P. R. McCrory, J. D. Wark, and P. D. Brukner. 1995. Risk factors for stress fractures in female track-and-field athletes: a retrospective analysis. *Clin J Sport Med* 5 (4):229–35.

Blackman, A. J., J. J. Blevins, B. J. Sangeorzan, and W. R. Ledoux. 2009. Cadaveric flatfoot model: ligament attenuation and Achilles tendon overpull. *J Orthop Res* 27 (12):1547–54.

Brilakis, E., E. Kaselouris, F. Xypnitos, C. G. Provatidis, and N. Efstathopoulos. 2012. Effects of foot posture on fifth metatarsal fracture healing: a finite element study. *J Foot Ankle Surg* 51 (6):720–8.

Burg, J., K. Peeters, T. Natsakis, G. Dereymaeker, J. Vander Sloten, and I. Jonkers. 2013. In vitro analysis of muscle activity illustrates mediolateral decoupling of hind- and mid-foot bone motion. *Gait Posture* 38 (1):56–61.

Cheung, J. T., and M. Zhang. 2005. A 3-dimensional finite element model of the human foot and ankle for insole design. *Arch Phys Med Rehabil* 86 (2):353–8.

Cheung, J. T., M. Zhang, A. K. Leung, and Y. B. Fan. 2005. Three-dimensional finite element analysis of the foot during standing—a material sensitivity study. *J Biomech* 38 (5):1045–54.

Coetzee, J. C., and D. Wickum. 2004. The Lapidus procedure: a prospective cohort outcome study. *Foot Ankle Int* 25 (8):526–31.

Cook, K. D., L. C. Jeffries, J. P. O'Connor, and D. Svach. 2009. Determining the strongest orientation for "Lisfranc's screw" in transverse plane tarsometatarsal injuries: a cadaveric study. *J Foot Ankle Surg* 48 (4):427–31.

Cottom, J. M., C. F. Hyer, and G. C. Berlet. 2008. Treatment of Lisfranc fracture dislocations with an interosseous suture button technique: a review of 3 cases. *J Foot Ankle Surg* 47 (3):250–8.

Dul, J. 1983. Development of a minimum-fatigue optimization technique for predicting individual muscle forces during human posture and movement with application to the ankle musculature during standing and walking. Ph.D. thesis, Vanderbilt University.

Eleftheriou, K. I., P. F. Rosenfeld, and J. D. Calder. 2013. Lisfranc injuries: an update. *Knee Surg Sports Traumatol Arthrosc*.

Garrido, I. M., J. C. Deval, M. N. Bosch, D. H. Mediavilla, V. P. Garcia, and M. S. Gonzalez. 2010. Treatment of acute Achilles tendon ruptures with Achillon device: clinical outcomes and kinetic gait analysis. *Foot Ankle Surg* 16 (4):189–94.

Gayzik, F. S., D. P. Moreno, C. P. Geer, S. D. Wuertzer, R. S. Martin, and J. D. Stitzel. 2011. Development of a full body CAD dataset for computational modeling: a multi-modality approach. *Ann Biomed Eng* 39 (10):2568–83.

Ghate, S. D., V. M. Sistla, V. Nemade, D. Vibhute, S. M. Shahane, and A. D. Samant. 2012. Screw and wire fixation for Lisfranc fracture dislocations. *J Orthop Surg (Hong Kong)* 20 (2):170–5.

Goetz, J., J. Beckmann, F. Koeck, J. Grifka, S. Dullien, and G. Heers. 2013. Gait analysis after tibialis anterior tendon rupture repair using z-plasty. *J Foot Ankle Surg*.

Graf, A., K. W. Wu, P. A. Smith, K. N. Kuo, J. Krzak, and G. Harris. 2012. Comprehensive review of the functional outcome evaluation of clubfoot treatment: a preferred methodology. *J Pediatr Orthop B* 21 (1):20–7.

Hetsroni, I., M. Nyska, D. Ben-Sira, Y. Arnson, C. Buksbaum, E. Aliev, G. Mann, S. Massarwe, G. Rozenfeld, and M. Ayalon. 2011. Analysis of foot and ankle kinematics after operative reduction of high-grade intra-articular fractures of the calcaneus. *J Trauma* 70 (5):1234–40.

Kamiya, T., E. Uchiyama, K. Watanabe, D. Suzuki, M. Fujimiya, and T. Yamashita. 2012. Dynamic effect of the tibialis posterior muscle on the arch of the foot during cyclic axial loading. *Clin Biomech (Bristol, Avon)* 27 (9):962–6.

Keyak, J. H., and S. A. Rossi. 2000. Prediction of femoral fracture load using finite element models: an examination of stress- and strain-based failure theories. *Journal of Biomechanics* 33 (2):209–214.

Kido, M., K. Ikoma, K. Imai, D. Tokunaga, N. Inoue, and T. Kubo. 2013. Load response of the medial longitudinal arch in patients with flatfoot deformity: in vivo 3D study. *Clin Biomech (Bristol, Avon)*.

Kuo, R. S., N. C. Tejwani, C. W. Digiovanni, S. K. Holt, S. K. Benirschke, S. T. Hansen, Jr., and B. J. Sangeorzan. 2000. Outcome after open reduction and internal fixation of Lisfranc joint injuries. *J Bone Joint Surg Am* 82-A (11):1609–18.

Kyu-Jung K. H. B. Kitaoka, Z.-P. Luo, S. Ozeki, L. J. Berglund, K. R. Kaufman. 2001. An in vitro simulation of the stance phase in human gait. *J Musculoskeletal Res.* 05 (02):113–121.

Liang, J., Y. Yang, G. Yu, W. Niu, and Y. Wang. 2011. Deformation and stress distribution of the human foot after plantar ligaments release: a cadaveric study and finite element analysis. *Sci China Life Sci* 54 (3):267–71.

Lins, C., A. F. Ninomya, C. K. Bittar, A. E. de Carvalho, Jr., and A. Cliquet, Jr. 2013. Kinetic and kinematic evaluation of the ankle joint after achilles tendon reconstruction with free semitendinosus tendon graft: preliminary results. *Artif Organs* 37 (3):291–7.

Lui, T. H., L. K. Chan, and K. B. Chan. 2013. Medial subtalar arthroscopy: a cadaveric study of the tarsal canal portal. *Knee Surg Sports Traumatol Arthrosc* 21 (6):1279–82.

Milgrom, C., M. Giladi, R. Chisin, and R. Dizian. 1985. The long-term follow-up of soldiers with stress fractures. *Am J Sports Med* 13 (6):398–400.

Myerson, M. 1989. The diagnosis and treatment of injuries to the Lisfranc joint complex. *Orthop Clin North Am* 20 (4):655–64.

Myerson, M. S. 1999. The diagnosis and treatment of injury to the tarsometatarsal joint complex. *J Bone Joint Surg Br* 81 (5):756–63.

Myerson, M. S., R. T. Fisher, A. R. Burgess, and J. E. Kenzora. 1986. Fracture dislocations of the tarsometatarsal joints: end results correlated with pathology and treatment. *Foot Ankle* 6 (5):225–42.
Nester, C. J. 2009. Lessons from dynamic cadaver and invasive bone pin studies: do we know how the foot really moves during gait? *J Foot Ankle Res* 2:18.
Panchbhavi, V. K., S. Vallurupalli, J. Yang, and C. R. Andersen. 2009. Screw fixation compared with suture-button fixation of isolated Lisfranc ligament injuries. *J Bone Joint Surg Am* 91 (5):1143–8.
Philbin, T., G. Rosenberg, and J. J. Sferra. 2003. Complications of missed or untreated Lisfranc injuries. *Foot Ankle Clin* 8 (1):61–71.
Rammelt, S., W. Schneiders, H. Schikore, M. Holch, J. Heineck, and H. Zwipp. 2008. Primary open reduction and fixation compared with delayed corrective arthrodesis in the treatment of tarsometatarsal (Lisfranc) fracture dislocation. *J Bone Joint Surg Br* 90 (11):1499–506.
Sawacha, Z., G. Cristoferi, G. Guarneri, S. Corazza, G. Dona, P. Denti, A. Facchinetti, A. Avogaro, and C. Cobelli. 2009. Characterizing multisegment foot kinematics during gait in diabetic foot patients. *J Neuroeng Rehabil* 6:37.
Shin, J., N. Yue, and C. D. Untaroiu. 2012. A finite element model of the foot and ankle for automotive impact applications. *Ann Biomed Eng* 40 (12):2519–31.
van Rijn, J., D. M. Dorleijn, B. Boetes, S. Wiersma-Tuinstra, and S. Moonen. 2012. Missing the Lisfranc fracture: a case report and review of the literature. *J Foot Ankle Surg* 51 (2):270–4.
Watanabe, K., H. B. Kitaoka, T. Fujii, X. Crevoisier, L. J. Berglund, K. D. Zhao, K. R. Kaufman, and K. N. An. 2012. Posterior tibial tendon dysfunction and flatfoot: Analysis with simulated walking. *Gait Posture*.
Weatherall, J. M., C. B. Chapman, and S. L. Shapiro. 2013. Postoperative second metatarsal fractures associated with suture-button implant in hallux valgus surgery. *Foot Ankle Int* 34 (1):104–10.

# 4 First Ray Model Comparing Normal and Hallux Valgus Feet

*Duo Wai-Chi Wong, Ming Zhang, and Aaron Kam-Lun Leung*

## CONTENTS



## SUMMARY

Hallux valgus has been reported as one of the most common foot problems. Besides its negative impact on quality of life, it also increases the risk of falling in elderly patients and imposes additional health risks on patients with diabetic or neuropathic feet. Pathoanatomy, pathomechanics, and hypermobility studies have been conducted to evaluate the features of hallux valgus. However, hallux valgus is not yet well understood, and this is reflected by the ineffectiveness of current treatments, complications, and recurrence rates. The current study constructed first ray finite element (FE) models of a normal foot and a hallux valgus foot to examine the sole effect of extrinsic loading on the bony structure. The model showed that loading at initial push-off could lead to metatarsus primus varus and hallux abductor valgus. The hallux valgus foot also showed impaired loading transfer from the distal phalanx, which could hinder the windlass mechanism. This result showed that extrinsic loading could predispose the patient to the risk of developing hallux valgus and the intrinsic stabilization structure, such as ligaments and muscles, could be important in preventing the development of hallux valgus.

## 4.1 INTRODUCTION

### 4.1.1 Background

Hallux valgus is an acquired deformity of the toes, characterized by lateral deviation of the hallux (hallux abducto valgus) and medial deviation of the first metatarsal (metatarsal primus varus). Another term, bunion, is often used to describe the same condition, more prominent on the enlarged and chronically swollen medial projected eminence. The condition could lead to disruption of the alignment of the first metatarsophalangeal joint. Eventually, the patient complains of pain and, in severe cases, joint dislocation may occur. Impaired gait and poor balance are also common symptoms in elderly patients (Abhishek et al. 2010; Cho et al. 2009; Koski et al. 1996; Mann and Coughlin 1981; Menz and Lord 2001a, 2001b; Tinetti, Speechley, and Ginter 1988).

Genetics and sexual dimorphism are two important intrinsic factors leading to hallux valgus, and family history shows a strong association with the prevalence of hallux valgus (Piqué-Vidal, Solé, and Antich 2007; Wu and Lobo Louie 2010). Genetics can have a substantial influence on metatarsal shape, arch height, and hypermobility (Bonney and Macnab 1952), with women being two to three times more susceptible than men (Nguyen et al. 2010; Roddy, Zhang, and Doherty 2008). Gender imposes differences in anatomy (Gutiérrez Carbonell, Sebastia Forcoda, and Betoldi Lizer 1998), bone morphology (Ferrari and Malone-Lee 2002), bone alignment (Ferrari, Hopkinson, and Linney 2004), ligamentous laxity (Wilkerson and Mason 2000), and first ray hypermobility (Coughlin and Shurnas 2003). Other publications have focused on extrinsic factors, such as footwear, occupation, and obesity (Coughlin and Jones 2007; Frey and Zamora 2007; Greer 1938).

Hallux valgus is one of the most common foot complaints (Vanore et al. 2003). Research undertaken in the United Kingdom reported a high prevalence of 28.4% in adults (Roddy, Zhang, and Doherty 2008). Owoeye et al. (2011) surveyed secondary school and undergraduate students and found a prevalence of 15.4%, with 9% reporting pain and 14% reporting an inability to walk for a prolonged period. Research targeting the elderly also demonstrated a high prevalence of 74% (Menz and Lord 2005). The high prevalence of hallux valgus imposes an unnecessary economic burden on the health care system and society in general. Thompson and Coughlin (1994) estimated that there were 56,500 bunionectomies performed in the United States in 1991, with 27% of all forefoot surgeries being undertaken to correct hallux valgus deformities. A study conducted in Australia (Australian Bureau of Statistics 2008) reported an average of 22 absent days following first metatarsophalangeal joint surgery, with a productivity loss of AUS$ 3,852 and the average hospitalization costing AUS$ 3,764.16 (Access Economics 2008; Courtney, Matz, and Webster 2002; Grimm and Fallat 1999; Mathers et al. 2001).

Though hallux valgus and its surgical correction are routine occurrences, the associated failures, complications, and recurrences remain. Complication rates range from 10% to 55% (Scioli 1997), and recurrence rates could be as high as 16% (Caminear et al. 2012; Lehman 2003). Failure rates could surpass 75% with soft tissue correctional procedures, such as the McBride procedure (Coughlin and Mann 2012). Osteotomy and arthrodesis could lead to over-correction (Easly et al. 1996), metatarsalgia (Wanivenhaus and Feldner-Busetin 1988), avascular necrosis, and arthrosis (Coughlin and Mann 2012).

### 4.1.2 Biomechanical Research on Hallux Valgus and Its Intervention

Biomechanical research on hallux valgus and its correction has been conducted with the aim of better understanding the pathogenesis of the disease and improving treatment outcomes. The concept of first ray hypermobility, introduced by Morton (1928) and Lapidus (1956), has been hotly debated in explaining hallux valgus and in selecting interventions (Faber et al. 2001; Glasoe et al. 2002). The importance of stability was stressed, as advocated by the arthrodesis procedure (Lapidus 1956). The evaluation of stability was also included in other studies on the osteotomy of hallux valgus (Coughlin et al. 2004; Kim et al. 2008). Currently, mobility and stability are assessed by manual dorsal excursion (Smith and Coughlin 2008), load-bearing radiographs (Coughlin and Jones 2008), or custom-made mechanical devices (Klaue, Hansen, and Masquelet 1994). Yet the quantification of hypermobility has been believed to be subjective and confined to static measurement (Martin et al. 2012; Wukich, Donley, and Sferra 2005; Faber et al. 2001).

Plantar pressure distribution is another set of parameters commonly used to classify foot types and deformities (Hillstrom et al. 2012). Various studies have demonstrated reduced pressure at the hallux region (Blomgren, Turan, and Agadir 1991; Hutton and Dhanendran 1981; Kernozek, Elfessi, and Sterriker 2003). Kernozek, Elfessi, and Sterriker (2003) found increased peak pressure at the central forefoot region peak and heightened pressure time integrals, while Stokes et al. (1979) discovered a lateral loci of peak pressure in the hallux valgus forefoot. However, converse findings on the medial shift of pressure were also detailed (Martínez-Nova et al. 2008; Mickle et al. 2011). Wen et al. (2012) commented that the medial metatarsal region may not be directly related to hallux valgus, whereas a reduction in hallux loading and increased loading on the central metatarsal would be more persistent in hallux valgus patients.

Surgical interventions have also been evaluated by means of plantar pressure assessment. Mittal, Raja, and Geary (2006) indicated that the McBride procedure could improve hallux function by increasing the contact area under the hallux. Saro et al. (2007) compared plantar pressure results between the operated and non-operated foot and demonstrated a significant reduction in peak pressure under the hallux and heel region. Dhukaram, Hullin, and Senthil Kumar (2006) compared the differences between a Mitchell and Scarf osteotomy by their differences in plantar pressure distribution.

Inasmuch as hallux valgus is not well-understood, numerous studies have examined the pathomechanism of hallux valgus and the biomechanical outcome of interventions by means of manual examinations and planar pressure studies to determine the altered load transfer pattern. Recently, FE analysis or simulations have been used to examine the internal stress/strain and load transfer behavior of the foot in the clinical field (Cheung and Nigg 2008). Yu et al. (2008) suggested the contribution of high-heeled shoes to the development of hallux valgus by simulating a foot shod with varying heel heights. Kai et al. (2006) undertook a primary FE analysis of the first ray to evaluate the relationship between first ray hypermobility and hallux valgus. In the present study, first ray models of a normal subject and hallux valgus patient were constructed and simulated at the initial push-off phase. The stress/strain and joint loading were studied to evaluate the alteration in kinematics and stability with the structural change of hallux valgus.

## 4.2 MODEL DEVELOPMENT

### 4.2.1 Geometry Construction

The models of the normal foot and hallux valgus foot were constructed from radiographic images of two women. The subject of the normal foot model was aged 28, 165 cm tall, and weighed 54 kg. The subject of the hallux valgus model had asymptomatic hallux valgus, was aged 28, 165 cm tall, and weighed 56 kg. Both participants reported no other musculoskeletal pathology, pain, or lower limb trauma or surgery within the past six months.

An ankle-foot orthosis was fabricated to keep the foot in the neutral position with minimum compression on the encapsulated soft tissue. The neutral position upon scanning was defined by the Society of Biomechanics, based on the joint coordinate system (Wu et al. 2002). The alignment of the participant was considered normal with a 25° calcaneal inclination angle (DiGiovanni and Smith 1976). A pad was also placed beneath the calcaneal and talus body to maintain foot alignment with respect to the scanning machine.

The geometry of the normal foot model was constructed via coronal magnetic resonance images with a 3.0T scanner (Siemens Medical Solutions, Erlangen, Germany), while that of the hallux valgus foot was constructed via coronal computer tomography (Aquilion, Toshiba Medical System, Japan).

The right foot images were obtained in neutral and non-weight-bearing states and scanned in 1-mm intervals. The clinical images were segmented in commercial software, MIMICS v10 (Materialise, Leuven, Belgium). A mask of the bones and soft tissue was distinguished and a three-dimensional mask was constructed. The segmented mesh was exported to the preprocessing software, Rapidform XOR2 (INUS Technology Ltd., Seoul, Korea). In this software, the mesh and geometry of different parts were refined and converted into an analytical surface representation before inputting into the finite element software, ABAQUS v6.11 (Dassault Systèmes Technologies, RI, USA).

### 4.2.2 Material Properties

The first ray model consisted of the medial cuneiform, the first metatarsal, the sesamoids, and the first proximal and distal phalanges. The sesamoids were merged with the first metatarsal, while the proximal and distal phalanges were fused together. An extract of the encapsulated soft tissue surrounding the extracted bones was created with the proximal boundary lying between the medial cuneiform. The geometry and the associated mesh are shown in Figure 4.1.

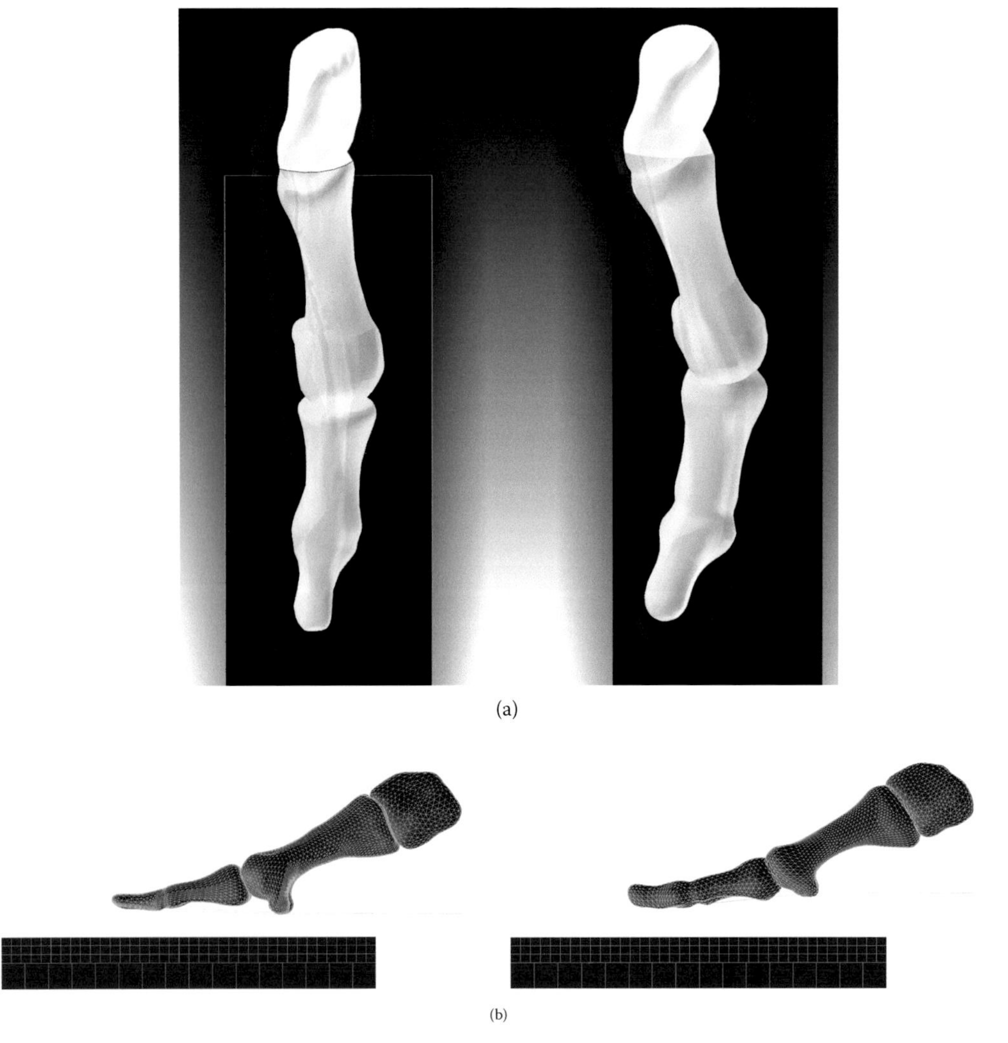


**FIGURE 4.1** (a) The geometry of the bone and encapsulated soft tissue of the first ray of the normal (left) and hallux valgus subject (right). (b) The mesh of the bone segment of the normal (left) and hallux valgus subject (right).

The elastic modulus of bone was assigned 7300 MPa and Poisson's ratio of 0.3 (Nakamura et al. 1981). The bone was assumed isotropic and homogeneous. Similar to the bones, the encapsulated soft tissue was also modeled as a three-dimensional solid. The hyperelastic behavior of the encapsulated soft tissue was referenced from Lemmon et al. (1997) and used a second-order energy potential Mooney-Rivlin equation. A layer of skin was modeled on the surface of the encapsulated soft tissue with a thickness of 1.2 mm (Pailler-Mattei, Bec, and Zahouani 2008) and assigned hyperelastic material properties (Gu et al. 2010a) using an Ogden formula. Two connectors that represented the plantar fascia were constructed with their proximal ends fixed. The distal ends were inserted into the boundary between the proximal and distal phalanges, while a pivot point was set at the sesamoids. The connector behavior was modeled as a slip-ring, in which material flow was allowed at the sesamoids that could mimic the windlass mechanism. The elasticity of the connector was assigned 203.3 MPa (Kitaoka et al. 1994).

### 4.2.3 Load and Boundary Conditions

An initial push-off instance was simulated with 30° metatarsophalangeal inclination (Kristen et al. 2005) that was governed by the anterior inclination of the floor plate. A vertical ground reaction force of 30 N, which was measured by pedobarography about the corresponding phase on the normal subject, was applied under the floor plate, with the proximal end of the medial cuneiform encrusted. The bone contact was frictionless, while the tissue-to-floor contact was assigned a coefficient of friction of 0.6 (Zhang and Mak 1999).

The FE analysis simulated an initial push-off instance with passive loading that ruled out the effects of both muscles and ligaments. The objective was to study the influence of gait loading on the bone configurations of a normal and a hallux valgus foot. The changes of bone alignment, stress, and loading were investigated.

## 4.3 IMPLICATIONS OF THE FINITE ELEMENT MODEL

### 4.3.1 Bone Alignment and Displacement

The change of bone alignment was evaluated by the intermetatarsal angle (IMA) and the hallux valgus angle (HVA). Axes of the bone shaft were approximated by a cylindrical regression axis algorithm via processing software, Rapidform XOR2 (INUS Technology Ltd., Seoul, Korea), and the axes were projected on the ground plane. The IMA was the angle between the first and second metatarsal axes, while the HVA was the angle between the first metatarsal and phalanx axes. Figure 4.2 shows the IMAs and HVAs of the normal foot and the hallux valgus foot before the simulation (undeformed state) and after the simulation (deformed state).

The HVA and IMA of the normal subject were 9.1° and 15.0°, respectively, while those of the hallux valgus patient were 10.4° and 25.7°, respectively. The HVAs and IMAs of both the normal subject and the hallux valgus patient increased after applying a forefoot loading. For the normal subject, both the IMA and HVA increased by about 1.2 times; for the hallux valgus patient, the IMA increased by about 140% and the HVA increased by about 89%. The increase of IMA was higher for the hallux valgus patient compared with the normal subject, while the increase of HVA was comparatively less. Figure 4.3 shows the simulation results indicating the displacement and change of bone alignment after applying a forefoot loading.

### 4.3.2 Influence on First Ray Joints

The passive tensile force of the plantar fascia and the joint force were also evaluated. The fascia forces of the normal foot and the hallux valgus foot were similar, with magnitudes of 25.80 N and 24.24 N, respectively. The metatarsocuneiform joint (MC) force and metatarsophalangeal joint

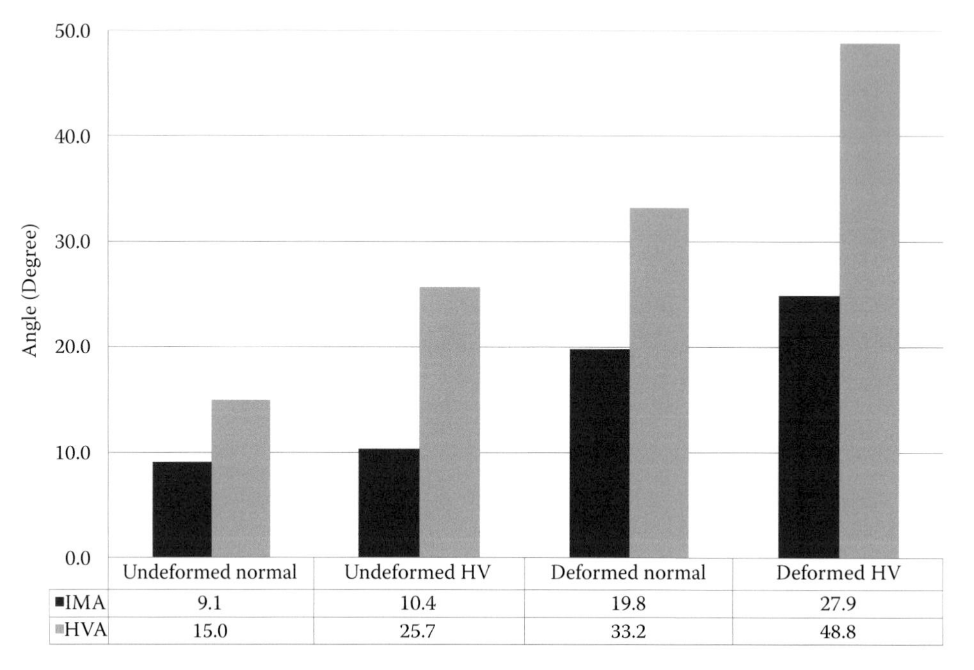


| | Undeformed normal | Undeformed HV | Deformed normal | Deformed HV |
|---|---|---|---|---|
| ■IMA | 9.1 | 10.4 | 19.8 | 27.9 |
| ■HVA | 15.0 | 25.7 | 33.2 | 48.8 |

**FIGURE 4.2** The hallux valgus angle (HVA) and intermetatarsal angle (IMA) of the normal and hallux valgus foot before the simulation (undeformed state) and after the simulation (deformed state).

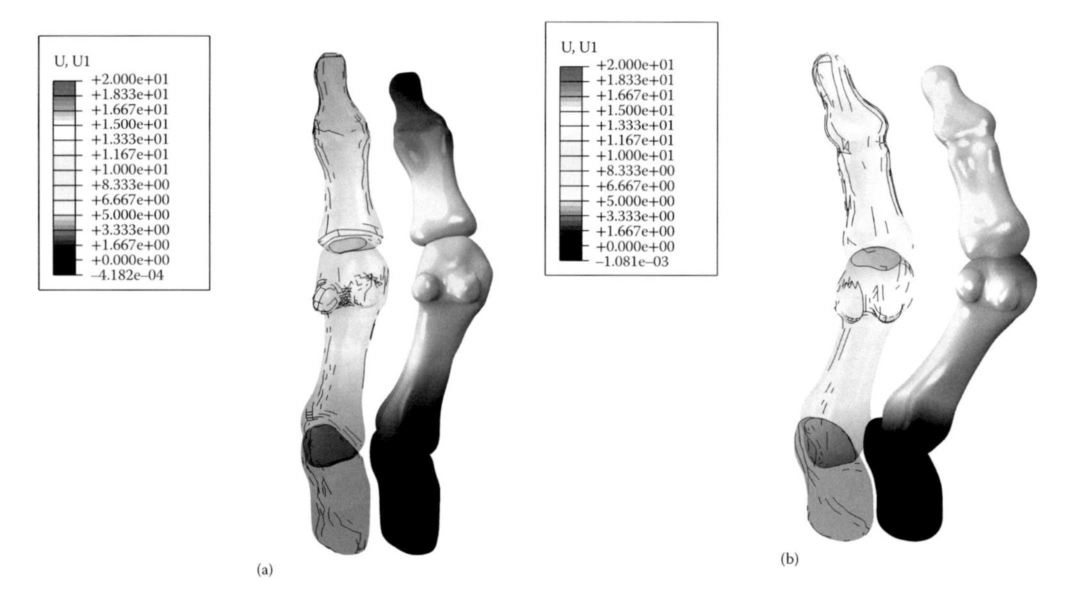


**FIGURE 4.3** **(See color insert.)** Simulation results on the displacement in medial-lateral direction indicating change of bone alignment after application of forefoot loading. (a) Normal foot; (b) hallux valgus foot.

(MTP) force are presented in Figure 4.4. The magnitude of the MC force of the normal foot was about 61 N. The deviation was less than 5 N in comparison to the hallux valgus foot. However, the MTP force of the hallux valgus foot was about 43% less than that of the normal foot. The MTP forces were 64 N and 36 N, respectively, for the normal foot and the hallux valgus foot.

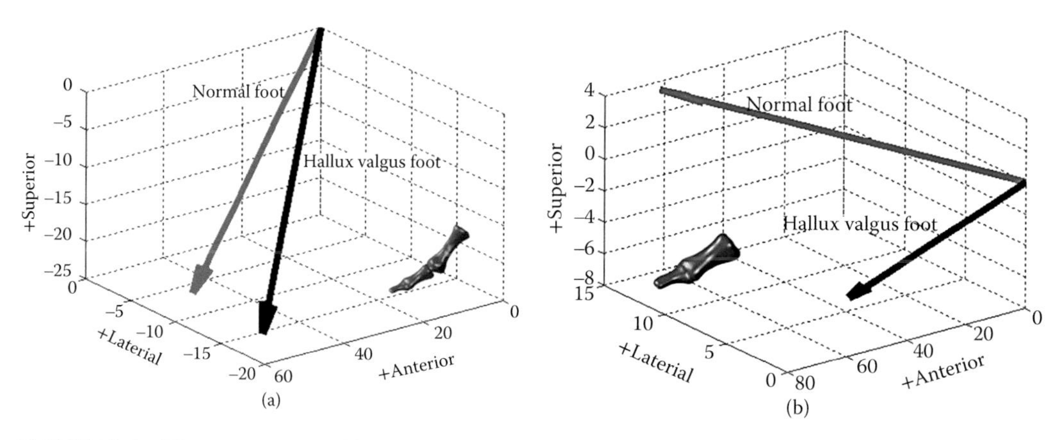


**FIGURE 4.4** The metatarsocuneiform (a) and metatarsophalangeal (b) joint force of the normal and hallux valgus foot. The direction of joint force is imposed from the proximal segment on the distal segment.

### 4.3.3 Plantar Pressure Distribution

The plantar pressure pattern on the supporting plate is shown in Figure 4.5. The localized pressure of the phalanges segment and the distal metatarsal segment can be seen clearly. The normal foot exhibited a concentrated pressure on the more distal portion of the metatarsal, while the hallux foot presented a more evenly distributed pattern with concentrated pressure at a more proximal location. The contact area of the hallux was also smaller in the hallux valgus foot. The peak contact pressure of the normal foot was 0.094 MPa, which was larger than the 0.069 MPa for the hallux valgus foot.

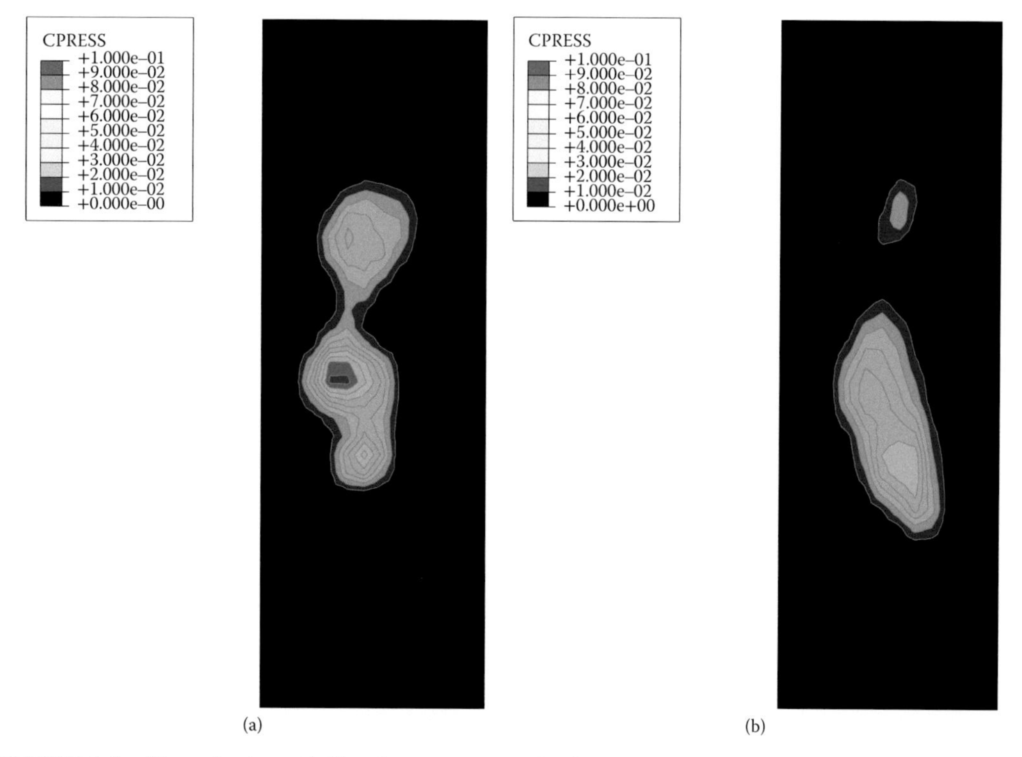


**FIGURE 4.5** **(See color insert.)** The plantar pressure distribution on the supporting plate of the normal (a) and hallux valgus (b) foot. Unit: MPa.

This study conducted FE analysis on normal and hallux valgus feet under passive loading conditions at initial push-off. The objective of this study was to evaluate the influence of loading conditions solely on the biomechanics of the skeletal structure, ruling out any effects of muscles and ligaments. Differences between a normal foot and the deranged bone alignment of the hallux valgus foot were studied.

### 4.3.4 Validation of the Model

Plantar pressure is one of the most common validation metrics used in FE foot models. Yu et al. (2008) reported that forefoot plantar pressure ranged from about 0.08 MPa to 0.12 MPa in their FE analysis and pedobarographic measurements using a flat support. Cheung et al. (2005) predicted the plantar pressure under the first metatarsal head at 0.097 MPa, which slightly deviated from the 0.06 MPa recorded through pedobarographic measurement. The FE result of this study found a peak plantar pressure of 0.094 MPa on the normal foot and 0.069 MPa on the hallux valgus foot, both of which are in general agreement with existing experimental results.

Bone stress is another commonly used parameter for model validation. The peak von Mises stress of the first metatarsal has been reported at 2 to 3 MPa (Cheung et al. 2005; Gu et al. 2010b). This simulation reported a relatively smaller peak first metatarsal shaft stress of 1.81 MPa in the normal foot and 2.61 MPa in the hallux valgus foot. The small deviation could be due to the differences in geometry, loading conditions, simulated gait instants, and the exclusion of muscle forces.

### 4.3.5 Deformity with Load Bearing

This simulation showed an increase in IMA and HVA upon the application of forefoot loading, which corresponded to metatarsus primus varus and hallux abductor valgus found in clinical settings. The reason for these findings should be due to the saddle shape of the joint facets. The dorsal-plantar excursion of the first ray is coupled with tri-plantar medial-lateral motion and rotation (Smith and Coughlin 2008). Lacking other stabilizing structures, the first metatarsal could tend to spray apart from the neutral line, leading to metatarsus primus varus, as described by the impaired tie-bar mechanism (Stainsby 1997). The withholding of the phalanx by the fascia would cause a secondary hallux abductor valgus deformity, as described by the bow-spring mechanism (Stainsby 1997).

### 4.3.6 Load Transfer across the First Ray

The joint force predicted in this simulation showed that the foot with hallux valgus had a weak force transfer capability at the MTP joint. This fact was further supported by the plantar pressure distribution, shown in Figure 4.5. The pressure under the hallux was reduced and accompanied by a posterior shift in the center of pressure. In fact, the arch of the foot formed a rigid lever arm during the push-off phase to shift the center of pressure from the lateral side to the medial side, which was demonstrated by the concentrated pressure of the normal foot under the hallux and the first metatarsal head (Figure 4.5). The hallux valgus foot impaired the arch function by hindering the windlass mechanism, possibly leading to secondary metatarsalgia (Hutton and Dhanendran 1981; Van Beek and Greisberg 2011).

Numerous publications on the hallux valgus have focused on studying the kinematic properties (Allen et al. 2004; Dietze et al. 2013) of the hypermobile first ray and some have attributed the cause to generalized ligament laxity (McNerney and Johnston 1979), deep transverse metatarsal ligament insufficiency (Stainsby 1997), and other extrinsic causes (Perera, Lyndon Mason, and Stephens 2011). The biomechanical cause of hypermobility or hallux valgus deformity could be multifactorial, and information on the role of different sources could aid in designing optimal treatment interventions. Computer simulation provides a valuable platform for examining specific factors in a controlled environment. The result in this study demonstrate that the extrinsic loading factor alone could affect the bone alignment and stress the importance of the intrinsic stabilizing structures. The results also illustrate the weakened windlass mechanism arising from the hallux valgus deformity. Future work could study the influence of ligaments and muscles on the load transfer mechanism of the first ray.

## ACKNOWLEDGMENT

The study was financed by the Research Studentship of Hong Kong Polytechnic University.

## REFERENCES

Abhishek, A., E. Roddy, W. Zhang, and M. Doherty. 2010. Are hallux valgus and big toe pain associated with impaired quality of life? A cross-sectional study. *Osteoarthritis Cartilage* 18 (7):923–6.

Access Economics. 2008. The economic impact of podiatric surgery. In *Report for the Australasian College of Podiatric Surgeons*. Sydney, Australia: Access Economics Pty Ltd.

Allen, M. K., T. J. Cuddeford, W. M. Glasoe, L. M. DeKam, P. J. Lee, K. J. Wagner, and H. J. Yack. 2004. Relationship between static mobility of the first ray and first ray, midfoot, and hindfoot motion during gait. *Foot Ankle Int* 25 (6):391–6.

Australian Bureau of Statistics. 2008. *Australian National Accounts: National Income, Expenditure and Product*. Canberra.

Blomgren, M., I. Turan, and M. Agadir. 1991. Gait analysis in hallux valgus. *J Foot Surg* 30 (1):70–1.

Bonney, G., and I. Macnab. 1952. Hallux valgus and hallux rigidus: a critical survey of operative results. *J Bone Joint Surg Br* 34-B (3):366–85.

Caminear, D.S., E. Addis-Thomas, A.W. Brynizcka, and A. Saxena. 2012. Revision hallux valgus surgery. In *International Advances in Foot and Ankle Surgery,* edited by A. Saxena, pp. 71–82. London: Springer-Verlag.

Cheung, J. T., M. Zhang, A. K. Leung, and Y. B. Fan. 2005. Three-dimensional finite element analysis of the foot during standing—a material sensitivity study. *J Biomech* 38 (5):1045–54.

Cheung, J.T.M., and B.M. Nigg. 2008. Clinical applications of computational simulation of foot and ankle. *Sports Orthop Traumatol* 23 (4):264–71.

Cho, N. H., S. Kim, D. J. Kwon, and H. A. Kim. 2009. The prevalence of hallux valgus and its association with foot pain and function in a rural Korean community. *J Bone Joint Surg Br* 91 (4):494–8.

Coughlin, M.J., and C.P. Jones. 2007. Hallux valgus: demographics, etiology, and radiographic assessment. *Foot Ankle Int* 28 (7):759–77.

Coughlin, M.J., and C.P. Jones. 2008. Hallux valgus and first ray mobility. A prospective study. *J Bone Joint Surg* 90 (5):1166–7.

Coughlin, M.J., C.P. Jones, R. Viladot, P. Glano, B.R. Grebing, M.J. Kennedy, P.S. Shurnas, and F. Alvarez. 2004. Hallux valgus and first ray mobility: a cadaveric study. *Foot Ankle Int* 25:537–44.

Coughlin, M. J., and R. A. Mann. 2012. Hallux valgus. In *Surgery of the Foot and Ankle,* 8th edition, edited by M. J. Coughlin, C. L. Saltzman, and R. A. Mann. Pennsylvania: Mosby Elsevier.

Coughlin, M.J., and P.S. Shurnas. 2003. Hallux valgus in men. Part II: First ray mobility after bunionectomy and factors associated with hallux valgus deformity. *Foot Ankle Int* 24 (1):73–8.

Courtney, T.K., S. Matz, and B.S. Webster. 2002. Disabling occupational injury in the U.S. construction industry, 1996. *J Occup Environ Med* 44 (12):1161–8.

Dhukaram, V., M.G. Hullin, and C. Senthil Kumar. 2006. The Mitchell and Scarf osteotomies for hallux valgus correction: a retrospective, comparative analysis using plantar pressures. *J. Foot Ankle Surgery* 45 (6):400-409.

Dietze, A., U. Bahlke, H. Martin, and T. Mittlmeier. 2013. First ray instability in hallux valgus deformity: a radiokinematic and pedobarographic analysis. *Foot Ankle Int* 34 (1):124–30.

DiGiovanni, J.E., and S.D. Smith. 1976. Normal biomechanics of the adult rearfoot: a radiographic analysis. *J Am Podiatry Assoc* 66 (11):812–24.

Easly, M.E., G.M. Kiebzak, W.H. Davis, and R.B. Anderson. 1996. Prospective, randomized comparison of proximal crescentic and proximal chevron osteotomies for correction of hallux valgus deformity. *Foot Ankle Int*17:307–16.

Faber, F.W.M., G.J. Kleinrensink, P.G.H. Mulder, and J.A.N. Verhaar. 2001. Mobility of the first tarsometatarsal joint in hallux valgus patients: a radiographic analysis. *Foot Ankle Int* 22 (12):965–9.

Ferrari, J., D.A. Hopkinson, and A.D. Linney. 2004. Size and shape differences between male and female foot bones: is the female foot predisposed to hallux abducto valgus deformity? *J Am Podiatric Med Assoc* 94 (5):434–52.

Ferrari, J., and J. Malone-Lee. 2002. The shape of the metatarsal head as a cause of hallux abductovalgus. *Foot Ankle Int* 23 (3):236–42.

Frey, C., and J. Zamora. 2007. The effects of obesity on orthopaedic foot and ankle pathology. *Foot Ankle Int* 28 (9):996–9.

Glasoe, W.M., M.K. Allen, C.L. Saltzman, P.M. Ludewig, and S.H. Sublett. 2002. Comparison of two methods used to assess first-ray mobility. *Foot Ankle Int* 23 (3):248–52.

Greer, W.S. 1938. Clinical aspect: relation to footwear. *Lancet* 232 (6017):1482–3.

Grimm, D.J., and L. Fallat. 1999. Injuries of the foot and ankle in occupational medicine: a 1-year study. *J Foot Ankle Surg* 38 (2):102–8.

Gu, Y., J. Li, X. Ren, M.J. Lake, and Y. Zeng. 2010a. Heel skin stiffness effect on the hind foot biomechanics during heel strike. *Skin Res Technol* 16 (3):291–6.

Gu, Y.D., X.J. Ren, J.S. Li, M.J. Lake, Q.Y. Zhang, and Y.J. Zeng. 2010b. Computer simulation of stress distribution in the metatarsals at different inversion landing angles using the finite element method. *Int Orthop* 34 (5):669–76.

Gutiérrez Carbonell, P., E. Sebastia Forcoda, and G. Betoldi Lizer. 1998. Factores morfológicos que influyen en el hallux valgus. *Revista de ortopedia y traumatologia* 42(5):356–62.

Hillstrom, H.J., J. Song, A.P. Kraszewski, J.F. Hafer, R. Mootanah, A.B. Dufour, B. Shingpui Chow, and J.T. Deland III. 2013. Foot type biomechanics part 1: Structure and function of the asymptomatic foot. *Gait Posture* 37(3):445–51.

Hutton, W.C., and M. Dhanendran. 1981. The mechanics of normal and hallux valgus feet: a quantitative study. *Clin Orthop Relat Res* 157:7–13.

Kai, T., W. Cheng-tao, W. Dong-mei, and W. Xu. 2006. Primary analysis of the first ray using a 3-dimension finite element foot model. Paper read at Engineering in Medicine and Biology Society, 2005. IEEE-EMBS 2005. 27th Annual International Conference.

Kernozek, T.W., A. Elfessi, and S. Sterriker. 2003. Clinical and biomechanical risk factors of patients diagnosed with hallux valgus. *J Am Podiatric Med Assoc* 93 (2):97–103.

Kim, J., J.S. Park, K.H. Seung, K.W. Young, and I.L.H. Sung. 2008. Mobility changes of the first ray after hallux valgus surgery: clinical results after proximal metatarsal chevron osteotomy and distal soft tissue procedure. *Foot Ankle Int* 29 (5):468–72.

Kitaoka, H.B., Z.P. Luo, E.S. Growney, L.J. Berglund, and K.N. An. 1994. Material properties of the plantar aponeurosis. *Foot Ankle Int* 15 (10):557–60.

Klaue, K., S.T. Hansen, and A.C. Masquelet. 1994. Clinical, quantitative assessment of first tarsometatarsal mobility in the sagittal plane and its relation to hallux valgus deformity. *Foot Ankle Int* 15 (1):9–13.

Koski, K., H. Luukinen, P. Laippala, and S. L. Kivela. 1996. Physiological factors and medications as predictors of injurious falls by elderly people: a prospective population-based study. *Age Ageing* 25 (1):29–38.

Kristen, K.H., K. Berger, C. Berger, W. Kampla, W. Anzbock, and S.H. Weitzel. 2005. The first metatarsal bone under loading conditions: a finite element analysis. *Foot Ankle Clin* 10 (1):1–14.

Lapidus, P.W. 1956. A quarter of a century of experience with the operative correction of the metatarsus varus primus in hallux valgus. *Bull Hosp Joint Dis* 17 (2):404–21.

Lehman, D.E. 2003. Salvage of complications of hallux valgus surgery. *Foot Ankle Clin* 8 (1):15–35.

Lemmon, D., T.Y. Shiang, A. Hashmi, J.S. Ulbrecht, and P.R. Cavanagh. 1997. The effect of insoles in therapeutic footwear—a finite element approach. *J Biomech* 30 (6):615–20.

Mann, R.A., and M.J. Coughlin. 1981. Hallux valgus—etiology, anatomy, treatment and surgical considerations. *Clin Orthop Relat Res* (157):31–41.

Martin, H., U. Bahlke, A. Dietze, V. Zschorlich, K.-P. Schmitz, and T. Mittlmeier. 2012. Investigation of first ray mobility during gait by kinematic fluoroscopic imaging-a novel method. *BMC Musculoskeletal Disorders* 13 (1):14.

Martínez-Nova, A., J.C. Cuevas-García, R. Sánchez-Rodríguez, J. Pascual-Huerta, and E. Sánchez-Barrado. 2008. Study of plantar pressure patterns by means of instrumented insoles in subjects with hallux valgus. *Revista española de cirugía ortopédica y traumatología (English edition)* 52(2):94–8.

Mathers, C.D., E.T. Vos, C.E. Stevenson, and S.J. Begg. 2001. The burden of disease and injury in Australia. *Bulletin World Health Organization* 79 (11):1076–84.

McNerney, J.E., and W.B. Johnston. 1979. Generalized ligamentous laxity, hallux abducto valgus and the first metatarsocuneiform joint. *J Am Podiatry Assoc* 69 (1):69–82.

Menz, H.B., and S.R. Lord. 2001a. The contribution of foot problems to mobility impairment and falls in community-dwelling older people. *J Am Geriatr Soc* 49 (12):1651–6.

Menz, H.B., and S.R. Lord. 2001b. Foot pain impairs balance and functional ability in community-dwelling older people. *J Am Podiatric Med Assoc* 91 (5):222–9.

Menz, H.B., and S.R. Lord. 2005. The contribution of foot problems to mobility impairment and falls in community-dwelling older people. *J Am Geriatr Soc* 49 (12):1651–6.

Mickle, K.J., B.J. Munro, S.R. Lord, H.B. Menz, and J.R. Steele. 2011. Gait, balance and plantar pressures in older people with toe deformities. *Gait Posture* 34 (3):347–51.

Mittal, D., S. Raja, and N.P.J. Geary. 2006. The modified McBride procedure: clinical, radiological, and pedobarographic evaluations. *J Foot Ankle Surg* 45 (4):235–9.

Morton, D.J. 1928. Hypermobility of the first metatarsal bone: the interlinking factor between metatarsalgia and longitudinal arch strains. *J Bone Joint Surg (Am)* 10 (2):187–96.

Nakamura, S., R.D. Crowninshield, and R.R. Cooper. 1981. An analysis of soft tissue loading in the foot—a preliminary report. *Bull Prosthetics Res* 10:27–34.

Nguyen, U.S.D.T., H.J. Hillstrom, W. Li, A.B. Dufour, D.P. Kiel, E. Procter-Gray, M.M. Gagnon, and M.T. Hannan. 2010. Factors associated with hallux valgus in a population-based study of older women and men: the MOBILIZE Boston Study. *Osteoarthritis and Cartilage* 18 (1):41–6.

Owoeye, B.A., S.R. Akinbo, A.L. Aiyegbusi, and M.O. Ogunsola. 2011. Prevalence of hallux valgus among youth population in Lagos, Nigeria. *Nigerian Postgraduate Med J* 18 (1):51–5.

Pailler-Mattei, C., S. Bec, and H. Zahouani. 2008. In vivo measurements of the elastic mechanical properties of human skin by indentation tests. *Med Eng Phys* 30 (5):599–606.

Perera, A.M., L. Mason, and M.M. Stephens. 2011. The pathogenesis of hallux valgus. *J Bone Joint Surg* 93 (17):1650–61.

Piqué-Vidal, C., M.T. Solé, and J. Antich. 2007. Hallux valgus inheritance: pedigree research in 350 patients with bunion deformity. *J Foot Ankle Surg* 46 (3):149–54.

Roddy, E., W. Zhang, and M. Doherty. 2008. Prevalence and associations of hallux valgus in a primary care population. *Arthritis Care Res* 59 (6):857–62.

Saro, C., B. Andrén, L. Felländer-Tsai, U. Lindgren, and A. Arndt. 2007. Plantar pressure distribution and pain after distal osteotomy for hallux valgus: a prospective study of 22 patients with 12-month follow-up. *Foot* 17 (2):84–93.

Scioli, M.W. 1997. Complications of hallux valgus surgery and subsequent treatment options. *Foot Ankle Clin* 2:719–40.

Smith, B.W., and M.J. Coughlin. 2008. The first metatarsocuneiform joint, hypermobility, and hallux valgus: what does it all mean? *Foot Ankle Surg* 14 (3):138–41.

Stainsby, G.D. 1997. Pathological anatomy and dynamic effect of the displaced plantar plate and the importance of the integrity of the plantar plate-deep transverse metatarsal ligament tie-bar. *Annals of the Royal College of Surgeons of England* 79 (1):58–68.

Stokes, I.A., W.C. Hutton, J.R. Stott, and L.W. Lowe. 1979. Forces under the hallux valgus foot before and after surgery. *Clin Orthop Related Res* (142):64–72.

Thompson, F.M., and M.J. Coughlin. 1994. The high price of high-fashion footwear. *J Bone Joint Surg (Am)* 76 (10):1586–93.

Tinetti, M.E., M. Speechley, and S.F. Ginter. 1988. Risk factors for falls among elderly persons living in the community. *N Engl J Med* 319 (26):1701–7.

Van Beek, C., and J. Greisberg. 2011. Mobility of the first ray: review article. *Foot Ankle Int* 32 (9):917–22.

Vanore, J.V., J.C. Christensen, S.R. Kravitz, J.M. Schuberth, J.L. Thomas, L.S. Weil, H.J. Zlotoff, R.W. Mendicino, and S.D. Couture. 2003. Diagnosis and treatment of first metatarsophalangeal joint disorders. Section 1: Hallux valgus. *J Foot Ankle Surg* 42 (3):124–36.

Wanivenhaus, A.H., and H. Feldner-Busetin. 1988. Basal osteotomy of the first metatarsal for the correction of metatarsus primus varus associated with hallux valgus. *Foot Ankle* 8 (6):337–43.

Wen, J., Q. Ding, Z. Yu, W. Sun, Q. Wang, and K. Wei. 2012. Adaptive changes of foot pressure in hallux valgus patients. *Gait Posture* 36 (3):344–9.

Wilkerson, R.D., and M.A. Mason. 2000. Differences in men's and women's mean ankle ligamentous laxity. *Iowa Orthop J* 20:46–8.

Wu, D., and D.P.E. Lobo Louie. 2010. Does wearing high-heeled shoe cause hallux valgus? A survey of 1,056 Chinese females. *Foot Ankle Online J* 3 (5): 3.

Wu, G., S. Siegler, P. Allard, C. Kirtley, A. Leardini, D. Rosenbaum, M. Whittle, D.D. D'Lima, L. Cristofolini, and H. Witte. 2002. ISB recommendation on definitions of joint coordinate system of various joints for the reporting of human joint motion, part I: ankle, hip, and spine. *J Biomech* 35 (4):543–8.

Wukich, D.K., B.G. Donley, and J.J. Sferra. 2005. Hypermobility of the first tarsometatarsal joint. *Foot Ankle Clin* 10 (1):157–66.

Yu, J., J.T. Cheung, Y. Fan, Y. Zhang, A.K. Leung, and M. Zhang. 2008. Development of a finite element model of female foot for high-heeled shoe design. *Clin Biomech (Bristol, Avon)* 23 Suppl 1:S31–8.

Zhang, M., and A.F.T. Mak. 1999. In vivo friction properties of human skin. *Prosthetics Orthotics Int* 23 (2):135–41.

# 5 Dynamic Foot Model for Impact Investigation

*Jia Yu, Duo Wai-Chi Wong, and Ming Zhang*

## CONTENTS



## SUMMARY

The foot acts as a shock absorber upon contact with the ground. Repetitive impact on the foot is considered one of the main etiological agents in foot pain and problems. Footwear or foot support designed for shock absorption has been identified as one of the important strategies for prevention of foot/ankle injuries. Simulations of foot impact are necessary for a better understanding of foot support design. This chapter reviews experimental and computational methods of existing investigations on foot impact and introduces implicit dynamic finite element (FE) methods. As an example of potential applications of such an implicit dynamic model for prevention of heel pain, results of the dynamic FE foot model during heel strike are presented.

## 5.1 INTRODUCTION

The foot is a shock absorber for the entire body during locomotion. Many methods are employed to assess the dynamic characteristics of the foot and ankle, such as clinical assessment, gait analysis, cadaveric examination, and dynamic computational analysis.

Repetitive impact forces during walking, running, and jumping impose substantial loading on the foot. The ground reaction force may reach up to three times the body weight during walking (Nigg, Cole, and Bruggemann 1995; Whittle 1999; Chi and Schmitt 2005) and about ten times during jumping (Cavanagh and Lafortune 1980; McClay et al. 1994). This repetitive impact force is considered one of the main etiological agents in degenerative joint diseases and acute injuries of the musculoskeletal system (Collins and Whittle 1989; Jorgensen 1985).

The shock-absorbing feature in the soles of footwear has been used effectively in the prevention of foot/ankle injuries. Sports shoes should be equipped with an effective cushioning system to attenuate potentially injurious foot-ground impact forces. A wide range of footwear sole materials is available, differing in density, elasticity, and viscoelasticity. The optimal design of cushioning materials and structures is imperative in order to meet complicated and individual functional and clinical demands (Fong, Hong, and Li 2007, 2009). The efficiency and effectiveness of a design are the main concerns for designers and manufactures. The following evaluation relies on information obtained from available experiments, such as materials tests, cadaveric experiments, kinematic and kinetic analysis, plantar pressure measurement, and muscle activity monitoring.

### 5.1.1 Experimental Approach

Experimental results show that the heel pad has the capacity to attenuate the heel strike peak acceleration transmitted to the lower leg by 80% (Noe et al. 1993). Polymeric shock-absorbing materials can augment this capacity depending on the compliance of the foot/shoe structures. The reduction in the peak accelerations due to shock-absorbing inserts may be around 18% in the presence of an intact compliant heel pad, but may rise to 83% in the absence of a heel pad. The rearfoot peak ground reaction force is significantly greater for a harder midsole during a step-off landing with basketball shoes (Zhang et al. 2005). Shock absorption capability has also been reported to be significantly greater in the calcaneal heel pad than in external shock absorbers (Jorgensen and Bojsen-Moller 1989).

#### 5.1.1.1 Pendulum Impact Experiments

Human pendulum impact tests were widely used to quantify lower limb biomechanical responses during walking, running, and trauma injury. A pendulum approach was used to simulate and quantify the shod heel region under impact loading, and it was found that peak loadings for the soft- and hard-soled feet differed significantly (Aerts and De Clercq 1993). However, impact tests can offer heel-pad deformation information without providing a true stress-strain response (Aerts and De Clercq 1993). It was also found that changes in peak muscle forces with simulated footwear conditions did not reflect changes in the peak impact force, and those responses to footwear conditions varied widely between subjects (Wright et al. 1998). A significant decrease in peak tibial acceleration and acceleration slope following fatigue using a vertical force plate on 24 healthy women was reported (Flynn, Holmes, and Andrews 2004). The venous plexus was also found to contribute to the damping properties of the heel pad during walking using three different impact velocities (0.2 to 0.6 m/s) (Weijers, Kessels, and Kemerink 2005).

#### 5.1.1.2 Accelerometer Measurement

Tibial axial acceleration was shown to be more sensitive for gauging footwear cushioning than ground reaction force measurement (Lafortune and Hennig 1992). Accelerometers positioned on the tibia have been used to evaluate the impact attenuation properties of different footwear during locomotion (Hamill 1999). During running, marked reductions were found in the higher frequency components of tibial shock signals (>20 Hz) by using specific midsole materials in the footwear (Shorten, Valiant, and Cooper 1986). Higher mean power frequencies were associated with more severe impacts (Johnson 1986). It was shown in animal studies that bone tissue development was very sensitive to the frequency content of transmitted impact shocks (Rubin, Mcleod, and Bain 1990).

#### 5.1.1.3 Material Test

Previous material tests focused on the effectiveness of reducing the magnitude of peak impact force and shock wave caused by heel strike. Besides transient force reduction, the evaluation of cushioning properties included material responsiveness to repetitive impacts (Sun et al. 2008). Long-term shock absorption characteristics could be as important as initial shock absorption responses in shoe designs (Jarrah et al. 1997). Plantar soft tissue demonstrated nonlinear stress-strain characteristics, and subcalcaneal tissue was shown to be different than other plantar soft tissue by compressive material test in vitro (Ledoux and Blevins, 2007).

Although recent advances in experimental measurement techniques provide data for footwear evaluation and design improvements, some important information, such as how the internal foot structures respond to impact loading and where damage may occur, remains difficult to obtain from experiments alone. Because of the complexity of the structures, limited measurement techniques, large variation in shoe design, and individual subject differences, consistent results regarding the biomechanical performance of the foot and footwear cannot be achieved in terms of the performance of different therapeutic and functional footwear. Experimental evaluation of individual shoes and optimal designs cannot be run efficiently and effectively. It has been suggested that a shift in focus from external to internal loading in the understanding of sporting injuries, and placing greater emphasis on subject-specific investigations, may be beneficial (Miller and Hamill 2009).

### 5.1.2 Computational Approach

Researchers have identified that computational modeling, based on FE methods, can allow realistic simulation of foot structures and the footwear interface and offer in-depth biomechanical information on both the foot and footwear. We have comprehensively reviewed the FE models developed for foot biomechanics and footwear design (Cheung et al. 2009). Both experimentation and modeling are important in investigating the biomechanics of foot impact. Finite element analysis (FEA) may be the most effective, combined with experiments.

The early FE models started from simplified two-dimensional or partial foot structures to investigate the effects of sole materials defined with either linear or hyperelastic or hyperfoam materials (Nakamura, Crowninshield, and Cooper 1981; Lemmon et al. 1997). Several studies used two-dimensional heel models to understand the loading response of the plantar heel pad (Verdejo and Mills 2004; Goske et al. 2006; Spears et al. 2007). Most of those models focused on the effects of sole design with varying materials or shapes on plantar pressure distribution in static conditions.

Several three-dimensional FE models were developed for footwear sole design and evaluation (Even-Tzur et al. 2006; Antunes et al. 2008; Cheung and Zhang 2008). FE models were used to investigate the effects of different sole materials, including silicone gel, plastazote, polyfoam, EVA, air and water pockets, and reinforcement bars within the sole, on plantar foot pressure and bone stresses during balanced standing, stance phases, or impact. Most models were analyzed under static or quasi-static loading, while nonlinear and viscoelastic properties were added to the FE model to study EVA midsole viscous damping of the shoe-heel interaction to heel pad stress and strain attenuation during heel strike in running (Even-Tzur et al. 2006).

The viscoelastic effect is often implemented for the footwear sole and soft tissues under dynamic loading. Recent computational studies on the viscoelastic material behavior of elastomeric materials and human soft tissue using the FE method were reported (Eskandari et al. 2008; Liu, Van Landingham, and Ovaert 2009; Fontanella et al. 2012). Such viscoelastic and hyperelastic models may provide a physically based simulation of tissue deformations for dynamic phenomena. A number of studies were undertaken to measure the nonlinear elastic or viscoelastic material properties of foot plantar soft tissues using mechanical or ultrasound indentation techniques (Erdemir et al. 2006; Gefen, Megido-Ravid, and Itzchak 2001; Zheng et al. 2000). The subject-specific plantar heel pad behavior could be obtained using the indenting test and analyzing by inverse FEA with a partially three-dimensional computational model. Optimized heel pad material coefficients were 0.001084 MPa ($\mu$), 9.780 ($\alpha$) (with an effective Poisson's ratio ($\nu$) of 0.475), for a first-order nearly incompressible Ogden material model (Chokhandre et al. 2012).

During a frontal car crash accident, the driver's foot and ankle may be injured due to the intrusion of the brake pedal. In car crash injury simulations, impact FE models of the foot and lower limb using explicit analysis were often developed to address the impact injury mechanism (Takahashi et al. 2000; Untaroiu, Darvish, and Crandall 2005; Cardot et al. 2006). A validated FE model of human detailed anatomical structures could be a useful tool for establishing injury criteria based on evaluation of the failure level of bone and ligament (Untaroiu, Darvish, and Crandall 2005).

Explicit techniques were widely used because they were easier to compute than implicit techniques. However, simulations based on explicit formulation without considering physiological muscle action may not accurately predict internal stress/strain of the foot and footwear.

Most existing FE models are based on static or quasi-static simulation. Since sole materials, such as elastomers, and biological tissues exhibit both viscoelasticity and nonlinear elasticity, an FE model with an implicit dynamic simulation can aid in evaluating these responses, based on time-dependent or velocity-related material properties. It is necessary to establish a dynamic FE model simulation of the foot for heel impact, using an implicit dynamic FE solver. A comprehensive dynamic three-dimensional FE model of the foot and footwear implemented by implicit formulation, incorporating realistic geometrical properties of bone and soft tissues, taking the viscoelasticity of soft tissue and sole into consideration, is still lacking and the biomechanically dynamic response of the foot to external impact forces has not been well addressed.

## 5.2 IMPLICIT DYNAMIC METHOD

Both explicit and implicit methods are algorithms used in numerical analysis for obtaining solutions of time-dependent ordinary and partial differential equations. The explicit method calculates the state of a system at a later time from the state of the system at the current time, while the implicit method finds a solution by solving an equation involving both the current state of the system and the later one. In mathematics, if $Y(t)$ is the current system state and $Y(t + \Delta t)$ is the state at the later time ($\Delta t$ is a small time step), then, for an explicit method:

$$Y(t+\Delta t) = F(Y(t)) \tag{5.1}$$

while for an implicit method one solves the equation

$$G(Y(t), Y(t+\Delta t)) = 0 \tag{5.2}$$

to find $Y(t + \Delta t)$.

It is clear that mplicit methods may require extra computational efforts to solve the above equations, which can be much harder to implement. The implicit method is used because many problems arising in practice are stiff, for which the use of an explicit method would require impractically small time steps $\Delta t$ to keep the error in the result tolerance. Therefore, for such problems, to achieve a given accuracy, it may take much less computational time to use an implicit method with larger time steps.

The implicit method is unconditionally stable, but it will encounter some difficulties when a complicated three-dimensional model is considered. Two main reasons are as follows: (1) with continuing reduction of the time increment, the computational costs in the tangent stiffness matrix will dramatically increase and even cause divergence; and (2) local instabilities will make it difficult to achieve equilibrium. Only direct integration methods, including implicit dynamics and explicit methods, are suitable for nonlinear problems. Most of the reported works on the comparison of implicit and explicit methods are on quasi-static nonlinear problems (Sun, Lee, and Lee 2000).

### 5.2.1 Solution Procedures

The implicit solution procedure uses an automatic increment method based on the success rate of a full Newton iterative solution method (ABAQUS manual):

$$\Delta \mathbf{u}^{(i+1)} = \Delta \mathbf{u}^{(i)} + \mathbf{K}_t^{-1} \cdot (\mathbf{F}^{(i)} - \mathbf{I}^{(i)}) \tag{5.3}$$

where $\Delta\mathbf{u}$ is the increment of displacement, $\mathbf{K}_t$ is the current tangent stiffness matrix, **F** the vector of applied loads, **u** the displacement vector, and **I** the internal force vector.

For an implicit dynamic procedure, the algorithm is defined as

$$\mathbf{M}\ddot{\mathbf{u}}^{(i+1)} + (1+\alpha)\mathbf{K}\mathbf{u}^{(i+1)} - \alpha\mathbf{K}\mathbf{u}^{(i)} = \mathbf{F}^{(i+1)} \tag{5.4}$$

where **M** is the mass matrix, **K** the stiffness matrix, **F** the vector of applied loads, and **u** the displacement vector:

$$\mathbf{u}^{(i+1)} = \mathbf{u}^{(i)} + \Delta t\dot{\mathbf{u}}^{(i)} + \Delta t^2\left(\left(\frac{1}{2}-\beta\right)\ddot{\mathbf{u}}^{(i+1)} + \beta\,\ddot{\mathbf{u}}^{(i+1)}\right) \tag{5.5}$$

and

$$\dot{\mathbf{u}}^{(i+1)} = \dot{\mathbf{u}}^{(i)} + \Delta t((1-\gamma)\ddot{\mathbf{u}}^{(i)} + \gamma\mathbf{u}^{(i+1)}) \tag{5.6}$$

with

$$\beta = \frac{(1-\alpha^2)}{4}, \gamma = 0.5 - \alpha \tag{5.7}$$

$\alpha = 0.05$ is often chosen by default as a small damping term to quickly remove the high-frequency noise without having a significant effect on the meaningful, lower-frequency response.

## 5.3 DYNAMIC FINITE ELEMENT MODEL FOR PLANTAR HEEL PAIN DURING HEEL STRIKE

Plantar heel pain in adults is the most common complaint among all foot ailments, accounting for about 15% of adult foot problems (Alshami, Souvlis, and Coppieters 2008). Plantar heel pain caused by atrophy of the fat pad often occurs in the elderly, because of thinning of the fat pad in the inferior heel or reduced shock absorption capability. Heel pain can be attenuated by wearing soft-soled shoes. These effects are difficult to understand through experimental techniques, such as plantar pressure and soft tissue deformation. FE methods could be an efficient tool to provide more information on internal stress/strain and joint movements.

### 5.3.1 Finite Element Model Development

The original FE foot model was previously developed from a female right foot; the development steps were described in Chapter 2. The plantar fat pad was further separated from other soft tissues based on magnetic resonance (MR) images (Figure 5.1) and assigned different material properties, including elastic (Gu et al. 2010), hyperelastic (Lemmon et al. 1997), and viscoelastic characteristics (Ledoux and Blevins 2007). Skin tissue was assigned as hyperelastic (Gu et al. 2010) by using a membrane element.

The plantar fat pad was separated using MIMICS (Materialise, Leuven, Belgium). The surface model of the plantar fat pad was imported into Rapidform XOR2 (3D Systems Korea Inc., Seoul, Korea) to generate the solid model. Thereafter, the FE package, ABAQUS v6.11 (Simulia, Providence, Rhode Island) was used for the creation of the FE mesh and subsequent analysis. In this study, a dynamic-implicit FE approach was employed to study foot impact during heel strike. In this FE prediction, the impact velocity was set at 0.5 m/s, in accordance with experimental results.

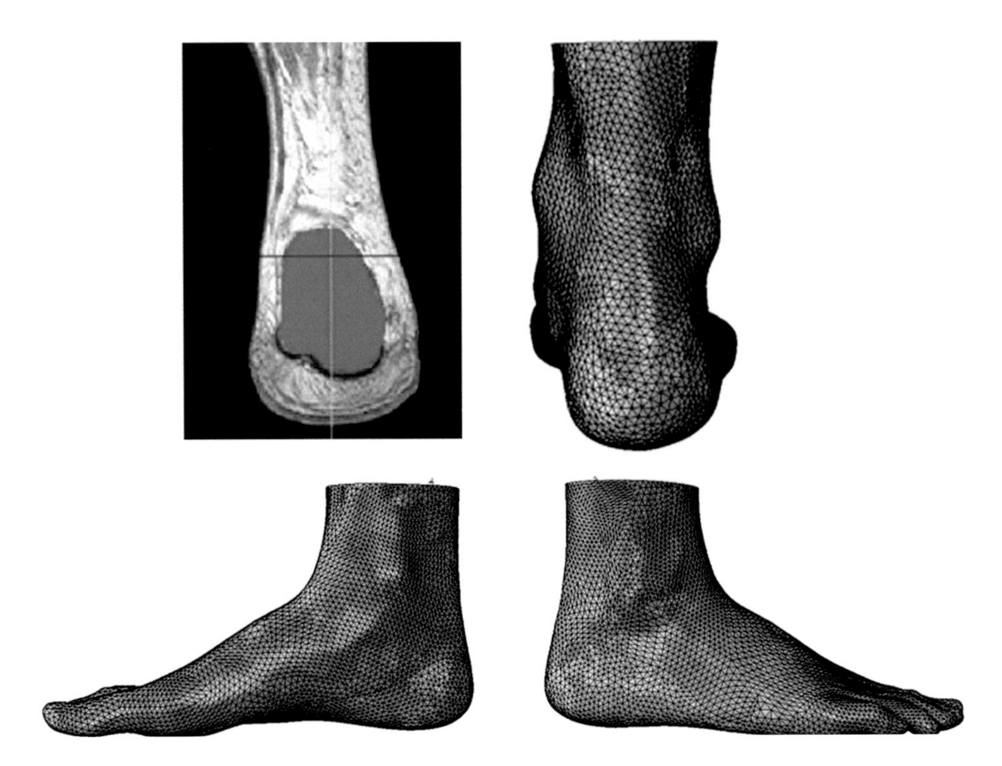

**FIGURE 5.1** **(See color insert.)** Finite element foot model with segmented plantar fat pad.

The implementation of a dynamic model was different from the traditional quasi-static model and is summarized in Table 5.1. The heel strike boundary condition is shown in Figure 5.2. A shock attenuation period (0 to 2% of gait cycle) from initial contact to loading response was simulated to investigate the biomechanical response of the foot and ankle.

### 5.3.2 Experimental Validation

Ground reaction force (GRF) during barefoot walking was recorded using a force platform (Advanced Mechanical Technology Inc., Newton, USA) at 2000 Hz (Figure 5.3). A high-speed camera (Casio EXILIM Pro EX-F1, Japan) at 1200 Hz was used to visualize fat pad deformation.

**TABLE 5.1**
**Comparison between Quasi-Static and Dynamic Models**

| Setting | Quasi-Static Simulation | Dynamic Simulation |
|---|---|---|
| Solver | Static, general | Dynamic, implicit |
| Boundary condition | Ground-striking<br>Foot-encrusted | Foot-striking<br>Ground-encrusted |
| Material properties | Young's modulus,<br>Poisson's ratio | Young's modulus,<br>Poisson's ratio, density,<br>time-related properties |
| Loading conditions | Force,<br>displacement | Initial velocity, change of any loading conditions with time |

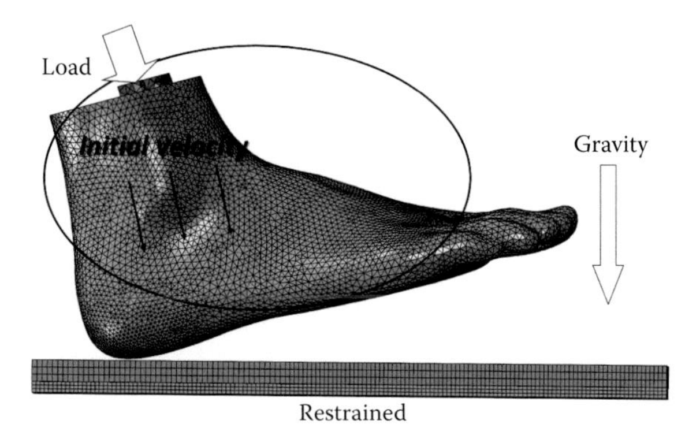


FIGURE 5.2 Impact boundary condition of the dynamic foot model.

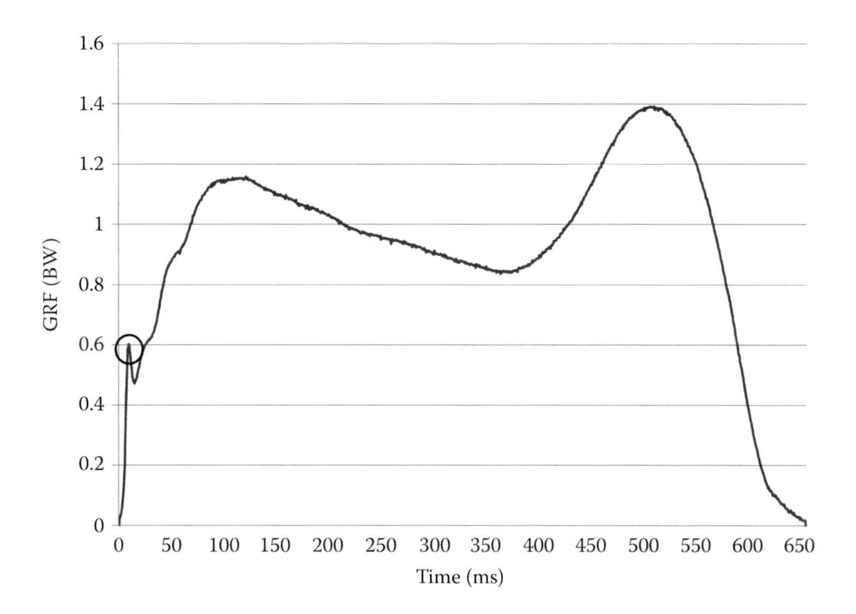


FIGURE 5.3 Vertical ground reaction force during walking.

Markers were placed on the lateral and medial posterior side of the rear foot (Figure 5.4). The data from GRF measurements and the high-speed camera were synchronized.

### 5.3.3 Results

The FE model's prediction of a peak GRF of about 60% body weight (BW) within 0.015 seconds was in agreement with experimental data (Figure 5.5). The maximal plantar heel pad compressive deformation was 5.64 mm by experimental measurement and 5.79 mm by FE prediction. Deformation and compressive strain at heel strike are shown in Figure 5.6.

Chi and Schmitt (2005) reported an impact force during walking at heel strike of 0.79 BW (from 0.37 to 1.41), time to impact peak of 14.61 ms (from 10.33 to 24.00), and compressive deformation of 5.24 mm (from 2.87 to 8.29). Our experimental and FE predicted data were consistent with these results. The impact force magnitude is more strongly affected by impact velocity than walking speed (Whittle 1999; Chi and Schmitt 2005).

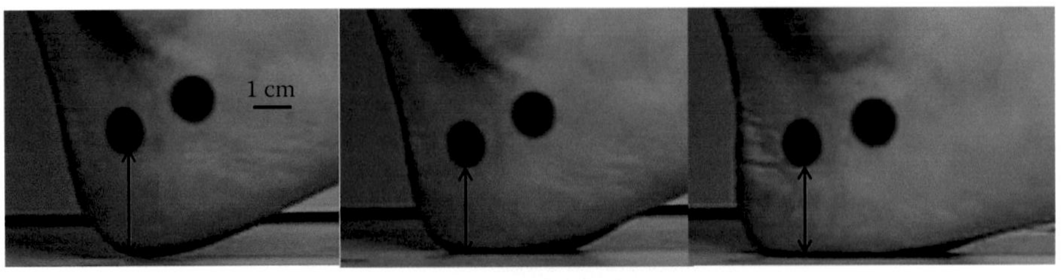


Lateral view of right foot

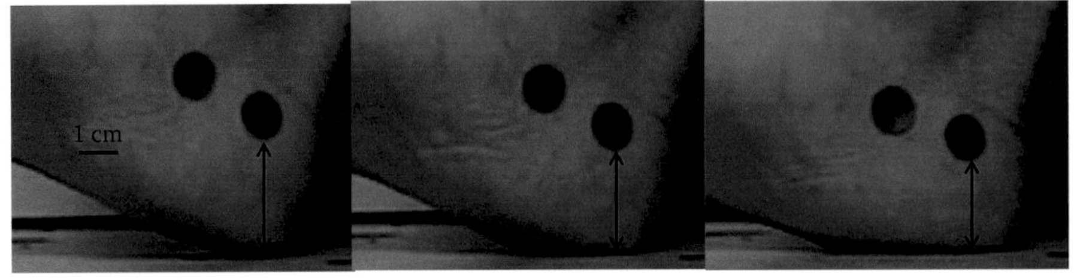


Medial view of right foot

**FIGURE 5.4** Lateral and medial views of heel pad deformation at heel strike.

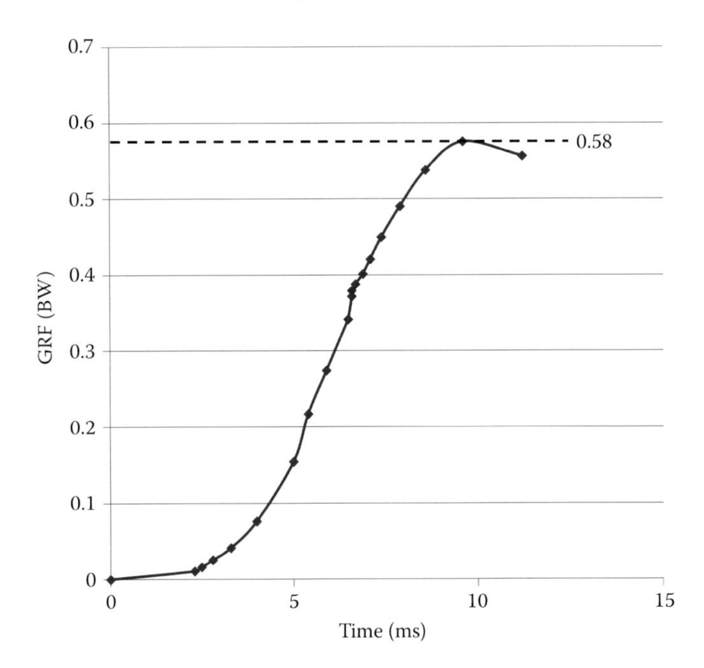


**FIGURE 5.5** Predicted plantar fat pad deformation curve during heel strike.

In fat pad atrophy patients, hard-soled shoes and walking on hard surfaces can aggravate pain. The central portion of the heel fat pad is the most intense region (Alshami, Souvlis, and Coppieters 2008). In this study, the largest strain of the heel fat pad during the heel strike was predicted around the central portion, demonstrated in Figure 5.6. The medial calcaneal nerve usually divides into anterior and posterior branches and offers sensory innervation to most of the heel fat pad (Louisia and Masquelet 1999). The medial calcaneal nerve can be irritated and compressed following atrophy of the heel fat pad (Davidson and Copoloff 1990).

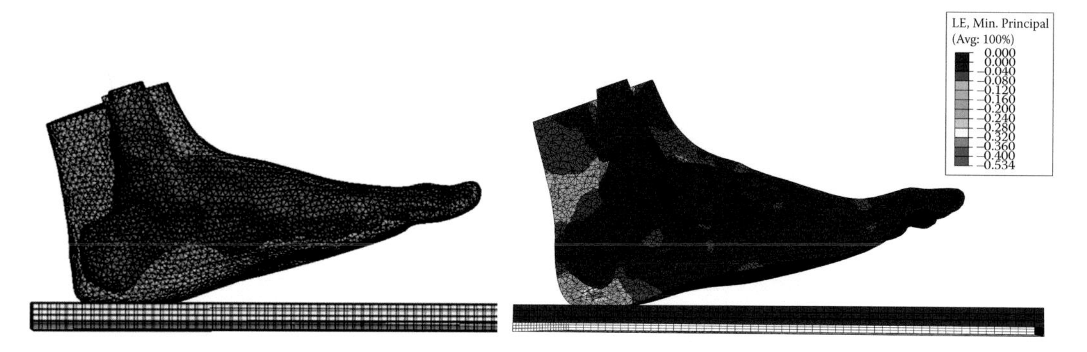


**FIGURE 5.6** **(See color insert.)** Predicted maximal plantar fat pad deformation and compressive strain at heel strike.

The indentation test combined with the computational approach can provide patient-specific fat pad deformation and internal stress/strain information for foot pathologies and therapeutic footwear designs (Erdemir et al. 2006). In this study, quantitative analysis of internal and contact stress of fat pad tissue using the dynamic FE foot model could help to address heel pain during walking or running.

FE analysis of shock attenuation period using dynamic-implicit algorithms is challenging and still undergoing constant improvement. Preliminary predicted results showed that the heel fat pad under impact force could induce large deformation and stress concentration on soft tissues under the calcaneus, which could affect the medial calcaneal nerve during heel strike, while similar heel pain location could have different pathophysiologic mechanisms and biomechanics patterns (Saggini et al. 2007). Moreover, in this study, we did not consider the venous plexus effect on fat pad damping stiffness, which could decrease impact force about 1.8% at 0.4 m/s (Weijers, Kessels, and Kemerink 2005). The simulation could further be tested and parametrically studied with different fat pad and sole material stiffness and thickness, striking velocities, striking angles, and proximal loads.

## ACKNOWLEDGMENTS

This work was supported by the Research Grant Council of Hong Kong (GRF Project no. PolyU5326/11E) and the National Natural Science Foundation of China (11272273, 11120101001).

## REFERENCES

Aerts, P., D. De Clercq. 1993. Deformation characteristics of the heel region of the shod foot during a simulated heel strike: The effect of varying midsole hardness, *J. Sports Sci.* 11:449–61.

Alshami, A.M., T. Souvlis, and M.W. Coppieters. 2008. A review of plantar heel pain of neural origin: Differential diagnosis and management. *Manual Therapy* 13:103–111.

Antunes, P.J., G.R. Dias, A.T. Coeiho, F. Rebeio, and T. Pereira. 2008. Hyperelastic modelling of cork polyurethane gel composites: Non-linear FEA implementation in 3D foot model. *Mater. Sci. Forum* 587–8.

Cardot, J., C. Masson, P.J. Arnoux, and C. Brunet. 2006. Finite element analysis of cyclist lower limb response in car-bicycle accident. *UCrash* 11(20):115–29.

Cavanagh, P.R., and M.A. Lafortune. 1980. Ground reaction forces in distance running. *J. Biomech.* 13:397–406.

Cheung, J.T., J. Yu, W.C. Wong, and M. Zhang. 2009. Current methods in computer-aided engineering for footwear design. *Footwear Sci.* 1:31–46.

Cheung, J.T.M., and M. Zhang. 2008. Parametric design of pressure-relieving foot orthosis using statistics-based finite element method. *Med. Eng. Phys.* 30:269–77.

Chi, K.J., and D. Schmitt. 2005. Mechanical energy and effective foot mass during impact loading of walking and running. *J. Biomech.* 38(7):1387–95.

Chokhandre, S., J.P. Halloran, A.J. van den Bogert, and A. Erdemir. 2012. A three-dimensional inverse finite element analysis of the heel pad. *J. Biomech. Eng.* 134(3):031002.
Collins, J.J., and M.W. Whittle. 1989. Impulsive forces during walking and their clinical implications. *Clin. Biomech.* 4:179–87.
Davidson, M.R., and J.A. Copoloff. 1990. Neuromas of the heel. *Clin. Podiatric Med. Surg.* 7(2):271–88.
Erdemir, A., M.L. Viveiros, J.S. Ulbrecht, and P.R. Cavanage. 2006. An inverse finite-element model of heel pad indentation. *J. Biomech.* 39(7): 1279–86.
Eskandari, H., S.E. Salcudean, R. Rohling, and J. Ohayon. 2008. Viscoelastic characterization of soft tissue from dynamic finite element models. *Phys. Med. Biol.* 53(22):6569–90.
Even-Tzur, N., E. Weisz, Y. Hirsch-Falk, and A. Gefen. 2006. Role of EVA viscoelastic properties in the protective performance of a sport shoe: Computational studies. *Bio-Med. Mater. Eng.* 16:289–99.
Flynn, J.M., J.D. Holmes, and D.M. Andrews. 2004. The effect of localized leg muscle fatigue on tibial impact acceleration. *Clin. Biomech.* 19(7):726-32.
Fong, D.T.P., Y. Hong, and J.X. Li. 2007. Cushioning and lateral stability functions of cloth sport shoes. *Sports Biomech.* 6(3):407–17.
Fong, D.T.P., Y. Hong, and J.X. Li. 2009. Human walks carefully when the ground dynamic coefficient of friction drops below 0.41. *Safety Sci.* 47(10):1429–33.
Fontanella, C.G., S. Matteoli, E.L. Carniel, J.E. Wilhjelm, A. Virga, A. Corvi, and A.N. Natali. 2012. Investigation on the load-displacement curves of a human healthy heel pad: In vivo compression data compared to numerical results. Med. Eng. Phys. 34(9):1253–9.
Gefen, A., M. Megido-Ravid, and Y. Itzchak. 2001. In vivo biomechanical behavior of the human heel pad during the stance phase of gait. *J. Biomech.* 34:1661–5.
Goske, S., A. Erdemir, M. Petre, S. Budhabhatti, and P.R. Cavanagh. 2006. Reduction of plantar heel pressures: Insole design using finite element analysis. *J. Biomech.* 39:2363–70.
Gu, Y., J. Li, X. Ren, M.J. Lake, and Y. Zeng. 2010. Heel skin stiffness effect on the hind foot biomechanics during heel strike. *Skin Res. Technol.* 16:291–6.
Hamill, J. 1999. Evaluation of shock attenuation, *Proceedings of the Fourth Symposium of Footwear Biomechanics*, Canmore, Canada, August 5–7, p. 7–8.
Jarrah, M., W. Qassem, M. Othman, and M. Gdeisat. 1997. Human body model response to mechanical impulse. *Med. Eng. Phys.* 19:308–16.
Johnson, G.R. 1986. The use of spectral analysis to assess the performance of shock absorbing footwear. *J. Eng. Med.* 15:117–22.
Jørgensen, U. 1985. Achillodynia and loss of heel pad shock absorbency. *Am. J. Sports Med.* 13:128–32.
Jørgensen, U., and F. Bojsen-Møller. 1989. Shock absorbency of factors in the shoe/heel interaction—with special focus on the role of the heel pad. *Foot Ankle* 9(6):294–9.
Lafortune, M.A., and E.M. Hennig. 1992. Cushioning properties of footwear during walking: Accelerometer and force platform measurements. *Clin. Biomech.* 7:181–4.
Ledoux, W.R., and J.J. Blevins. 2007. The compressive material properties of the plantar soft tissue. *J. Biomech.* 40L:2975–81.
Lemmon, D., T.Y. Shiang, A. Hashmi, J.S. Ulbrecht, and P.R. Cavanagh. 1997. The effect of insoles in therapeutic footwear: A finite element approach. *J. Biomech.* 30:615–20.
Liu, K., M.R. Van Landingham, and T.C. Ovaert. 2009. Mechanical characterization of soft viscoelastic gels via indentation and optimization-based inverse finite element analysis. *J. Mech. Behav. Biomed. Mater.* 2:355–62.
Louisia, S., and A.C. Masquelet. 1999. The medial and inferior calcaneal nerves: An anatomic study. *Surg. Radiol. Anat.* 21(3):169–73.
McClay, I.S., J.R. Robinson, T.P. Andriacchi, E.C. Frederick, T. Gross, P. Martin, G. Valiant, K.R. Williams, and P.R. Cavanagh. 1994. A profile of ground reaction forces in professional basketball players. *J. Appl. Biomech.* 10:222–36.
Miller, R.H., and J. Hamill. 2009. Computer simulation of the effects of shoe cushioning on internal and external loading during running impacts. *Comput. Methods Biomech. Biomed. Eng.* 12(4):481–90.
Nakamura, S., R.D. Crowninshield, and R.R. Cooper. 1981. An analysis of soft tissue loading in the foot: A preliminary report. *Bull. Prosthetics Res.* 18:27–34.
Nigg, B.M., G.K. Cole, and G.P. Bruggemann. 1995. Impact forces during heel–toe running. *J. Appl. Biomech.* 11:407–32.
Noe, D.A., S.J. Voto, M.S. Hoffmann, M.J. Askew, and I.A. Gradisar. 1993. Role of the calcaneal heel pad and polymeric shock absorbers in attenuation of heel strike impact. *J. Biomed. Eng.* 15(1):23–6.

Rubin, C., K. Mcleod, and S. Bain. 1990. Osteoregulatory nature of mechanical stimuli. *J. Biomech.* 23(Suppl.):43–54.

Saggini, R., R.G. Bellomo, G. Affaitati, D. Lapenna, and M.A. Giamberardino. 2007. Sensory and biomechanical characterization of two painful syndromes in the heel. *J. Pain* 8(3):215–22.

Shorten, M., G. Valiant, and L. Cooper. 1986. Frequency analysis of the effects of shoe cushioning on dynamic shock in running. *Med. Sci. Sports Exercise* 18(Suppl.):S80–S81.

Spears, I.R., J.E. Miller-Young, J. Sharma, R.F. Ker, and F.W. Smith. 2007. The potential influence of the heel counter on internal stress during static standing: A combined finite element and positional MRI investigation. *J. Biomech.* 40:2774–80.

Sun, P.C., H.W. Wei, C.H. Chen, C.H. Wu, H.C. Kao, and C.K. Cheng. 2008. Effects of varying material properties on the load deformation characteristics of heel cushions. *Med. Eng. Phys.* 30(6):687–92.

Sun, J.S., K.H. Lee, and H.P. Lee. 2000. Comparison of implicit and explicit finite element methods for dynamic problems. *J. Mater. Process. Technol.* 105:110–18.

Takahashi, Y., Y. Kikuchi, A. Konosu, and H. Ishikawa. 2000. Development and validation of the finite element model for the human lower limb of pedestrians. *Stapp Car Crash J.* 44:335–55.

Untaroiu, C., K. Darvish, and J. Crandall. 2005. A finite element model of the lower limb for simulating pedestrian impacts. *Stapp Car Crash J.* 49:157–81.

Verdejo, R., and N.J. Mills. 2004. Heel-shoe interactions and the durability of EVA foam running-shoe midsoles. *J. Biomech.* 37:1379–86.

Weijers, R.E., A.G. Kessels, and G.J. Kemerink. 2005. The damping properties of the venous plexus of the heel region of the foot during simulated heelstrike. *J. Biomech.* 38(12):2423–30.

Whittle, M.W. 1999. Generation and attenuation of transient impulsive forces beneath the foot: A review. *Gait Posture* 10: 264–75.

Wright, I.C., R.R. Neptune, A.J. van den Bogert, and B.M. Nigg. 1998. Passive regulation of impact forces in heel-toe running. *Clin. Biomech.* 13(7):521–31.

Zhang, S., K. Clowers, C. Kohstall, and Y.J. Yu. 2005. Effects of various midsole densities of basketball shoes on impact attenuation during landing activities. *J. Appl. Biomech.* 21(1):3–17.

Zheng, Y.P., Y.K. Choi, K. Wong, S. Chan, and A.F. Mak. 2000. Biomechanical assessment of plantar foot tissue in diabetic patients using an ultrasound indentation system. *Ultrasound Med. Biol.* 26(3):451–6.

# Section II

## Knee Joint

# 6 Knee Joint Model for Anterior Cruciate Ligament Reconstruction

*Jie Yao, Ming Zhang, and Yubo Fan*

## CONTENTS



## SUMMARY

Tearing of the anterior cruciate ligament (ACL) is a common knee injury, and is frequently treated through ACL reconstruction. However, sequelae such as tunnel enlargement and tibia fracture have been reported. The creation of bone tunnels or implantation of interference devices may interrupt the normal loading transmission and potentially contribute to long-term sequelae. This chapter aims to (1) develop a three-dimensional finite element (FE) model of the human knee joint with ACL reconstructions; (2) quantify the change of strain energy density (SED) distribution induced by the tunnel creation and interference screw; and (3) investigate the influence of screw material on the SED changes in the tibia. The bone SED distribution was derived from the validated FE model under compressive loading. The numerical results confirmed that the bone SED distribution and stress orientation changed after surgery. These changes occurred around the bone tunnel, and could produce abnormal bone remodeling. The consequential bone resorption and micro-damage may serve as a predisposing factor for tunnel enlargement and osteoarthritis. The material property of the screw could also influence the postoperative SED distribution in the tibia. On the premise of achieving sufficient fixation strength, using a screw with a modulus similar to the bone could decrease the risk of stress shielding. These findings together with histology factors could help us to understand the pathomechanism of the sequelae, and help to improve surgical techniques.

## 6.1 INTRODUCTION

The ACL is an essential component for the stability of the human knee joint. It provides constraint on excessive translation and rotation of the tibia. However, injuries to the ACL are common, particularly in sports activities (more than 120,000 cases per year in the United States; Kim et al. 2011). A torn ACL causes knee instability and loss of extension during the early stages of injury (Muneta et al. 1996), and may lead to osteoarthritis if left untreated (Louboutin et al. 2009). Since an injured ACL cannot heal spontaneously, ACL reconstruction is often required, to replace the torn ACL with an autogenous or allogenous tendon graft. ACL reconstruction can successfully restore short-term knee stability (Yagi et al. 2002), and can also lower but not eliminate the risk of developing osteoarthritis (Louboutin et al. 2009). A possible explanation is that the surgery could restore the natural kinematics and interaction of the tibiofemoral joint, thus resulting in a beneficial mechanical environment on the articular surface.

However, there are still several long-term sequelae, including bone tunnel enlargement, tunnel communication, and bone fracture (Delcogliano et al. 2001; Zerahn et al. 2006; Siebold 2007; Nyland et al. 2010; Gobbi et al. 2011). The exact mechanism of the sequelae, however, remains unclear. A possible predisposing factor is the creation of bone tunnels and implantation of an interference screw, which are utilized to accommodate and fix the graft ends in ACL reconstruction. The bone tunnels and interference screw may interrupt the normal loading transmission in the knee and result in undesirable stress distribution adjacent to the tunnel wall. Consequently, abnormal bone remodeling, including bone resorption or micro-damage, may be activated, and contribute to collapse of the bone adjacent to the tunnel wall.

Changes in the internal mechanical environment and the consequential histological response are difficult to anticipate from in vivo or in vitro experiments. An alternative option is the FE method, which can visualize the distribution of mechanical stimuli (e.g., stress, strain, and SED) in the knee under complex loading conditions. With a precisely validated FE model of the human knee joint, the mechanical stimuli after ACL reconstruction can be quantified. A better understanding of the postoperative mechanical environment may also improve treatment methods and lower the risk of secondary injury.

## 6.2 DEVELOPMENT OF A COMPREHENSIVE KNEE MODEL

The development of the FE model of the knee joint involved geometry acquisition, mesh discretization, and model validation. The details of this model development process are described next.

### 6.2.1 Geometry Acquisition

Since the present study focused on SED distribution in both bone and cartilage, the geometry of the knee joint was obtained with magnetic resonance (MR) scanning, which can identify both hard and soft tissues. A male subject (weight 65 kg, height 172 cm, age 30) volunteered for this study. Knee pathology was ruled out with physical and MR image examinations. The right knee of the subject was maintained in an extended but relaxed position during scanning. Since any movement during scanning will produce image artifact, a knee brace with low-temperature thermoplastic sheets (ORFIT® Eco, ORFIT, Inc., Belgium) was utilized to keep the knee static. The right knee joint was scanned from 20 cm proximal to 20 cm distal of the articular line with the MR machine (Siemens, Sonata 1.5T, Germany). Images were acquired at 2-mm intervals, 0.47 × 0.47 pixel resolution, 43-ms echo time, and 7170-ms repetition time. Ethical approval was granted prior to image capture and the subject was detailed on the procedures and gave written consent.

Three-dimensional models of the femur, tibia, fibula, and meniscus were reconstructed from the MR images using the clinical images processing software MIMICS (Materialise, Inc., Belgium). Segmentations of cortical bone, cancellous bone, cartilage, and meniscus were performed based on the different grayscales of tissues with the guidance of a surgeon. The insertion sites of the ACL, posterior cruciate ligament (PCL), medial collateral ligament (MCL), lateral collateral ligament (LCL), and four meniscal attachments on the bone were marked for ligament modeling in the FE model. Since the subchondral plate is difficult to identify on MR images, a uniform thickness of 1.5 mm was assumed, in accordance with previous research (Milz and Putz 1994).

### 6.2.2 Finite Element Discretization

The FE model of the human knee joint was developed based on a three-dimensional geometrical model using the finite element software ABAQUS (Simulia, Inc., USA). Discretizations were implemented in the cortical bone, cancellous bone, cartilage, and meniscus with three-dimensional four-node tetrahedral elements. Four ligaments (ACL, PCL, MCL, and LCL) were modeled as bundles of two-node truss elements. The nodes of the truss elements were uniformly fixed within the ligament insertion sites on the bone. Four meniscal horn attachments were also modeled as truss elements, with one side fixed to the meniscal horn and the other side to bone.

The ligament was assumed to be a nonlinear hyperelastic material with the force-strain relationship:

$$f = \begin{cases} 0, & \varepsilon \le 0, \\ \frac{1}{4} k\varepsilon^2 / \varepsilon_1, & 0 \le \varepsilon \le 2\varepsilon_1, \\ k(\varepsilon - \varepsilon_1), & 2\varepsilon_1 \le \varepsilon, \end{cases} \tag{6.1}$$

where *f* is the in situ force of the ligament, ε is the strain on the ligament, ε*l* is the reference strain, assumed to be 0.03, and *k* is the stiffness coefficient. The *k* of ACL, PCL, MCL, and LCL are 10,000, 18,000, 6,000, and 8,250, respectively (Blankevoort et al. 1991; Li et al. 2002; Netravali et al. 2011). The meniscal attachments were assumed to be a noncompressive elastic material with an elastics modulus of 600 MPa. The medial and lateral menisci were assumed to be transversely isotropic, linear elastic materials. The moduli in circumferential ($E_\theta$), radial ($E_R$), and axial ($E_Z$) directions were 125, 27.5, and 27.5 MPa; the Poisson ratios ($\nu_{\theta R}$, $\nu_{\theta Z}$, and $\nu_{RZ}$) were 0.1, 0.1, and 0.33, and the shear moduli ($G_{\theta R}$ and $G_{\theta Z}$) were 2.0 MPa (Yao et al. 2006). Cortical bone, cancellous bone, subchondral bone, and cartilage were assumed to be homogeneous linear elastic materials. The elastic moduli and Poisson ratios are shown in Table 6.1 (Taylor et al. 1998; Au et al. 2005; Yao et al. 2006). Since the friction on the articular surface is extremely small, a frictionless and finite sliding algorithm was assigned on cartilage-cartilage and cartilage-meniscus contact pairs.

### 6.2.3 Model Validation

For an effective FE simulation, validation is critical to ensure that the model can accurately predict the realistic deformation of tissue. In the present study, validation was implemented by comparing realistic meniscal deformation with the deformation predicted by the FE model. The realistic three-dimensional shape of the meniscus (including information on meniscal movement and deformation)

**TABLE 6.1**
**Material Properties of Bone and Cartilage**

| Tissue | Young's Modulus | Poisson Ratio |
|---|---|---|
| Cortical bone | 17 GPa | 0.33 |
| Subchondral bone | 1.15 GPa | 0.25 |
| Cancellous bone | 0.4 GPa | 0.33 |
| Cartilage | 5 MPa | 0.35 |

*Source:* Taylor et al., *J Biomech*, 31, 4, 303–310, 1998; Au et al., *J Biomech* 38, 4, 827–832, 2005; Yao et al., *J Biomech Eng*, 128, 1, 135–141, 2006.

at 45 degrees of knee flexion was obtained from MR scanning. The virtual three-dimensional shape of the meniscus was obtained by flexing the FE model of the knee joint to 45 degrees. After overlapping the realistic meniscus and the FE meniscus, the mismatch volumes on the medial and lateral sides were 22% and 26.5%, respectively. These mismatches were within the deviation of MR image pixel resolution. The result of this validation could ensure the precision of the following applications.

## 6.3 APPLICATION OF THE FINITE ELEMENT MODEL

The sequelae following ACL reconstruction, such as tunnel enlargement and osteoarthritis, have been correlated to changes in the bone material and morphology. According to Wolff's law, the postoperative mechanical environment plays an important role in these bone alterations, and could influence the long-term surgical outcome. Therefore, characterization of the postoperative mechanical environment and its potential influence on bone remodeling could help to understand the pathomechanism of the sequelae, and may help advance the surgical procedure. In this light, the following sections developed an FE model of the human knee joint with ACL reconstruction, quantified the alteration of bone SED distribution in comparison with the intact knee, and analyzed the effect of the interference screw material on the SED.

### 6.3.1 Model of ACL Reconstruction

An anatomic single-bundle ACL reconstruction was performed on the validated model of the human knee joint. A routine procedure was undertaken with the guidance of a surgeon. After the removal of the natural ACL, a tibial bone tunnel was drilled from the medial side of the tibial tubercle through the natural footprint of the ACL on the tibial plateau. The tunnel diameter was 9 mm, and the length was approximately 40 mm. The femoral bone tunnel was drilled from the natural ACL footprint on the femoral notch and through the lateral side of the femoral cortical bone. Specifically, an Endobutton fixation requires the extra-articular part of the femoral tunnel to be 4.5 mm in diameter and 10 mm in length. The intra-articular part of the tunnel was 9 mm in diameter and approximately 25 mm in length.

Bundles of nonlinear springs were used to simulate the graft in the FE model. The virtual graft was placed within the bone tunnels and served as a connector between the femur and tibia. In reality, the material properties of the tendon graft and Endobutton tape were different from the natural ACL. However, this study focused on the SED redistribution induced by the tunnel and interference screw. To eliminate the side effects of graft and Endobutton tape material on SED distribution, their material properties were assumed to be equal to that of the ACL. On the intra-articular tunnel aperture, the graft was free to move along the tunnel axis, whereas it was constrained in the direction perpendicular to the tunnel axis. To simulate this motion pattern, a slip ring algorithm was utilized in the interaction of "tunnel aperture-graft."

An interference screw was utilized to fix the graft end within the tibial tunnel. As shown in Figure 6.1, a screw 9 mm in diameter and 25 mm in length was placed in the distal tibial tunnel, whereby its head engaged the extra-articular tunnel aperture. The screw was meshed with a four-node tetrahedral element and assigned with a homogeneous linear elastic material. To quantify the influence of the screw material on the SED distribution, an elastic modulus ranging from 0 to 113.5 GPa (modulus of titanium alloy) was analyzed, with a Poisson ratio of 0.27. The FE model of the knee following ACL reconstruction is shown in Figure 6.1. Loading and boundary conditions were as follows: distal tibia was rigidly fixed in all directions, and a compressive loading equal to the body weight was applied at the proximal femur in the direction of the femoral shaft. Knee extension was maintained during the loading.

### 6.3.2 Change of SED Distribution after ACL Reconstruction

After ACL reconstruction, the SED distributions changed in both femur and tibia, and these changes occurred primarily around the tunnels (Figure 6.2). Adjacent to the femoral tunnel, a reduction in SED occurred at both proximal and distal regions of the tunnel wall. The minimum SED was

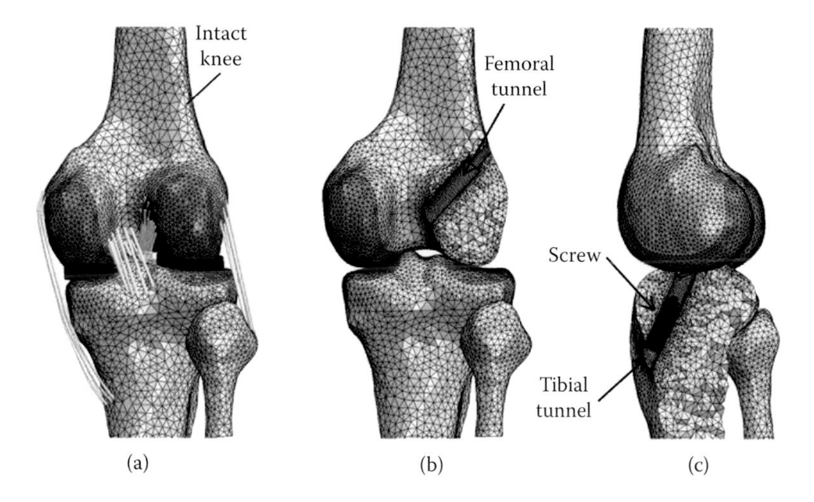


**FIGURE 6.1 (See color insert.)** FE model of the human knee joint. (a) Intact knee. (b) Cross-section of the femoral tunnel. (c) Cross-section of the tibial tunnel and interference screw.

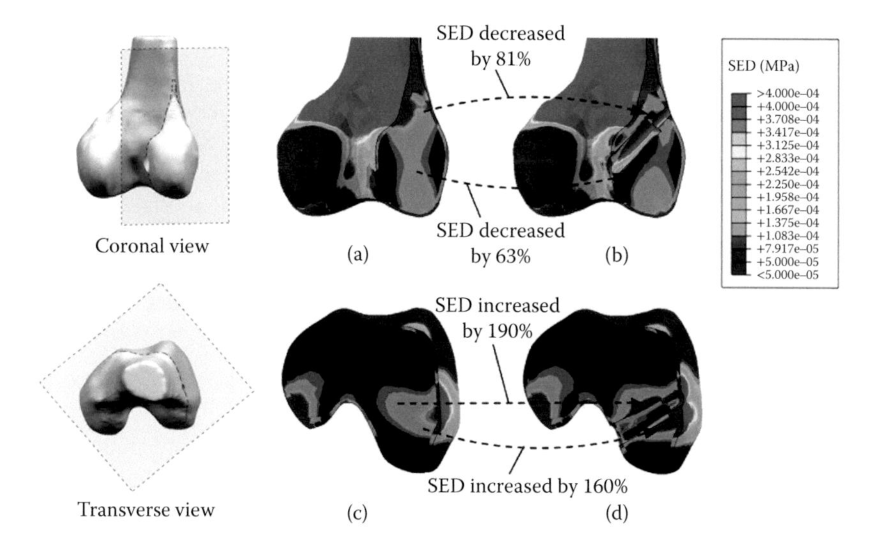


**FIGURE 6.2 (See color insert.)** SED distributions in the femur. (a) SED distribution in the intact femur (coronal view). (b) SED distribution in the femur after ACL reconstruction (coronal view). (c) SED distribution in the intact femur (transverse view). (d) SED distribution in the femur after ACL reconstruction (transverse view).

$2.7 \times 10^{-5}$ MPa, and the SED per density (SED/ρ) was $1.7 \times 10^{-5}$ Joule per gram (J/g), which is lower than the threshold for bone resorption (Mellal et al. 2004; Lin et al. 2010). The low SED may decrease the bone density with time. Since compressive loadings (e.g., standing and walking) are typical daily activities, the postoperative reduction in SED may frequently activate bone resorption and contribute to gradual tunnel enlargement. On the other hand, a dramatic increase in SED was observed in the cortical bone close to the extra-articular tunnel aperture. This SED increase could be caused by the Endobutton graft fixation at the aperture. The maximum SED was 0.017 MPa, and the SED/ρ was 0.0085 J/g, which is greater than the threshold for bone micro-damage. The accumulated micro-damages may contribute to long-term loosening of the graft and postoperative bone fracture.

SED distribution in the cartilage was also influenced by ACL reconstruction. In the femoral cartilage, the SED increased by approximately 280% at the posterolateral site. In the medial tibial cartilage, the SED increased by approximately 83% at the anterolateral site. In the lateral tibial cartilage, the SED increased by approximately 45% at the anteromedial site. The increased SED may affect the cartilage signal pathways and potentially contribute to osteoarthritis (Gosset et al. 2008).

### 6.3.3 Change of Stress Direction after ACL Reconstruction

ACL reconstruction further influenced the direction of stress in the bone. As shown in Figure 6.3, tensile stress can be seen close to the femoral tunnel wall after ACL reconstruction, and reduced compressive stress is apparent around the tunnel. From the trajectories of the tensile and compressive stress vectors, it was confirmed that the creation of the bone tunnel interrupted the normal load transmission in the knee and was responsible for the abnormal SED after surgery. The normal compressive stress trajectory was blocked by the tunnel and the tensile stress increased around the tunnel wall. The changes in stress orientation occurred near the intra-articular tunnel aperture, which is a high-risk region for tunnel enlargement and osteoarthritis. Furthermore, these changes in stress also influenced the mechanical environment within the cartilage. In both femoral and tibial cartilages, SEDs were increased around the cartilage edge close to the intra-articular tunnel apertures. Since there is a close relationship between the mechanical environment and cartilage metabolism (Gosset et al. 2008), the postoperative abnormal SED may contribute to long-term cartilage degradation.

### 6.3.4 Effect of Screw Material on SED Distribution

The influence of screw material on the redistribution in tibial SED was also investigated. Figure 6.4 indicates the coupled effect of both the tunnel and screw in altering the SED. When the modulus of the screw is greater than that of cancellous bone, the screw played a dominant role in the redistribution of the tibial SED. Conversely, when the modulus of the screw is lower than that of cancellous bone, the bone tunnel became the dominant factor. Figure 6.4 shows that the abnormal SED alteration was minimized when the screw modulus was equal to that of cancellous bone. This result implies that, on the premise of sufficient fixation strength, using a screw with a modulus similar to that of bone could decrease the risk of stress shielding, and thus could be positive for the surgical outcome.

### 6.3.5 Distribution of SED in Interference Screw

The SED distribution in the screw was also nonuniform. The SED was low around the anterolateral shaft of the screw, whereas it was high at the posteromedial shaft and the head of the screw (Figure 6.5). Recently, there has been increasing interest in biodegradable interference screws, which

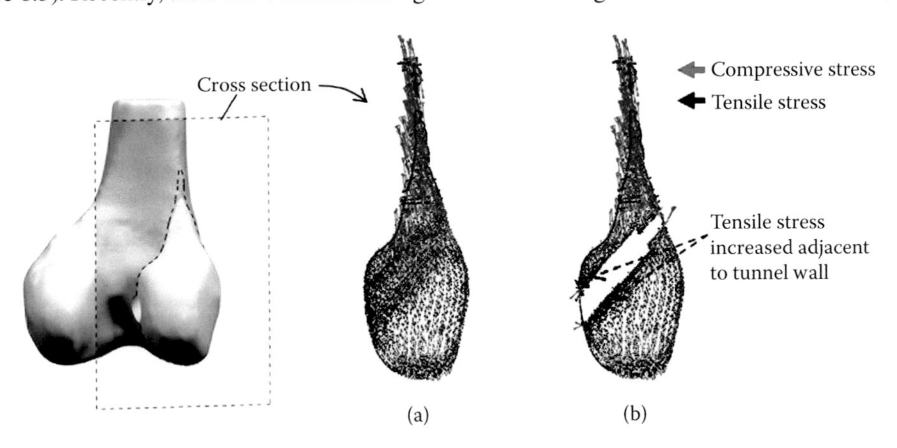


**FIGURE 6.3 (See color insert.)** Compressive and tensile stress trajectories in a coronal section of the femur. The red arrow indicates compressive stress, and the black arrow indicates tensile stress. (a) Intact knee. (b) Knee after ACL reconstruction.

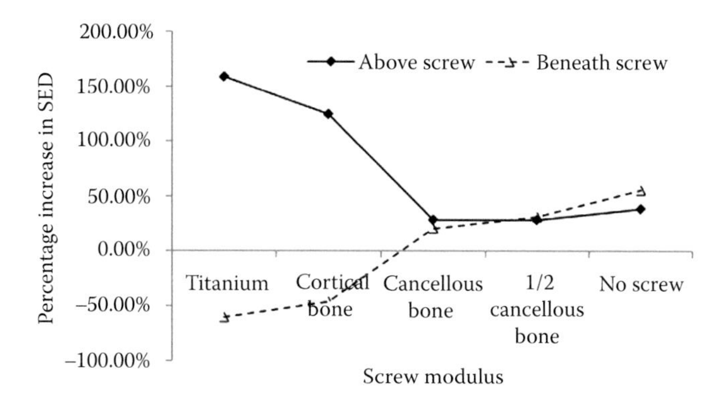


**FIGURE 6.4** Influence of the screw modulus on bone SED alteration above and beneath the screw.

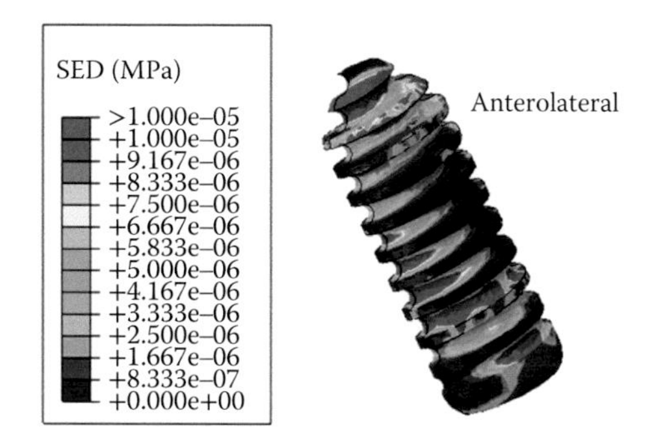


**FIGURE 6.5** SED distribution in the interference screw.

can offer initial graft fixation and gradually be replaced by the host tissue. However, the interaction between the biodegradation process and bone remodeling remains unclear. Consequently, complications caused by undesirable biodegradation have been frequently reported (Konan and Haddad 2009). The present study indicated that the in vivo SED distribution in the screw was nonuniform (Figure 6.5). However, the material used in most biodegradable screws is homogeneous. Since mechanical environments can influence the biodegradation process, the mechanical strength and morphology of the screw may change nonuniformly and undesirably in the long term, which may serve as a predisposing factor for surgical failure. In this light, the present study could provide preliminary knowledge for the screw-bone interaction, and contribute to the optimal design of a biodegradable screw.

In conclusion, this chapter developed an FE model of the human knee joint with an ACL reconstruction. The numerical results confirmed that the bone SED distribution and stress orientation were altered after surgery. These changes occurred around the bone tunnel, and could activate abnormal bone remodeling. The consequential bone resorption and micro-damage may serve as a predisposing factor for tunnel enlargement and osteoarthritis. The material properties of the screw could also influence the postoperative SED distribution in the tibia. On the premise of sufficient fixation strength, using a screw with a modulus similar to that of bone could decrease the risk of stress shielding, and thus could offer a more favorable surgical outcome. These findings together with histology factors could help us to understand the pathomechanism of the sequelae, and help to improve surgical techniques.

## ACKNOWLEDGMENTS

This work was supported by the National Natural Science Foundation of China (Nos. 10925208, 11120101001), the National Science and Technology Pillar Program (Nos. 2012BAI18B05, 2012BAI18B07), and the 111 Project (No. B13003).

## REFERENCES

Au, A. G., V. J. Raso, A. B. Liggins, D. D. Otto and A. Amirfazli (2005). "A three-dimensional finite element stress analysis for tunnel placement and buttons in anterior cruciate ligament reconstructions." *J Biomech* **38**(4): 827–832.

Blankevoort, L., J. H. Kuiper et al. (1991). "Articular contact in a three-dimensional model of the knee." *J Biomech* **24**(11): 1019–1031.

Delcogliano, A., S. Chiossi et al. (2001). "Tibial plateau fracture after arthroscopic anterior cruciate ligament reconstruction." *Arthroscopy* **17**(4): E16.

Gobbi, A., V. Mahajan et al. (2011). "Tibial plateau fracture after primary anatomic double-bundle anterior cruciate ligament reconstruction: a case report." *Arthroscopy* **27**(5): 735–740.

Gosset, M., F. Berenbaum et al. (2008). "Mechanical stress and prostaglandin E2 synthesis in cartilage." *Biorheology* **45**(3–4): 301–320.

Kim, S., J. Bosque et al. (2011). "Increase in outpatient knee arthroscopy in the United States: a comparison of National Surveys of Ambulatory Surgery, 1996 and 2006." *J Bone Joint Surg Am* **93**(11): 994–1000.

Konan, S., and F. S. Haddad (2009). "A clinical review of bioabsorbable interference screws and their adverse effects in anterior cruciate ligament reconstruction surgery." *Knee* **16**(1): 6–13.

Li, G., J. Suggs et al. (2002). "The effect of anterior cruciate ligament injury on knee joint function under a simulated muscle load: a three-dimensional computational simulation." *Ann Biomed Eng* **30**(5): 713–720.

Lin, C. L., Y. H. Lin et al. (2010). "Multi-factorial analysis of variables influencing the bone loss of an implant placed in the maxilla: prediction using FEA and SED bone remodeling algorithm." *J Biomech* **43**(4): 644–651.

Louboutin, H., R. Debarge et al. (2009). "Osteoarthritis in patients with anterior cruciate ligament rupture: a review of risk factors." *Knee* **16**(4): 239–244.

Mellal, A., H. W. Wiskott et al. (2004). "Stimulating effect of implant loading on surrounding bone. Comparison of three numerical models and validation by in vivo data." *Clin Oral Implants Res* **15**(2): 239–248.

Milz, S. and R. Putz (1994). "Quantitative morphology of the subchondral plate of the tibial plateau." *J Anat* **185**(Pt 1): 103–110.

Muneta, T., Y. Ezura et al. (1996). "Anterior knee laxity and loss of extension after anterior cruciate ligament injury." *Am J Sports Med* **24**(5): 603–607.

Netravali, N. A., S. Koo et al. (2011). "The effect of kinematic and kinetic changes on meniscal strains during gait." *J Biomech Eng* **133**(1): 011006.

Nyland, J., B. Fisher et al. (2010). "Osseous deficits after anterior cruciate ligament injury and reconstruction: a systematic literature review with suggestions to improve osseous homeostasis." *Arthroscopy* **26**(9): 1248–1257.

Siebold, R. (2007). "Observations on bone tunnel enlargement after double-bundle anterior cruciate ligament reconstruction." *Arthroscopy* **23**(3): 291–298.

Taylor, M., K. E. Tanner and M. A. Freeman (1998). "Finite element analysis of the implanted proximal tibia: a relationship between the initial cancellous bone stresses and implant migration." *J Biomech* **31**(4): 303–310.

Yagi, M., E. K. Wong et al. (2002). "Biomechanical analysis of an anatomic anterior cruciate ligament reconstruction." *Am J Sports Med* **30**(5): 660–666.

Yao, J., J. Snibbe et al. (2006). "Stresses and strains in the medial meniscus of an ACL deficient knee under anterior loading: a finite element analysis with image-based experimental validation." *J Biomech Eng* **128**(1): 135–141.

Zerahn, B., A. O. Munk et al. (2006). "Bone mineral density in the proximal tibia and calcaneus before and after arthroscopic reconstruction of the anterior cruciate ligament." *Arthroscopy* **22**(3): 265–269.

# 7 Knee Joint Models for Kneeling Biomechanics

*Yuxing Wang, Yubo Fan, and Ming Zhang*

## CONTENTS



## SUMMARY

Kneeling is a common movement required for both occupational and cultural reasons and has been shown to be associated with an increased risk of knee disorders. Since excessive stress is considered to be a possible aggressor, it is important to determine the mechanical features of this posture. Models of the knee joint can provide this information, particularly the stress distribution within the inner region, which is very difficult to measure through experimental approaches. In this chapter, a three-dimensional finite element (FE) model of the human knee joint in a kneeling position, flexed through 90 degrees, was developed from magnetic resonance images. The bones, cartilages, and ligaments were incorporated into the model, with relevant interaction between cartilage, ligaments, and bone.

## 7.1 INTRODUCTION

Kneeling is required by many occupations, such as mining, baggage handling, building construction, and agricultural work. It is also a critical part of life in the Middle East and Asia for religious or cultural reasons. Since kneeling is an important and common activity, it is natural that people have a desire to investigate its effects on the health of the knee joint. Previous studies have shown that kneeling and crouching can increase the risk of knee injuries (Jensen and Eenberg 1996; Coggon et al. 2000; McMillan and Nichols 2005). A dose-response relationship between kneeling and knee disorders has also been reported (Jensen 2005; Klussmann et al. 2010). While the exact mechanisms for these diseases are not clear, excessive cartilage stress could be one explanation. Therefore, the determination of stress distribution plays an important role in understanding and preventing knee injuries associated with kneeling.

FE analysis is a useful tool for investigating the mechanical status of joints. Using this approach, many researchers have studied the stress within and around the knee under different situations. For example, investigators have reported on the stress and strain values of loaded knee cartilage, the effects of different loading conditions, and run parametric analyses after changing the material properties

(Li, Lopez, and Rubash 2001; Donahue et al. 2002). Other studies have investigated the changes of cartilage stress associated with different flexion angles (Moglo and Shirazi-Adl 2003; Shirazi-Adl and Mesfar 2005), calculated from full extension to deep flexion. The stress and strain field of ligaments was also predicted when solid models of these parts were introduced (Pena et al. 2006).

In this chapter, we present an FE model of a knee in a kneeling position to investigate the stress distribution of cartilage, menisci, and bones. Based on this case, we discuss the FE modeling procedure and show how these models can help us to better understand the relevant biomechanical issues.

## 7.2 DEVELOPMENT OF A KNEELING MODEL

MR images were obtained from a 26-year-old healthy male volunteer. Scanning was carried out on his right knee flexed through 90 degrees. During scanning, the volunteer was asked to remain relaxed in order to eliminate the influence of pre-stressing. The resolution of these images was 0.75 mm, with slice distance of 0.94 mm.

The images were used to build a geometrical model using the commercial software MIMICS, including the femur, fibula, tibia, patella, cartilage, medial and lateral menisci, patellar tendon, and ligaments. The points of attachment of the biceps and semimembranosus were also identified. When performing the mask editing process, instead of directly forming the precise contour for each part, an overlap of the masks was carefully created using a Boolean operation. For example, when we constructed the femur and its cartilage, we precisely shaped their outer surfaces and created an overlap by manually moving the inner surface of the cartilage into the femur. The reason for this procedure was that if we created masks such as that shown in Figure 7.1b, then there would be both overlap and gaps at the interface, that could not be eliminated by a simple Boolean subtract. Because forming thin objects is usually more difficult, the cartilage rather than the femur was used as the subtracted part. The .stl format files of all parts, which described the surface as triangles, were imported into Rapidform XOR for improvement of triangle qualities and further modification.

Following these procedures, all parts were imported into ABAQUS. A Boolean operation was carried out at this stage to avoid any possible errors caused by shifting among different software. Then all parts were meshed by hexahedral elements, which can provide more accurate results over tetrahedral elements with the same element size. This meshing strategy requires manual partition of the parts. The method used can be seen in Figure 7.2.

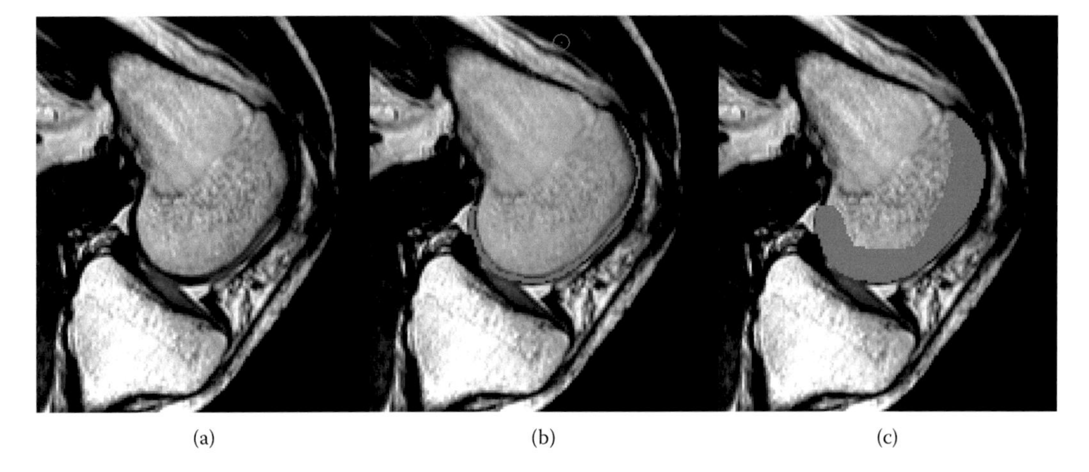

(a) (b) (c)

**FIGURE 7.1** Geometric models were created based on magnetic resonance images. (a) MR images that are used to create the geometric model. (b) Direct contouring of the cartilage, which may cause gaps between cartilage and bone of the 3D models. (c) Manually moving the inner surface of the cartilage into the femur to cause an overlap while keeping its outer surface precisely shaped. The overlapped part will be subtracted by Boolean operation, by which all the possible gaps between cartilage and bone can be eliminated.

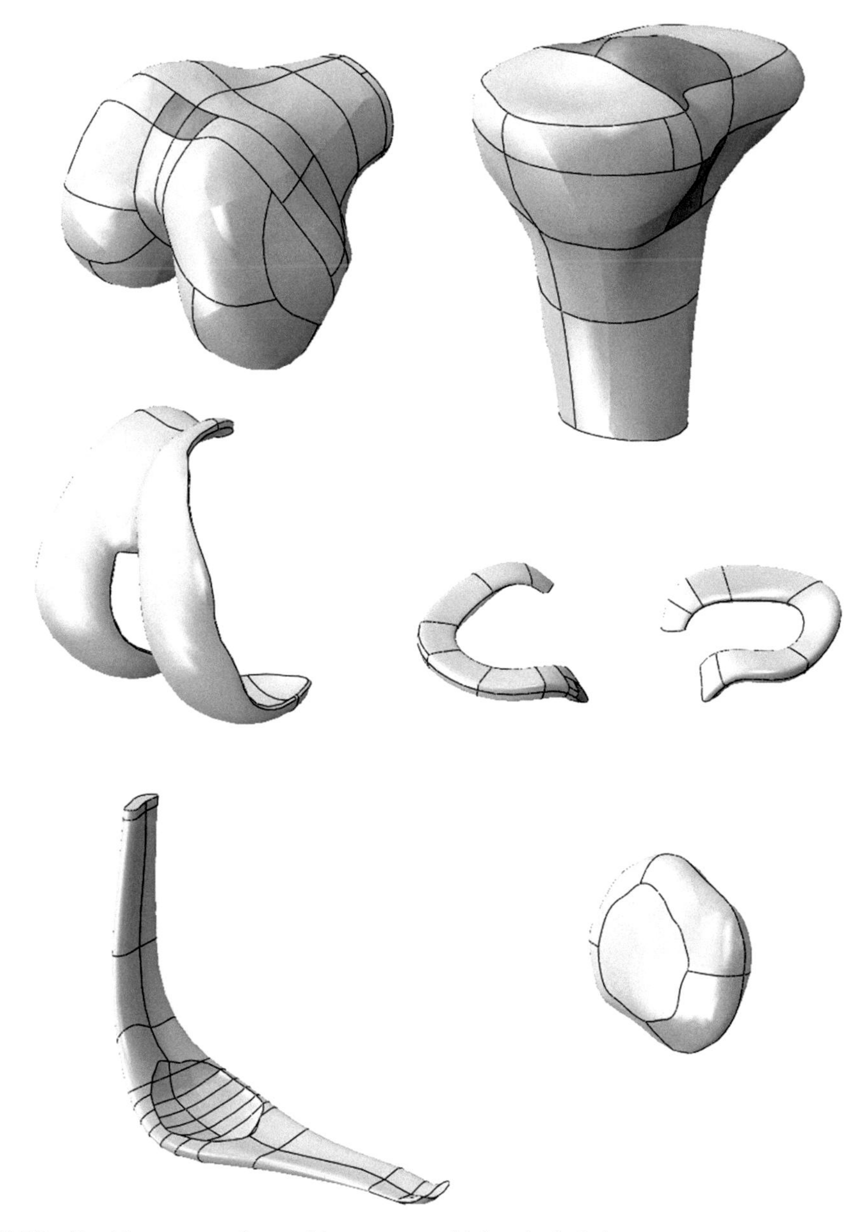

**FIGURE 7.2** Partition strategy for meshing process with hexahedral elements.

The number of elements is an important factor that needs be considered when developing FE models. Too many elements will slow down the calculation process, while too few elements will result in loss of accuracy. Convergence tests are typically run to overcome this problem, whereby the element density is increased incrementally until the difference in results for each successive model becomes negligible. However, since there have already been a great number of models developed for the knee joint, meshing details can be referenced from the literature. In our model, the element size for cartilage ranged from 0.8 mm to 1.6 mm, and was around 2 mm for the remaining parts (Figure 7.3). This mesh density has been proven to be accurate enough for convergence (Donahue et al. 2002).

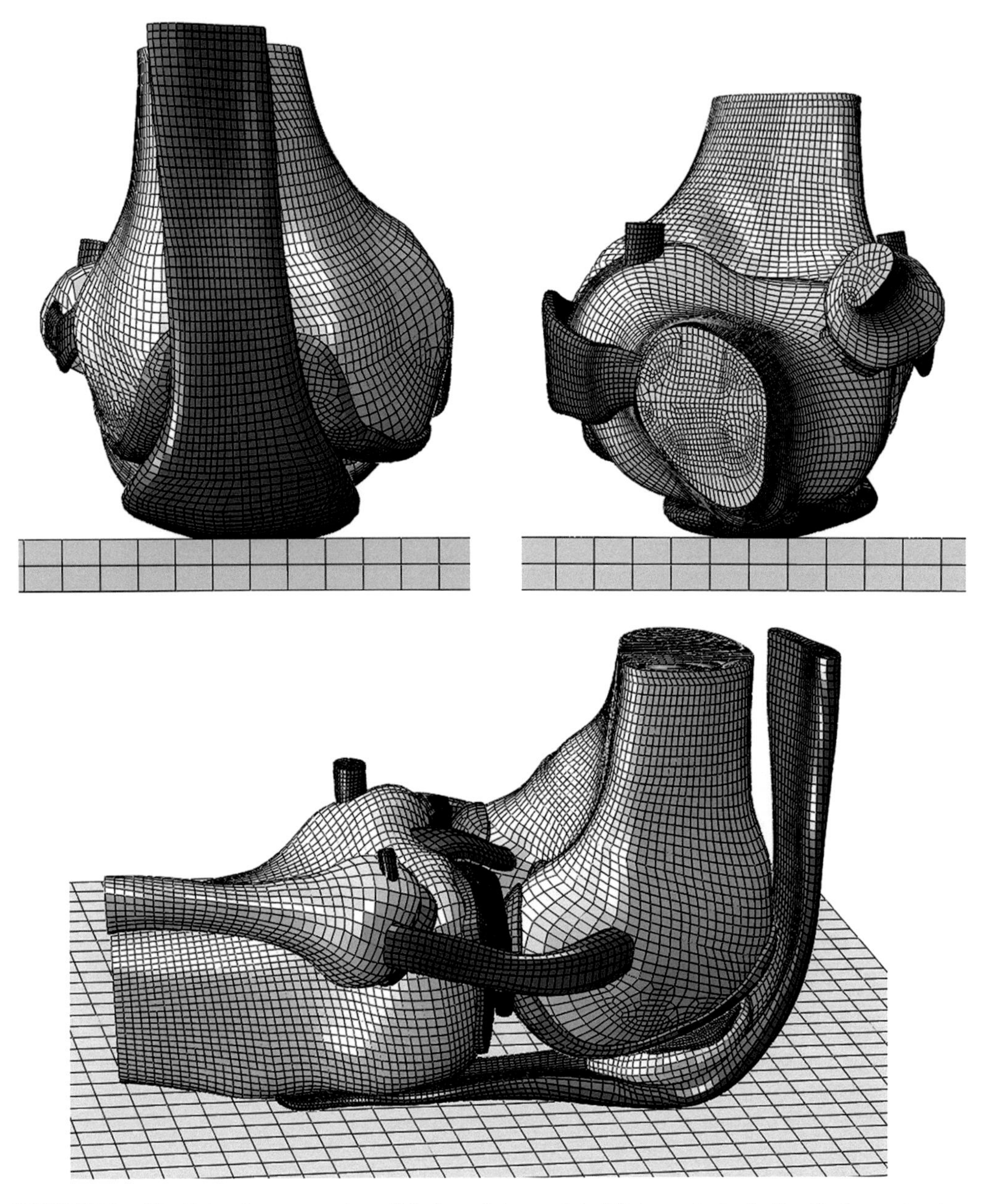

**FIGURE 7.3** The finite element meshes of the knee joint model. All the parts, including the bony structures, cartilages, menisci, ligaments, and patellar tendon, were meshed by 8-node hexahedral elements.

### 7.2.1 Material Properties

The material properties can also influence the simulation results, and were sourced from previous publications on validated knee models. The bony parts were set as cortical bone with an elastic modulus of 20,000 MPa and Poisson's ratio of 0.3 (Rho, Ashman, and Turner 1993; Zysset et al. 1999). The inner structures, such as trabecular bone, were ignored since they would not interact directly with the cartilage. The cartilage was assumed to be isotropic elastic with an elastic modulus of 10 MPa and a Poisson's ratio of 0.45. The menisci were modeled as transversely isotropic materials,

**TABLE 7.1**
**Material Properties and Element Types Defined in the Finite Element Model**

| Component | Element Type | Young's Modulus *E* (MPa) | Poisson's Ratio *v* |
|---|---|---|---|
| Bony structures | 8-node hexahedra | 20,000 | 0.3 |
| Cartilage | 8-node hexahedra | 10 | 0.45 |
| Menisci | 8-node hexahedra | Circumferential: 140<br>Radial and axial: 20 | In-plane: 0.2<br>Out-of-plane: 0.3 |
| Ligaments and patellar tendon | 8-node hexahedra | Hyperelastic | 0.49 |
| Ground support | 8-node hexahedra | Upper layer: 20,000<br>Lower layer: rigid | 0.1<br>- |

with a radial and axial modulus of 20 MPa and a circumferential modulus of 140 MPa. The in-plane Poisson's ratio was 0.2 while the out-of-plane Poisson's ratio was 0.3 (Donahue et al. 2002; Netravali et al. 2011). The ligaments were assumed to behave as an isotropic hyperelastic material. Although attempts have been made to involve some sophisticated anisotropic hyperelastic descriptions (Pena et al. 2006), in this chapter the ligament behavior was assumed to be homogeneous and isotropic to avoid the coding process in ABAQUS's user-defined subroutine. To simply the process, the coefficients form of input was not used. Instead, test data on the stress-strain relationship of each ligament obtained from previous FE and experimental studies was inputted (Butler, Kay, and Stouffer 1986; Shirazi-Adl and Mesfar 2005). The patellar tendon was set as compressible since it would bear contact loads from the ground, while the ligaments were assumed to be incompressible. In the literature, the testing data usually only includes tension since ligaments are considered to be incompressible. Thus, in ABAQUS, the compressive side of the stress-strain curve needed to be added. In practice, a completely incompressible material may lead to problems with convergence, so a material with properties approaching incompressibility is preferable. By this approach, the input process may be hastened, but attention should be paid to the results of the material evaluation to determine whether the material description can fit the results. In this chapter, the Ogden model with second-order energy potential was used. The summary of material properties and element types can be seen in Table 7.1.

### 7.2.2 Loading and Boundary Conditions

The inner surfaces of cartilages and the connecting faces of ligaments were assumed to be rigidly fixed to corresponding bones, while for menisci only the two horns of each one were fixed. The interaction between cartilages and menisci was defined with frictionless contact. The boundary conditions for this kneeling model had the femur fixed in space with the tibia set totally free. The ground plane was allowed to move only perpendicularly, with the other five degrees of freedom limited. The end surface of the patellar tendon was constrained, as it can only have displacement parallel to the direction of the femur. Muscle forces (quadriceps 215 N, biceps 31 N, and semimembranosus 54 N) were used to simulate the physiological loading of the knee joint (Wickiewicz et al. 1983; Hofer et al. 2011). An incrementally increased load up to 1000 N was applied to the ground, which can only move perpendicularly towards the femur. Contact interaction was defined for cartilages, menisci, and possible contact regions between ligaments and bony structures. According to the literature, prestrain for ligaments was also applied (Table 7.2), and ligaments were divided into anterior and posterior parts.

**TABLE 7.2**
**Initial Strain of the Ligaments (Percent)**

| aACL | pACL | aPCL | pPCL | aLCL | pLCL | aMCL | pMCL |
|---|---|---|---|---|---|---|---|
| 0.06 | 0.1 | 0.0 | 0.0 | 0.0 | 0.0 | 0.04 | 0.04 |

## 7.3 DETERMINING THE MECHANICAL STATUS OF KNEELING

Because the purpose of this model was to determine the mechanical status of the knee joint under kneeling, the stress distributions of cartilages, menisci, and ligaments were determined.

### 7.3.1 STRESS DISTRIBUTION OF CARTILAGE

The largest von Mises stress occurred at the point of contact between the tibia and patellar cartilages (Figure 7.4). The peak von Mises stress reached 1.875 MPa under a 300-N compressive load. It then increased to 2.008 MPa at 600 N and 2.471 MPa at 1000 N. Li, Lopez, and Rubash (2001) calculated von Mises stress distribution and ran a parametric analysis on their standing model. They found the peak von Mises stress to be 1.57 MPa for a standing position. Using the same cartilage material properties (elastic modulus of 10 MPa and Poisson's ratio of 0.45) and similar boundary conditions (1000-N compression force with 300-N muscle force), the von Mises stress was found to be 2.471 MPa. This indicated that kneeling would increase the von Mises stress by approximately 57%.

The trend for the peak cartilage contact pressure is similar, reaching 4.358 MPa under 300-N compressive loads, and increasing to 4.805 MPa at 600 N and 5.581 MPa at 1000 N. The contact area also increased from 315.34 mm$^2$ to 379.104 mm$^2$ and to 424.842 mm$^2$, with the loads changing from 300 N to 600 N and to 1000 N. With increasing loads, the location of maximum stress also changed, migrating toward the lower part of the patella.

Some researchers have used the largest contact pressure to illustrate the risks to the cartilage. Donahue et al. (2002) showed the peak contact pressure on knee cartilage while standing fluctuated from 2.10 to 2.51 MPa and was influenced by different material properties and boundary conditions. In this research, the peak pressure reached 5.581 MPa, suggesting that kneeling carries a greater risk of injuries than normal standing. If we use a 4.5-MPa peak contact pressure as the criterion for

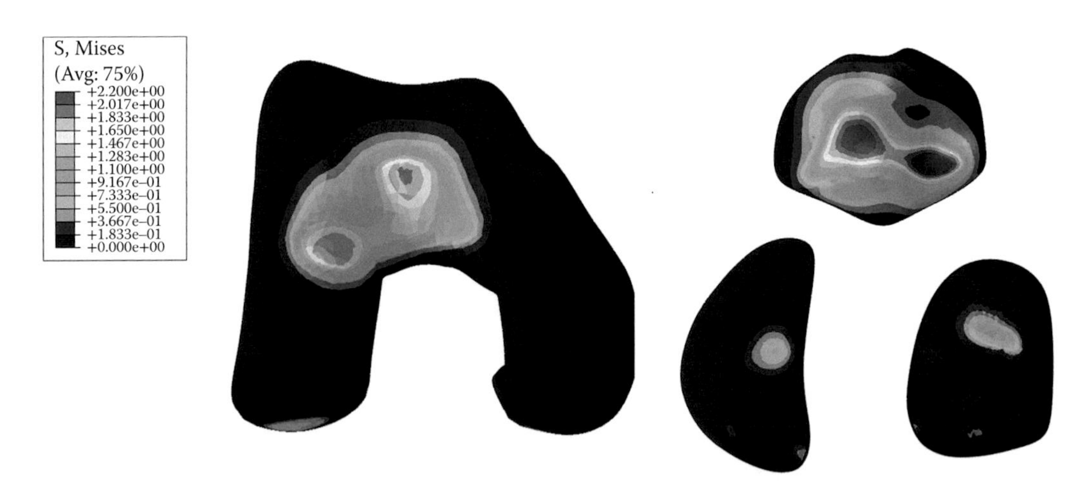


**FIGURE 7.4** **(See color insert.)** The von Mises stress on the cartilages of the knee joint in a kneeling posture under 1000 N.

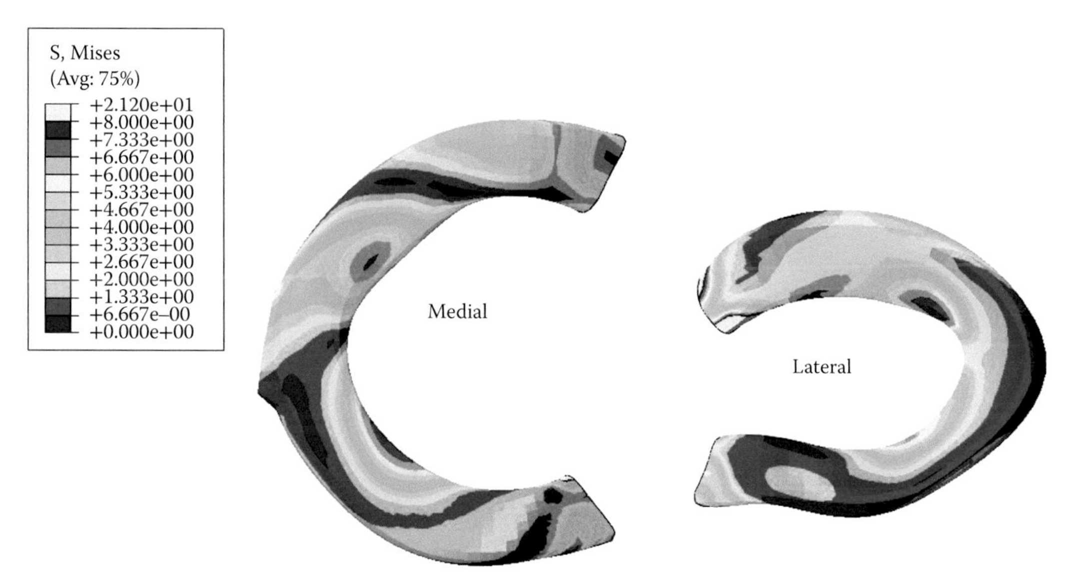


**FIGURE 7.5** **(See color insert.)** The von Mises stress of menisci under a 1000 N compressive load.

cartilage damage, according to Grodzinsky et al. (2000), then we can estimate the safe criterion for kneeling. According to our results (Figure 7.5), 600-N compressive loads (causing 4.805-MPa peak contact pressure) for each knee have the potential to cause knee disorders. In other words, single stance kneeling or double stance kneeling with loads greater than one times body weight could be dangerous to the knee joint. Therefore, when in a kneeling position, the loads on the knee should be kept below 600 N. According to the results, contact area alone is not adequate for determining the risk of injury.

### 7.3.2 Stress Distribution of Menisci and Bones

Besides the cartilage, the stress distribution of menisci and bones was also calculated from this model. The von Mises stress of the lateral meniscus is approximately equal to that of the medial one, which indicates that there should be an equivalent risk of menisci injury for both sides (Figure 7.5). The maximum von Mises stress of bones is 26.15 MPa, located at the contact point between the patella and the ground (Figure 7.6). This suggests an injury risk for patella disorders.

The FE model of a kneeling position provided a detailed stress distribution of the knee joint. However, there are still several limitations. First, the material properties used in our model are based on previous literature that does not correspond to a specific subject. Therefore, the exact value should be used cautiously. Second, here we only simulated a case of 90 degrees of flexion. It would be important to investigate the influence of the flexion angle and make comparisons between crouching and kneeling. In addition, we only used a static model to investigate the consequence of constant loading. The result of impacts also needs to be considered using a dynamic model with more sophisticated consecutive equations.

In spite of these limitations, the von Mises stress, contact pressure, and contact area were determined for a kneeling position. The region of high risk for injury was identified and a comparison made with other studies of the standing position. All of these findings help us better understand the mechanical influence of kneeling and provide information that could be used to prevent associated injuries.

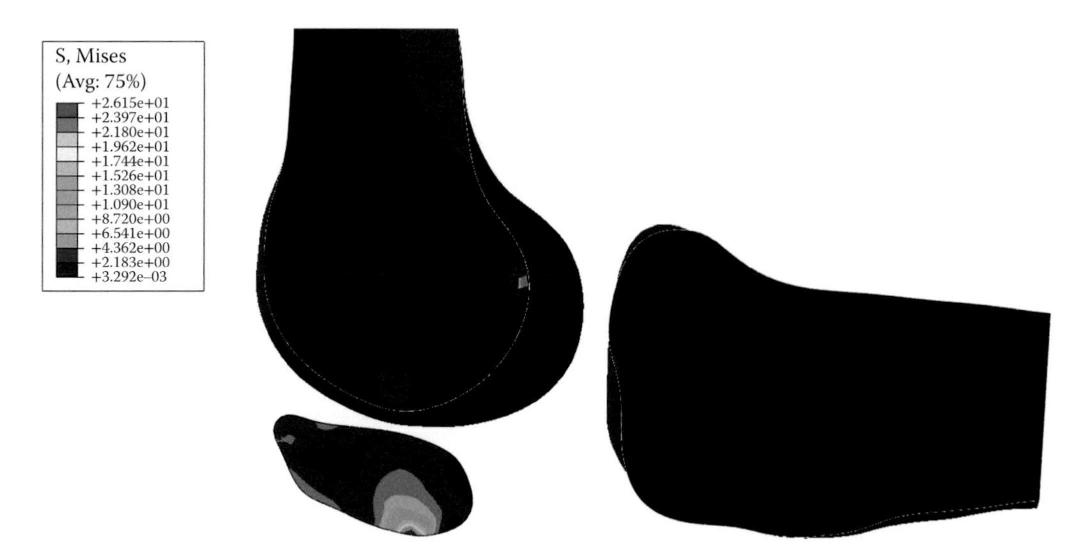


**FIGURE 7.6 (See color insert.)** Sagittal view of the von Mises stress distribution of bones and patellar tendon under 1000 N compressive loads.

## ACKNOWLEDGMENTS

This study was supported by grants from the National Natural Science Foundation of China (Nos. 11272273, 10925208, and 11120101001), and research grants from The Hong Kong Polytechnic University (G-U796, G-YL72) and the National Key Lab of Virtual Reality Technology. Grants were also received from National Science and Technology Pillar Program (Nos. 2012BAI18B05, 2012BAI18B07), 111 Project (No. B13003).

## REFERENCES

Butler, D. L., M. D. Kay, and D. C. Stouffer. 1986. Comparison of material properties in fascicle-bone units from human patellar tendon and knee ligaments. *J Biomech* 19: 425–32.

Coggon, D., P. Croft, S. Kellingray, D. Barrett, M. McLaren, and C. Cooper. 2000. Occupational physical activities and osteoarthritis of the knee. *Arthritis Rheum* 43: 1443–49.

Donahue, T. L. H., M. L. Hull, M. M. Rashid, and C. R. Jacobs. 2002. A finite element model of the human knee joint for the study of tibio-femoral contact. *J Biomech Eng Trans ASME* 124: 273–80.

Grodzinsky, A. J., A. M. Loening, I. E. James, M. E. Levenston, A. M. Badger, E. H. Frank et al. 2000. Injurious mechanical compression of bovine articular cartilage induces chondrocyte apoptosis. *Arch Biochem Biophys* 381: 205–12.

Hofer, J. K., R. Gejo, M. H. McGarry, and T. Q. Lee. 2011. Effects on tibiofemoral biomechanics from kneeling. *Clin Biomech* 26: 605–11.

Jensen, L. K. 2005. Knee-straining work activities, self-reported knee disorders and radiographically determined knee osteoarthritis. *Scand J Work Environ Health* 31: 68–74.

Jensen, L. K., and W. Eenberg. 1996. Occupation as a risk factor for knee disorders. *Scand J Work Environ Health* 22: 165–75.

Klussmann, A., H. Gebhardt, M. Nubling, F. Liebers, E. Q. Perea, W. Cordier, L. V. von Engelhardt, M. Schubert, A. David, B. Bouillon, and M. A. Rieger. 2010. Individual and occupational risk factors for knee osteoarthritis: results of a case-control study in Germany. *Arthritis Res Ther* 12(3): R88.

Li, G., O. Lopez, and H. Rubash. 2001. Variability of a three-dimensional finite element model constructed using magnetic resonance images of a knee for joint contact stress analysis. *J Biomech Eng Trans ASME* 123: 341–46.

McMillan, G., and L. Nichols. 2005. Osteoarthritis and meniscus disorders of the knee as occupational diseases of miners. *Occup Environ Med* 62: 567–75.

Moglo, K. E., and A. Shirazi-Adl. 2003. On the coupling between anterior and posterior cruciate ligaments, and knee joint response under anterior femoral drawer in flexion: a finite element study. *Clin Biomech* 18: 751–59.

Netravali, N. A., S. Koo, N. J. Giori, and T. P. Andriacchi. 2011. The effect of kinematic and kinetic changes on meniscal strains during gait. *J Biomech Eng Trans ASME* 133, 001106-1-6.

Pena, E., B. Calvo, M. A. Martinez, and M. Doblare. 2006. A three-dimensional finite element analysis of the combined behavior of ligaments and menisci in the healthy human knee joint. *J Biomech* 39: 1686–701.

Rho, J. Y., R. B. Ashman, and C. H. Turner. 1993. Young's modulus of trabecular and cortical bone material: ultrasonic and microtensile measurements. *J Biomech* 26: 111–9.

Shirazi-Adl, A., and W. Mesfar. 2005. Biomechanics of the knee joint in flexion under various quadriceps forces. *Knee* 12: 424–34.

Wickiewicz, T. L., R. R. Roy, P. L. Powell, and V. R. Edgerton. 1983. Muscle architecture of the human lower limb. *Clin Orthop Relat Res*: 275–83.

Zysset, P. K., X. E. Guo, C. E. Hoffler, K. E. Moore, and S. A. Goldstein. 1999. Elastic modulus and hardness of cortical and trabecular bone lamellae measured by nanoindentation in the human femur. *J Biomech* 32: 1005–12.

# 8 Knee Implant Model

## *A Sensitivity Study of Trabecular Stiffness on Periprosthetic Fracture*

*Duo Wai-Chi Wong and Ming Zhang*

## CONTENTS



## SUMMARY

Periprosthetic fracture is one of the leading causes of knee implant failure. Risk factors are associated with a reduction in bone density or stock. The purpose of this study was to evaluate the reduction in bone stiffness with the risk of fracture using finite element (FE) analysis. The femur geometry was acquired from a three-dimensional reconstruction of magnetic resonance imaging (MRI) scans and the knee implant was digitized and imported into FE software. The anterior flange, screw region, and posterior supracondylar regions were crucial regions in determining the risk of fracture. The risk of fracture was accessed by compressive stress yielding criteria. With an increasing flexion angle and decreasing bone stiffness, the compressive stress of the trabecular bone approached its yielding point. This outcome also suggested that normal bone stiffness/stock at high flexion angles could also contribute to stress yielding, notwithstanding the small yielding volume. Yet, osteoporotic patients could have impaired bone reparability and accumulated bone yielding with a reduction of bone stiffness and sustainability. Physicians should be careful in their consideration of knee replacements for osteoporotic patients to reduce the possible risk of periprosthetic fracture.

## 8.1 INTRODUCTION

### 8.1.1 Background

Knee replacements generally boast good longevity and high survivorship and are commonly prescribed for patients with severe knee problems, predominantly osteoarthritis (National Institutes of Health 2003). The disease is more prevalent in the elderly, due to accumulative overuse and degeneration. With an ever-aging population, the number of people using joint replacement is expected to increase. It is logical to assume that most people would like to get an implant and remain mobile rather than be confined to a wheelchair.

Despite the high demand and success of knee implants, some failure is inevitable (National Institutes of Health 2003). Though uncommon, periprosthetic fracture is one cause of failure, with a prevalence ranging from 0.3% to 2.5% (Su, DeWal, and Di Cesare 2004). In fact, the consequences of periprosthetic fracture are devastating, with complication rates of 25% to 75% after treatment (Dennis 2001). The majority of the fractures are associated with elderly subjects with osteoporosis and rheumatoid arthritis patients receiving steroid therapy. These risk factors are related to a reduction of bone stock, with a consequential loss in bone strength and stress alterations within the bone.

### 8.1.2 Notching and Stress Concentration

Anterior notching of implants is believed to contribute to the risk of periprosthetic fracture. Culp et al. (1987) conducted a theoretical analysis and suggested that a 3-mm notching reduced bone torsional strength by 29.2%. This result was further supported by Lesh et al. (2000), who reported a decrease of 18% in bending strength and 39.2% in torsional strength after notching. Zalzal et al. (2006) characterized local stress by notching and believed that a femoral notch greater than 3 mm with sharp corners would produce high stress concentration. However, Gujarathi et al. (2009) suggested that there was no relationship between minimal notching and fracture. Though the vulnerable location was identified, the mechanism of fracture and the association with loading conditions are not well understood. Su, DeWal, and Di Cesare (2004) suggested that fractures occurred most frequently from low-velocity falls, while high-energy trauma contributed to a smaller proportion of fractures. With the growing number of knee replacements and increased activity of elderly patients, an investigation of the fracture mechanism could aid physicians in determining viable knee replacements.

Computational methods are efficient tools to investigate the parametric effects of the bone-implant interface in a well-controlled environment. In the present study, a three-dimensional FE bone-implant model was developed to study the effect of reducing bone strength on periprosthetic stress failure.

## 8.2 MODEL DEVELOPMENT

### 8.2.1 Model Construction

The geometry of the distal femur and proximal tibia was acquired from MRI of a normal adult female subject (age 28, height 165 cm, weight 54 kg), and reconstructed using the segmentation software MIMICS (Materialise, Leuven, Belgium). The knee implant (Advance® Primary System, Wright Medical Technology Inc., Arlington, Virginia) was digitized and reconstructed using the reverse engineering software Rapidform XOR2 (INUS Technology, Seoul, South Korea). The bone-implant model was then imported into the FE software ABAQUS (Dassault Systèmes Technologies, RI, USA) for FE meshing and analysis.

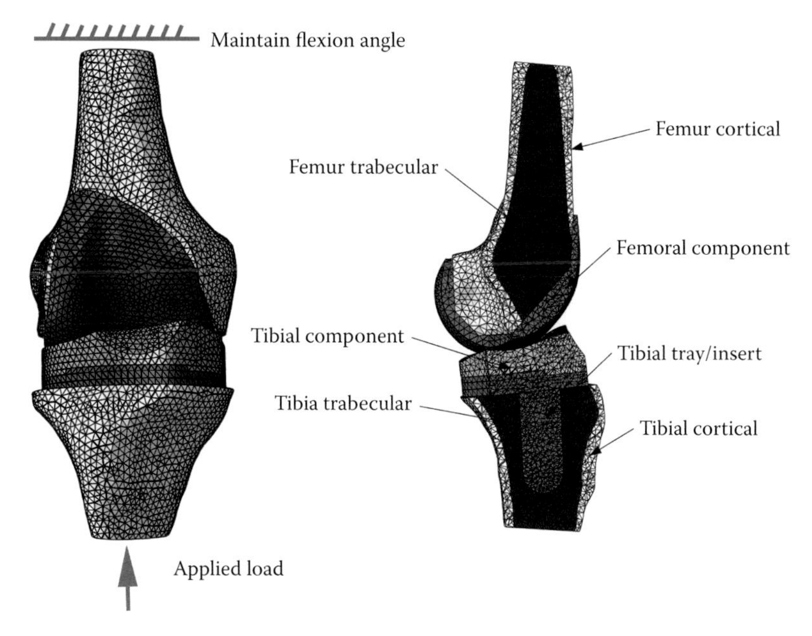


**FIGURE 8.1** **(See color insert.)** Finite element model of the knee implant and bones simulating compression with flexion.

### 8.2.2 Material Properties

The FE model, as shown in Figure 8.1, consisted of the distal femur and proximal tibia with sectioned trabecular and cortical regions, and the implant, including the femoral component, tibial insert, and tibial tray. The FE model consisted of about 2.36 million tetrahedral elements and was idealized with homogeneous material properties. Orthotropic material properties were assigned to the cortical bone (Ashman et al. 1984). The trabecular bone of the femur and tibia was assigned with a Young's modulus ranging from half to two times the suggested elastic moduli of 389 MPa (femoral) and 445 MPa (tibial) (Linde, Hvid, and Pongsoipetch 1989; Rohlmann et al. 1980). The femoral component and tibia tray of the implant was assigned an elastic modulus and Poisson's ratio of 110 GPa and 0.34, respectively (Chu 1999), while the ultra-high-molecular-weight polyethylene (UHMWPE) insert was assigned values of 8.1 GPa and 0.46 (Miyoshi et al. 2002). The coefficient of friction between the femoral components and the tibial insert was set to be 0.07 (Godest et al. 2002). The implant and bones were tied together. The material properties of the model are summarized in Table 8.1. Because the surgical resection process depended on surgical experience and the conditions of the patients, the optimal position of the implant and knee was established by a correlation operation that gave a relative objective alignment. The tibial tray was tilted 10 degrees posteriorly (Singerman et al. 1996).

### 8.2.3 Boundary and Loading Conditions

The model was assigned a compressive load of 2000 N on the distal tibia with the proximal femur restrained (Kim, Kwon, and Kim 2008; Villa et al. 2004). Typical knee examination angles of 0, 5, 30, 45, and 60 degrees of flexion were simulated. The predicted principal compressive stress was evaluated against the suggested yield to assess the risk of bone yielding failure. Three targeted locations of interest were identified via stress concentration on the anterior flange (AF), the posterior supracondylar region (PR), and the screw apex (SR).

**TABLE 8.1**
**Material Properties and Element Types of the Finite Element Model**

| Component | Element Type | Elastic Modulus | Poisson's Ratio | Number of Elements |
|---|---|---|---|---|
| Femoral component (titanium) | Three-dimensional (3D)-tetrahedral | 110 Pa | 0.34 | 35,452 |
| Tibial tray (titanium) | 3D-tetrahedral | 110 GPa | 0.34 | 25,194 |
| Tibial insert (UHMWPE) | 3D-tetrahedral | 8.1 GPa | 0.46 | 15,966 |
| Cortical bone | 3D-tetrahedral | Orthotropic $E_1$, $E_2$, $E_3$, 12, 13.4, 20, GPa | $\nu_{12}$,$\nu_{13}$,$\nu_{23}$ 0.38, 0.22, 0.24 | 18,958 (femur) 8,787 (tibia) |
| Trabecular bone—femur | 3D-tetrahedral | 389 MPa 194.5–1167 MPa | 0.2 | 76,589 |
| Trabecular bone—tibia | 3D-tetrahedral | 445 MPa 222.5–1335 MPa | 0.2 | 54,882 |

*Note:* Subscript 1 = radial direction relative to the long axis of the bone; 2 = tangential direction; 3 = longitudinal direction.

## 8.3 RESULTS

### 8.3.1 Contact Pressure of the Tibial Insert

Typical contact pressure of the tibial insert is illustrated in Figure 8.2 with normal bone stiffness. In both full extension and 60-degree flexion, the localized pressure of the tibial insert was located at the concave zone, which is designed to have a large contact surface with the condyles of the femoral component. The localized pressure exhibited a circular pattern, except for a crescent pattern apparent at the lateral region in the flexed knee. The peak pressure of 26.46 MPa was located in the medial region upon full extension. With 60-degree flexion, the peak pressure was 25.68 MPa and located in the lateral region. In addition, the peak pressure was 24.72 MPa and 23.28 MPa, respectively, for the 30- and 45-degree flexion angles. The peak contact pressure of the tibial component deviated by less than 10% across all flexion angles.

### 8.3.2 Compressive Stress of the Trabecular Femur

Figure 8.3 demonstrates the stress contour in the trabecular portion of the femur (Linde, Hvid, and Pongsoipetch 1989; Rohlmann et al. 1980). In full extension, stress was concentrated on PR and SR, with a peak compressive stress of about 1.13 MPa. When flexed to 60 degrees, the stress concentrated on the AF, with a magnitude of 3.18 MPa.

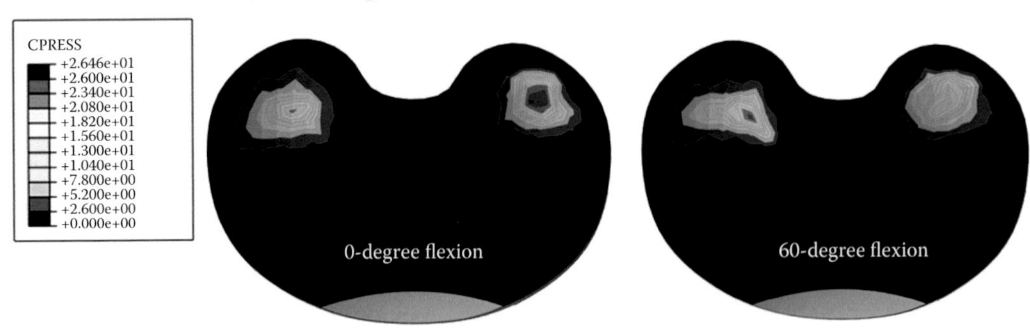


**FIGURE 8.2** **(See color insert.)** Contact pressure of the tibial insert at 0-degree flexion (left) and 60-degree flexion (right). Unit: MPa.

The principal compressive stress against various flexion angles is illustrated in Figure 8.4. There was a relatively small increase in compressive stress on the PR and SR regions over the flexion range, especially at small flexion angles. The maximum change was 0.16 MPa and 0.57 MPa for the PR and SR, respectively, whereas the maximum values were about 1.28 MPa and 1.2 MPa at

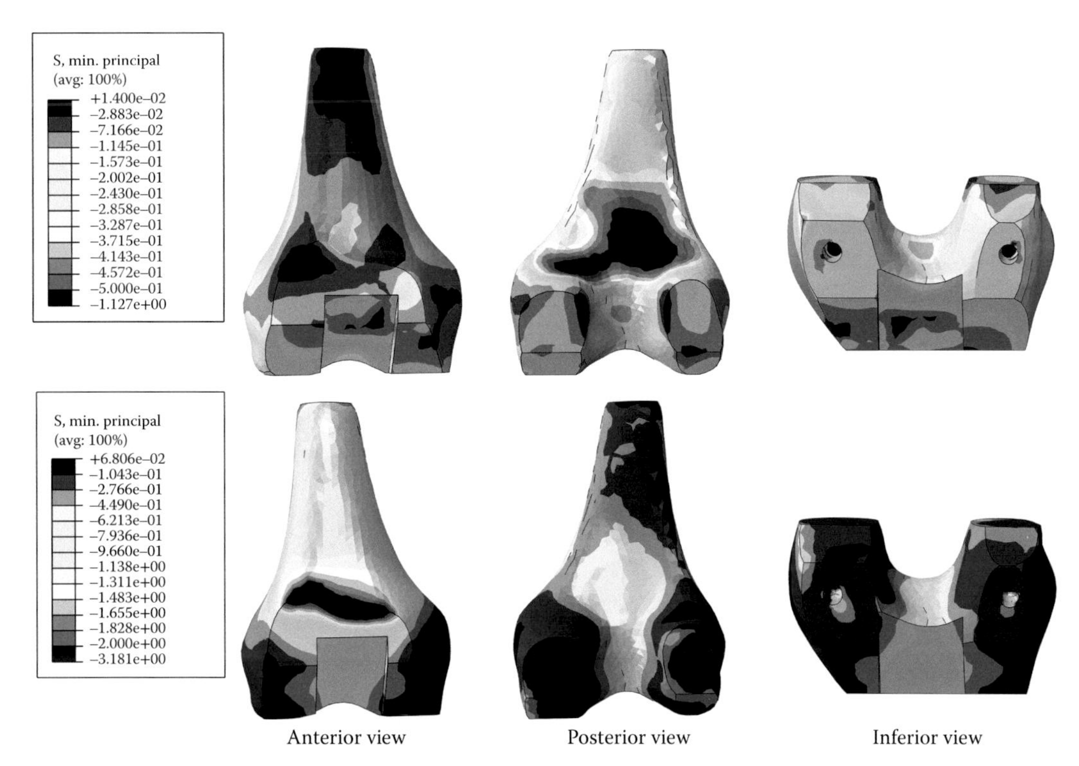


**FIGURE 8.3 (See color insert.)** Concentrated compressive stress on the trabecular femur at the anterior flange, screw sites, and posterior supracondylar location at neutral position (top) and 60-degree flexion (bottom). Unit: MPa.

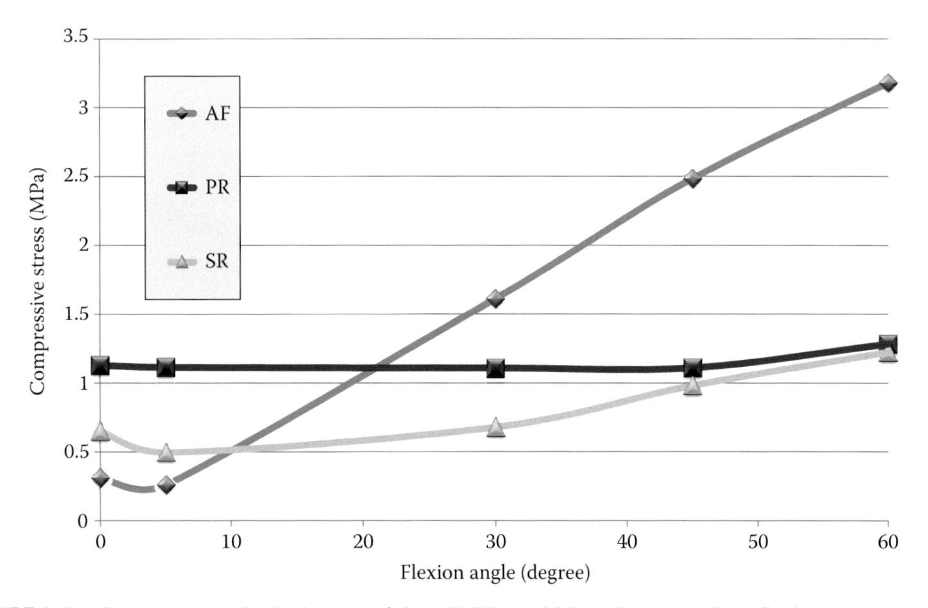


**FIGURE 8.4** Compressive principal stress of the AF, PR, and SR regions at various flexion angles.

60-degree flexion. Conversely, the change at the AF was relative drastic. The compressive stress increased from about 0.31 MPa at full extension to 3.18 MPa at 60-degree flexion, which is about a 10-fold increase.

The stiffness of the trabecular bone was parameterized and divided into the AF, PR, and SR regions (Figure 8.5). Around the AF region, the predicted compressive stress generally increased

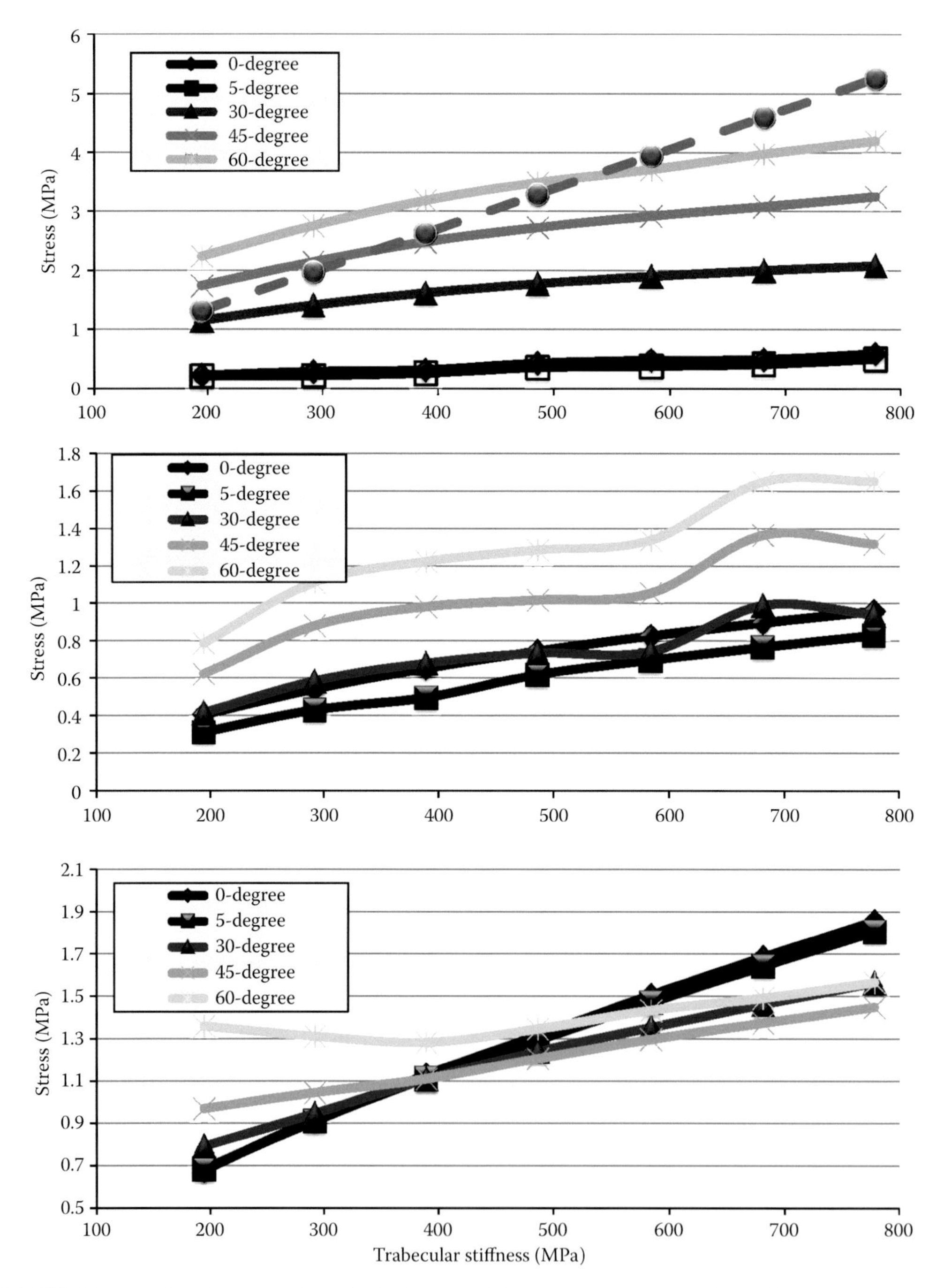


**FIGURE 8.5** Predicted compressive principal stress of the AF (top), PR (center), and SR (bottom) regions with varying trabecular stiffness. The constitutive equation of yielding stress (dotted line) was suggested by Burgers et al. (2008).

with increasing trabecular stiffness. The rise was more apparent with greater flexion angles. The stress increased from 0.22 MPa to 0.59 MPa in full extension and from 2.24 MPa to 4.20 MPa at 60-degree flexion. The trend of the PR was similar to that of the AF. There was an increase of 0.56 MPa and 0.86 MPa in compressive stress when flexed at 0 degrees and 60 degrees, despite the fact that the gain was apparently smaller than that of the AF. Around the SR site, the change was not consistent in spite of the diminishing stress with decreasing trabecular stiffness. The range of change was more persistent at lower flexion angles, while indeed the smaller flexion angles displayed a lower compressive stress than high flexion angles. The compressive stress increased from 0.68 MPa to 1.86 MPa at 0 degrees and from 1.36 MPa to 1.57 MPa at 60 degrees.

### 8.3.3 Parametric Study on Trabecular Stiffness and Yield

The yielding stress of the trabecular bone (Burgers et al. 2008) is incorporated in Figure 8.5 to evaluate possible material failure upon reducing trabecular stiffness. The range of yielding stress was between 1.32 MPa and 5.26 MPa for the given range of trabecular stiffness, indicating that yielding at the PR and SR was unlikely. The predicted compressive stress from full extension to 30-degree flexion did not exceed the suggested yielding stress in general. Yet, the predicted compressive stress overlapped the line of yielding stress for flexion angles greater than 45 degrees. The indication of over-yielding became more apparent with lower trabecular stiffness and greater flexion angles.

The yielded volume in the cross-section and anterior views of the trabecular femur are shown in Figure 8.6 under 45-degree and 60-degree flexion stiffness ranging from 25% to 125%. The magnitude of yielded volume is also shown; it should be noted that a better discretization was used for calculation and a slight inconsistency may be apparent between the values and preceding plot/graph. The region of yield is located along the superior line of the AF. The yielding volume was 52 mm$^3$ under 45-degree flexion with 125% trabecular stiffness. It was gradually increased and reached approximately 450 mm$^3$ under 60-degree flexion with 50% trabecular stiffness. A deeper flexion angle and smaller trabecular stiffness generally preceded a larger yield volume, which was also supported by the wider gap between the predicted stress and yield in Figure 8.5.

In this study, the parametric effects of trabecular bone stiffness on compressive principal stress at different susceptible regions were quantified with suggested yielding stresses. Pure load compression with different flexion angles was adopted in the simulations. The boundary condition mimicked highly physical work with knee flexion to accommodate factors that may contribute to knee disorders that may exaggerate the risk of periprosthetic fracture (Jensen and Eenberg 1996; McMillan and Nichols 2005). In fact, this study applied an axial load in different flexion angles, which is also a common practice as it complies with the material and standard testing for implant. Chu (1999) applied 2200 N to evaluate the contact stress on the tibial component with different flexion angles. Similar protocols were adopted by Villa et al. (2004) using different force-angle combinations. Pure compression in the neutral position was also widely used. Liau et al. (2002) and Kim, Kwon, and Kim (2008) applied 3000-N and 2000-N forces, respectively, despite the fact that most of the implant studies focused on the response of the tibial component due to the importance of wear and tear.

### 8.3.4 Validation of the Model

The implant stress from the current study was compared to existing works. The contact stress of the tibial component in our studies ranged from 23.28 MPa to 25.68 MPa, with less than 10% deviation. Miyoshi et al. (2002) and Godest et al. (2002) presented a contact pressure of about 24 MPa in their tibial components. An experiment-validated simulation conducted by Villa et al. (2004) reported a contact pressure of 15 MPa at 15-degree flexion and 27.7 MPa at 45-degree flexion. The current prediction was generally agreeable with the existing literature in terms of the magnitude of contact pressure on the tibial component, with deviations arising from differences in implant designs, loading conditions, and alignment protocols.

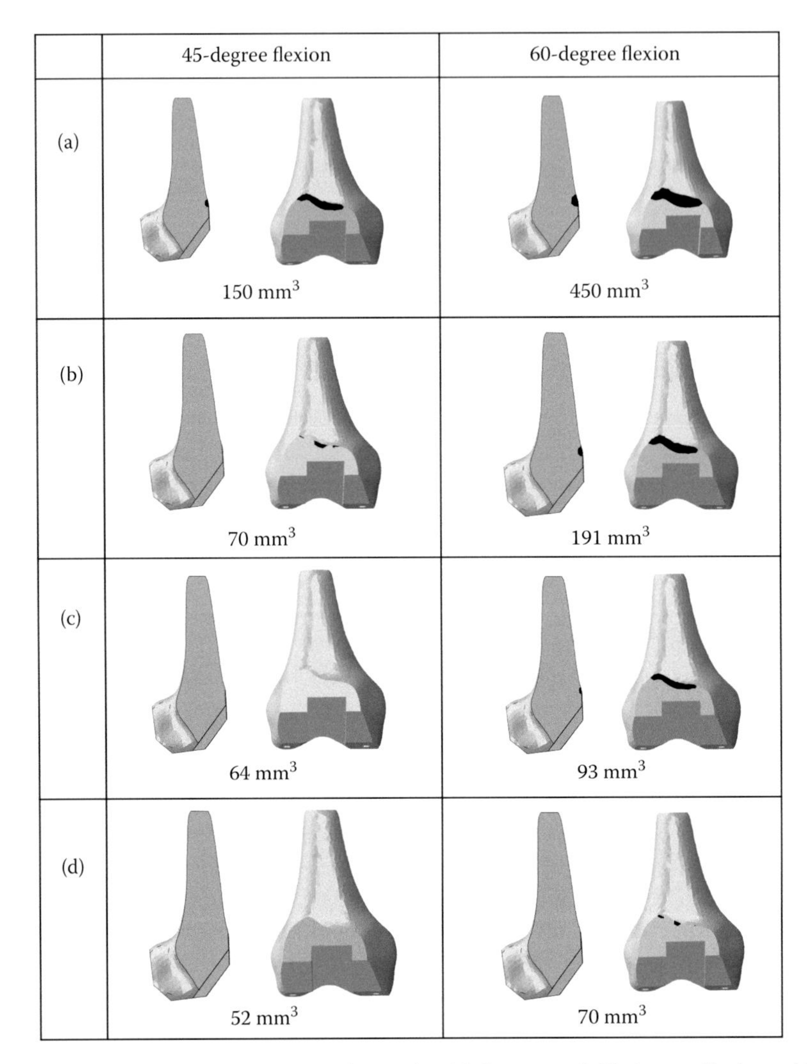


**FIGURE 8.6** Yielded volume of the trabeculae under 45-degree and 60-degree flexion with trabecular stiffness: (a) 194.50 MPa; (b) 291.75 MPa; (c) 389.00 MPa; (d) 486.25 MPa.

## 8.4 RISK OF PERIPROSTHETIC FRACTURE

Extensive efforts have been made to understand the stress of the implant due to the importance of wear and tear. There have been few studies to investigate the relationship between the weakening of bone (which could be due to osteoporosis or rheumatoid arthritis) and the risk of knee arthroplasty. This study conducted a parametric study on bone stiffness to evaluate the risk of periprosthetic fracture via a computational platform.

High flexion angles predispose the patient to the risk of periprosthetic fracture, as presented by the elevated compressive principal stress. Although the SR and PR presented a higher stress magnitude at full extension, the compressive stress of the AF was three times higher in 60-degree flexion. These stress concentration regions were also pronounced in an FE study conducted by Tissakht, Ahmed, and Chan (1996), while the high stress in the AF also corresponded to a typical periprosthetic fracture site (Culp et al. 1987; Zalzal et al. 2006). A sensitivity study on trabecular stiffness was conducted to imitate the effect of osteoporotic degeneration, which imposes additional risks to elderly patients. The association between trabecular stiffness and bone degeneration was supported by the linear and power law relationship of elastic modulus and bone mineral apparent density (Burgers et al. 2008).

Despite the predicted compressive stress being reduced upon a reduction in stiffness due to stress shielding, this will not circumvent the risk of periprosthetic fracture. In fact, the reduction of trabecular stiffness leads to abatement in sustainability in terms of yielding stress (Burgers et al. 2008). The compressive stress experienced in the FE model surpassed the yielding stress for stiffness below 500 MPa in 45-degree and 60-degree flexion, notwithstanding the relatively small yielding volume. The predicted yielding volume reached approximately 450 $mm^3$ for 25% trabecular stiffness under 60-degree flexion. It was interesting that yielding occurred for 125% trabecular stiffness, which was greater than the suggested normal trabecular stiffness. It could be interpreted that the installation of a knee implant would heavily influence the stress at the AF. It could also be due to the variety and inconsistency of trabecular stiffness in different sites and subjects (Burgers et al. 2008; Yeni et al. 2011). Regardless, the yielding volume and fatigue may not necessarily trigger fracture due to the bone undergoing constant repair and regeneration. However, osteoporotic patients may experience inadequate bone resorption and bone stiffness that may lead to accumulated yield and subsequent fracture.

## 8.5 LIMITATIONS OF THE STUDY

The boundary conditions applied in this study were based on the protocol of material testing. Literally, a physiological gait boundary condition could accurately present the response of the most frequent loading conditions. However, the installation of the implant could inevitably compromise normal knee kinematics, and the designated motion provided by the implant should be considered. Also, the patello-femoral joint was not taken into account. Similarly, the design of the femoral component and the patellar implant could change the trajectory of patellar motion and the direction of force. Neglecting the patello-femoral joint could lead to underestimating the predicted stress experienced by the trabecular bone.

## 8.6 CONCLUSION

In conclusion, this research implemented a sensitivity study of trabecular stiffness with the aim to evaluate periprosthetic fracture in terms of yielding criterion. The outcome showed that high flexion angles and weakened bone would predispose the bone to yielding and could lead to subsequent fracture. The condition could be exaggerated in the case of osteoporotic patients, who are presented with lower bone density and compromised bone repair.

## ACKNOWLEDGMENTS

This study was financed by the Research Committee of The Hong Kong Polytechnic University (GHP/052/06).

## REFERENCES

Ashman, R. B., S. C. Cowin, W. C. Van Buskirk, and J. C. Rice. 1984. A continuous wave technique for the measurement of the elastic properties of cortical bone. *J Biomech* 17 (5):349–61.

Burgers, T. A., J. Mason, G. Niebur, and H. L. Ploeg. 2008. Compressive properties of trabecular bone in the distal femur. *J Biomech* 41 (5):1077–85.

Chu, T. 1999. An investigation on contact stresses of New Jersey low contact stress (NJLCS) knee using finite element method. *J Systems Integration* 9 (2):187–199.

Culp, R. W., R. G. Schmidt, G. Hanks, A. Mak, J. L. Esterhai, Jr., and R. B. Heppenstall. 1987. Supracondylar fracture of the femur following prosthetic knee arthroplasty. *Clin Orthop Relat Res* (222):212–22.

Dennis, D. A. 2001. Periprosthetic fractures following total knee arthroplasty. *Instr Course Lect* 50:379–89.

Godest, A. C., M. Beaugonin, E. Haug, M. Taylor, and P. J. Gregson. 2002. Simulation of a knee joint replacement during a gait cycle using explicit finite element analysis. *J Biomech* 35 (2):267–75.

Gujarathi, N., A. B. Putti, R. J. Abboud, J. G. MacLean, A. J. Espley, and C. F. Kellett. 2009. Risk of periprosthetic fracture after anterior femoral notching. *Acta Orthop* 80 (5):553–6.

Jensen, L. K., and W. Eenberg. 1996. Occupation as a risk factor for knee disorders. *Scand J Work Environ Health* 22 (3):165–75.

Kim, Y. H., O. S. Kwon, and K. Kim. 2008. Analysis of biomechanical effect of stem-end design in revision TKA using Digital Korean model. *Clin Biomech (Bristol, Avon)* 23 (7):853–8.

Lesh, M. L., D. J. Schneider, G. Deol, B. Davis, C. R. Jacobs, and V. D. Pellegrini, Jr. 2000. The consequences of anterior femoral notching in total knee arthroplasty. A biomechanical study. *J Bone Joint Surg Am* 82-A (8):1096–101.

Liau, J. J., C. K. Cheng, C. H. Huang, and W. H. Lo. 2002. The effect of malalignment on stresses in polyethylene component of total knee prostheses: a finite element analysis. *Clin Biomech (Bristol, Avon)* 17 (2):140–6.

Linde, F., I. Hvid, and B. Pongsoipetch. 1989. Energy absorptive properties of human trabecular bone specimens during axial compression. *J Orthop Res* 7 (3):432–9.

McMillan, G., and L. Nichols. 2005. Osteoarthritis and meniscus disorders of the knee as occupational diseases of miners. *Occup Environ Med* 62 (8):567–75.

Miyoshi, S., T. Takahashi, M. Ohtani, H. Yamamoto, and K. Kameyama. 2002. Analysis of the shape of the tibial tray in total knee arthroplasty using a three dimension finite element model. *Clin Biomech (Bristol, Avon)* 17 (7):521–5.

National Institutes of Health. 2003. Consensus Statement on total knee replacement. *NIH Consens State Sci Statements* 20 (1):1–34.

Rohlmann, A., H. Zilch, G. Bergmann, and R. Kolbel. 1980. Material properties of femoral cancellous bone in axial loading. Part I: Time independent properties. *Arch Orthop Trauma Surg* 97 (2):95–102.

Singerman, R., J. C. Dean, H. D. Pagan, and V. M. Goldberg. 1996. Decreased posterior tibial slope increases strain in the posterior cruciate ligament following total knee arthroplasty. *J Arthroplasty* 11 (1):99–103.

Su, E. T., H. DeWal, and P. E. Di Cesare. 2004. Periprosthetic femoral fractures above total knee replacements. *J Am Acad Orthop Surg* 12 (1):12–20.

Tissakht, M., A. M. Ahmed, and K. C. Chan. 1996. Calculated stress-shielding in the distal femur after total knee replacement corresponds to the reported location of bone loss. *J Orthop Res* 14 (5):778–85.

Villa, T., F. Migliavacca, D. Gastaldi, M. Colombo, and R. Pietrabissa. 2004. Contact stresses and fatigue life in a knee prosthesis: comparison between in vitro measurements and computational simulations. *J Biomech* 37 (1):45–53.

Yeni, Y. N., M. J. Zinno, J. S. Yerramshetty, R. Zauel, and D. P. Fyhrie. 2011. Variability of trabecular microstructure is age-, gender-, race- and anatomic site-dependent and affects stiffness and stress distribution properties of human vertebral cancellous bone. *Bone* 49 (4):886–94.

Zalzal, P., D. Backstein, A. E. Gross, and M. Papini. 2006. Notching of the anterior femoral cortex during total knee arthroplasty characteristics that increase local stresses. *J Arthroplasty* 21 (5):737–43.

# Section III

## Hip and Pelvis

# 9 Femur Model for Predicting Strength and Fracture Risk

*He Gong, Yubo Fan, and Ming Zhang*

## CONTENTS



## SUMMARY

Hip fractures often occur as a result of bone fragility due to osteoporosis, particularly in the elderly, and have been recognized as a major public health problem given the ever-increasing number of elderly people. In clinics, osteoporosis is often defined in terms of bone mineral density. However, density alone cannot accurately determine bone strength in practice. Many hip fractures occur in people whose hip bone mineral density (BMD) is not severely reduced. Other factors, such as bone geometry, bone internal architecture, and tissue properties, may affect bone strength. Accurate evaluation of hip fracture risk in patients can identify those at high risk so that preventive measures can be taken. It is very important to develop better measures to assess femoral strength and fracture risk using clinically available information. Subject-specific finite element (FE) analysis is a very powerful tool to obtain more sophisticated evaluations of bone strength and the related fracture risk. In this chapter, subject-specific, image-based, nonlinear FE modeling of the proximal femur will be introduced to predict proximal femoral strength and locations of failure.

## 9.1 INTRODUCTION

Whole bone strength is mainly determined by architecture, geometry, and material properties. Architecture and geometry include the gross morphology (size and shape) and the arrangement of the internal trabecular architecture. Bone material properties depend on the degree of mineralization, collagen characteristics, and micro-damage (Seeman and Delmas, 2006). For a more accurate and reliable diagnosis, it is necessary to know the mechanical parameters of bone, especially stiffness and strength, because these parameters reflect bone fracture risk.

To obtain more sophisticated evaluations of bone strength and the related fracture risk, subject-specific FE analysis is particularly useful (Keyak et al., 2005; Bessho et al., 2007; Viceconti et al., 2005). If developed properly while taking the most relevant information into consideration, subject-specific FE modeling has the following advantages: physical properties of bone measured noninvasively can derive the nonlinear mechanical properties of bone; parametric studies can be performed easily; heterogeneity and anisotropy can be addressed; and the magnitudes and directions of loading can be varied easily. The strategies to develop subject-specific FE models are summarized below.

### 9.1.1 Linear versus Nonlinear Finite Element Analysis

Initially, most subject-specific FE models for prediction of femoral fracture load and stiffness were established under linear assumptions (Keyak et al., 1998; Cody et al., 2000). That means linear elastic material properties were assigned to bone tissue and small deformations were assumed. The linear model could save a lot of computational time. The correlations predicted were statistically comparable to those reported previously for fracture load and simpler, density-based measures because of the inability of linear models to represent the nonlinear mechanical behaviors that occur during bone failure. In spite of this limitation of linear analysis, linear models can still be useful if the aim is not to investigate the whole fracture process, but to verify if a certain failure criterion can reproduce the conditions at the onset of failure (Schileo et al., 2008).

Nonlinear FE analyses were developed (Keyak et al., 2005; Bessho et al., 2007), with two kinds of constitutive relationships developed to describe the nonlinearity of bone material. One is a bilinear elastoplastic constitutive relationship, in which the asymmetric strength characteristics between tension and compression can be taken into consideration (Bayraktar et al., 2004; Bessho et al., 2007; Gong et al., 2012). The other is to describe the post-yield material behavior as an initial perfectly plastic phase, then a strain-softened phase, followed by an indefinite perfectly plastic phase (Keyak et al., 2005; Keyak, 2001).

### 9.1.2 Isotropic versus Anisotropic Bone Material Model

In most FE models an isotropic material property was assigned to bone tissue. Bone tissue at the trabecular level can be assumed to be isotropic, although bone at the apparent level is anisotropic (van Rietbergen et al., 1998; Gong et al., 2007). There was also orthotropic material property assignment adopted in some studies (Wirtz et al., 2003; Peng et al., 2006).

### 9.1.3 Bone Tissue Heterogeneity

To consider bone heterogeneity, the mechanical properties of each element should be calculated from the image grey levels or Hounsfield unit values from computed tomography (CT) scanning. The mechanical properties can be discretized into a number of sets of material properties. There are generally two strategies to transfer the grey level into mechanical properties. One is based on equal density intervals (Peng et al., 2006) and the other uses a logarithmic increment of modulus with the aim of obtaining a finer discrimination of mechanical properties at a lower stiffness value (Perillo-Marcone, Alonso-Wazquez, and Taylor, 2003).

### 9.1.4 Effects of the Loading Condition

Identifying the loading condition under which the femur is most likely at risk may aid in the prevention of hip fracture. The effect of force direction on fracture load is a factor inherently associated with fracture risk. There are generally two types of loading conditions investigated in the literature: atraumatic loading, that is, loading similar to joint loading during daily activities (Keyak, Skinner, and Fleming, 2001; Bessho et al., 2007; Schileo et al., 2008), and fall loading, simulating impact from a fall (Keyak et al., 1998; Verhulp, van Rietbergen, and Huiskes, 2008). Keyak, Skinner, and Fleming (2001) investigated the effect of force direction and found that for the fall configuration, the force direction with the lowest fracture load corresponded to an impact to the posterolateral aspect of the great trochanter, and for atraumatic loading, the lowest fracture loads for the force directions occurred when the load was similar to conditions while standing on one leg or climbing stairs. In this chapter, subject-specific, image-based, nonlinear FE modeling of the proximal femur is introduced to predict proximal femoral strength and failure location.

## 9.2 DEVELOPMENT OF FEMUR MODELS TO PREDICT STRENGTH AND FRACTURE RISK

### 9.2.1 Quantitative CT Scanning Procedure

Quantitative CT (QCT) scans were made of the hip region. The settings for the QCT scanning were 80 KVp, 280 mA, and 512 × 512 matrix in spiral reconstruction mode with 0.9375 mm pixel size and 2.5 mm increments (GE Medical Systems/lightspeed 16, Wakesha, Wisconsin). To calibrate CT Hounsfield units to equivalent bone mineral concentration, a calibration phantom containing known hydroxyapatite concentrations was included in each scan (Image Analysis, Columbia, Kentucky). The calibration phantom extending from the L1 vertebral body to the mid-femoral shaft was positioned under each subject. The phantom contained calibration cells of 0, 0.075, and 0.15 g/cm$^3$ equivalent concentration of calcium hydroxyapatite (Gong et al., 2012).

### 9.2.2 Three-Dimensional Modeling of the Proximal Femur

Three-dimensional reconstruction and surface meshing of the proximal femur was performed in MIMICS software (Materialise Inc., Leuven, Belgium) from the femoral head to 1 cm below the lesser trochanter; then it was imported into ABAQUS software (Simulia Inc., Waltham, Massachusetts) to convert the triangular surface mesh into a four-node tetrahedral elements mesh. The largest average element edge length was 1.76 mm, which could achieve a sufficiently precise prediction.

### 9.2.3 Bone Tissue Heterogeneity

To account for bone tissue heterogeneity, bone material in the whole proximal femur was divided into approximately 170 discrete sets of materials so that the modulus of each material increased in an increment of 5% over the modulus of the previous material (Keyak et al., 1998; Perillo-Marcone, Alonso-Wazquez, and Taylor, 2003). All bone materials in the proximal femur were set with a Young's modulus of 22.5 GPa. Relationships between ash density and elastic modulus can be obtained from the literature (see Table 9.1). The apparent density of each element can be determined from the linear regression relationship of the Hounsfield units of the calibration phantom to their apparent density, and its ash density can be obtained from its apparent density using a relationship between ash density and apparent density reported in the literature (Keyak et al., 1998). Then, which bone material an element was assigned with can be determined.

A specific nonlinear constitutive relationship was assigned to each bone material. A four-parameter bilinear constitutive model was used to describe the nonlinear constitutive relationship of each bone material (Gong, Zhang, and Fan, 2011; Gong et al., 2012). The four parameters in the constitutive model were tensile yield strain ($\varepsilon_t^T$), compressive yield strain ($\varepsilon_c^T$), pre-yield Young's modulus ($E$), and post-yield modulus ($E_u$, assuming that the post-yield modulus in compression was equal to that in tension) (Gong, Zhang, and Fan, 2011). The schematic of the bilinear constitutive model of one bone material is shown in Figure 9.1.

**TABLE 9.1**
**Relationships between Ash Density and Elastic Modulus**

| Ash Density (g/cm³) | Young's Modulus (MPa) |
|---|---|
| $\rho = 0$ | 0.001 |
| $0 < \rho \leq 0.27$ | $33{,}900\ \rho^{2.20}$ |
| $0.27 < \rho \leq 0.6$ | $5307\ \rho + 469$ |
| $0.6 < \rho$ | $10\ 200\ \rho^{2.01}$ |

*Source:* J. H. Keyak, S. A. Rossi, K. A. Jones, and H. B. Skinner, *J Biomech,* 31, 125–33, 1998. With permission.

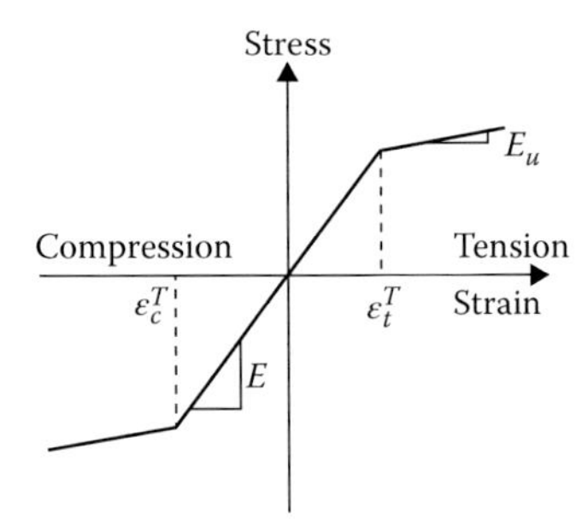


**FIGURE 9.1** Schematic of the bilinear constitutive model of one bone material (not to scale).

### 9.2.4 Nonlinear Finite Element Analysis for the Estimation of Femoral Strength

In this chapter, two typical proximal femur models (model A and model B) were built to estimate their strengths and failure locations and types. Nonlinear FE analyses were performed using ABAQUS finite element software. The models were meshed with 261,784 and 289,362 four-node tetrahedral elements, respectively (Figure 9.2). A distributive pressure load with a maximum magnitude of 6.5 N/mm² was applied on the femoral head, tilting the specimen 8° in the frontal plane (Gong et al., 2012). The distal end of the model was constrained. Figure 9.2 shows the FE models with boundary and loading conditions.

The load when at least one element in the outer cortical surface yielded was used to describe femoral strength (Bessho et al., 2007). The maximum principal strain criterion was used to judge bone failure (Schileo et al., 2008), that is, failure of a surface element would occur in tension when the maximum principal strain exceeded 1.174%, or in compression when the minimum principal strain was less than –0.73% (Gong et al., 2012).

## 9.3 RESULTS FROM THE MODEL ANALYSIS

The femoral strengths predicted from nonlinear FE analyses for models A and B in Figure 9.2 were 3033.85 N and 2056.00 N, respectively. An initial tensile yield in the superolateral aspect of the outer cortical surface of the femoral neck was predicted for model A. The initial plastic strain in this model is shown in Figure 9.3a. For model B, an initial compressive yield in the femoral neck was predicted and the initial plastic strain in this model is shown in Figure 9.3b.

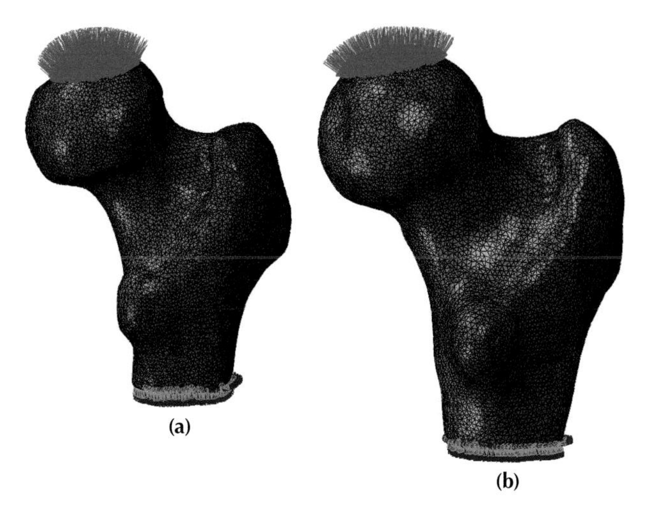


**FIGURE 9.2** **(See color insert.)** Finite element models of two typical proximal femurs with boundary and loading conditions: (a) model A; (b) model B.

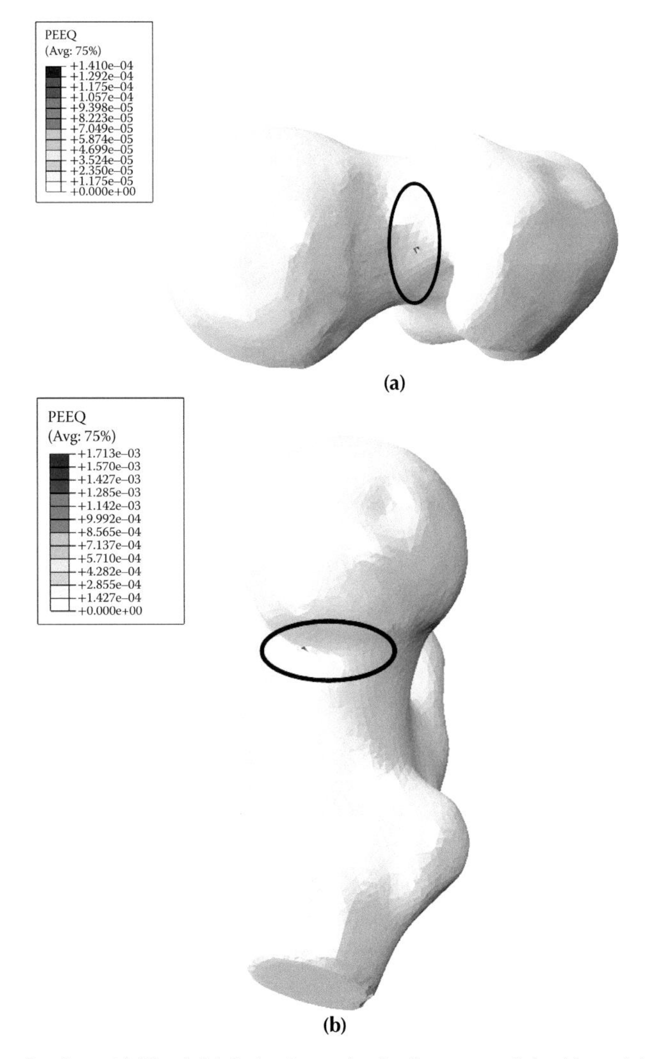


**FIGURE 9.3** **(See color insert.)** The initial plastic strains in the two models: (a) model A; (b) model B.

## 9.4 APPLICATIONS OF THE MODEL

FE analysis of QCT scans is the most advanced method for noninvasive clinical assessment of femoral strength, and was regarded as biomechanical CT (Keaveny et al., 2010). Identifying the associations between the subject-specific femoral strength predicted by image-based nonlinear FE analysis and clinically available information may have the potential for noninvasive clinical assessment of individual hip fracture risk.

Material distribution can provide more details about the material properties of the proximal femur than BMD, and it can be easily obtained from QCT data. Figure 9.4 shows the material distributions of the two models evaluated using MIMICS software, from which it can be clearly seen that the material distributions of the two models were significantly different, with model A having more materials with greater elastic moduli.

It is also convenient to obtain the geometrical parameters from QCT data (Lv et al., 2012), which included neck length (NL), diameter of femoral head (HD), height of femoral head (HH), offset (OFF), neck shaft angle (NSA), height of the top of the greater trochanter (TRH), thickness of femur (TOF), and neck diameter (ND). A schematic of the geometric parameters of a proximal femur (model A) is shown in Figure 9.5. The three-dimensional geometric parameters of the two models in this chapter are listed in Table 9.2, from which it can be seen that the geometries of the two models were quite different.

Recently, three independent factors (principal components, or PCs) were extracted from BMD, material distribution from QCT data, height, weight, and geometric parameters. A high correlation was found between the three PCs and the strength predicted from FE analysis. Besides BMD, other parameters, such as material distribution and geometric parameters, also contributed significantly to femoral strength (Gong et al., 2012).

If developed properly, while taking the most relevant information into consideration, subject-specific FE modeling has the following advantages: the physical properties of bone measured noninvasively can derive the nonlinear mechanical properties of the bone; parametric studies can be performed easily; heterogeneity and anisotropy can be addressed; and the magnitudes and directions of loading can be varied easily. Accordingly, subject-specific, image-based, nonlinear FE analysis can quantitatively predict the form and magnitude of loads that may bring about fracture, as well as the possible location and type of fracture. It offers insight into the fracture mechanism and may help to predict brittle fracture risk and determine feasible exercises for the elderly to avoid falls. Subject-specific bone strength and evaluation of the related fracture risk can be obtained noninvasively based on a patient's clinical information.

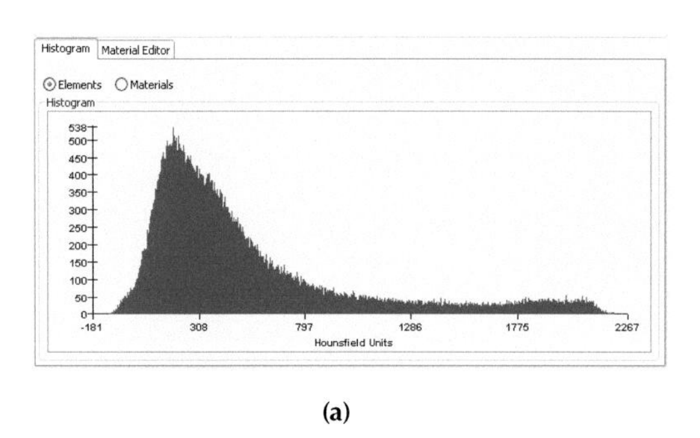


(a)

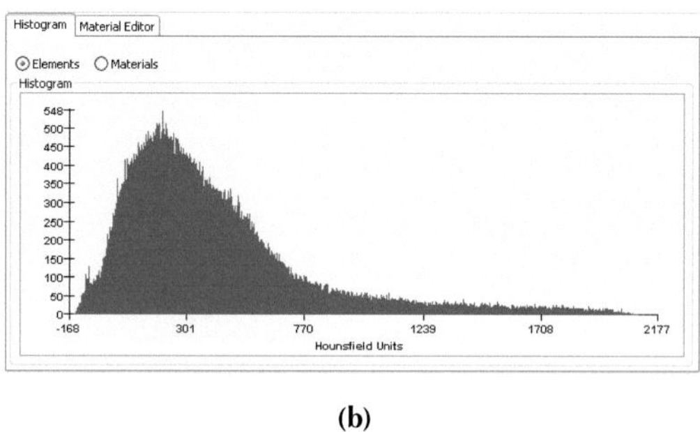


(b)

**FIGURE 9.4** Material distributions of the two models evaluated using MIMICS software: (a) model A; (b) model B.

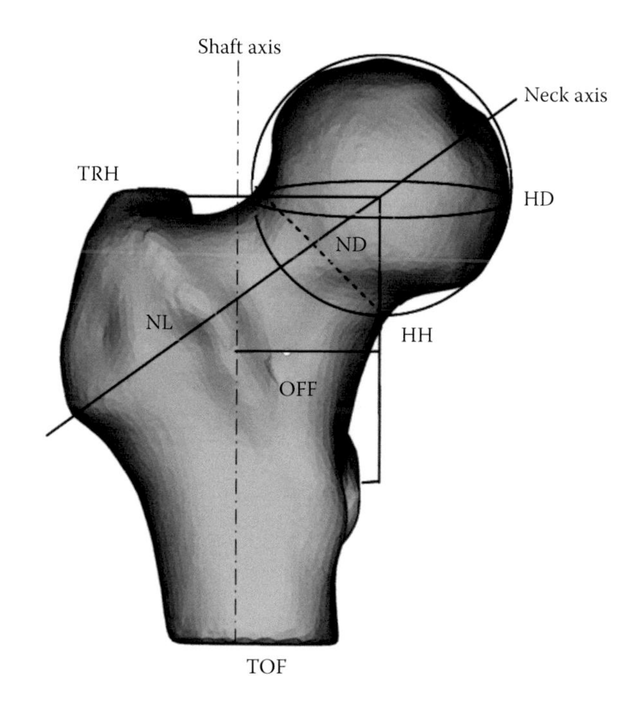


**FIGURE 9.5** Schematic of the geometric parameters of a proximal femur (model A).

**TABLE 9.2**
**Three-Dimensional Geometric Parameters, Age, Height, and Weight of Models A and B**

| Parameter | Model A | Model B |
|---|---|---|
| HD (mm) | 48.5549 | 48.2565 |
| HH (mm) | 55.1895 | 53.3638 |
| OFF (mm) | 34.522 | 35.9124 |
| NSA (°) | 128.3331 | 131.4655 |
| THR (mm) | 3.8922 | 7.4645 |
| TOF (mm) | 11.3688 | 8.2368 |
| ND (mm) | 37.707 | 40.0248 |
| NL (mm) | 90.8636 | 98.4625 |
| Age | 74 | 72 |
| Height (cm) | 161.75 | 166.2 |
| Weight (kg) | 61.5 | 54.8 |

## ACKNOWLEDGMENTS

This work is supported by a grant from the National Natural Science Foundation of China (No. 11120101001, 11322223, 11272273) and the Program for New Century Excellent Talents in University (NCET-12-0024).

## REFERENCES

Bayraktar, H. H., E. F. Morgan, G. L. Niebur, G. Morris, E. Wong, and T. Keaveny. 2004. Comparison of the elastic and yield properties of human femoral trabecular and cortical bone tissue. *J Biomech* 37:27–35.

Bessho, M., I. Ohnishi, J. Matsuyama, T. Matsumoto, K. Imai, and K. Nakamura. 2007. Prediction of strength and strain of the proximal femur by a CT-based finite element method. *J Biomech* 40:1745–53.

Cody, D. D., F. J. Hou, G. W. Divine, and D. P. Fyhrie. 2000. Femoral structure and stiffness in patients with femoral neck fracture. *J Orthop Res* 18:443–8.

Gong, H., M. Zhang, and Y. Fan. 2011. Micro-finite element analysis of trabecular bone yield behavior: effects of tissue non-linear material properties. *J Mech Med Biol* 11(3):563–80.

Gong, H., M. Zhang, Y. Fan, W. L. Kwok, and P. C. Leung. 2012. Relationships between femoral strength evaluated by nonlinear finite element analysis and BMD, material distribution and geometric morphology. *Ann Biomed Eng* 40(7):1575–85.

Gong, H., M. Zhang, L. Qin, and Y. Hou. 2007. Regional variations in the apparent and tissue-level mechanical parameters of vertebral trabecular bone with aging using micro-finite element analysis. *Ann Biomed Eng* 35(9):1622–31.

Keaveny, T. M., D. L. Kopperdahl, L. J. Melton III, P. F. Hoffmann, S. Amin, B. L. Riggs, and S. Khosla. 2010. Age-dependence of femoral strength in white women and men. *J Bone Miner Res* 25(5):994–1001.

Keyak, J. H. 2001. Improved prediction of proximal femoral fracture load using nonlinear finite element models. *Med Eng Phys* 23:165–73.

Keyak, J. H., T. S. Kaneko, J.Tehranzadeh, and H. B. Skinner. 2005. Predicting proximal femur strength using structural engineering models. *Clin Orthop Relat Res* 437:219–28.

Keyak, J. H., S. A. Rossi, K. A. Jones, and H. B. Skinner. 1998. Prediction of femoral fracture load using automated finite element modeling. *J Biomech* 31:125–33.

Keyak, J. H., H. B. Skinner, and J. A. Fleming. 2001. Effect of force direction on femoral fracture load for two types of loading conditions. *J Orthop Res* 19:539–44.

Lv, L., G. Meng, H. Gong, D. Zhu, and W. Zhu. 2012. A new method for the measurement and analysis of three-dimensional morphological parameters of proximal male femur. *Biomed Res* 23(2):219–26.

Peng, L., J. Bai, X. Zeng, and Y. Zhou. 2006. Comparison of isotropic and orthotropic material property assignment on femoral finite element models under two loading directions. *Med Eng Phys* 28:227–33.

Perillo-Marcone, A., A. Alonso-Wazquez, and M. Taylor. 2003. Assessment of the effect of mesh density on the material property discretisation within QCT based FE models: a practical example using the implanted proximal tibia. *Comput Meth Biomech Biomed Eng* 6(1):17–26.

Schileo, E., F. Taddei, L. Cristofolini, and M. Viceconti. 2008. Subject-specific finite element models implementing a maximum principal strain criterion are able to estimate failure risk and fracture location on human femurs tested in vitro. *J Biomech* 41: 356–67.

Seeman, E., and P. D. Delmas. 2006. Bone quality: the material and structural basis of bone strength and fragility. *New Engl J Med* 354:2250–61.

van Rietbergen, B., A. Odgaard, J. Kabel, and R. Huiskes. 1998. Relationship between bone morphology and bone elastic properties can be accurately quantified using high-resolution computer reconstructions. *J Orthop Res* 16:23–8.

Verhulp, E., B. van Rietbergen, and R. Huiskes. 2008. Load distribution in the healthy and osteoporotic human proximal femur during a fall to the side. *Bone* 42:30–5.

Viceconti, M., S. Olsen, L. P. Nolte, and K. Burton. 2005. Extracting clinically relevant data from finite element simulations. *Clin Biomech* 20:451–4.

Wirtz, D. C., T. Pandorf, F. Portheine, K. Radermacher, N. Schiffers, A. Prescher, D. Weichert, and F. U. Niethard. 2003. Concept and development of an orthotropic FE model of the proximal femur. *J Biomech* 36(2):289–93.

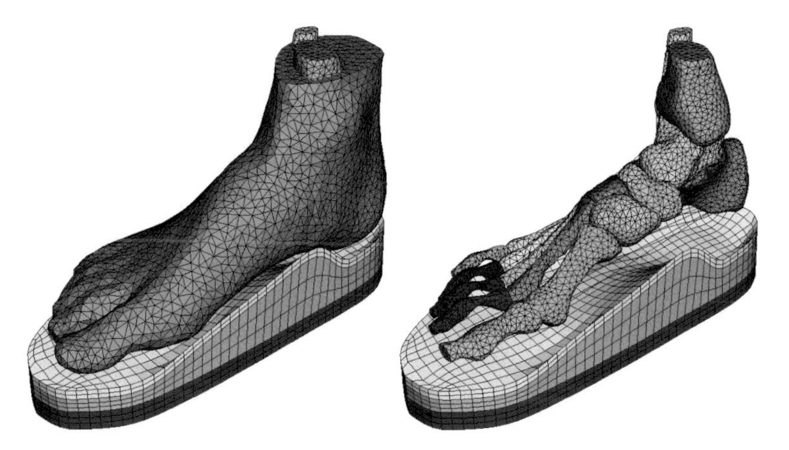

**FIGURE 1.4** Foot and foot orthosis model.

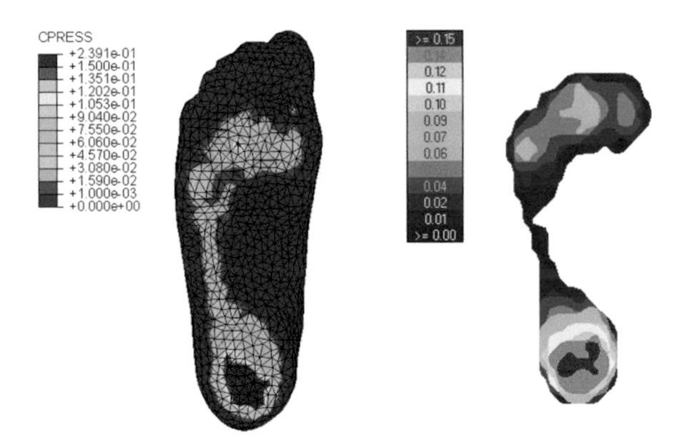


**FIGURE 1.5** Plantar pressure distributions predicted by finite element modeling and measured by Tekscan.

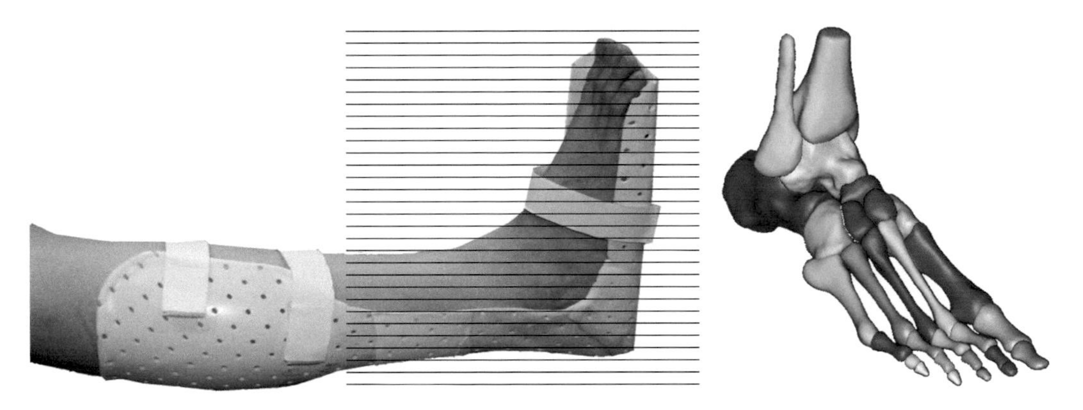

**FIGURE 2.1** Fixing foot and ankle in a neutral position by foot orthosis (left), and foot bones surface model (right).

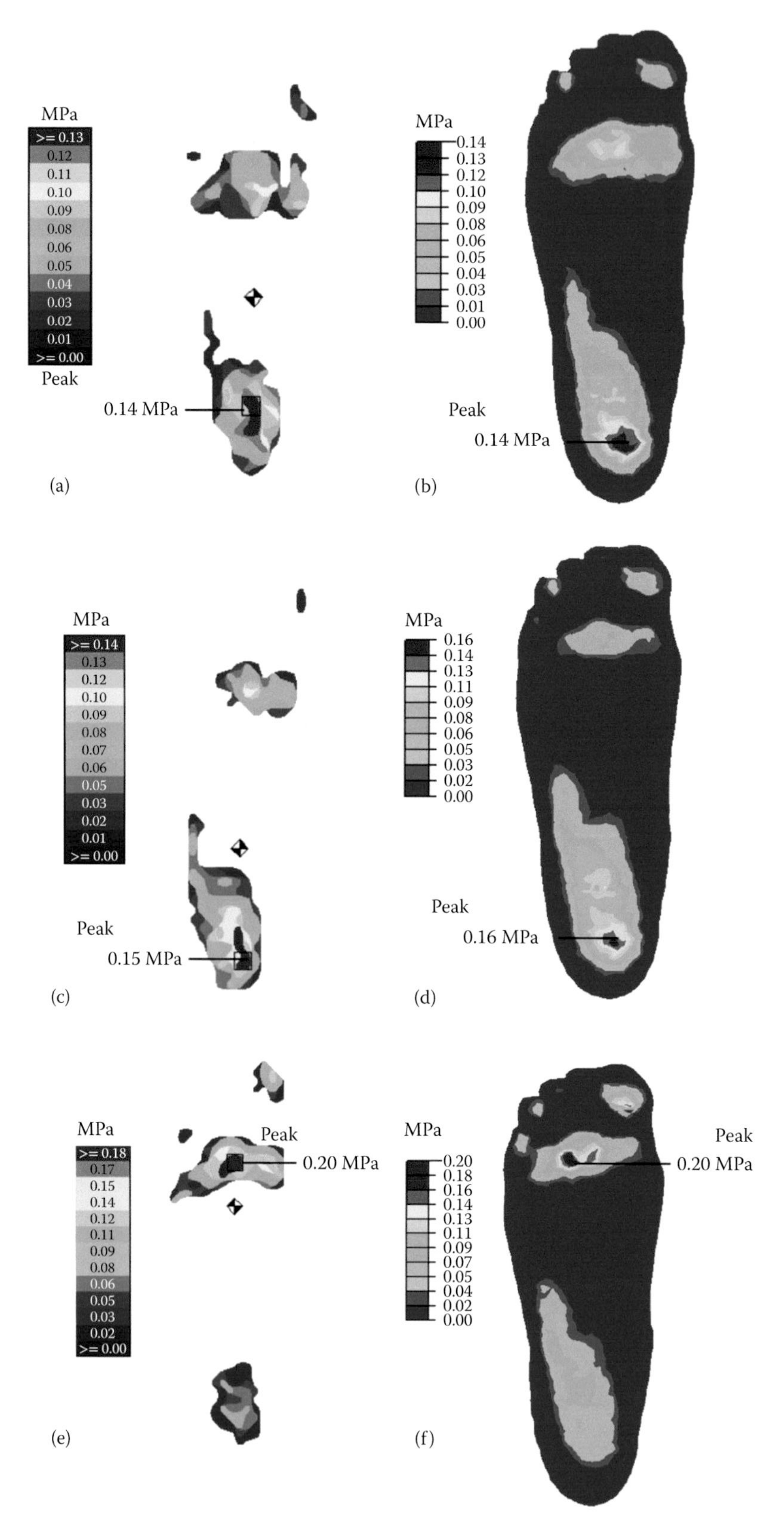


**FIGURE 2.6** Plantar pressure distributions: (a) 1-inch from F-scan measurement, (b) 1-inch from finite element (FE) prediction, (c) 2-inch from F-scan measurement, (d) 2-inch from FE prediction, (e) 3-inch from F-scan measurement, and (f) 3-inch from FE prediction.

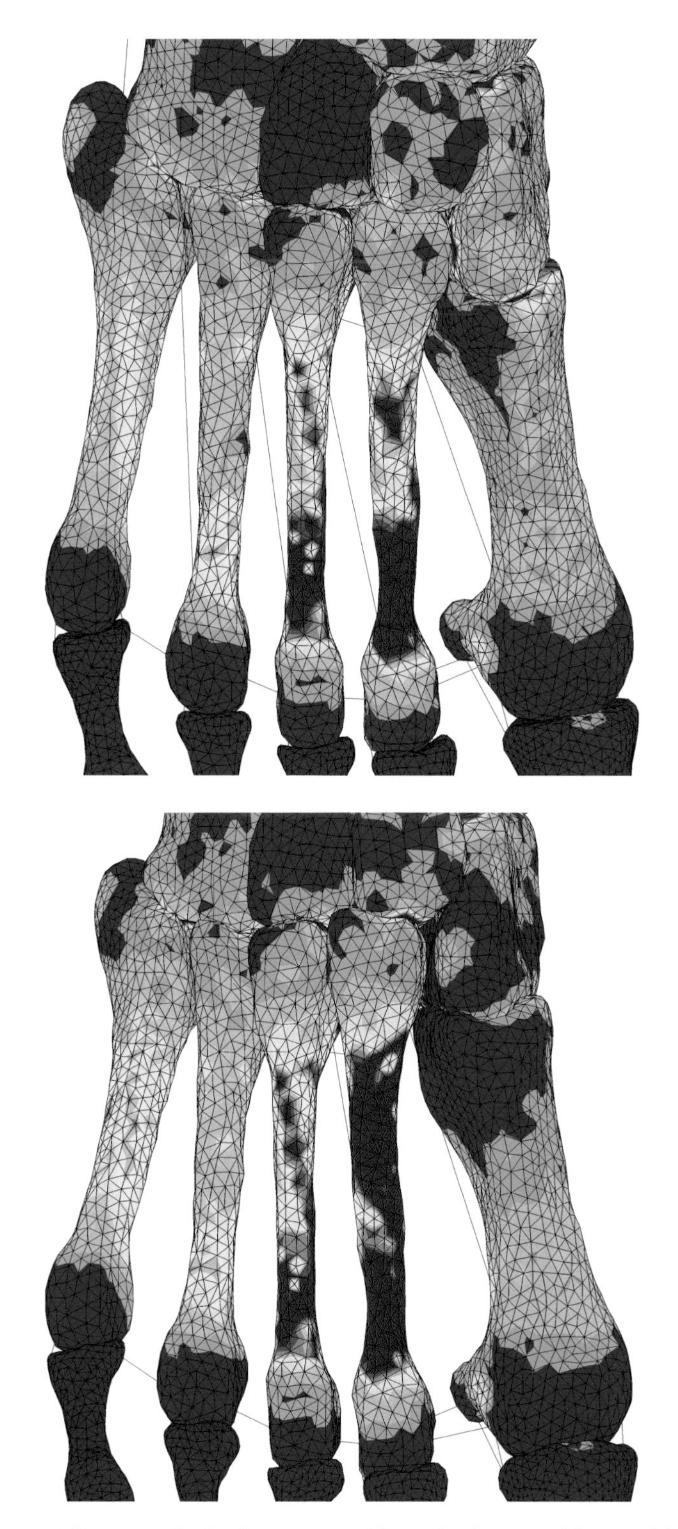

**FIGURE 3.4** The von Mises stress in the five metatarsal bones in the normal foot model (a) and model with the first and second TMT joints fused (b) in midstance.

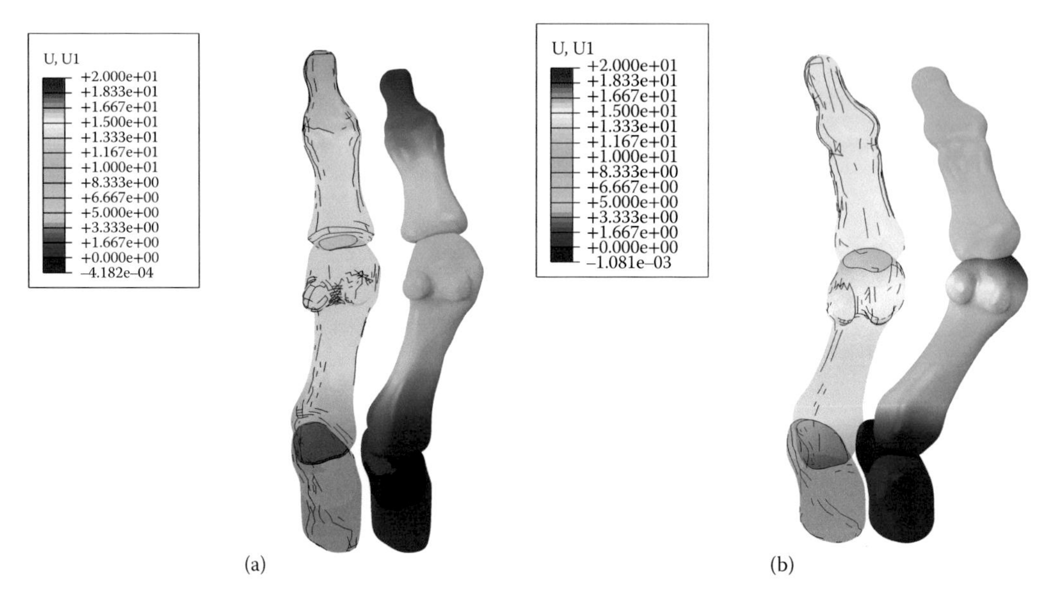


**FIGURE 4.3** Simulation results on the displacement in medial-lateral direction indicating change of bone alignment after application of forefoot loading. (a) Normal foot; (b) hallux valgus foot.

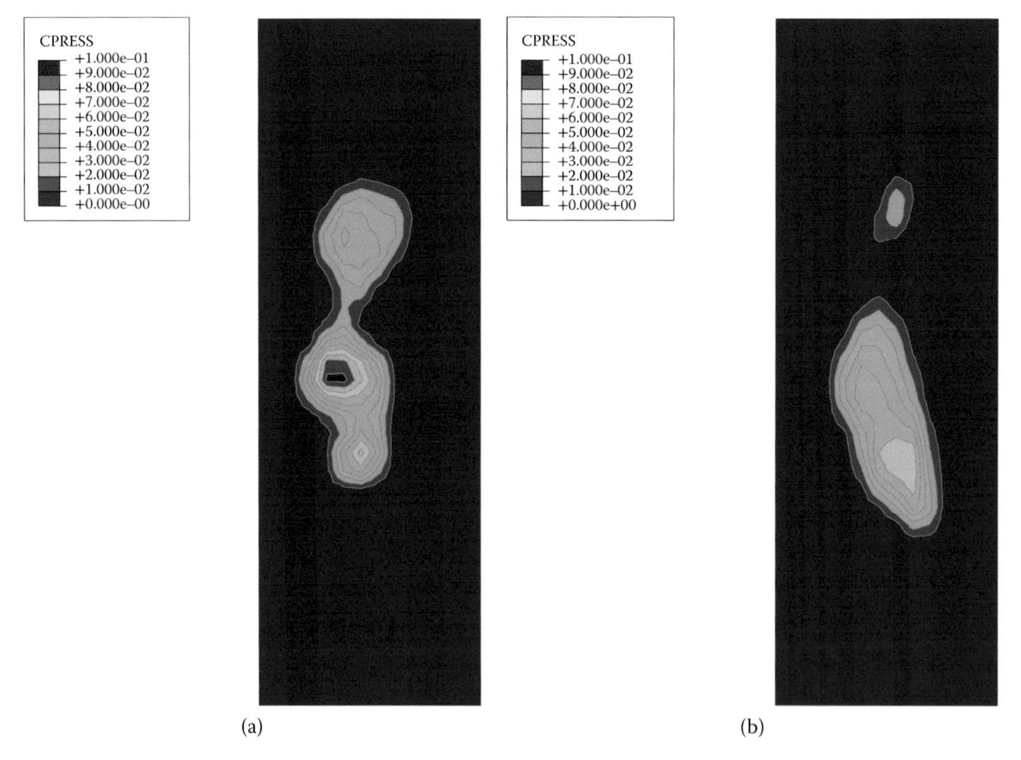


**FIGURE 4.5** The plantar pressure distribution on the supporting plate of the normal (a) and hallux valgus (b) foot. Unit: MPa.

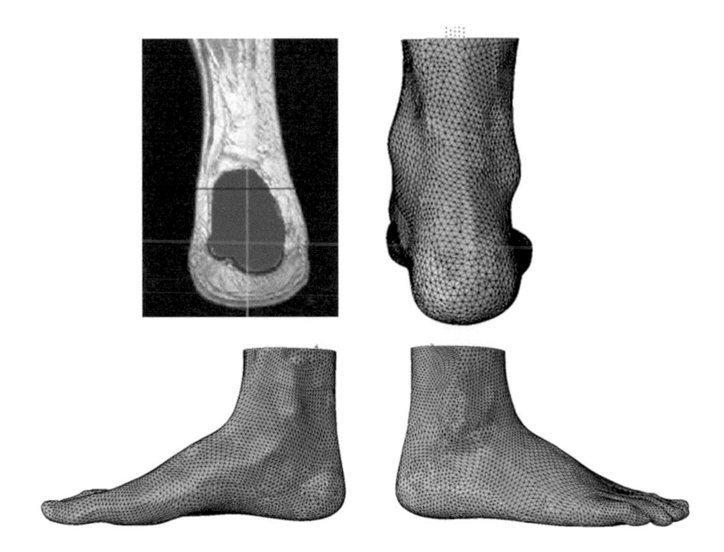

**FIGURE 5.1** Finite element foot model with segmented plantar fat pad.

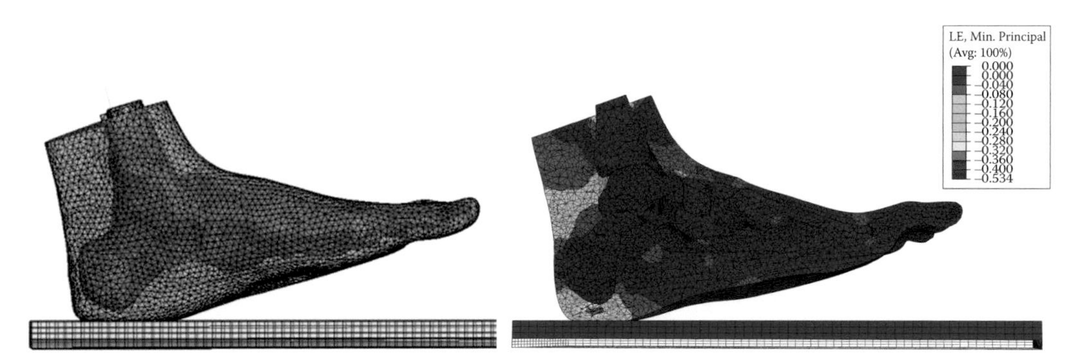


**FIGURE 5.6** Predicted maximal plantar fat pad deformation and compressive strain at heel strike.

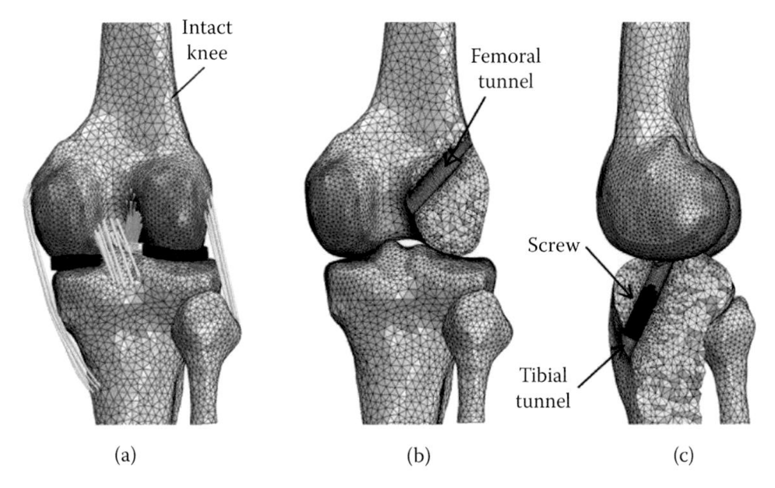


**FIGURE 6.1** FE model of the human knee joint. (a) Intact knee. (b) Cross-section of the femoral tunnel. (c) Cross-section of the tibial tunnel and interference screw.

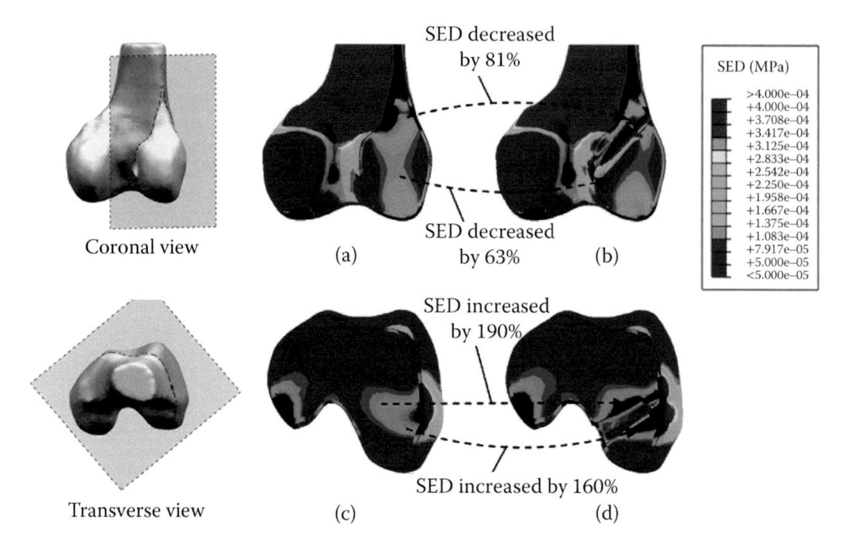


**FIGURE 6.2** SED distributions in the femur. (a) SED distribution in the intact femur (coronal view). (b) SED distribution in the femur after ACL reconstruction (coronal view). (c) SED distribution in the intact femur (transverse view). (d) SED distribution in the femur after ACL reconstruction (transverse view).

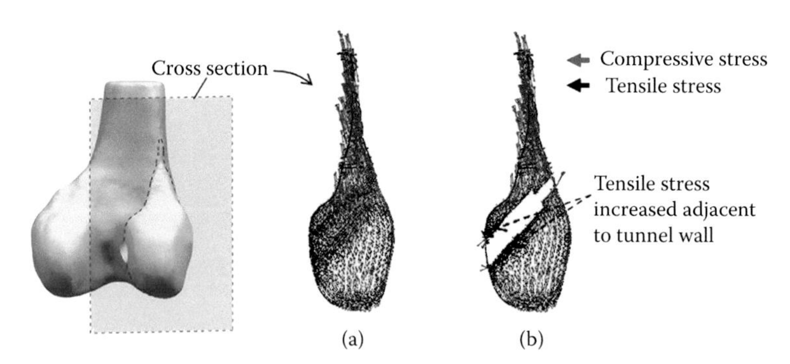


**FIGURE 6.3** Compressive and tensile stress trajectories in a coronal section of the femur. The red arrow indicates compressive stress, and the black arrow indicates tensile stress. (a) Intact knee. (b) Knee after ACL reconstruction.

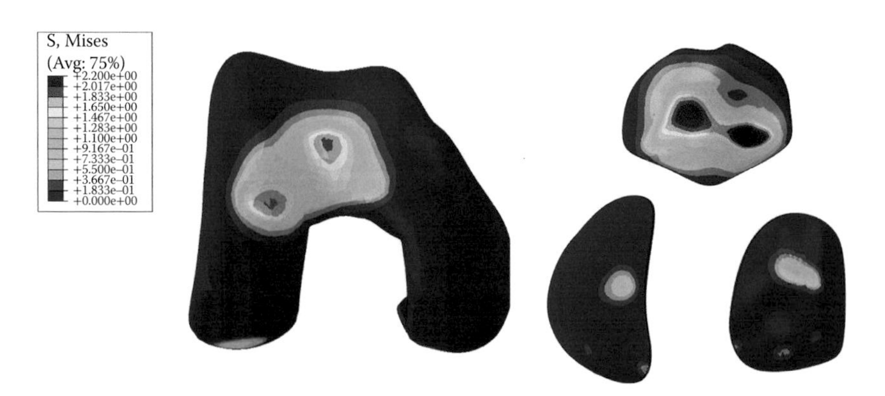


**FIGURE 7.4** The von Mises stress on the cartilages of the knee joint in a kneeling posture under 1000 N.

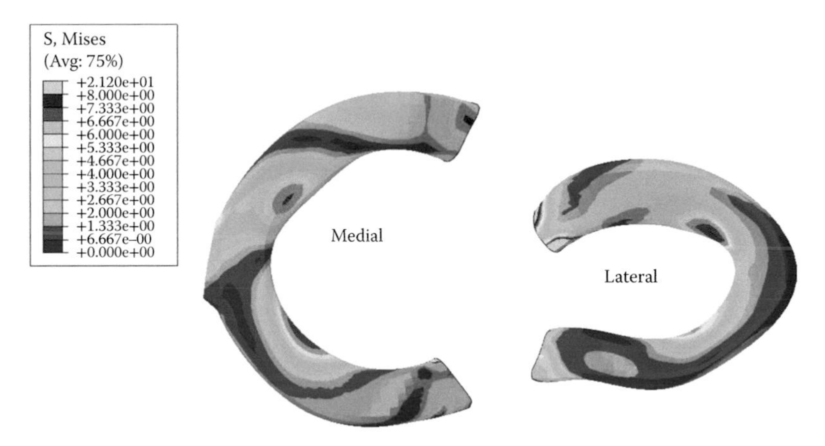


**FIGURE 7.5** The von Mises stress of menisci under a 1000 N compressive load.

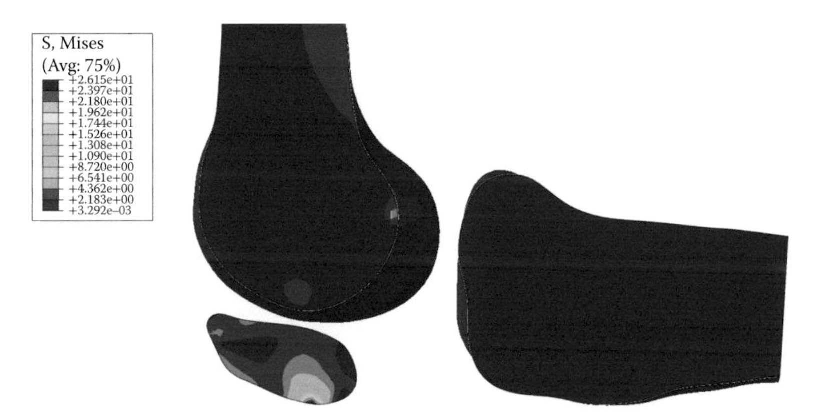


**FIGURE 7.6** Sagittal view of the von Mises stress distribution of bones and patellar tendon under 1000 N compressive loads.

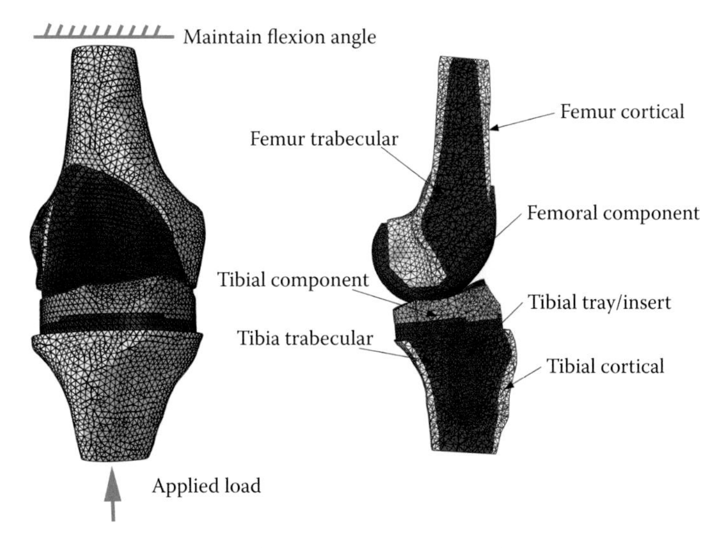


**FIGURE 8.1** Finite element model of the knee implant and bones simulating compression with flexion.

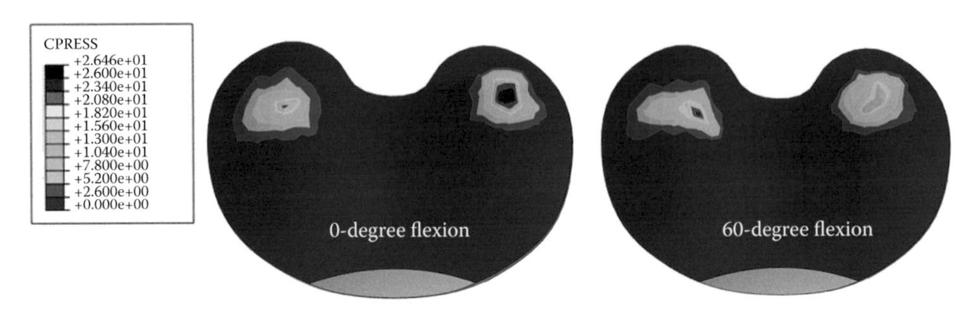


**FIGURE 8.2** Contact pressure of the tibial insert at 0-degree flexion (left) and 60-degree flexion (right). Unit: MPa.

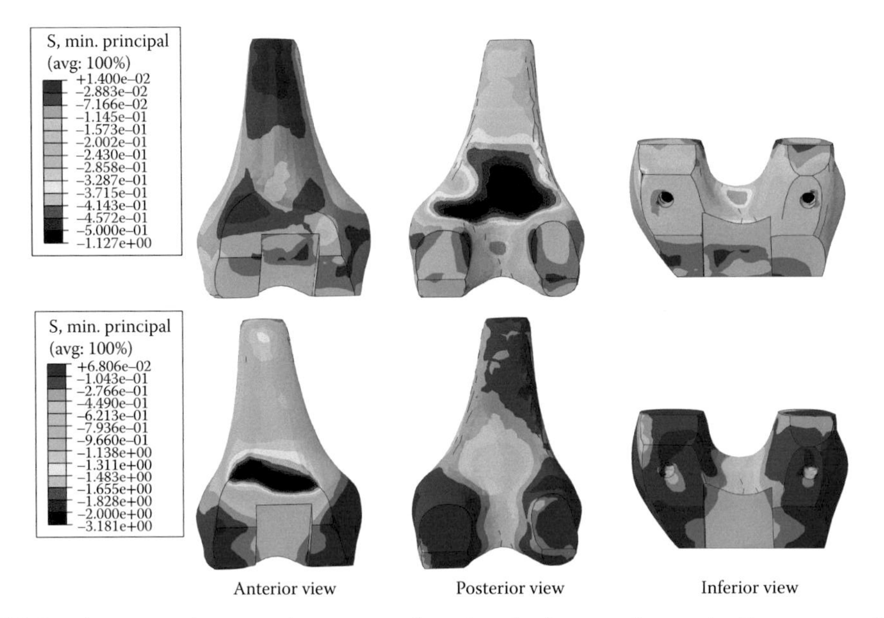


**FIGURE 8.3** Concentrated compressive stress on the trabecular femur at the anterior flange, screw sites, and posterior supracondylar location at neutral position (top) and 60-degree flexion (bottom). Unit: MPa.

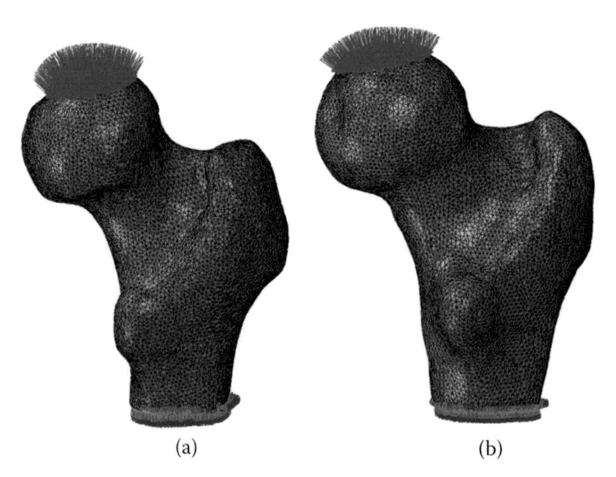


**FIGURE 9.2** Finite element models of two typical proximal femurs with boundary and loading conditions: (a) model A; (b) model B.

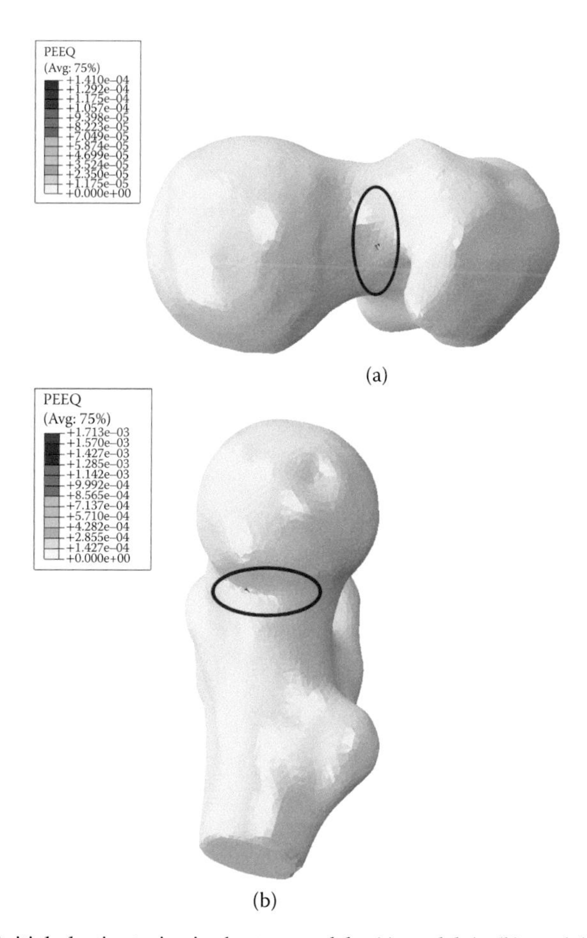


**FIGURE 9.3** The initial plastic strains in the two models: (a) model A; (b) model B.

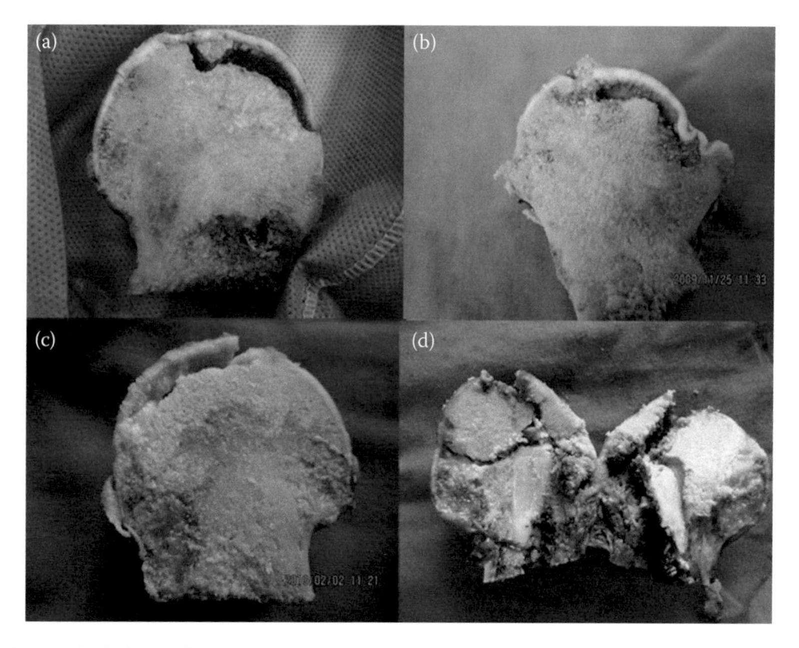


**FIGURE 10.1** Typical signs of ONFH: (a) crescent fracture; (b) femoral head collapse; (c) cortical fracture and osteosclerosis; (d) necrotic boundary fracture.

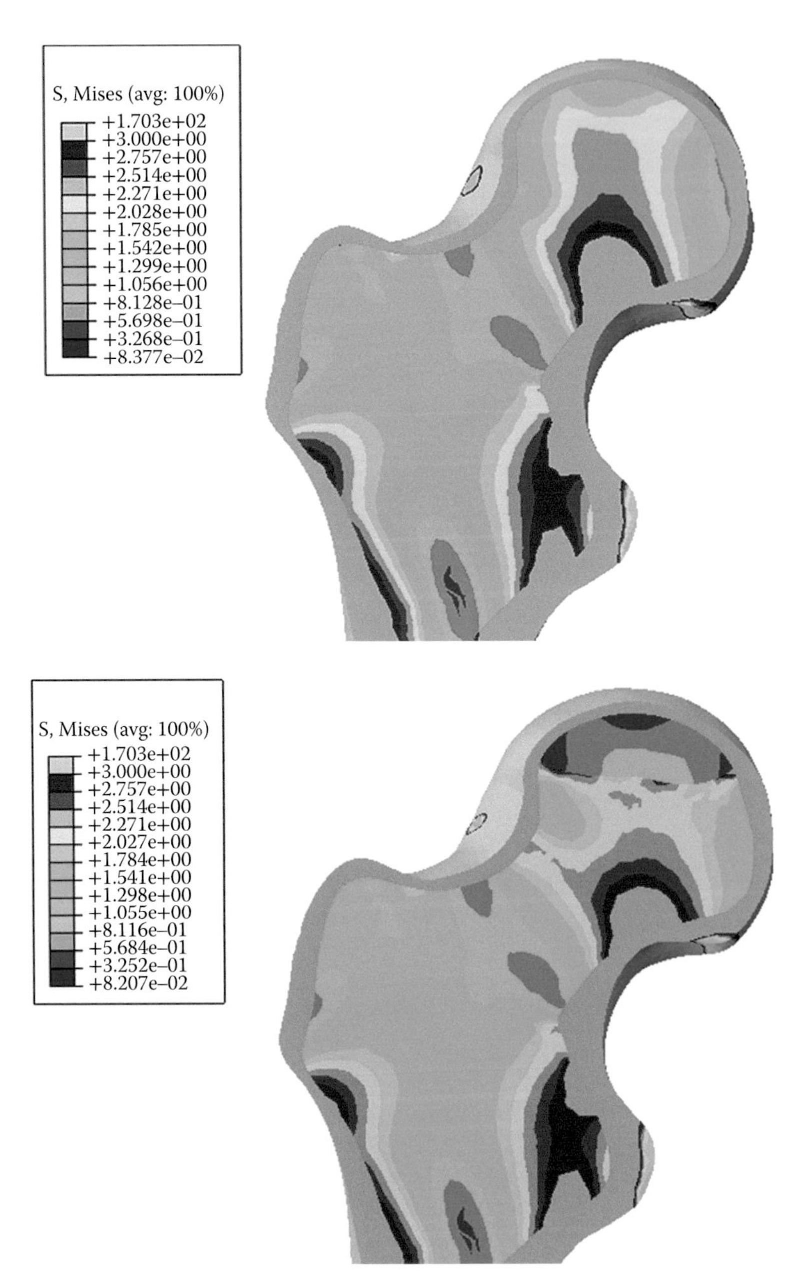


**FIGURE 10.2** The von Mises stress contour of the femur head for a normal femur (top) and a necrotic femur (bottom). Unit: MPa.

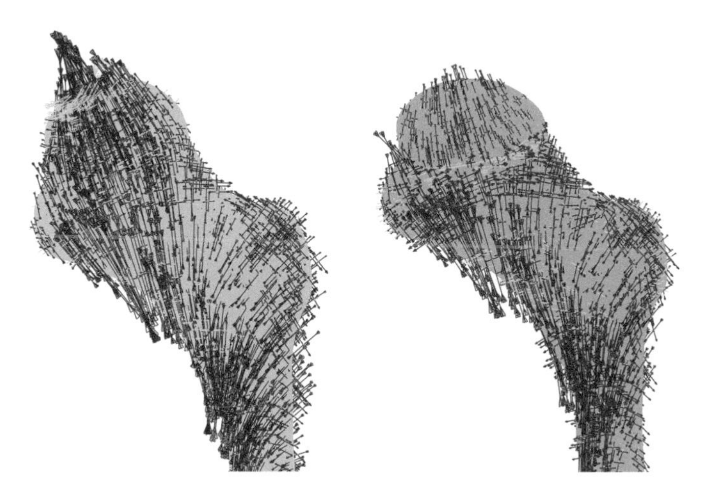

**FIGURE 10.4** Field vector of compressive principal stress (red) and tensile principal stress (blue) in normal (left) and necrotic condition (right).

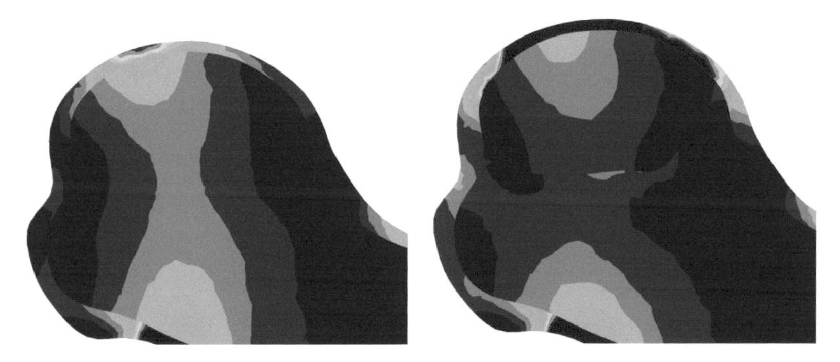

**FIGURE 10.5** Strain energy density of the cortical bone and trabecular bone at the weight-bearing region: (left) healthy; (right) necrotic condition.

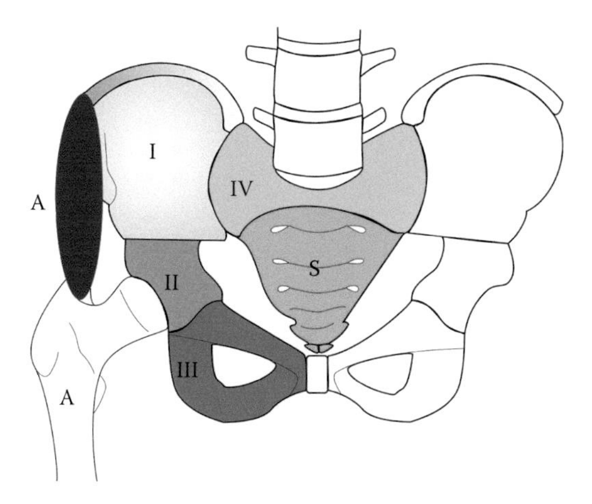


**FIGURE 11.1** Types of pelvic resection. The red oval indicates the gluteus medius muscle; Type I = resection of the ilium; Type II = resection of the periacetabular region; Type III = resection of the ischiopubic region; Type IV = en bloc excision of the sacral ala; A = aggressive resection; S = sacrum.

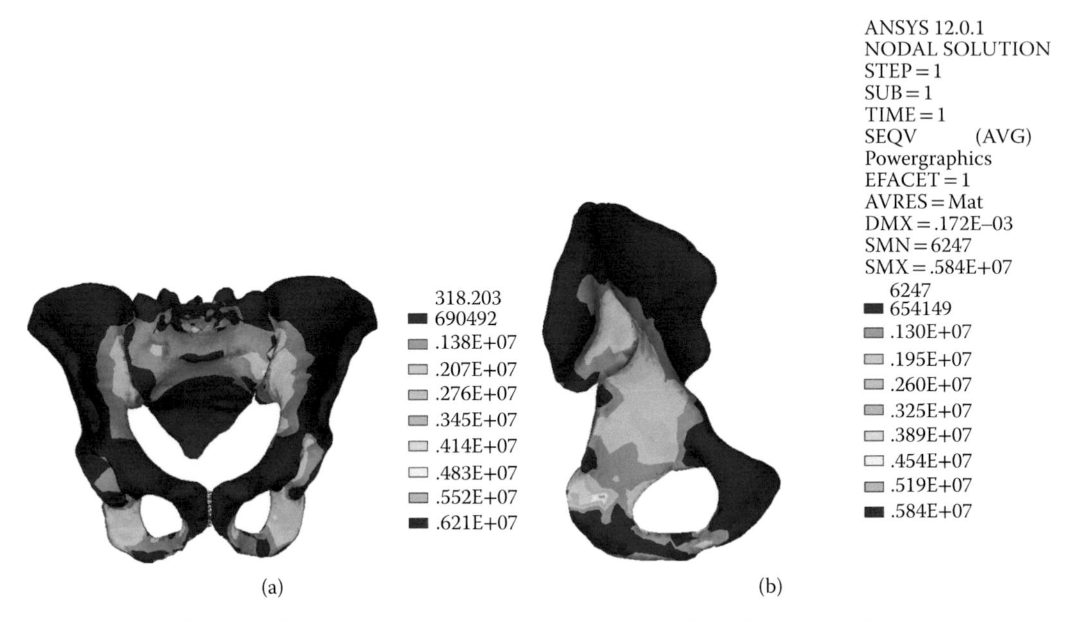


**FIGURE 11.6** The von Mises stress distribution of the intact model: (a) global model; (b) the left ilium.

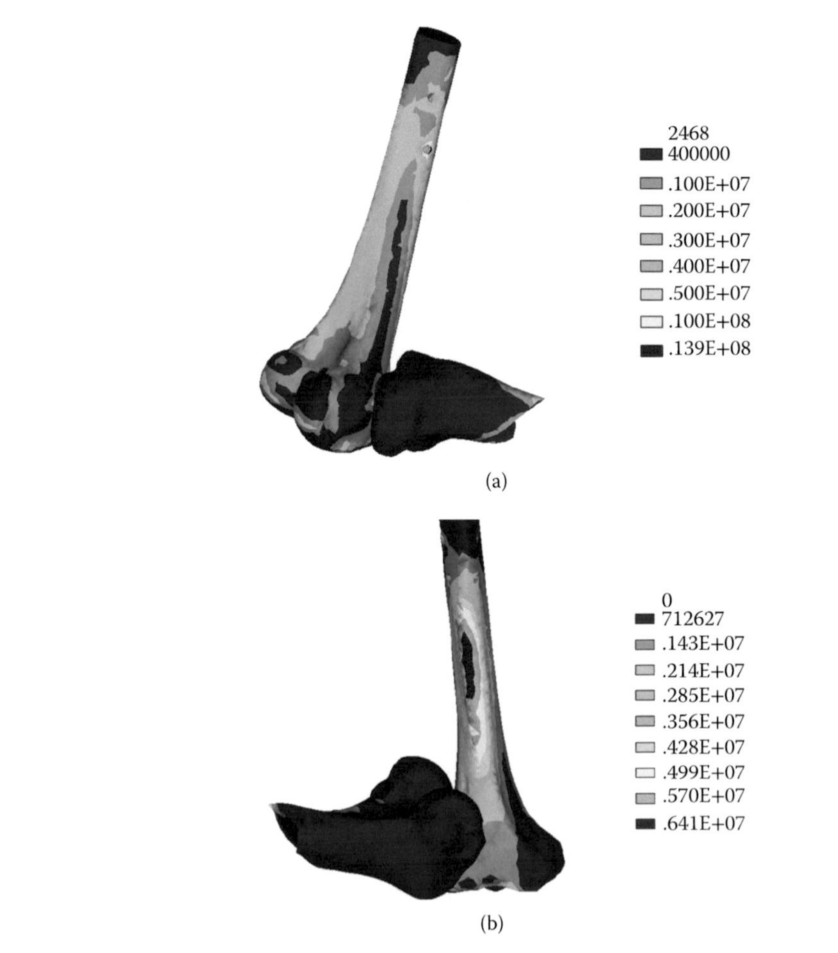


**FIGURE 11.9** The von Mises stress distribution of the autografts in the reconstruction models: (a) the femur in the FC-reconstructed model; (b) the tibia in the TP-reconstructed model.

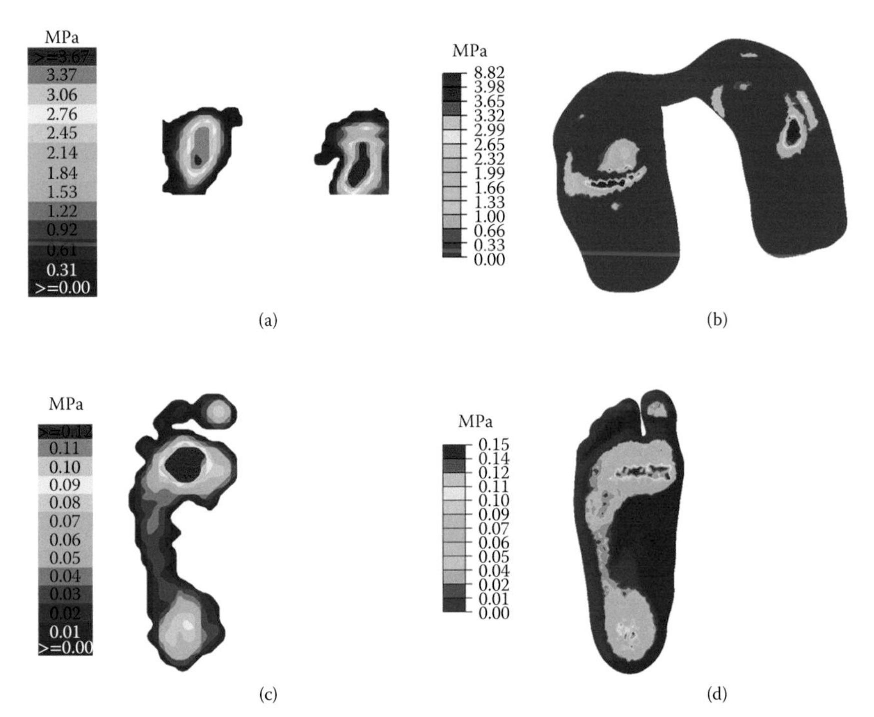


**FIGURE 12.3** Model validation: (a) tibiofemoral pressure measured by K-Scan; (b) FE predicted tibiofemoral pressure; (c) plantar pressure measured by F-Scan; (d) FE predicted plantar pressure.

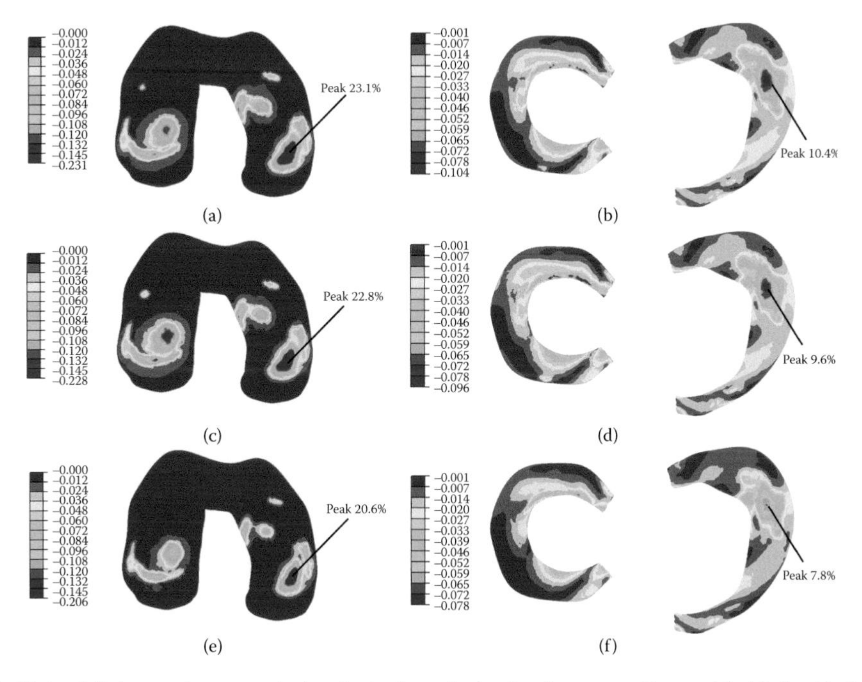


**FIGURE 12.4** Minimum (compressive) principal strain in the femur cartilage with (a) 0-, (c) 5-, and (e) 10-degree laterally wedged orthosis and minimum principal strain in the meniscus with (b) 0-, (d) 5-, and (f) 10-degree laterally wedged orthosis.

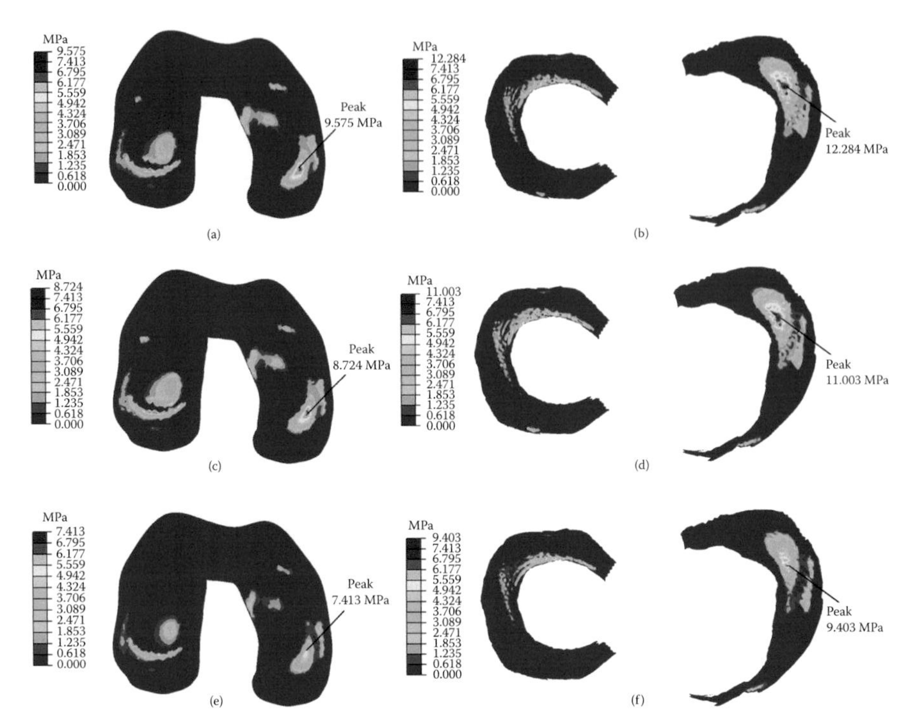


**FIGURE 12.5** Contact pressure on the articulation surface of the femur cartilage with (a) 0-, (c) 5-. and (e) 10-degree laterally wedged orthosis and contact pressure on articulation surfaces of the meniscus with (b) 0-, (d) 5-, and (f) 10-degree laterally wedged orthosis.

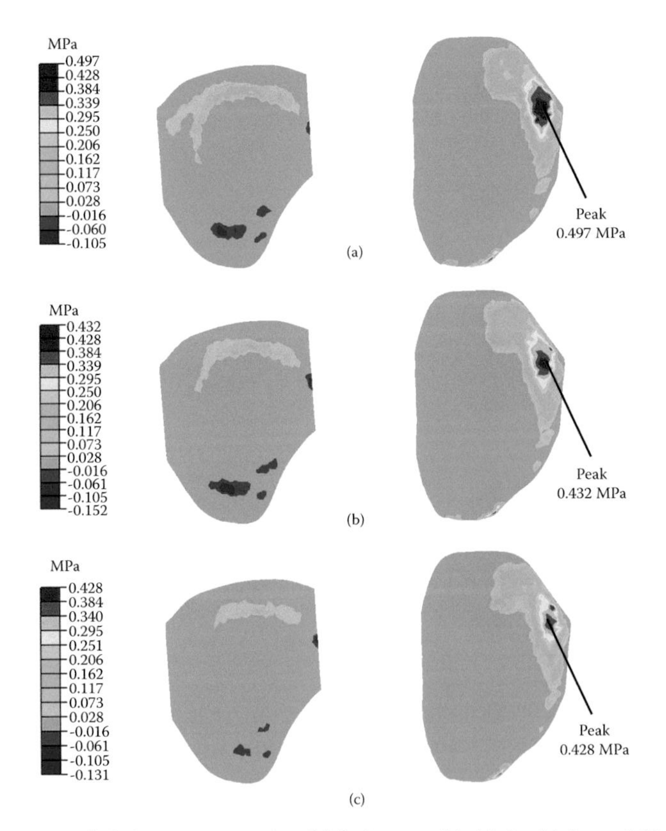


**FIGURE 12.6** Lateral-medial shear stress on the tibial plateau with (a) 0-, (b) 5-, and (c) 10-degree laterally wedged orthoses.

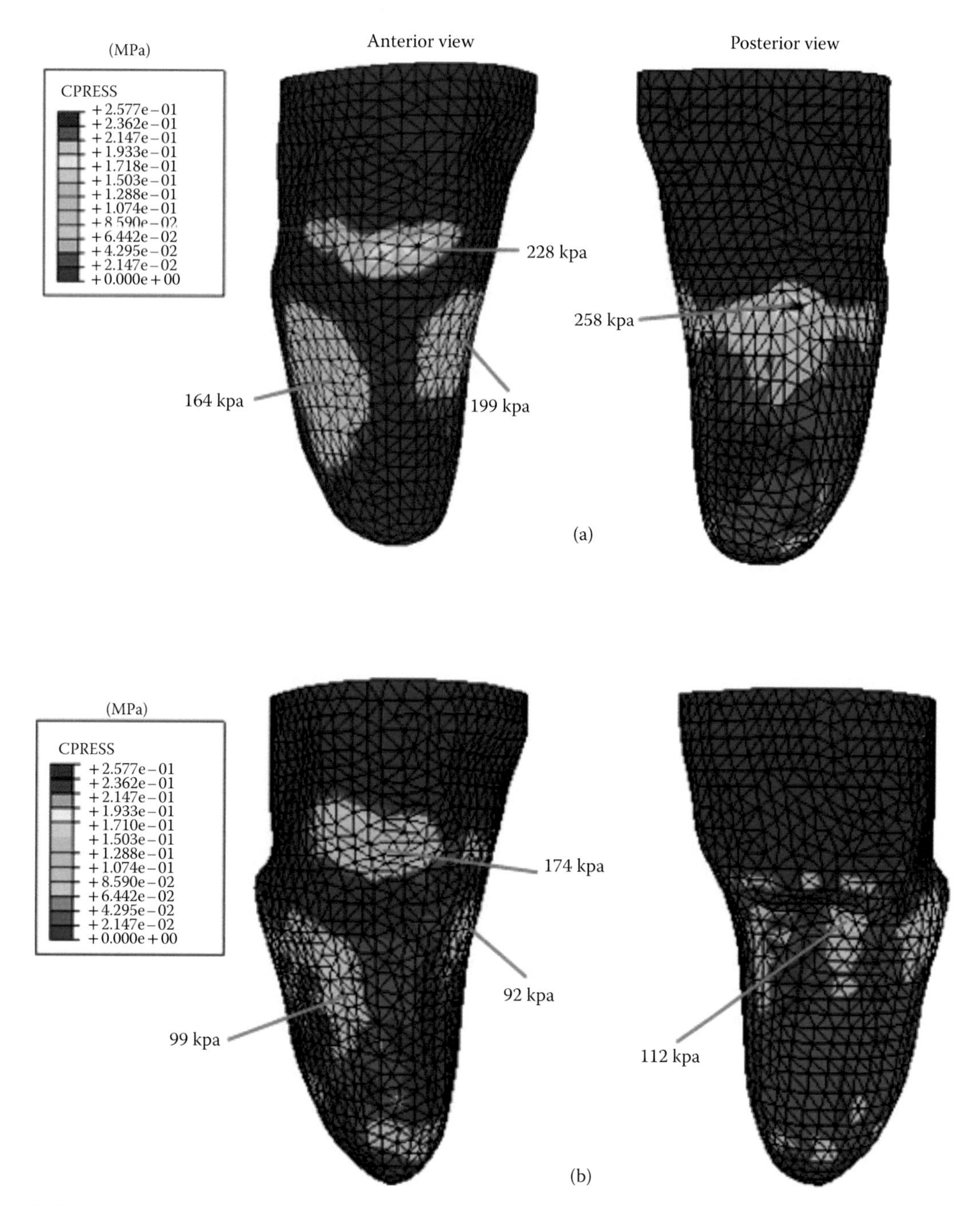


**FIGURE 13.4** Contact normal stress distribution (a) with pre-stress considered and (b) with pre-stress ignored upon application of 800 N.

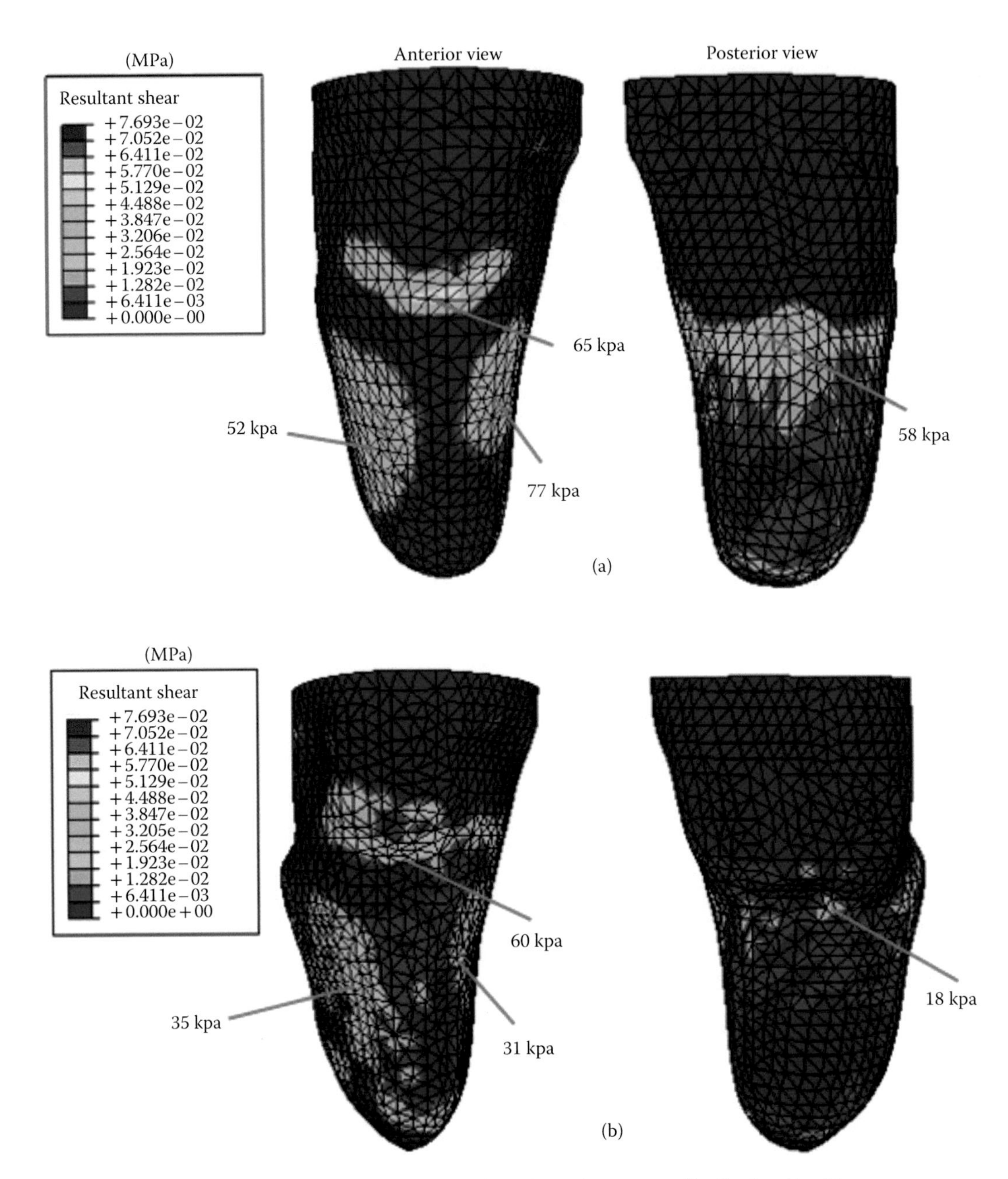


**FIGURE 13.5** The anterior and posterior views of resultant shear stress distribution (a) with pre-stress considered and (b) with pre-stress ignored upon application of 800 N.

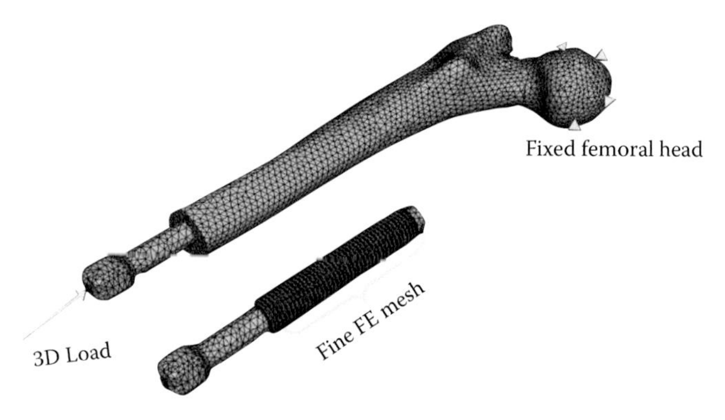


**FIGURE 14.1** The finite element model of the bone and the fixator (implant and abutment).

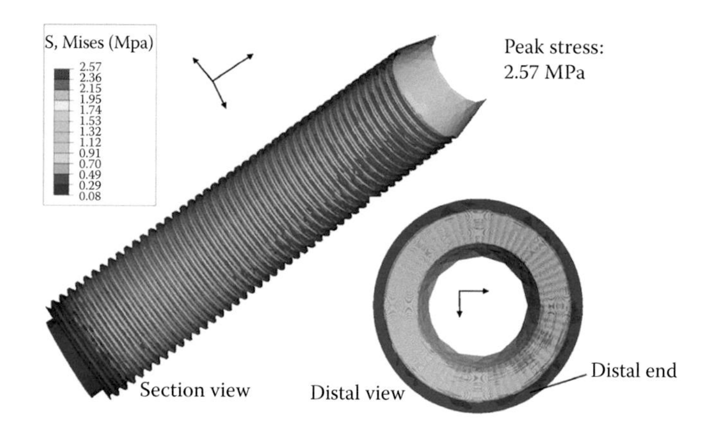


**FIGURE 14.2** Stress distribution at the bone adjacent to the implant at loading condition A.

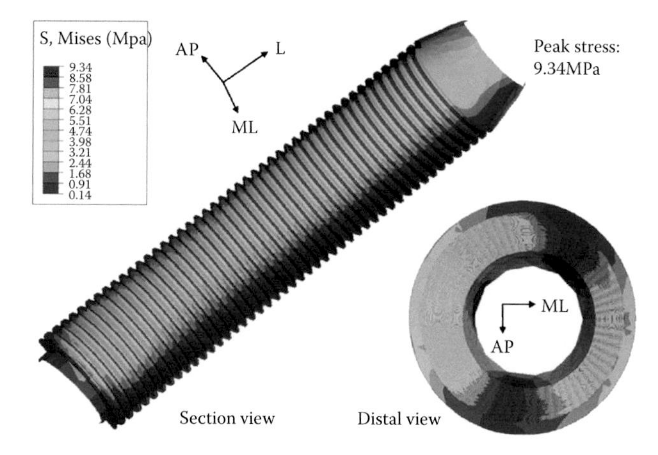


**FIGURE 14.3** Stress distribution at the bone adjacent to the implant at loading condition B.

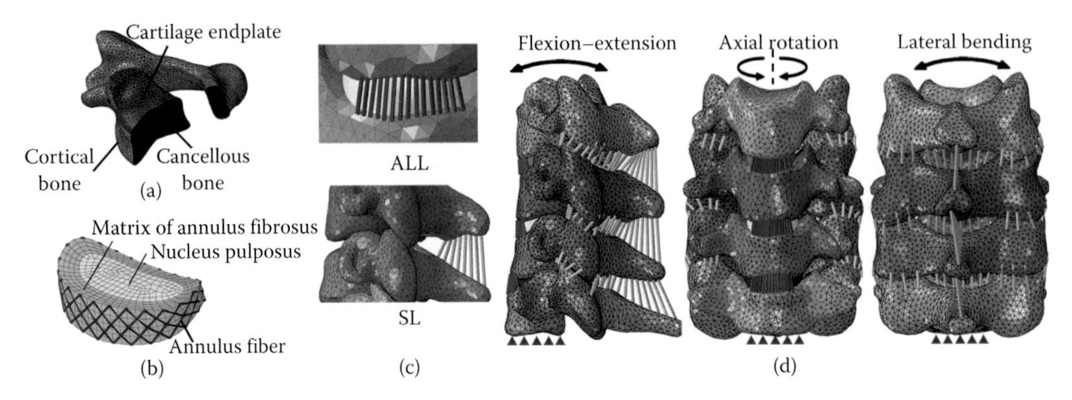


**FIGURE 16.2** Details of the finite element cervical model: (a) vertebra, (b) disc, (c) ligament insertion point, (d) entire cervical segment and illustration of spinal motion.

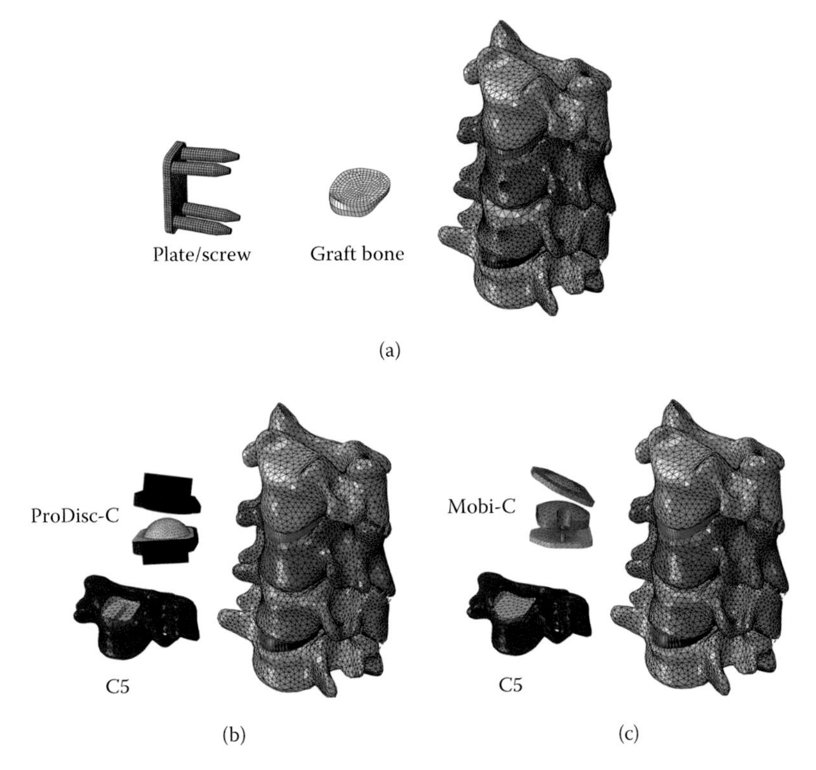


**FIGURE 16.4** Surgical details. (a) Anterior cervical decompression and fusion. (b) Total disc replacement with ProDisc-C. (c) Total disc replacement with Mobi-C.

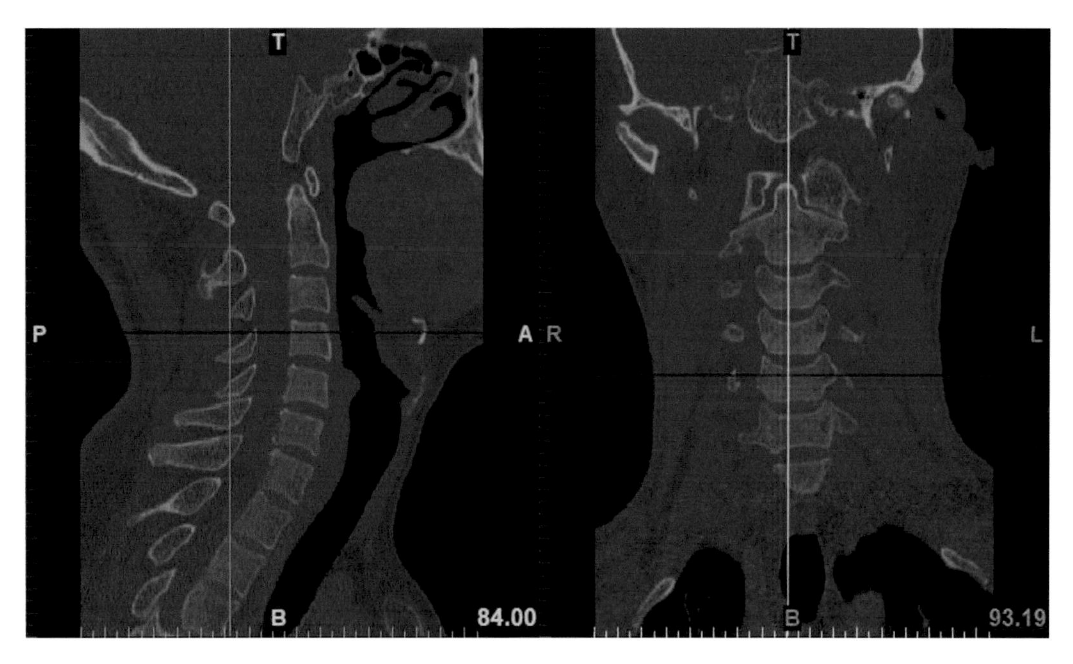


**FIGURE 17.5** CT images at 1-mm intervals on C3-C7 of a 35-year-old healthy male volunteer.

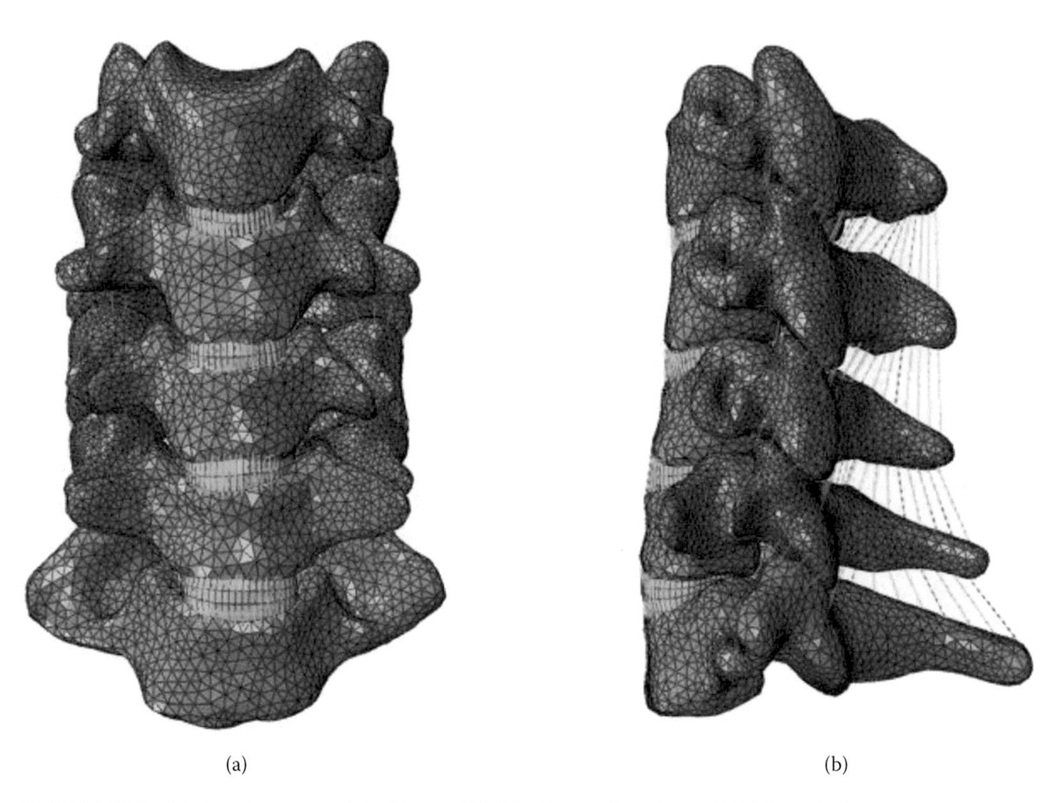


**FIGURE 17.6** Finite element model of intact C3-C7: (a) anterior view and (b) lateral view.

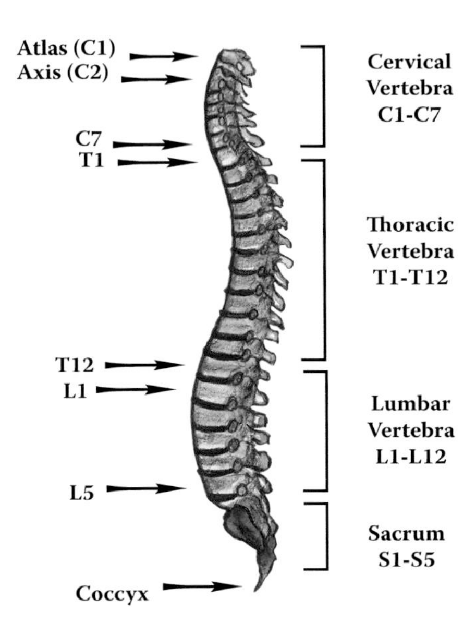


**FIGURE 18.1** Anatomical structure of the human spine.

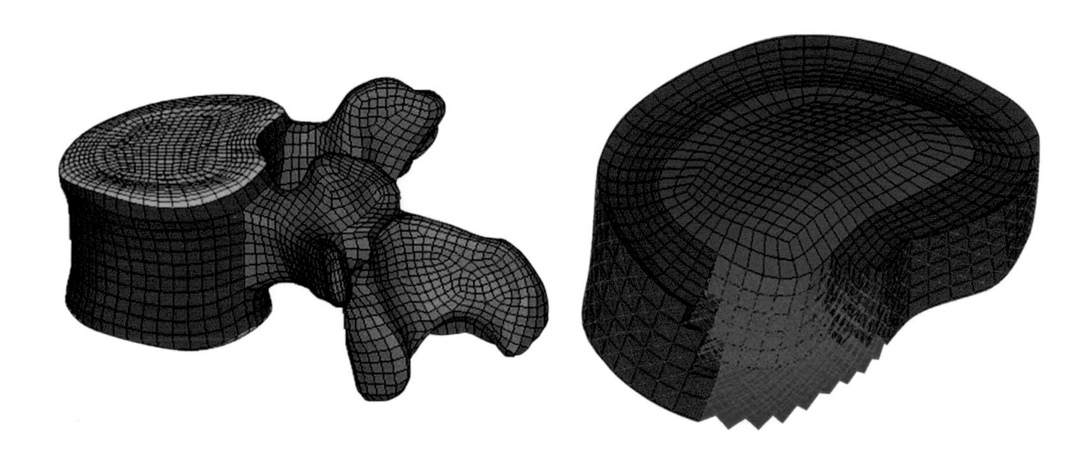

**FIGURE 18.7** Model of L2 vertebra (left) and disc of L2-L3 (right).

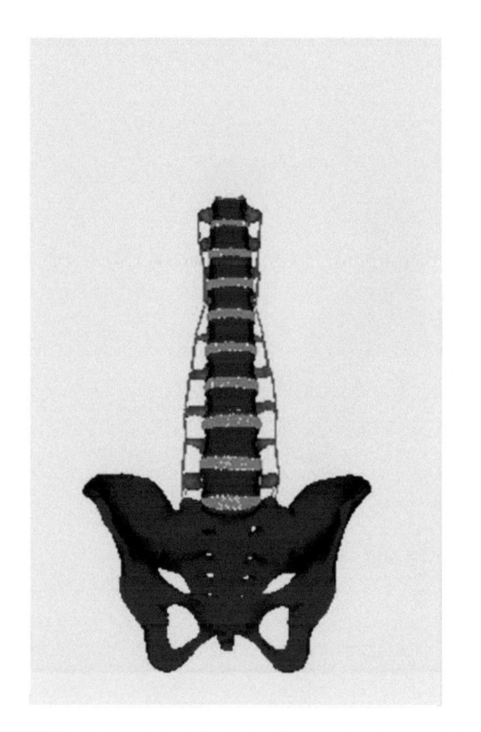

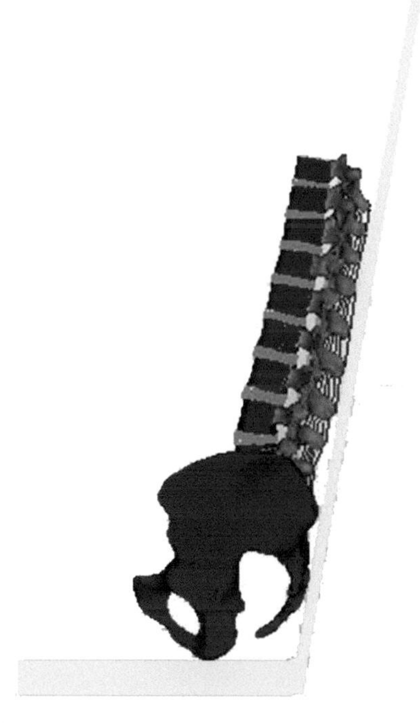

**FIGURE 18.8** Model of the thoracolumbar-pelvis complex along with the seat.

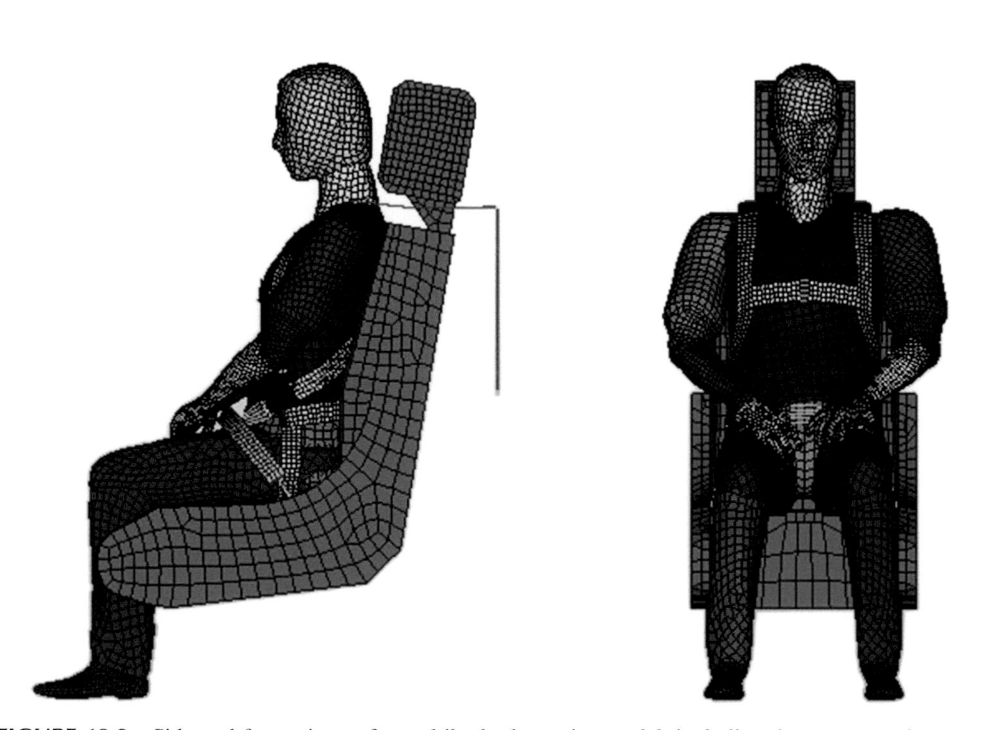

**FIGURE 18.9** Side and front views of a multibody dynamics model, including dummy, restraint systems, and seat.

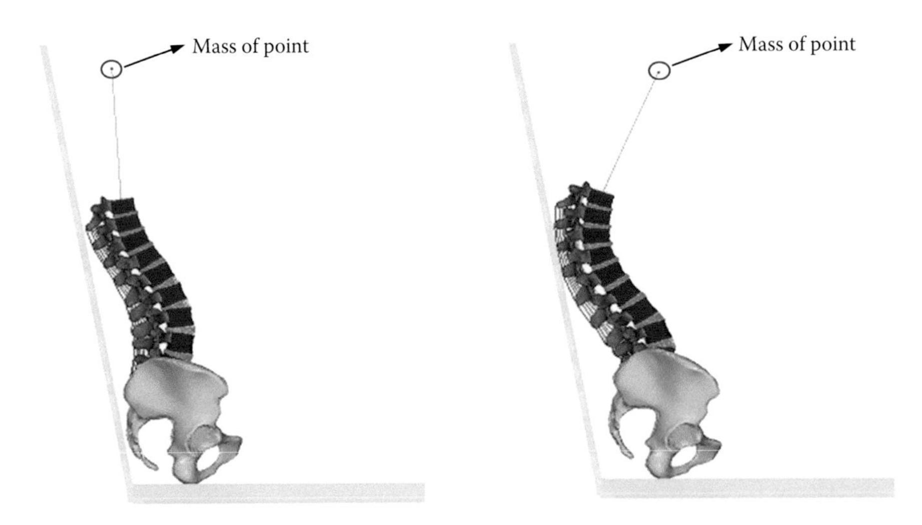


**FIGURE 18.14** Response of the thoracolumbar spine with (left) and without (right) the harness.

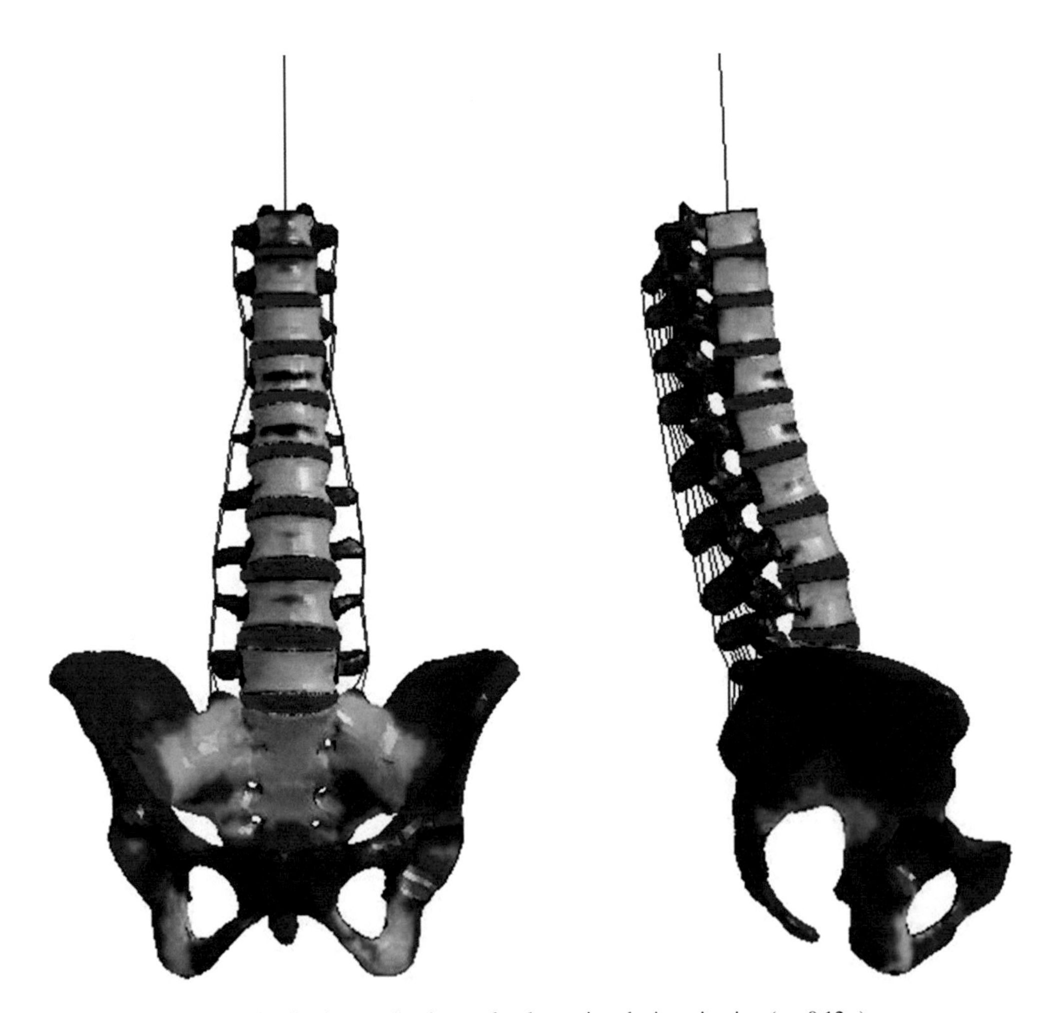

**FIGURE 18.16** Stress distribution on the thoracolumbar spine during ejection ($t$ = 0.12 s).

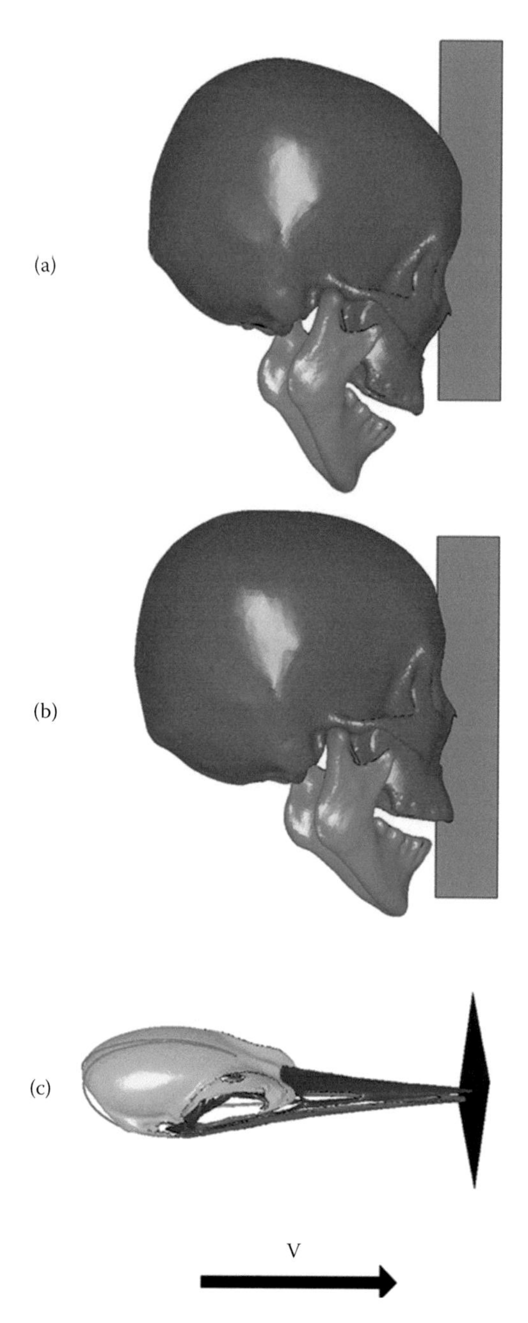


**FIGURE 19.4** Boundary conditions of the human and woodpecker head model: (a) human forehead collision, (b) human frontal collision, and (c) woodpecker pecking.

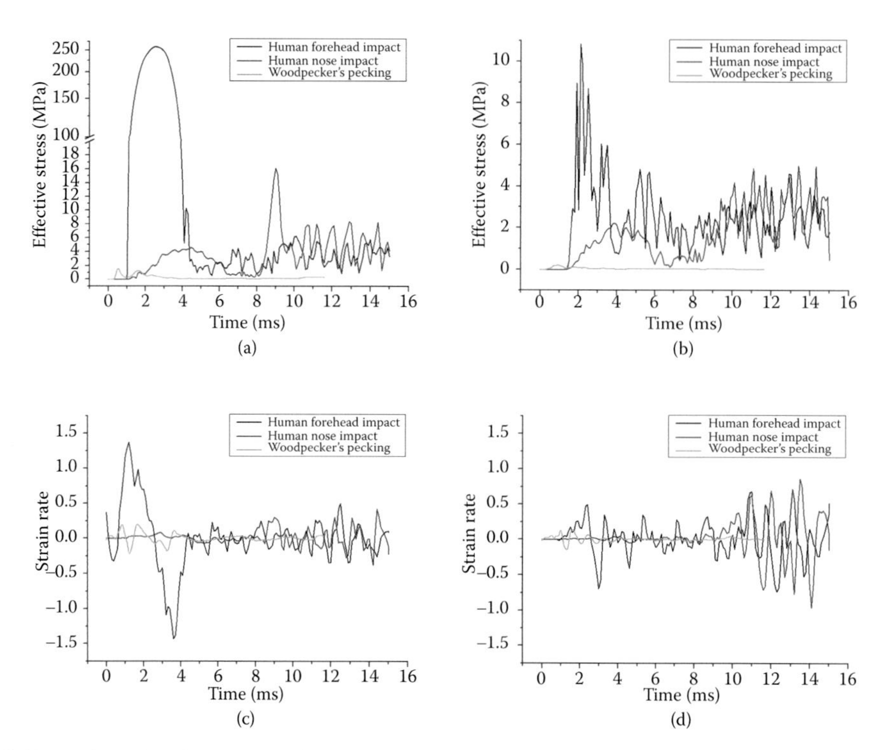


**FIGURE 19.5** Stress/strain rate time histories of the skull and brain for the human and woodpecker: (a) stress history on the forehead on the skull; (b) stress history on the occiput on the skull; (c) strain rate history on the forehead on the brain; and (d) strain rate history on the occiput on the brain.

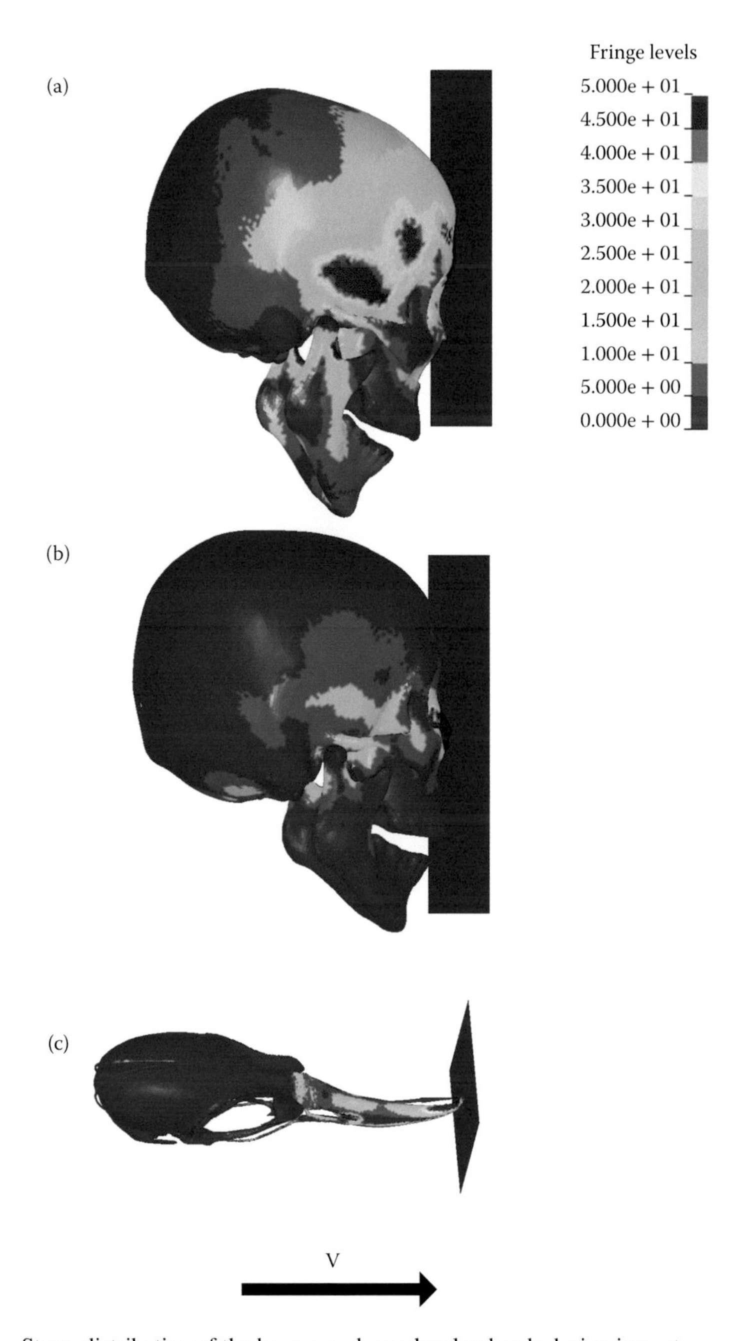


**FIGURE 19.6** Stress distribution of the human and woodpecker heads during impact.

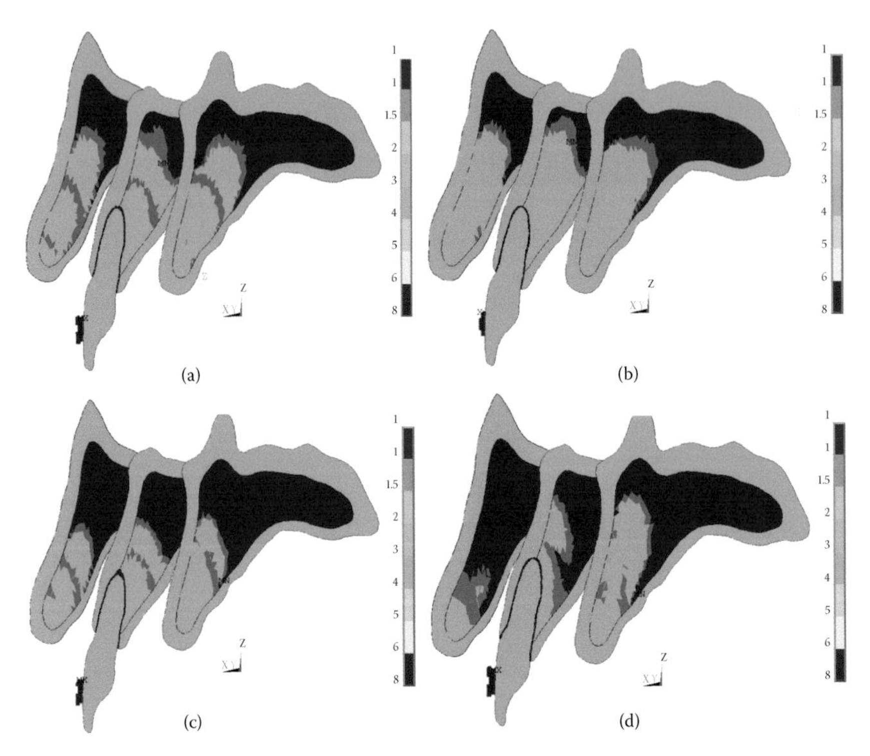


**FIGURE 20.5** Simulated distribution of apparent bone density (g/cm$^3$) under the different actions of orthodontic loading. (a) Tipping; (b) rotation; (c) extrusion; and (d) intrusion. The cross-section of the model is shown in each case.

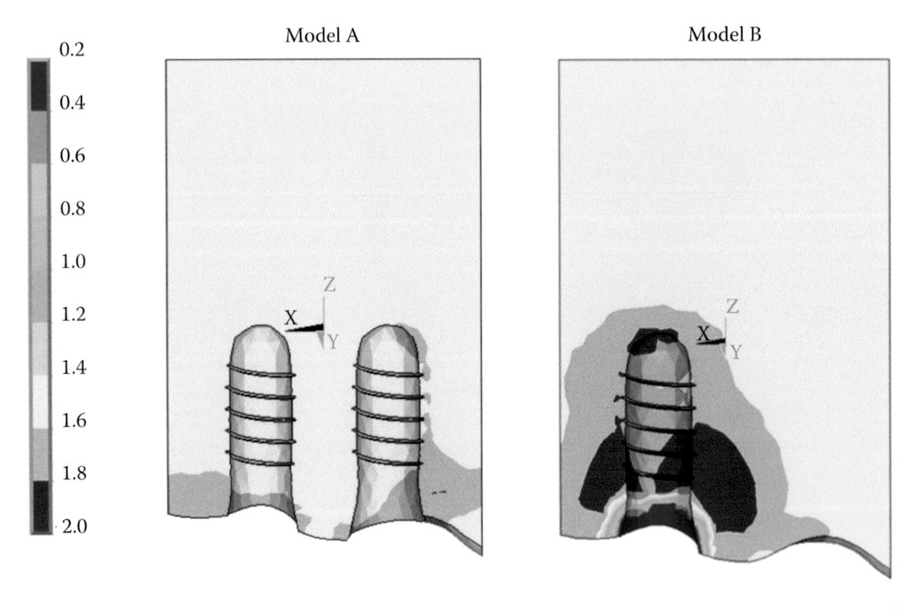


**FIGURE 20.6** Comparison of final results showing the variation of alveolar bone density (g/cm$^3$) under mechanical conditions. Model A: non-cantilever model; Model B: cantilever model.

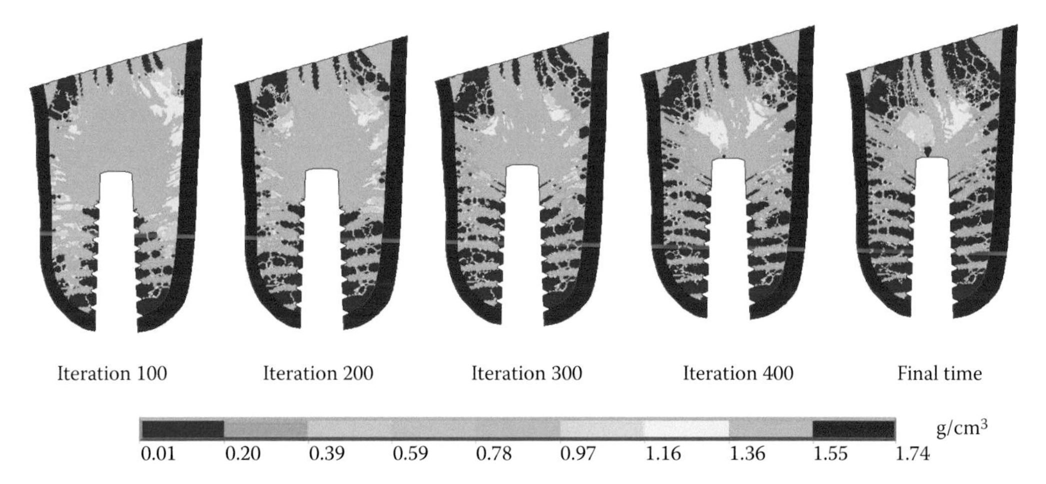


**FIGURE 20.7** Cancellous trabeculae morphology distribution after the 100th, 200th, 300th, 400th, and final time steps. Colors represent bone mineral density, shown in g/cm$^3$. Red is the cortical bone. The total number of iteration steps for this case to achieve the converged result is 510.

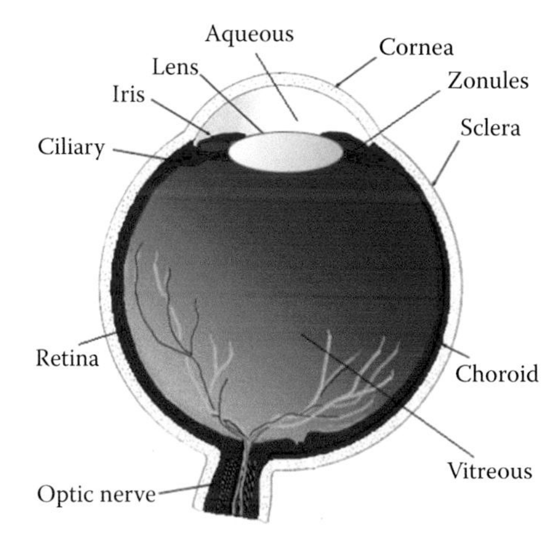


**FIGURE 21.1** Ocular structure: cornea, sclera, iris, ciliary body, zonules, lens, optic nerve, retina, choroid, and vitreous and aqueous humors.

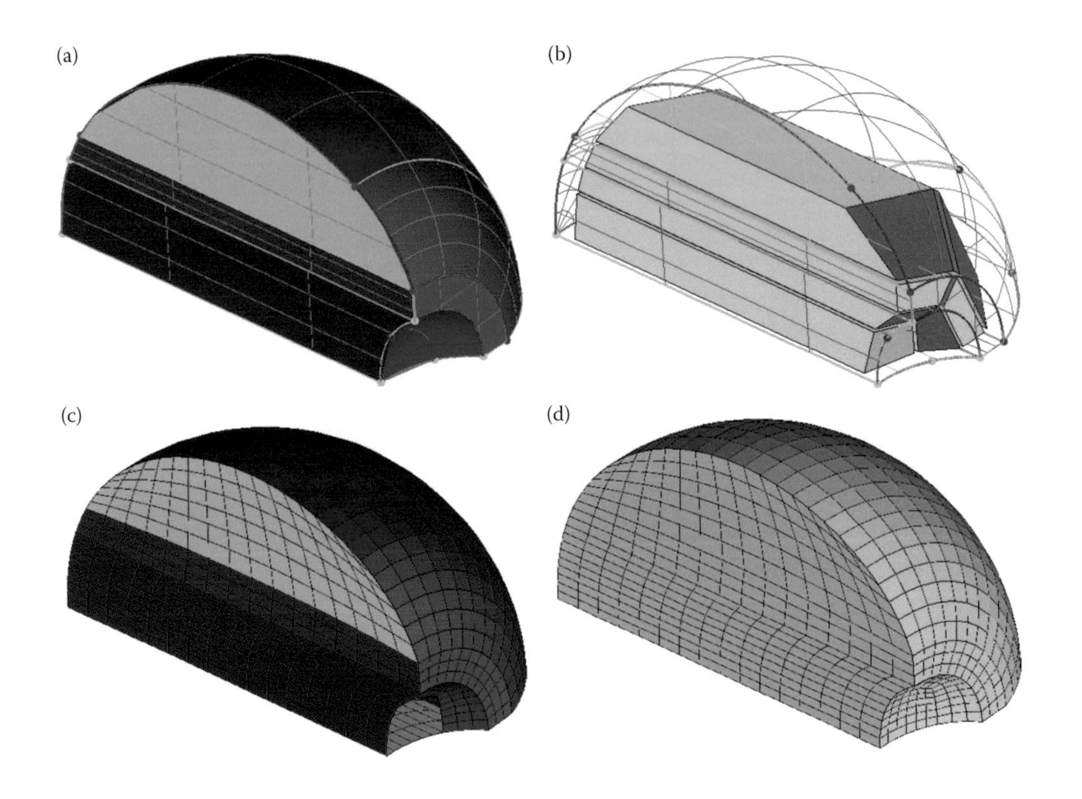


**FIGURE 21.5** Process of mesh generation for the vitreous model. (a) Surface creation. (b) Block division. (c) Mesh preview. (d) Mesh generation.

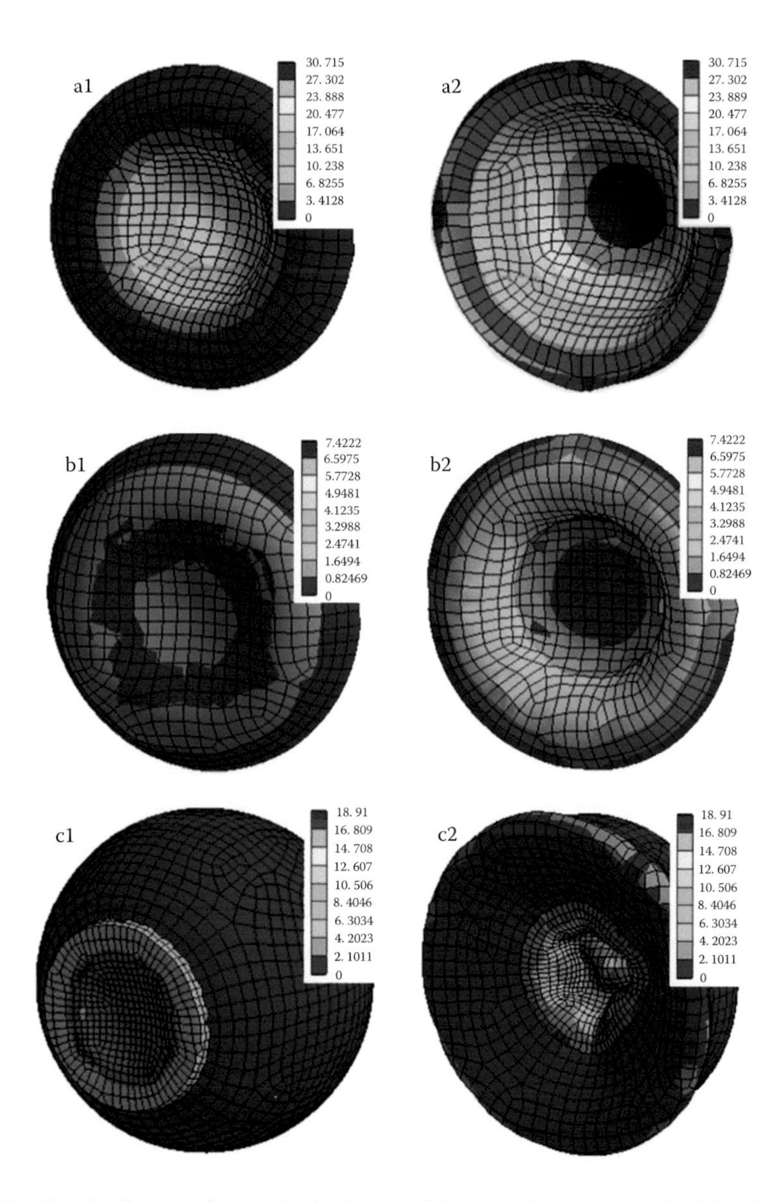


**FIGURE 21.7** The distribution of max principal stress of the globe in the six matched simulations: (a) stress distribution for low (left) and high (right) speed BB simulation; (b) stress distribution for low (left) and high (right) speed foam simulation; and (c) stress distribution for low (left) and high (right) speed baseball simulation.

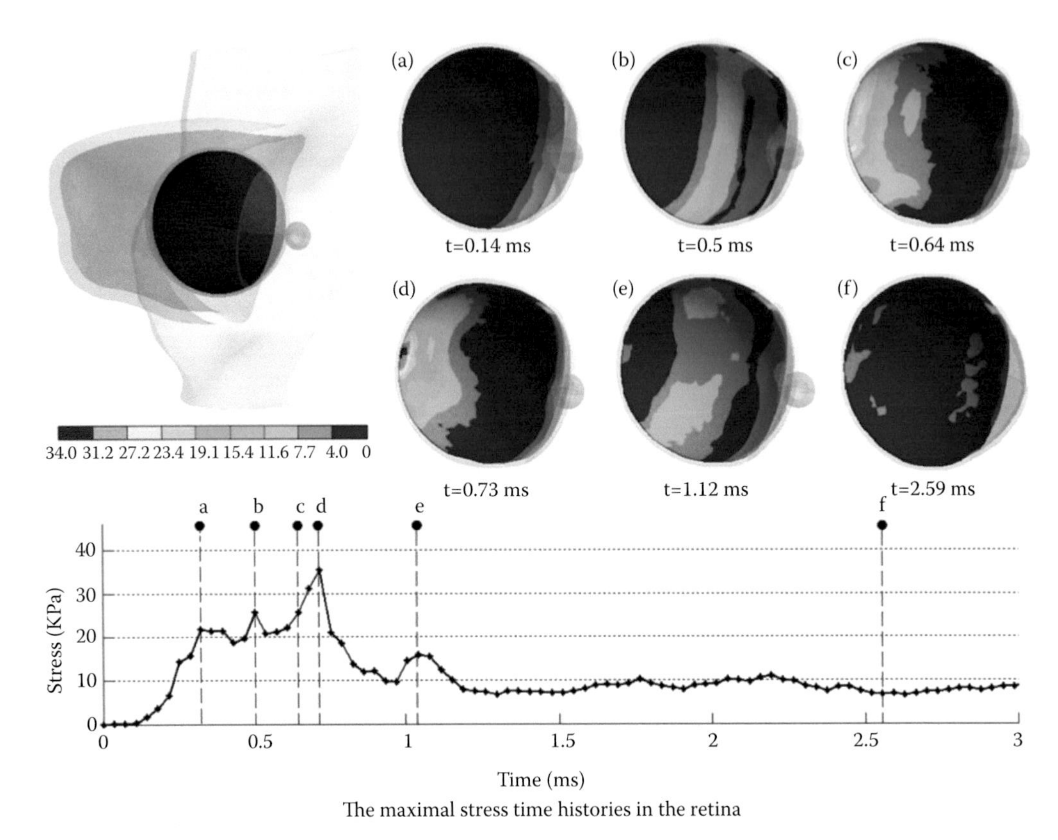


**FIGURE 21.9** The stress wave propagating in the retina (projectile speed of 62.5 m/s). The profile of the maximal stress time histories during the simulation from 0 to 3 ms was drawn. The stress distributions in the retina at six time nodes are illustrated.

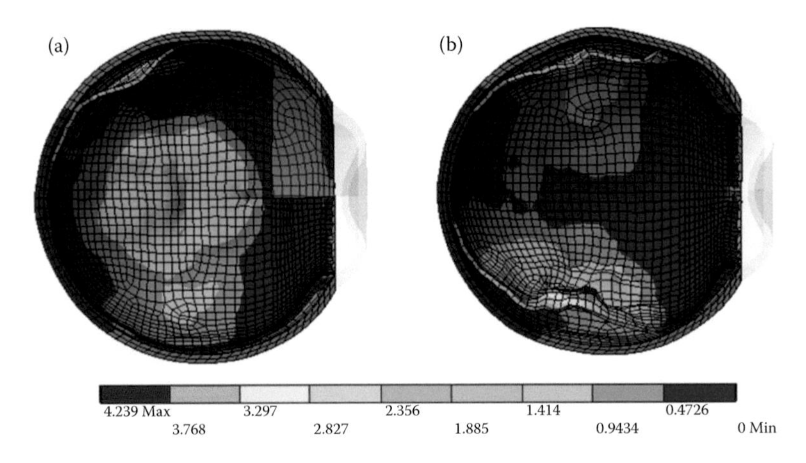


**FIGURE 21.10** Retinal detachments simulated at lower (a) and higher (b) impacting speeds by BB projectile.

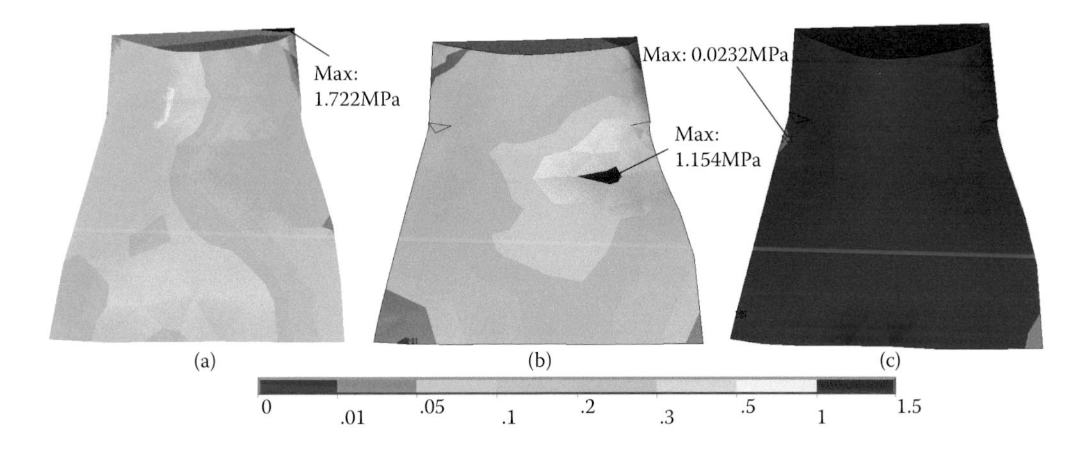


**FIGURE 22.2** The von Mises stresses in the left discs of the three models (the maximum stresses were listed): (a) the bond model, (b) the contact model, and (c) the gap model.

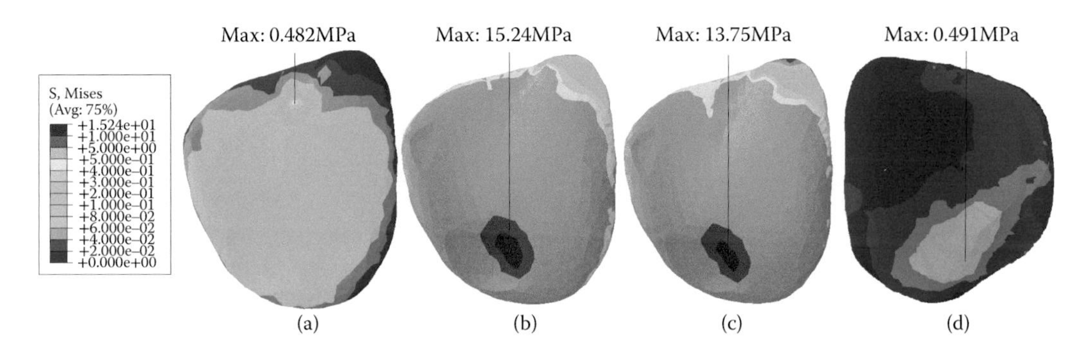


**FIGURE 22.5** The von Mises stress distributions of the disc under the three occlusions (the maximum stresses were listed): (a) central occlusion (left disc), (b) anterior occlusion (left disc), (c) right side molar occlusion (left disc), and (d) right side molar occlusion (right disc).

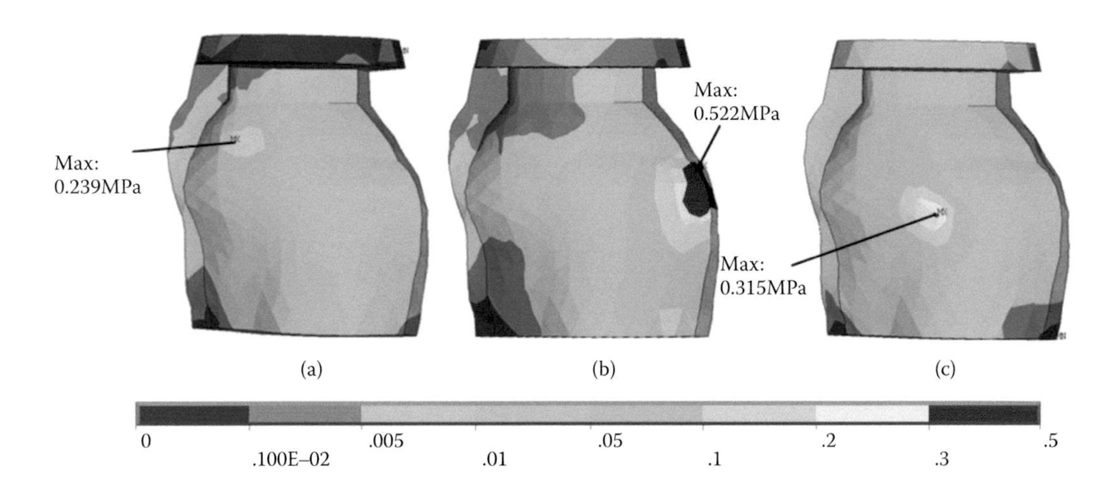


**FIGURE 22.7** The von Mises stress distributions of the normal disc and the discs with relaxed attachment (the maximum stresses were listed; units: MPa): (a) normal disc, (b) disc with relaxation of anterior attachments, and (c) disc with relaxation of bilaminar zones.

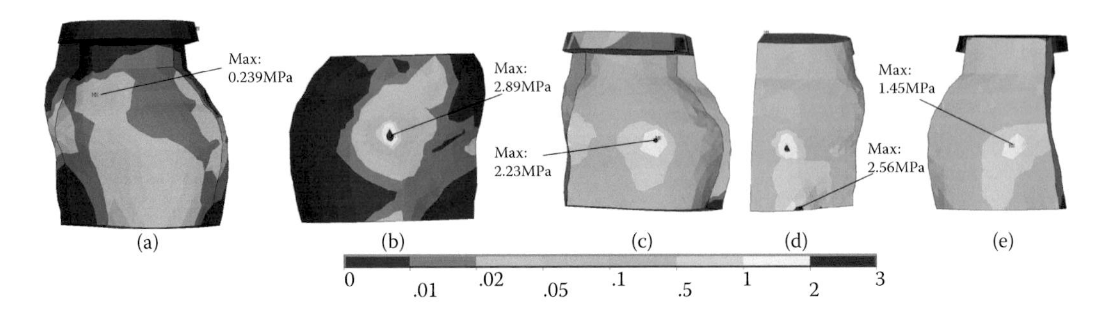


**FIGURE 22.8** The von Mises stress distributions of the discs in models with disc displacements (the maximum stresses were listed; units: MPa): (a) normal disc, (b) anteriorly displaced disc, (c) posteriorly displaced disc, (d) medially displaced disc, and (e) laterally displaced disc.

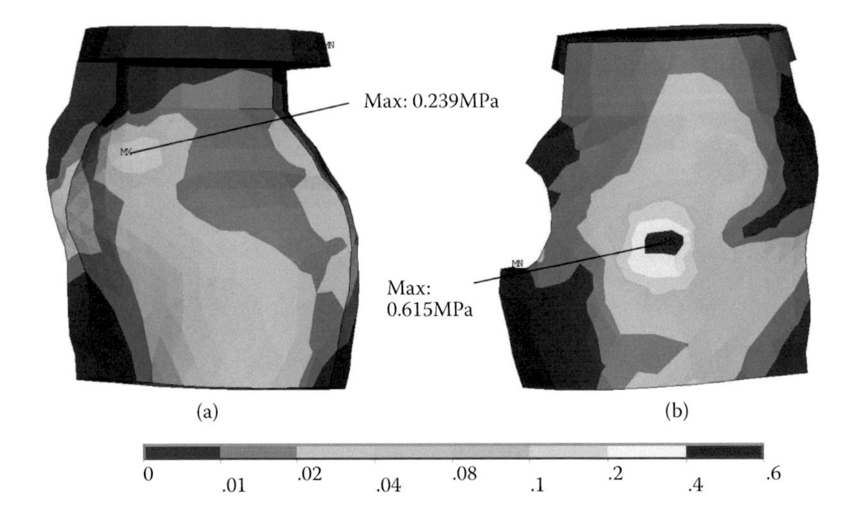


**FIGURE 22.9** The von Mises stress distributions of the normal and disc perforation models (the maximum stresses were listed; units: MPa): (a) normal disc and (b) perforated disc.

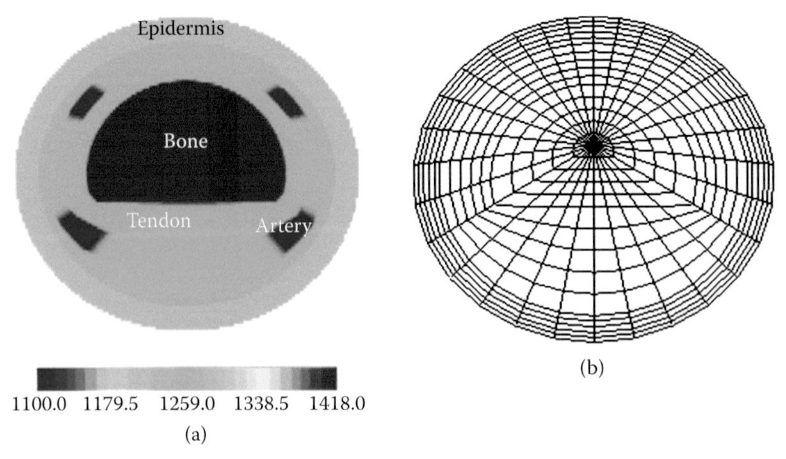


**FIGURE 23.2** (a) Different materials in the finger and (b) a mesh network of the cross section of the finger.

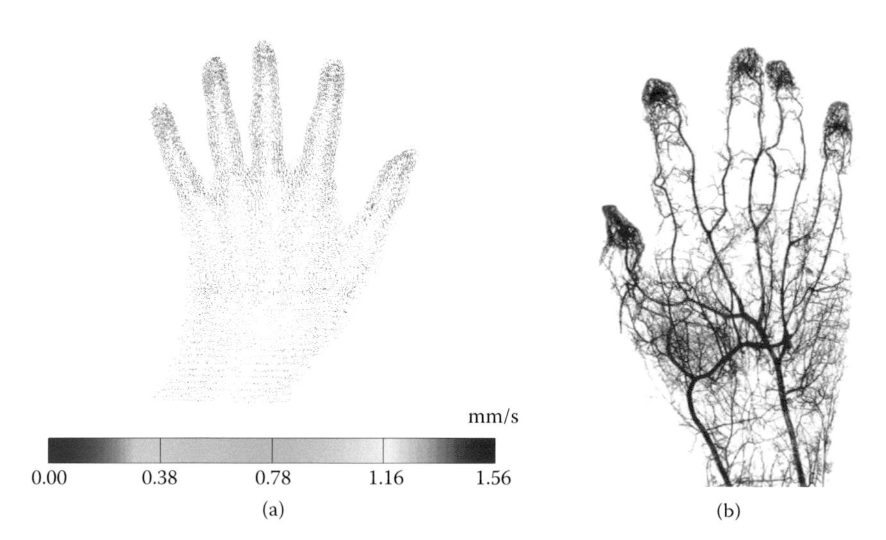


**FIGURE 23.8** (a) Simulated blood perfusion of the human hand. (Reproduced from Shao HW et al., *Computer Methods in Biomechanics and Biomedical Engineering*, 2012. With permission.) (b) Vasculature in the human hand. (From http://www.zcool.com.cn/ZMjEzNjg4.html.)

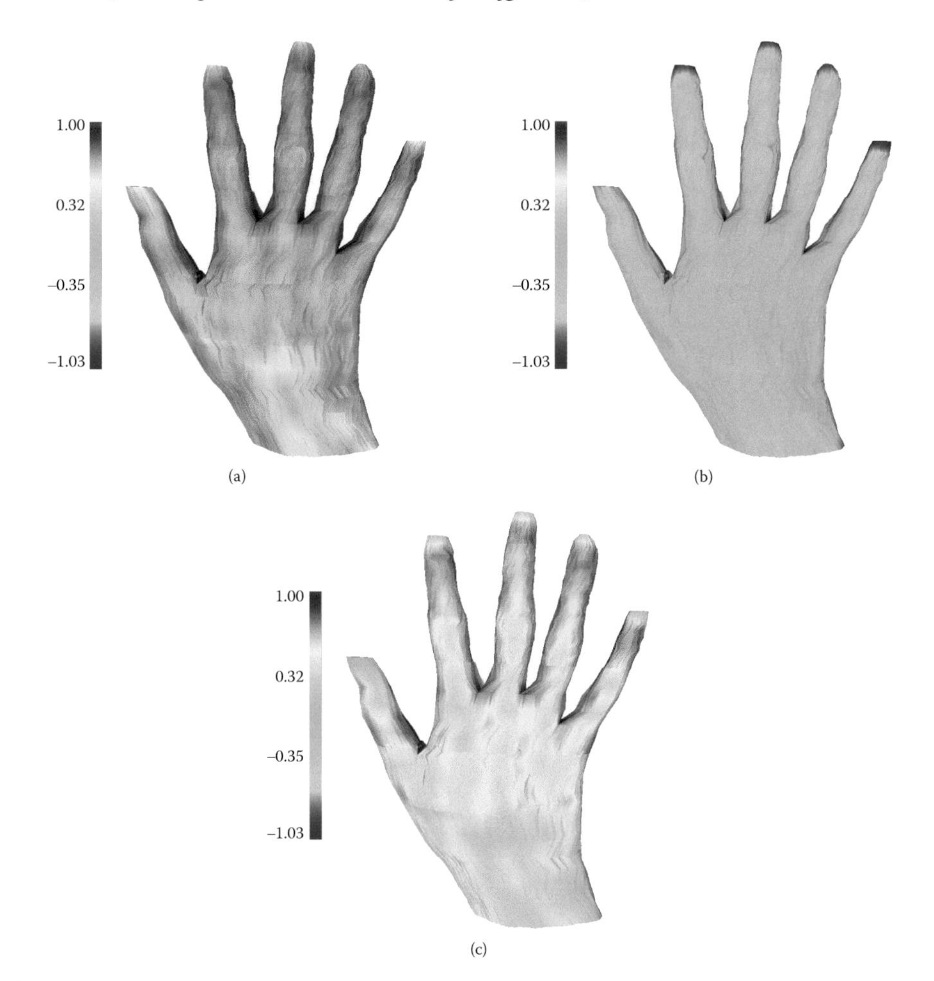


**FIGURE 23.13** Skin temperature distribution in different stages of the cooling test. (a) At 600 s. (b) At 780 s. (c) At 1350 s.

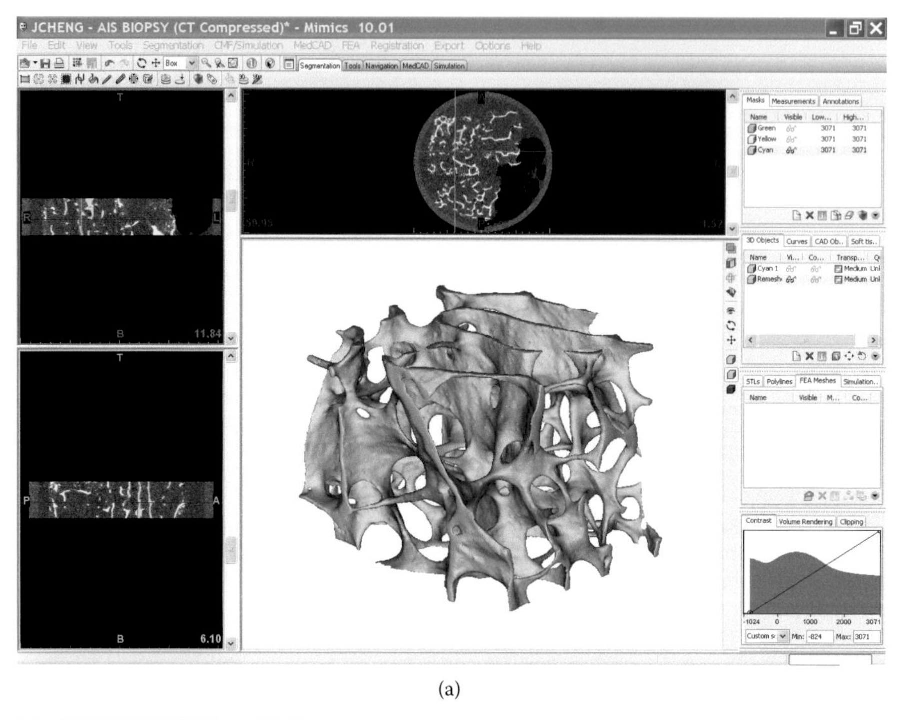


(a)

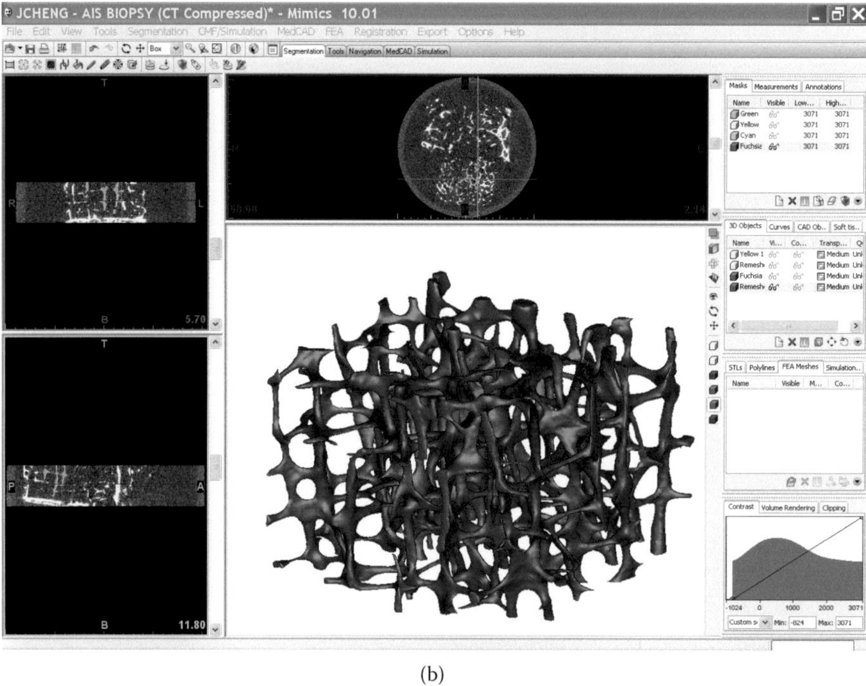


(b)

**FIGURE 24.1** The three-dimensional models generated in MIMICS software: (a) trabecular specimen A and (b) trabecular specimen B.

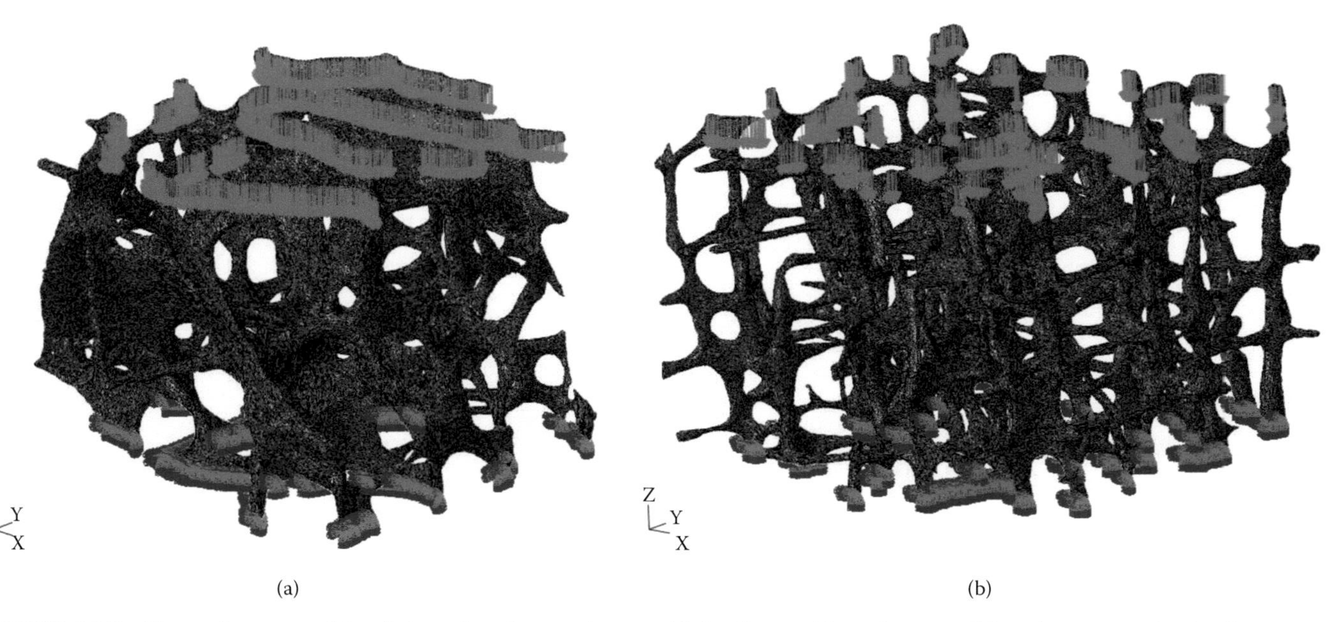


**FIGURE 24.2** Finite element models of the trabecular specimens with loading and boundary conditions in compressive loading conditions in the longitudinal direction, respectively: (a) trabecular specimen A and (b) trabecular specimen B.

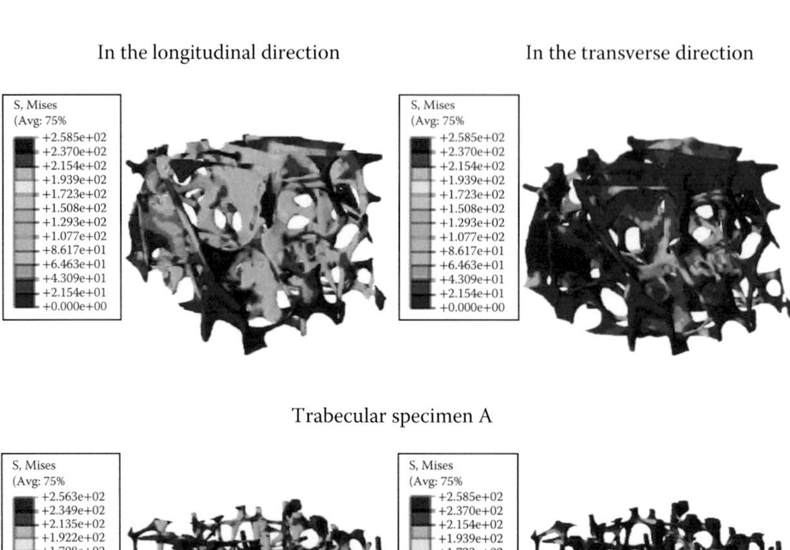


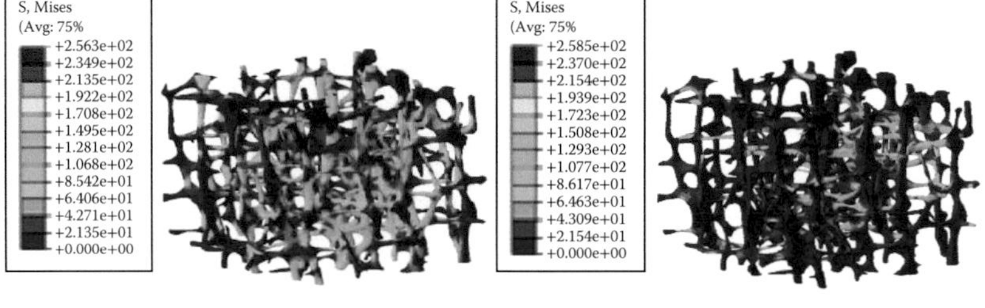


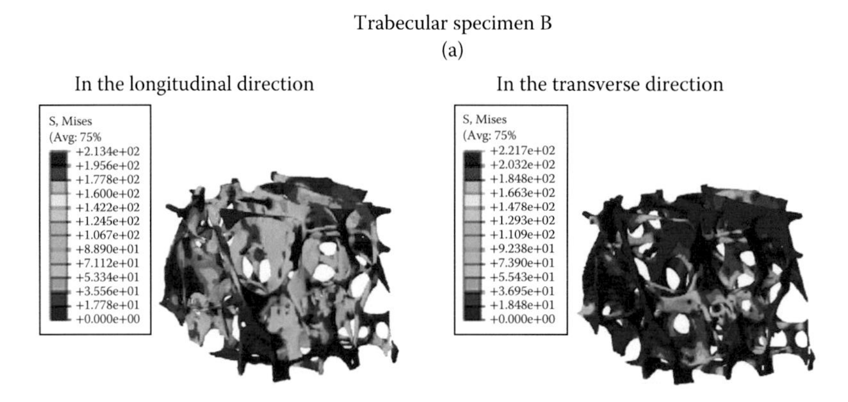


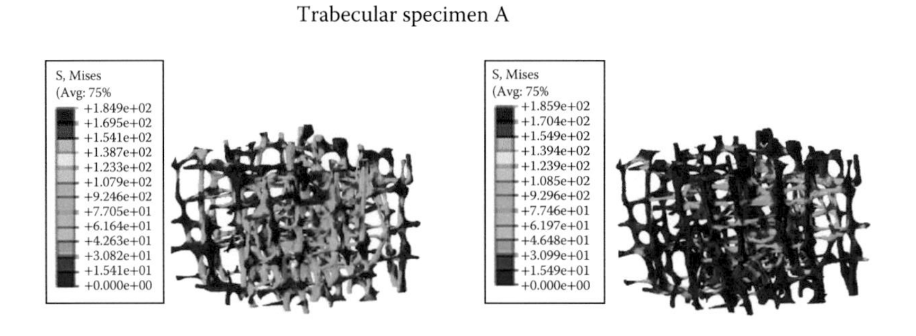


**FIGURE 24.4** von Mises stress distributions of the trabecular specimens at the yield point in apparent compressive and tensile loading conditions in the longitudinal and transverse directions, respectively: (a) compression and (b) tension.

# 10 Hip Model for Osteonecrosis

*Duo Wai-Chi Wong, Zhihui Pang, Jia Yu, Aaron Kam-Lun Leung, and Ming Zhang*

## CONTENTS



## SUMMARY

Osteonecrosis of the femoral head (ONFH) is one of the common pathologies of the hip joint (Hernigou and Beaujean 2002). Femoral head collapse is the most devastating stage of ONFH. The surgical failure rate of post-collapse patients remains very high. Nearly all femoral head collapse patients require additional surgery or total hip replacement (Hernigou and Beaujean 2002; Lieberman, Conduah, and Urist 2004; Wirtz, Rohrig, and Neuss 2003; Yoon et al. 2001). Joint-preserving modalities are not adequate to prevent and delay the deterioration of the collapsed hip (Lieberman et al. 2012).

Joint-preserving surgical treatment during the early stages of ONFH is often recommended (Mont et al. 2010), including deep compression, vascularized bone grafting, and tantalum rod insertion. Though satisfactory outcomes have been reported, studies have been inconsistent in their reporting of patient satisfaction and success (Chen et al. 2009; Fairbank et al. 1995; Markel et al. 1996).

In this section, the biomechanical nature of ONFH, such as lesion size, angle, material, and weight-bearing characteristics, and its associated operation, is investigated (Yoo et al. 2008; Mont et al. 1998; Yoshida et al. 2006; Brown, Way, and Ferguson 1981). Computational simulations were also undertaken on an osteonecrotic hip (Brown, Mutschler, and Ferguson 1982; Daniel et al. 2006; Yang et al. 2002). Nevertheless, the optimal modality and when to carry out the surgery for ONFH remain troubling points for clinicians (Hungerford and Jones 2004; Lieberman et al. 2003). A platform must be established to systematically evaluate the pathomechanism and intervention strategies for ONFH.

## 10.1 INTRODUCTION

### 10.1.1 Background

Osteonecrosis is also known as avascular necrosis or ischemic necrosis and is commonly found on the femoral head (Anderson 2006). ONFH is characterized by poorly localized pain at the buttocks, groin, and the medial side of the thigh. The problem results from temporary or permanent disruption of blood supply on the femoral head, causing ischemic death of osteocytes and bone marrow within 6 to 12 hours after insult. The phenomenon was also analogized as "the coronary disease of the hip" by Chandler (1948). Yet, avascular necrosis and subsequent femoral head collapse could take from several months to years to progress, making it difficult to diagnose until the disease has progressed to a severe state.

The strength of the femoral head is reduced as trabeculae thin out and the new favorability for bone resorption leaves the bone with inadequate mechanical strength. The situation is further exasperated as the reparability of the bone is impaired. Because the femoral head is one of the main weight-bearing regions of the body, the normal functioning of the joint could impose a high load on the femoral head and lead to subsequent fatigue or fracture of the enervated bone. It is also believed that the zone around the necrotic bone experiences abnormal loading that causes additional microfractures and lesions. Eventually, the femoral head collapses and, with the accompanying osteoarthritis, the patient becomes practically immobile. Total hip replacement is typically recommended at this stage. Nevertheless, such procedures impose a heavy economic and social burden for young patients.

In the pre-collapse stage, nonoperative interventions, such as physical therapy, reduced weight bearing, and medications, are prescribed, though proven to be ineffective (Amanatullah, Strauss, and Di Cesare 2011a; Mont and Hungerford 1995). Although studies on bisphosphonates prescription and extracorporeal shockwave therapy demonstrate reduced pain and delayed progression of ONFH, none of them could provide concrete evidence to prevent progression (Agarwala et al. 2005; Agarwala, Shah, and Joshi 2009; Massari et al. 2006; Wang et al. 2008). Surgical intervention, such as core decompression, vascularized bone grafting, and porous tantalum rod insertion, produce better clinical outcomes and survivorship (Amanatullah, Strauss, and Di Cesare 2011b).

### 10.1.2 Prevalence

ONFH is a common consequence of other diseases or complications. An estimation of prevalence is difficult to determine. Moreover, the onset of symptoms could take some time before a patient seeks medical advice (Fordyce and Solomon 1993).

An early prevalence study by Mont and coworkers estimated that there were about 100,000 to 200,000 new cases of ONFH in the United States each year (Mont and Hungerford 1995; Mont et al. 1998). Hirota et al. (1993) reported an annual incidence of 2500 to 3000 idiopathic cases of ONFH in Japan. Another nationwide epidemiology study conducted by Kang et al. (2009) concluded that there was a prevalence rate of 28.91 cases per 100,000 people. Kang's study analyzed the period from 2002 to 2006 and showed that the rate of ONFH increased each year. However, the author suggested that the increasing trend could also be due to earlier diagnosis being made possible through advances in health care. The prognosis for ONFH was poor and there was a high rate of total hip replacement; around 5% to 18% of patients required total hip arthroplasties (Mankin 1992; Mont, Carbone, and Fairbank 1996; Urbaniak and Harvey 1998). Some studies reported a rate as high as 50% to 60% (Kim and Rubash 2007).

Men are more susceptible to ONFH than women, by a ratio of seven to three, excluding the independent risk of systemic lupus erythematosus (Assouline-Dayan et al. 2002). More than 70% of ONFH cases will proceed to bilateral involvement within two years (Malizos et al. 2007). The disease can affect young adults (Lieberman et al. 2003, 2012; Petrigliano and Lieberman 2007) and imposes a heavy socioeconomic burden if total hip arthroplasty is undertaken. In nontraumatic ONFH, alcoholism and steroid use contribute a predominant risk to the disease. Excessive alcoholism was seen in 22% to 74% of ONFH patients, while 18% to 35% of the patients were enrolled in steroid therapy (Hirota et al. 1993; Merle D'Aubigne et al. 1965). Steroid use was the most prevalent risk factor, as presented in epidemiology studies in Japan and France (Kang et al. 2009; Merle D'Aubigne et al. 1965).

### 10.1.3 Etiology and Diagnosis

ONFH is not a specific kind of disease, but is a condition that can result from different diseases or problems. The causes of ONFH can be classified into two groups: traumatic and nontraumatic, with traumatic causes being more frequent (Anderson 2006). In traumatic ONFH, the blood circulation of the femoral head is affected by blunt force trauma to the area. For example, patients with a hip dislocation could suffer vessel interference from the ligamentum teres.

The source of nontraumatic ONFH remains unclear, though multifactorial origins have been suggested. A disorder in lipid metabolism is believed to be the major contributing etiology. Excessive alcoholism predisposes the patient to a risk of developing ONFH by blocking the blood vessels with lipids, while diseases, such as sickle cell disease, Gaucher's disease, and pancreatitis, contribute in a similar lipid coagulation pathway. Different pathomechanisms have been suggested to produce an insufficient vascular supply.

ONFH is ironically called the "disease by medical advancement" because of the high and growing prevalence of glucocorticosteroid-induced ONFH. Although it could similarly induce disorders in lipid metabolism, it is argued that embolism or thrombosis are not the major causative factors. Another theory suggests that the increased size of marrow cells induced by medication could increase intraosseous pressure and eventually lead to vessels rupturing, as demonstrated in Caisson disease-induced ONFH.

Radiographic evaluation could help to diagnose and classify the severity of the disease. Anterior-posterior and frog-leg plain x-rays are used to identity fracture, crescent sign, and collapse of the bone initially, while magnetic resonance imaging (MRI) can provide more evidence in symptomatic patients when x-ray scans cannot distinguish the problem. The size of the lesion is also another essential index to classify the stage of ONFH. The Ficat-Arlet classification is one of the common standards in the evaluation of ONFH.

### 10.1.4 Natural History

Femoral head collapse is the most devastating stage of ONFH. The surgical failure rate of patients post-collapse remains very high. Nearly all patients with a collapsed femoral head require additional surgery or total hip replacement (Hernigou and Beaujean 2002; Lieberman, Conduah, and Urist 2004; Wirtz, Rohrig, and Neuss 2003; Yoon et al. 2001). Some studies have reported failure rates exceeding two-thirds of patients (Chen et al. 2009; Kim and Rubash 2007; Langlais et al. 2004). Once the joint collapses, joint-preserving modalities cannot prevent or delay the deterioration of the collapsed hip (Lieberman et al. 2012). Thus, identifying the characteristics and progression pathway of ONFH is especially important. Though joint-preserving operations have been suggested as prophylactics (Amanatullah, Strauss, and Di Cesare 2011b), patients and surgeons are reluctant to resort to surgery because of the additional risks it carries, especially for asymptomatic patients.

In 1993, Takatori et al. reported about one-quarter of patients diagnosed with ONFH progressed to femoral head collapse in a follow-up study of 21 months. However, Bradway and Morrey (1993) in the same year showed that all patients with ONFH progressed to a symptomatic state and then all eventually underwent collapse. Nam et al. (2008) and Kubo et al. (1997) extended the follow-up study to 103 and 52 months, respectively, and presented a collapse rate of about one-third. A study with a larger sample size by Hernigou et al. (2006) showed a high collapse rate of up to 77%. A meta-analysis conducted by Mont et al. (2010) revealed that about 60% of ONFH patients would deteriorate. After the diagnosis of asymptomatic ONFH, it was suggested that the disease would progress to symptomatic within 39 months and in an additional 10 months could pose a risk of collapse (Mont et al. 2010).

The progression of ONFH could depend on the severity of the disease and the nature of the lesion. Hungerford and Jones (2004) reported that a small lesion of a size less than 15% of the femoral head was unlikely to progress. Another suggestion was that a large lesion of a size more than 30% of the femoral head was also unlikely to progress, or may be successfully treated by decompression. Conversely, Hernigou et al. (2006) examined patients with small asymptomatic lesions and reported 88% progressed further and 73% led to collapse. A systematic review of the modality of ONFH is difficult due to the different and wide variability in methodologies and the complicated nature of the disease. For example, asymptomatic ONFH has been presented with severe clinical findings (stage III collapse; Mont et al. 2010).

### 10.1.5 Review of Biomechanical Studies

The weight-bearing nature of the hip joint relates to the mechanical characteristics of the femoral head and the acetabulum. The surface of the joint is covered by articular cartilage, while the cartilage on the acetabulum is lunar in shape, surrounded by a thick labrum. This configuration, together with the collateral ligaments, maintains the stability of the hip joint. In low loading, the articular surfaces are not in total contact. However, the surfaces come into contact with high loading to redistribute the localized pressure.

The loading-bearing region of the femoral head was documented by Chen et al. (2005). In an approximated primitive femoral head model, the weight-bearing region during standing was presented as a cone, while the apex of the cone was the center of the spherical femoral head; the cone region was 40° to the femoral neck axis (Chen et al. 2005). Under gait, the weight-bearing region extended 40° in the flexion direction, 5° in the direction of extension, and 5° in the abduction and adduction directions (Chen et al. 2005).

The nature of the weight-bearing region was further studied by Yoshida et al. (2006), who made use of a computational model to evaluate the contact pressure on the acetabulum cartilage. Yoshida et al. (2006) divided the loaded region into four main regions: lateral, anterior, medial, and posterior.

Biomechanical evaluations have also been carried out on necrotic bone. Brown and coworkers (Brown, Baker, and Brand 1992; Brown and Hild 1983; Brown, Mutschler, and Ferguson 1982; Brown, Way, and Ferguson 1981) have extensively studied the material and mechanical properties of ONFH. A material examination of the necrotic and collapsed trabecular bone revealed a reduction of 52% in yielding strength and 72% in elastic modulus (Brown, Way, and Ferguson 1981). Brown, Mutschler, and Ferguson (1982) also conducted finite element (FE) analysis of the stress transfer of avascular necrotic bone and discovered that the stress in the center of the bone was apparently reduced, while stress concentration was localized at the boundary between the necrotic and healthy bone. This corresponds to the clinical hypothesis that abnormal loading in these regions leads to micro-fracture and a preference for bone resorption (mosaic stage). Another FE analysis by Brown, Baker, and Brand (1992) investigated the importance of the subchondral plate in ONFH. The results suggested that the subchondral plate under the articular cartilage was not of particular importance in the total integrity of the structure, although the authors commented that the unsustainability of necrotic bone under the subchondral region may contribute to structural collapse.

Kim et al. (1991) studied the morphological behavior of necrotic bone in different regions through mechanical testing and found that the bone at the necrotic boundary could be weakened from bone resorption. Yang et al. (2002) established a finite element platform for evaluating different lesion sizes by stress-to-strength ratio. They discovered that a ratio exceeding the physiological limit was unlikely and suggested that fatigue fracture could be a possible cause of material failure. Another computational model constructed by Daniel et al. (2006) investigated the effect of necrotic volume and location. With an increase in necrotic volume and lateralization of the site, the contact pressure within the hip joint increased. These findings suggested a possible source for the subsequent arthritis.

Grecu et al. (2010) extended the FE model to include the pelvis. With the application of ground reaction forces and gluteal medius muscle forces, a high strain area was demonstrated on a particular region of the cartilaginous surface which corresponded to the damaged site of the necrotic femoral head specimen.

Volokh et al. (2006) simulated seated and walking positions using FE analysis and reported that buckling was the most important mechanism leading to collapse in ONFH. A parametric analysis of the cortical thickness and cortical and trabecular elastic modulus was carried out. The buckling mode was assessed by the critical buckling pressure and predicted pressure. However, Volokh et al. did not rule out other possible mechanisms, such as fatigue, that could occur concurrently.

### 10.1.6 Investigation of the Pathomechanism

The pathomechanism suggested by various sources in the literature is quite general. Amanatullah, Strauss, and Di Cesare (2011a) reported that femoral head collapse was basically caused by bone resorption. After bone resorption, the weakened trabeculae could not accommodate stress concentration and would eventually fracture (Amanatullah, Strauss, and Di Cesare 2011a). In fact, the bone deterioration in ONFH is quite unique and progresses with different clinical symptoms, and different patterns of necrosis are speculated to occur in different regions. Figure 10.1 shows some of the signs of ONFH.

Crescent fractures are the most common findings in ONFH and are illustrated by a crescent shape on x-ray scans. It is believed that the crescent predisposes the femoral head to collapse, whereby a portion of the necrotic bone is lost (Figure 10.1b). Fracture of the cortical bone is also seen in some cases. Although it could result from preceding collapse and unsustainability, a precollapse condition also reveals the possibility of cortical fracture (Figure 10.1c). A large gray area beneath the necrotic regions is also seen; this is an area of osteosclerosis. Occasionally, fracture occurs on the boundary between the necrotic and healthy trabecular bone, as shown in Figure 10.1d.

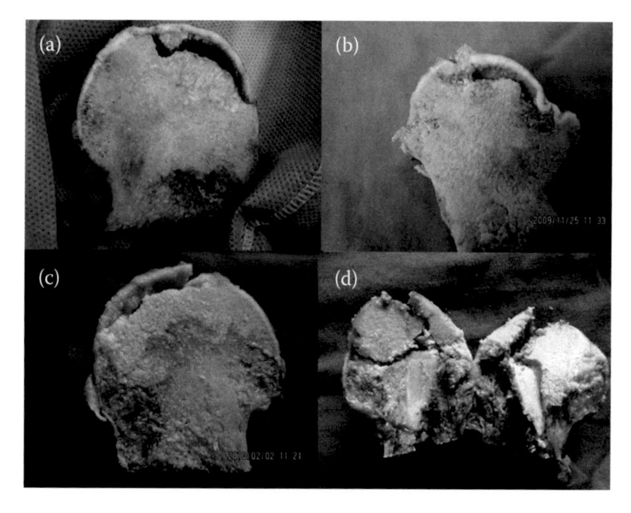


**FIGURE 10.1 (See color insert.)** Typical signs of ONFH: (a) crescent fracture; (b) femoral head collapse; (c) cortical fracture and osteosclerosis; (d) necrotic boundary fracture.

## 10.2 MODEL DEVELOPMENT

### 10.2.1 Geometry Acquisition

The model was developed from images of a 29-year-old male patient who was admitted in April 2012. He weighed 60 kg and was about 170 cm tall. The patient was diagnosed with bilateral femoral head necrosis (steroid type) and systemic lupus erythematosus.

Both MRI and computer tomography (CT) were used to construct the geometry for the FE model. MRI segmentation has the advantage of distinguishing bone structure. It was used to construct the trabecular bone core and the cortical bone shell, whereas CT has a higher resolution for differentiating the necrotic region from normal bone. The left side of the hip joint was constructed, including the hemi-pelvis, proximal femur, articular cartilage, and capsule. The femur was divided into trabecular, cortical, and necrotic regions. The necrotic volume accounted for about 31% of the trabecular volume in the femoral head.

### 10.2.2 Material Properties

The material properties of the trabecular and cortical bone were assigned orthotropically (Krone and Schuster 2006). The necrotic region was parameterized with 10% to 100% Young's modulus and shear modulus as the trabecular bone. The principal directions of the trabecular and necrotic regions were modeled by the trajectory suggested by Skedros and Baucom (2007). The pelvis was assigned a Young's modulus and Poisson's ratio of 15,100 MPa and 0.3, respectively (Brown and Hild 1983). The material properties of the cartilage and capsule were 10.5 MPa and 12.4 MPa, respectively (Grecu et al. 2010; Stewart et al. 2002).

### 10.2.3 Boundary Conditions

Midstance of the gait cycle was simulated with ground reaction forces and muscle forces adopted from the literature (Brand et al. 1982). The von Mises stress distribution of the bone is shown in Figure 10.2.

## 10.3 APPLICATIONS

The stress transfer within the femur was altered with the assignment of a necrotic region. Figure 10.2 shows the von Mises stress in a cross-sectional view. The stress at the superior surface of the femur head was gradually transferred to the inferior side of the femoral head. However, the reduction in stiffness in the necrotic region results in a sharp reduction in its ability to sustain stress. The stress was transferred to the exterior of the femoral head. This could be reflected by the von Mises stress of the cortical bone at the weight-bearing region. It increased from 15 MPa to 40 MPa when the necrotic region was reduced to 10% stiffness.

### 10.3.1 Trabecular Yield

As the bone degrades, the Young's modulus of the trabecular bone in the necrotic region decreases. Due to the effect of stress shielding, the load shared by the necrotic trabeculae also reduces. However, the reduction of experienced load is coupled with a reduction in sustainability. When the experienced load exceeds the sustainability, the bone will most likely fail. Figure 10.3 demonstrates the increase in volume of the yielding region as the degradation progresses.

### 10.3.2 Load Transfer Alteration

Figure 10.4 illustrates changes in load transfer as the bone progresses from a healthy to a necrotic state. The compressive principal stress of the healthy trabeculae at the load-bearing area is of greater

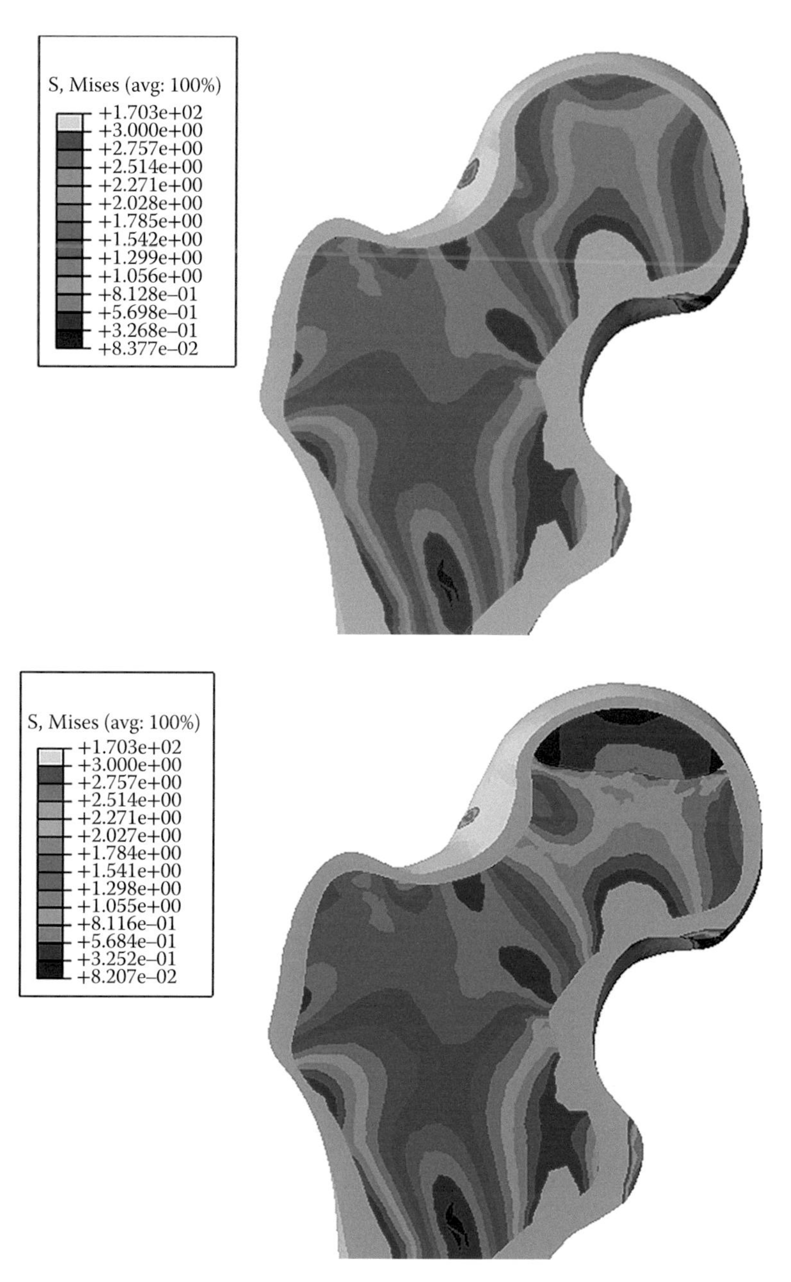


**FIGURE 10.2** **(See color insert.)** The von Mises stress contour of the femur head for a normal femur (top) and a necrotic femur (bottom). Unit: MPa.

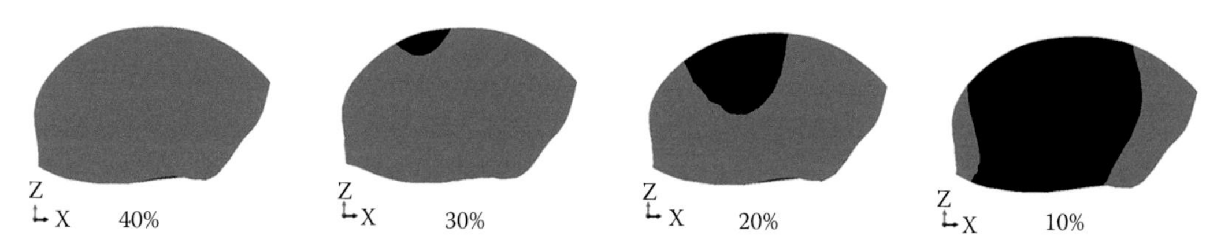


**FIGURE 10.3** Coronal cross-section views of the necrotic region at different stages of deterioration; the black region represents an over-yield region.

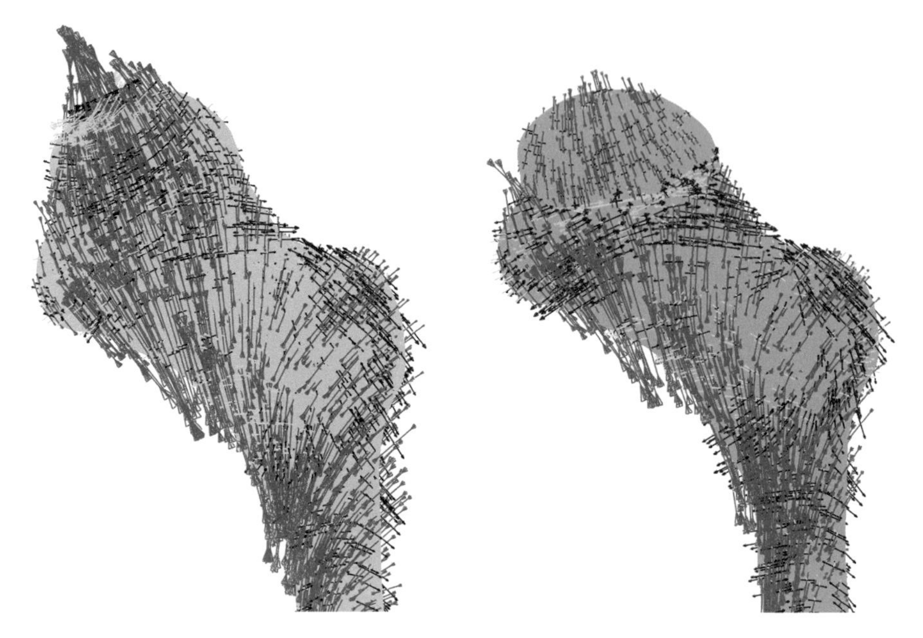

**FIGURE 10.4** **(See color insert.)** Field vector of compressive principal stress (red) and tensile principal stress (blue) in normal (left) and necrotic condition (right).

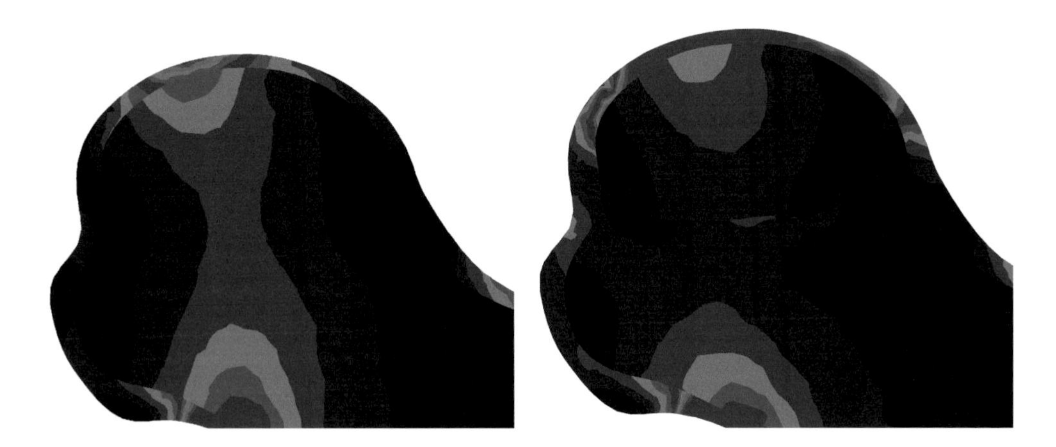

**FIGURE 10.5** **(See color insert.)** Strain energy density of the cortical bone and trabecular bone at the weight-bearing region: (left) healthy; (right) necrotic condition.

magnitude than that of the necrotic bone, while the stress within the necrotic bone is also less due to stress shielding. On the other hand, the transfer continuum of the necrotic bone is distorted along the necrotic line with abnormal tensile stress at the boundary.

### 10.3.3 Strain Energy Density Ratio

Since the load transfer is now more dependent on the cortical bone shell than the trabecular bone, the cortical bone absorbs more energy, as illustrated in Figure 10.5. As the severity of the necrosis continues, the energy contrast between the two layers becomes larger and causes a greater discontinuum. High strain energy density tends to stabilize through dilatation (change in volume) or distortion (change in shape). The discontinuum creates instability across the trabecular-cortical boundary as energy dissipation may not be consistent. When a flaw exists that could be caused by the micro-failure of the trabeculae in daily activities, the discontinuous energy rapidly dissipates, leading to crack propagation and delamination between the cortical shell and the trabecular core. Crack propagation along the laminated layer results in the crescent sign or fracture.

This study reviewed literature associated with clinical and biomedical engineering aspects of ONFH. A finite element simulation of ONFH was conducted and the associated pathomechanisms were examined. Further clinical investigations and validation should be carried out to verify these hypotheses and pathomechanisms.

## ACKNOWLEDGMENTS

This research was funded by the Natural Science Foundation of Guangdong Province (S2011040005966), the Science and Technology Planning Project of Guangdong Province (2011: 106), and the Research Fund for the Doctoral Program of Higher Education (S20104425120 012) of China.

## REFERENCES

Agarwala, S., D. Jain, V. R. Joshi, and A. Sule. 2005. Efficacy of alendronate, a bisphosphonate, in the treatment of AVN of the hip. A prospective open-label study. *Rheumatology (Oxford)* 44 (3):352–9.

Agarwala, S., S. Shah, and V. R. Joshi. 2009. The use of alendronate in the treatment of avascular necrosis of the femoral head: follow-up to eight years. *J Bone Joint Surg Br* 91 (8):1013–8.

Amanatullah, D. F., E. J. Strauss, and P. E. Di Cesare. 2011a. Current management options for osteonecrosis of the femoral head: part 1, diagnosis and nonoperative management. *Am J Orthop (Belle Mead NJ)* 40 (9):E186–92.

Amanatullah, D. F., E. J. Strauss, and P. E. Di Cesare. 2011b. Current management options for osteonecrosis of the femoral head: part II, operative management. *Am J Orthop (Belle Mead NJ)* 40 (10):E216–25.

Anderson, J. M. 2006. Avascular necrosis often is trauma-induced, but alcohol, steroid use play roles as well. *J Controversial Med Claims* Feb.

Assouline-Dayan, Y., C. Chang, A. Greenspan, Y. Shoenfeld, and M. E. Gershwin. 2002. Pathogenesis and natural history of osteonecrosis. *Semin Arthritis Rheum* 32 (2):94–124.

Bradway, J. K., and B. F. Morrey. 1993. The natural history of the silent hip in bilateral atraumatic osteonecrosis. *J Arthroplasty* 8 (4):383–7.

Brand, R. A., R. D. Crowninshield, C. E. Wittstock, D. R. Pedersen, C. R. Clark, and F. M. van Krieken. 1982. A model of lower extremity muscular anatomy. *J Biomech Eng* 104 (4):304–10.

Brown, T. D., K. J. Baker, and R. A. Brand. 1992. Structural consequences of subchondral bone involvement in segmental osteonecrosis of the femoral head. *J Orthop Res* 10 (1):79–87.

Brown, T. D., and G. L. Hild. 1983. Pre-collapse stress redistributions in femoral head osteonecrosis: a three-dimensional finite element analysis. *J Biomech Eng* 105 (2):171–6.

Brown, T. D., T. A. Mutschler, and A. B. Ferguson, Jr. 1982. A non-linear finite element analysis of some early collapse processes in femoral head osteonecrosis. *J Biomech* 15 (9):705–15.

Brown, T. D., M. E. Way, and A. B. Ferguson, Jr. 1981. Mechanical characteristics of bone in femoral capital aseptic necrosis. *Clin Orthop Relat Res* (156):240–7.

Chandler, F. A. 1948. Coronary disease of the hip. *J Int Coll Surg* 11 (1):34–6.

Chen, C. C., C. L. Lin, W. C. Chen, H. N. Shih, S. W. Ueng, and M. S. Lee. 2009. Vascularized iliac bone-grafting for osteonecrosis with segmental collapse of the femoral head. *J Bone Joint Surg Am* 91 (10):2390–4.

Chen, W. P., C. L. Tai, C. F. Tan, C. H. Shih, S. H. Hou, and M. S. Lee. 2005. The degrees to which transtrochanteric rotational osteotomy moves the region of osteonecrotic femoral head out of the weight-bearing area as evaluated by computer simulation. *Clin Biomech (Bristol, Avon)* 20 (1):63–9.

Daniel, M., S. Herman, D. Dolinar, A. Iglic, M. Sochor, and V. Kralj-Iglic. 2006. Contact stress in hips with osteonecrosis of the femoral head. *Clin Orthop Relat Res* 447:92–9.

Fairbank, A. C., D. Bhatia, R. H. Jinnah, and D. S. Hungerford. 1995. Long-term results of core decompression for ischaemic necrosis of the femoral head. *J Bone Joint Surg Br* 77 (1):42–9.
Fordyce, M. J., and L. Solomon. 1993. Early detection of avascular necrosis of the femoral head by MRI. *J Bone Joint Surg Br* 75 (3):365–7.
Grecu, D., I. Pucalev, M. Negru, D. N. Tarnita, N. Ionovici, and R. Dita. 2010. Numerical simulations of the 3D virtual model of the human hip joint, using finite element method. *Rom J Morphol Embryol* 51 (1):151–5.
Hernigou, P., and F. Beaujean. 2002. Treatment of osteonecrosis with autologous bone marrow grafting. *Clin Orthop Relat Res* (405):14–23.
Hernigou, P., A. Habibi, D. Bachir, and F. Galacteros. 2006. The natural history of asymptomatic osteonecrosis of the femoral head in adults with sickle cell disease. *J Bone Joint Surg Am* 88 (12):2565–72.
Hirota, Y., T. Hirohata, K. Fukuda, M. Mori, H. Yanagawa, Y. Ohno, and Y. Sugioka. 1993. Association of alcohol intake, cigarette smoking, and occupational status with the risk of idiopathic osteonecrosis of the femoral head. *Am J Epidemiol* 137 (5):530–8.
Hungerford, D. S., and L. C. Jones. 2004. Asymptomatic osteonecrosis: should it be treated? *Clin Orthop Relat Res* (429):124–30.
Kang, J. S., S. Park, J. H. Song, Y. Y. Jung, M. R. Cho, and K. H. Rhyu. 2009. Prevalence of osteonecrosis of the femoral head: a nationwide epidemiologic analysis in Korea. *J Arthroplasty* 24 (8):1178–83.
Kim, S. Y., and H. E. Rubash. 2007. Avascular necrosis of the femoral head: the Korean experience. In *The Adult Hip,* 2nd ed., edited by J. J. Callaghan, A. G. Rosenberg, and H. E. Rubash. Philadelphia, PA: Lippincot Williams and Wilkins.
Kim, Y. M., S. H. Lee, F. Y. Lee, K. H. Koo, and K. H. Cho. 1991. Morphologic and biomechanical study of avascular necrosis of the femoral head. *Orthopedics* 14 (10):1111–6.
Krone, R., and P. Schuster. 2006. An investigation on the importance of material anisotropy in finite-element modeling of the human femur. *SAE International* (2006-01-0064).
Kubo, T., S. Yamazoe, N. Sugano, M. Fujioka, S. Naruse, N. Yoshimura, T. Oka, and Y. Hirasawa. 1997. Initial MRI findings of non-traumatic osteonecrosis of the femoral head in renal allograft recipients. *Magn Reson Imaging* 15 (9):1017–23.
Langlais, F., J. Fourastier, J. E. Gedouin, M. Ropars, J. C. Lambotte, and H. Thomazeau. 2004. Can rotation osteotomy remain effective for more than ten years? *Orthop Clin North Am* 35:345–51.
Lieberman, J. R., D. J. Berry, M. A. Mont, R. K. Aaron, J. J. Callaghan, A. D. Rajadhyaksha, and J. R. Urbaniak. 2003. Osteonecrosis of the hip: management in the 21st century. *Instr Course Lect* 52:337–55.
Lieberman, J. R., A. Conduah, and M. R. Urist. 2004. Treatment of osteonecrosis of the femoral head with core decompression and human bone morphogenetic protein. *Clin Orthop Relat Res* (429):139–45.
Lieberman, J. R., S. M. Engstrom, R. M. Meneghini, and N. F. SooHoo. 2012. Which factors influence preservation of the osteonecrotic femoral head? *Clin Orthop Relat Res* 470 (2):525–34.
Malizos, K. N., A. H. Karantanas, S. E. Varitimidis, Z. H. Dailiana, K. Bargiotas, and T. Maris. 2007. Osteonecrosis of the femoral head: etiology, imaging and treatment. *Eur J Radiol* 63 (1):16–28.
Mankin, H. J. 1992. Nontraumatic necrosis of bone (osteonecrosis). *N Engl J Med* 326 (22):1473–9.
Markel, D. C., C. Miskovsky, T. P. Sculco, P. M. Pellicci, and E. A. Salvati. 1996. Core decompression for osteonecrosis of the femoral head. *Clin Orthop Relat Res* (323):226–33.
Massari, L., M. Fini, R. Cadossi, S. Setti, and G. C. Traina. 2006. Biophysical stimulation with pulsed electromagnetic fields in osteonecrosis of the femoral head. *J Bone Joint Surg Am* 88 Suppl 3:56–60.
Merle D'Aubigne, R., M. Postel, A. Mazabraud, P. Massias, J. Gueguen, and P. France. 1965. Idiopathic necrosis of the femoral head in adults. *J Bone Joint Surg Br* 47 (4):612–33.
Mont, M. A., J. J. Carbone, and A. C. Fairbank. 1996. Core decompression versus nonoperative management for osteonecrosis of the hip. *Clin Orthop Relat Res* (324):169–78.
Mont, M. A., and D. S. Hungerford. 1995. Non-traumatic avascular necrosis of the femoral head. *J Bone Joint Surg Am* 77 (3):459–74.
Mont, M. A., L. C. Jones, T. A. Einhorn, D. S. Hungerford, and A. H. Reddi. 1998. Osteonecrosis of the femoral head. Potential treatment with growth and differentiation factors. *Clin Orthop Relat Res* (355 Suppl):S314–35.
Mont, M. A., M. G. Zywiel, D. R. Marker, M. S. McGrath, and R. E. Delanois. 2010. The natural history of untreated asymptomatic osteonecrosis of the femoral head: a systematic literature review. *J Bone Joint Surg Am* 92 (12):2165–70.
Nam, K. W., Y. L. Kim, J. J. Yoo, K. H. Koo, K. S. Yoon, and H. J. Kim. 2008. Fate of untreated asymptomatic osteonecrosis of the femoral head. *J Bone Joint Surg Am* 90 (3):477–84.
Petrigliano, F. A., and J. R. Lieberman. 2007. Osteonecrosis of the hip: novel approaches to evaluation and treatment. *Clin Orthop Relat Res* 465:53–62.

Skedros, J. G., and S. L. Baucom. 2007. Mathematical analysis of trabecular "trajectories" in apparent trajectorial structures: the unfortunate historical emphasis on the human proximal femur. *J Theor Biol* 244 (1):15–45.
Stewart, K. J., R. H. Edmonds-Wilson, R. A. Brand, and T. D. Brown. 2002. Spatial distribution of hip capsule structural and material properties. *J Biomech* 35 (11):1491–8.
Sverdlova, N. S., and U. Witzel. 2010. Principles of determination and verification of muscle forces in the human musculoskeletal system: Muscle forces to minimise bending stress. *J Biomech* 43 (3):387–96.
Takatori, Y., T. Kokubo, S. Ninomiya, S. Nakamura, S. Morimoto, and I. Kusaba. 1993. Avascular necrosis of the femoral head. Natural history and magnetic resonance imaging. *J Bone Joint Surg Br* 75 (2):217–21.
Urbaniak, J. R., and E. J. Harvey. 1998. Revascularization of the femoral head in osteonecrosis. *J Am Acad Orthop Surg* 6 (1):44–54.
Volokh, K. Y., H. Yoshida, A. Leali, J. F. Fetto, and E. Y. Chao. 2006. Prediction of femoral head collapse in osteonecrosis. *J Biomech Eng* 128 (3):467–70.
Wang, C. J., F. S. Wang, K. D. Yang, C. C. Huang, M. S. Lee, Y. S. Chan, J. W. Wang, and J. Y. Ko. 2008. Treatment of osteonecrosis of the hip: comparison of extracorporeal shockwave with shockwave and alendronate. *Arch Orthop Trauma Surg* 128 (9):901–8.
Wirtz, D.C., Rohrig, H., Neuss, M. 2003. Core decompression for avascular necrosis of the femoral head. *Oper Orthop Traumatol* 15:288–303.
Yang, J. W., K. H. Koo, M. C. Lee, P. Yang, M. D. Noh, S. Y. Kim, K. I. Kim, Y. C. Ha, and M. S. Joun. 2002. Mechanics of femoral head osteonecrosis using three-dimensional finite element method. *Arch Orthop Trauma Surg* 122 (2):88–92.
Yoo, M. C., K. I. Kim, C. S. Hahn, and J. Parvizi. 2008. Long-term followup of vascularized fibular grafting for femoral head necrosis. *Clin Orthop Relat Res* 466 (5):1133–40.
Yoon, T. R., E. K. Song, S. M. Rowe, and C. H. Park. 2001. Failure after core decompression in osteonecrosis of the femoral head. *Int Orthop* 24 (6):316–8.
Yoshida, H., A. Faust, J. Wilckens, M. Kitagawa, J. Fetto, and E. Y. Chao. 2006. Three-dimensional dynamic hip contact area and pressure distribution during activities of daily living. *J Biomech* 39 (11):1996–2004.

# 11 Pelvis Model for Reconstruction with Autografted Long Bones following Hindquarter Amputation

*Wen-Xin Niu, Jiong Mei, Ting-Ting Tang, Yubo Fan, Ming Zhang, and Ming Ni*

## CONTENTS



## SUMMARY

It is anatomically feasible and potentially therapeutic to reconstruct a defective pelvis ring with antogenous long bones after hemipelvic amputation, but the biomechanical characteristics of the surgery are unclear to doctors and researchers. The objectives of this study were to analyze the stress distribution of two hemipelvic reconstruction surgeries using the finite element (FE) method and to find which surgery gave a more favorable outcome. An FE model of the intact pelvis was constructed through sequenced computer tomography (CT) images and validated with an in vitro experiment. Two operative schemes, the ischiadic tuberosity being replaced with the femur condyles or the tibial plateau, were modeled based on the intact model. A normal sitting posture was simulated, and von Mises stresses were calculated and compared among the three models. From a biomechanical perspective, reconstruction with the femur condyles should be considered preferential. In planning the operation, screws of larger than normal diameter should be used and their rigidity decreased to reduce the incidence of stress shielding.

## 11.1 INTRODUCTION

The human pelvis, the lower part of the trunk, includes the bony pelvis, cavity, floor, and the perineum. The pelvic skeleton is a complicated mechanical structure. It is formed by the sacrum and a pair of iliac bones, which connect the spine with the lower extremities. The two iliac bones are attached to the sacrum posteriorly with the sacroiliac joints (SIJs), articulated with the two femurs with the hip joints, and connected to each other anteriorly with the symphysis pubica.

Surgical treatment of highly aggressive and malignant tumors originating from the pelvis is always challenging due to the complex anatomy of the pelvis and the proximity of neurovascular structures and viscera (Campanacci et al. 2012; Cheng et al. 2011). Partial resection of the pelvis is still necessary in some patients (Grimer et al. 2013). Enneking and Dunham (1978) proposed a classification scheme for describing the various subtypes of pelvic resection. A modified scheme is shown in Figure 11.1. The resection locations were classified into several regions (Hosalkar and Dormans 2007). Depending on the lesion location and extent, Enneking and Dunham (1978) performed three types of procedures individually or in combination: (1) wide excision or radical resection of the iliac wing; (2) periacetabular wide excision or radical resection; and (3) wide excision or radical resection of the pubis.

However, partial resection of the pelvis can cause trouble sitting, standing, and fitting a prosthesis to the patient after amputation (Sneppen et al. 1978; Hoffmann et al. 2006). The continuity of the pelvic ring must be restored to achieve structural stability and mechanical function following hindquarter amputation. There are various options for reconstructing the pelvic ring after partial amputation, such as arthrodesis, prosthesis, autograft, and allograft prosthetic composite reconstructions (Campanacci et al. 2012). Each of these methods has its inherent advantages and limitations (Hugate and Sim 2006).

The biomechanical properties of any implant must be evaluated to meet the stability and mechanical requirements of reconstructive surgery. Some in vitro experiments have been undertaken to study the biomechanical behavior of various surgeries (Yu et al. 2010; Cheng et al. 2011; Mindea et al. 2012), but the measurable parameters in cadaveric experiments are very limited as the experiments are very costly and samples are not easily acquired. Biomechanists and surgeons need cheaper and more effective methods of studying relevant subjects.

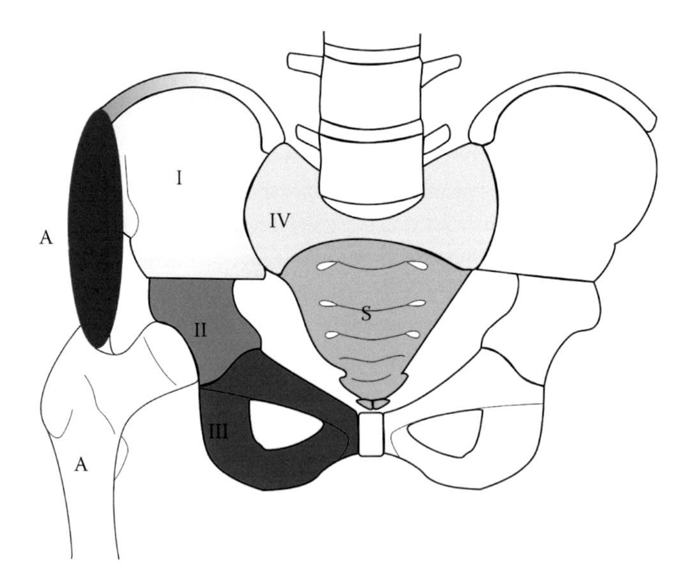


**FIGURE 11.1** **(See color insert.)** Types of pelvic resection. The red oval indicates the gluteus medius muscle; Type I = resection of the ilium; Type II = resection of the periacetabular region; Type III = resection of the ischiopubic region; Type IV = en bloc excision of the sacral ala; A = aggressive resection; S = sacrum.

The finite element method is an ideal biomechanical tool because it is able to provide the details of multiple parameters regarding stress, strain, displacement, and energy. With FE modeling and analysis, biomechanical information can be compared between a reconstructed and normal pelvis, and physicians and researchers can determine which operation is more biomechanically reliable or what can be done to improve current methods. In the last five years, the number of FE studies based on the pelvis has increased remarkably, as summarized in Table 11.1. However, studies on pelvic rcconstruction are still limited, as listed in Table 11.2.

**TABLE 11.1**
**Finite Element Models of the Pelvis and Applications in the Literature since 2009**

| Year | Author(s) | Model Components | Objective(s) |
|---|---|---|---|
| 2009 | Ivanov et al. | L4-S1 unites, pelvis and ligaments | To quantify the increase in sacrum angular motions and stress across SIJ as a function of fused lumbar spine |
| 2009 | Majumder, Roychowdhury, and Pal | Pelvis, femur, soft tissue | To evaluate the responses to changing body configurations during backward fall |
| 2009 | Leung et al. | Pelvic bones and ligaments, bilateral proximate femurs | To analyze pelvic strains as a function of interior and cortical surface bone density, and to compare high-strain regions with common insufficiency fracture sites |
| 2010 | Zhang et al. | Pelvic bones with a cemented acetabular cup | To develop a subject-specific FE pelvic bone model and study the bone-cement interfacial response in cemented acetabular replacements |
| 2011 | Eichenseer, Sybert, and Cotton | Pelvic bones and ligaments | To characterize the sacroiliac ligament strains in response to different loads and quantify the changes in SIJ stress and angular displacement in response to changes in the ligament stiffness |
| 2011 | Hao et al. | Pelvic bones and ligaments | To study the effect of boundary conditions on the pelvic biomechanics predictions |
| 2011 | Shim et al. | Fractured pelvis, fragment and femoral head | To measure interfragmentary movements in the conventional open approach with plate fixations in the acetabular fractures and compare them; and to develop a way of predicting interfragmentary movement |
| 2012 | Böhme et al. | Fractured pelvis and iliosacral screws | To test if patient-specific FE models can predict implant behavior under real conditions to avoid implant failure and secondary operation |
| 2012 | Bréaud et al. | Pelvis, fat, puboprostatic ligament, perineal membrane, prostate, the urinary system | To develop a computerized FE model by digitizing the male pelvis in order to understand the associated pelvic ring trauma and posterior urethral trauma |
| 2012 | Kiapour et al. | L3-S1, pelvic bones and ligaments, bilateral proximate femurs | To assess the relationship between leg length discrepancy and the load distribution across SIJ |
| 2012 | Kunze et al. | Pelvis, acetabular implant socket | To determine the muscle forces of the activity of getting up from different seat heights by multibody simulation and the evaluation of the micro-motions at the acetabular implant-bone interface during those activities |
| 2012; 2013 | Ghosh et al. | Intact and implanted composite hemipelvis | To assess the validity of the generation procedure of the FE model of intact and implanted artificial pelvises |
| 2013 | Small et al. | Implanted composite hemipelvis | To examine the combined effects of acetabular cup orientation and stiffness and on pelvic osseous loading |

*Note:* SIJ: sacroiliac joint; FE: finite element.

**TABLE 11.2**
**Finite Element Models of Pelvic Reconstruction and Applications**

| Year | Author(s) | Resection Location(s) | Reconstruction Method(s) |
|---|---|---|---|
| 2003 | Kawahara et al. | IV + S | A modified Galveston; a triangular frame; and a novel reconstruction |
| 2008 | Jia et al. | I | Fibular autograft with four different rod-screw systems |
| 2009 | Niu et al. | I + II + III | Autograft long bones |
| 2010 | Ji et al. | II + III | Modular hemipelvic prosthesis |
| 2012 | Zhu et al. | IV + S | Sacral-rod; 4-rod; bilateral fibular flaps; and improved compound Galveston techniques |
| 2013 | Zhou et al. | I + II + III | Modular hemipelvic prosthesis |
| 2013 | Hao et al. | II + III | Modular hemipelvic prosthesis |

## 11.2 HEMIPELVIC AUTOGRAFT RECONSTRUCTION

After hemipelvic amputation, ipsilateral autogenous bones are harvested as native materials for repairing wide bone defects. It is a purely biological reconstruction with perfect osteoinductive properties and can avoid the loosening and breakage often seen with metal implants. Rejection and potential for disease transmission can be ignored in this procedure. Because of the small number of complications and good functionality, autograft implantation was recommended by Hillmann et al. (2003).

Bramer and Taminiau (2005) reconstructed the pelvic ring and tuberosity of the ischium (TI) with the ipsilateral femur or tibia in three patients, two of whom were alive one and four years postoperatively. The authors believed that in selected cases with a wide defect in the hemipelvis after resection, this method could improve function and quality of life, but they failed to report on whether the femoral condyle (FC) used to replace TI was situated in the vacated space of the original TI.

Wang et al (2012) recently studied the anatomy of proximal femoral autografts of 13 fresh-frozen Chinese male cadavers. Based on their measurements, we calculated the diameter of the femoral head, the distance from the apex of the greater trochanter perpendicular to the medial cortex edge of the femoral neck, the length between the apex of the femoral head and the midpoint of the osteotomy line under the lesser trochanter, and the width of the greater trochanter from anterior to posterior. These parameters were 50.0 ± 2.7 mm, 56.9 ± 5.9 mm, 100.8 ± 5.7 mm, and 48.9 ± 2.7 mm, respectively. There was a positive correlation between subject height and each of these parameters. The authors thought that proximal femoral autograft reconstruction was a good option after hemipelvic resection.

Seventy cases have been investigated and measured retrospectively using radiographic data at the Tongji University School of Medicine to determine the maximum diameter that could be contained in the trochanter. Dual-energy x-ray absorptiometry was used to detect the bone mineral density (BMD) of the natural acetabulum and trochanter of eight volunteers to reflect BMD differences and determine whether postoperative protection was needed. The experiment was detailed by Gao et al. (2011). It was found that the section of the greater trochanter was round and had a maximum diameter of 44.2 ± 5.75 mm. The BMD was 1.224 ± 0.183 $g \cdot cm^{-3}$ in the natural acetabulum and 0.866 ± 0.132 $g \cdot cm^{-3}$ in the trochanter region. The hemipelvis and acetabulum could be reconstructed with an autograft from the ipsilateral proximal femur. Therefore, it is feasible to reconstruct the hemipelvis and hip joint with an autografted ipsilateral proximal femur combined with a normal-sized total hip replacement. Postoperative protection is mandatory for the new acetabulum as it would not be able to replace the natural one in terms of BMD.

Ten fresh cadaveric specimens of the pelvis and lower limb were analyzed in this study. Some of the recorded parameters are listed in Table 11.3. CT scans and cross-section dissections were performed on these specimens to observe the distribution of cancellous bone in the distal femur and the proximal tibia. Both the tibial plateau (TP) and FC were mainly occupied with cancellous bone, and the average thickness of the cancellous bone in these two regions was 6.24 and 5.64 cm, respectively. This suggested that the FC and TP are potential materials for reconstructing an incomplete hemipelvis, but no single one of them is capable of reconstructing the pelvic ring and TI in situ (Mei et al. 2008).

Two samples, using combinations of the femur and tibia, were used to reconstruct the hemipelvis. In one sample, TI was reconstructed with the FC and the proximal tibia was used as a bridging bone to connect the FC and contralateral pubic as a natural pubic arch. In the other sample, the TP was used to reconstruct TI and the distal femur was used as a bridging bone. FC-reconstructed (FCR) and TP-reconstructed (TPR) samples and their x-ray photographs are shown in Figures 11.2 and 11.3. The images indicate that the combined use of the distal femur and proximal tibia could regain the pelvic ring structure.

**TABLE 11.3**
**Anatomic Measurements of the Pelvis, Femur, and Tibia**

| Parameters | Measurement |
|---|---|
| The perpendicular distance from TI to the median sagittal plane | 4.52 ± 0.48 |
| The perpendicular distance from TI to the coronal plane through the AS center | 3.31 ± 0.57 |
| The distance from TI to the ipsilateral AS | 11.75 ± 6.19 |
| The distance from TI to the symphysis pubis | 15.72 ± 5.19 |
| The maximum oblique diameter of the FC | 7.93 ± 0.44 |
| The maximum oblique diameter of the TP | 7.19 ± 0.61 |

*Note:* TI: tuberosity of ischium; AS: auricular surface; FC: femoral condyle; TP: tibial plateau. Measurements in cm; mean ± one standard deviation.

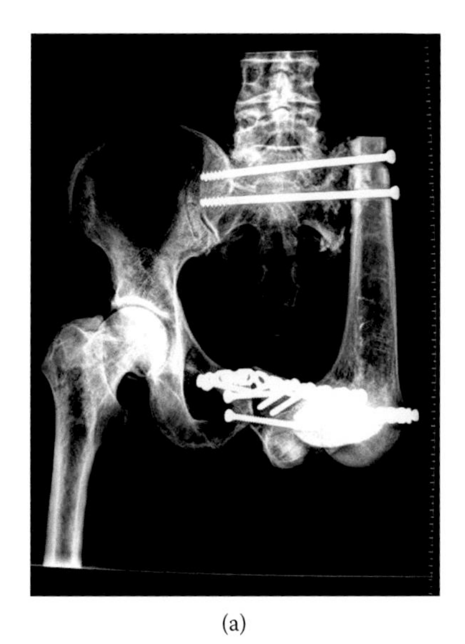

(a)

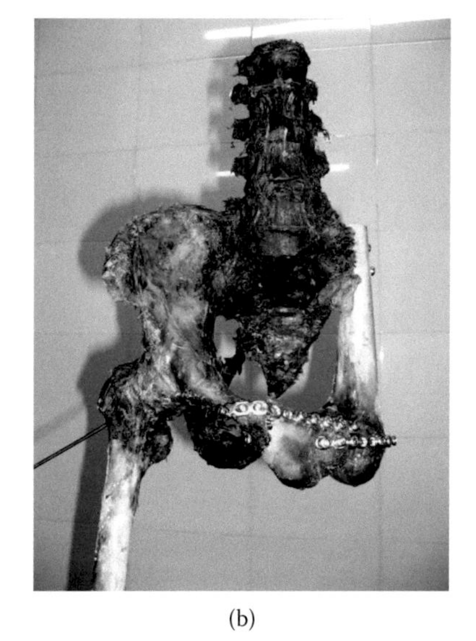

(b)

**FIGURE 11.2** Photograph (a) and x-ray (b) of pelvic reconstruction using the femoral condyle to replace the tuberosity of the ischium.

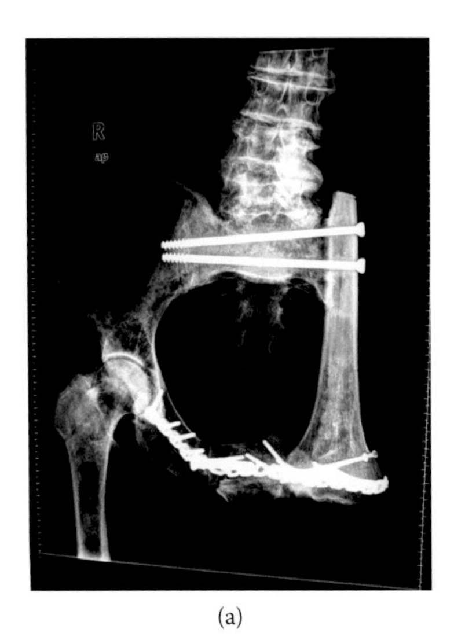


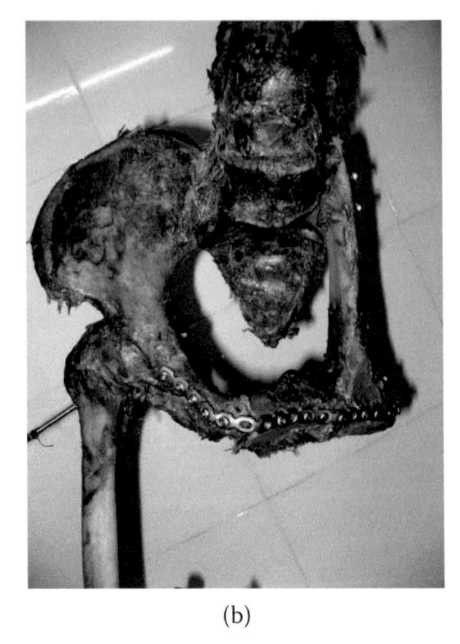

(a) (b)

**FIGURE 11.3** Photograph (a) and x-ray (b) of pelvic reconstruction using the tibial plateau to replace the tuberosity of the ischium.

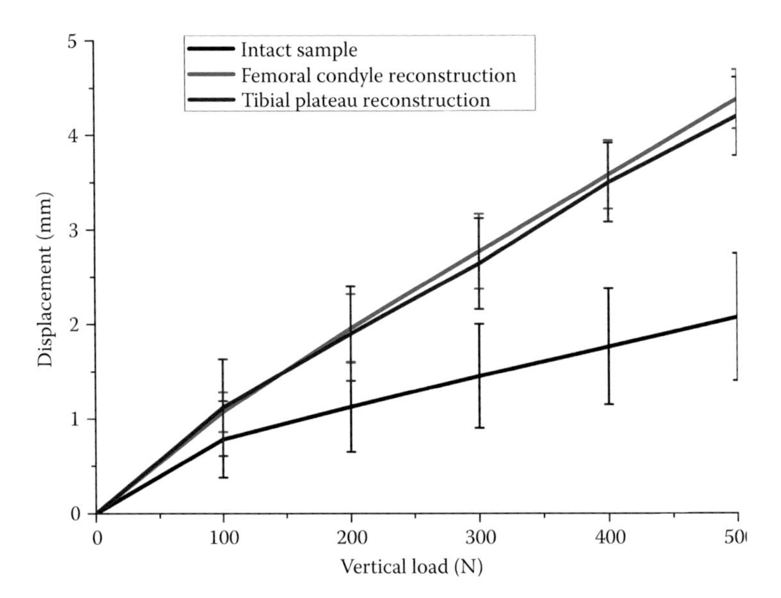


**FIGURE 11.4** Load-displacement curves of the intact and reconstructed pelvis.

Biomechanical testing was also performed on the intact and two reconstructed pelvises to compare stability. A material testing machine (CSS-44010, Chuangchun Research Institute for Mechanical Science Co. Ltd, Jilin, China) was used to load the samples five consecutive times with a vertical force of 500 N. Before each formal trial, a preconditioning trial was carried out to load and unload the sample several times with 300 N to get a steady load-displacement curve.

A comparison of load-displacement curves for the different samples is shown in Figure 11.4. The axial stiffness of the intact pelvis was significantly greater than that of any of the reconstructed samples, but the difference in axial stiffness between the two reconstructed samples was not significant. After unloading, all samples returned to their initial conditions. This indicated that the reconstruction methods were reliable, though their global stiffness was not at the same level as the intact pelvis. Also, the effects of these surgeries on the stress and strain distributions of both tissues and implants must be evaluated.

## 11.3 FINITE ELEMENT ANALYSIS OF THE INTACT PELVIS

### 11.3.1 Finite Element Modeling and Analysis

The pelvis is a complex and irregular structure that cannot be subject to oversimplification. In some previous studies, two-dimensional or quasi-three-dimensional models were constructed to tackle clinical problems (Huiskes 1987). The over-simplified structure was unable to simulate complex operations and provide accurate results. With the development of advanced medical imaging techniques and high-performance computers, any human organ can be accurately modeled in three dimensions (Liang et al. 2011; Cheung et al. 2005). A male cadaver (age 30 years; mass 65 kg; height 172 cm) free from any pathological abnormalities, trauma, or deformity in the pelvis or lower extremities was used in this study. Transverse CT images of this subject were acquired at 1.25-mm intervals from the third lumbar vertebra to the middle tibia. A total of 960 images were acquired, of which 620 were used to construct the FE model.

Model segmentation was carried out with MIMICS 12.0 (Materialise, Inc., Leuven, Belgium) on the sacrum and bilateral iliac bones. The cortical and cancellous bones were also distinguished effectively as different point clouds. The point cloud files were imported into Geomagic Studio 9.0 (Raindrop Geomagic, Research Triangle Park, North Carolina) and converted into polygonal surface models through the point phase, polygon phase, and shape phase.

The solid models were then imported into ANSYS 12.0 (ANSYS Inc., Canonsburg, Pennsylvania), in which a Boolean operation was used to produce cortical and cancellous models for various bones. All bones were meshed with three-dimensional 10-node tetrahedron structural solid elements (SOLID92). This element has a quadratic displacement behavior and is capable of modeling irregular meshes. To simplify the analysis, all tissues were idealized as homogeneous, linearly elastic, and isotropic. The Young's modulus and Poisson ratio of all tissues are listed in Table 11.4. The parameters were all adopted from Bodzay, Flóris, and Váradi (2011).

The symphysis pubica was modeled with three-dimensional link elements. The nodes at the opposite iliac bones were selected first and exported as two groups with the space coordinates. A C-language program was then used to search for a matching for each point in either group. A matching was defined as two points between which the displacement was the shortest than any other couples. This method has been previously used to simulate cartilage and soft tissues (Liang et al. 2011).

The FE model of the intact pelvis is shown in Figure 11.5. The TI was fixed in six degrees of freedom to simulate a sitting posture. A compressive force of 500 N was vertically loaded on the lumbosacral disc to simulate normal body weight (Jia et al. 2008). The FE analysis was carried out with ANSYS software. The von Mises stress of the intact model is detailed in Figure 11.6. Stress concentrations occurred mainly around the arcuate line, acetabulum, isciadic ramus, SIJ, and TI. The maximum von Mises stresses in the sacrum and right and left iliums were 5.40, 6.21, and 5.84 MPa, respectively. The stress was distributed symmetrically to a great extent. Considering the physiological asymmetry, modeling, and receivable computational error, the computation was encouraging.

**TABLE 11.4**
**Material Properties of Tissues in the Finite Element Models**

| | Young's Modulus (MPa) | Poisson Ratio |
|---|---|---|
| Cortical bone | 17,000 | 0.3 |
| Cancellous bone | 400 | 0.2 |
| Sacroiliac joint | 68 | 0.2 |
| Symphysis pubica | 50 | 0.2 |

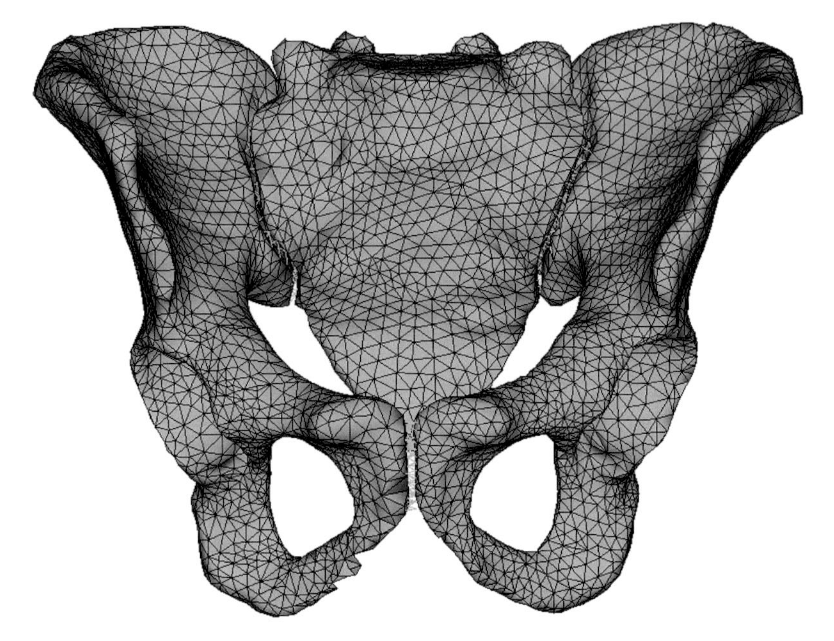

FIGURE 11.5 Finite element model of the intact pelvis.

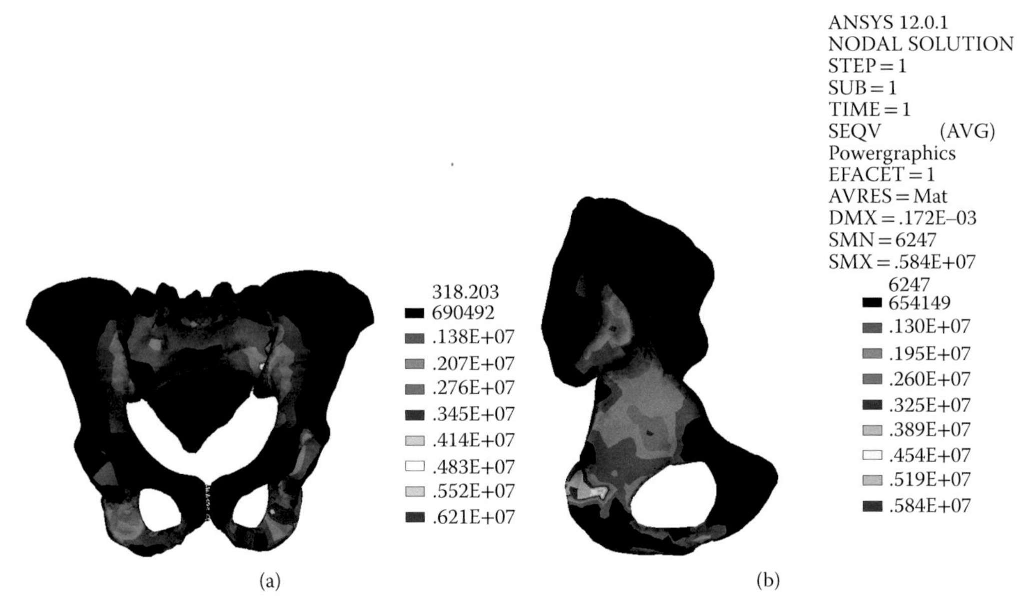


FIGURE 11.6 **(See color insert.)** The von Mises stress distribution of the intact model: (a) global model; (b) the left ilium.

### 11.3.2 Finite Element Model Validation

As shown in Table 11.1, there have been numerous recent publications on FE models of the pelvis, but only a few of them were experimentally validated. Model validation is extremely necessary in biomechanical research for gauging the accuracy of predictions. There are several methods for validating an FE model, such as in vivo and in vitro experiments and model-model validation (using a validated model to validate another new model).

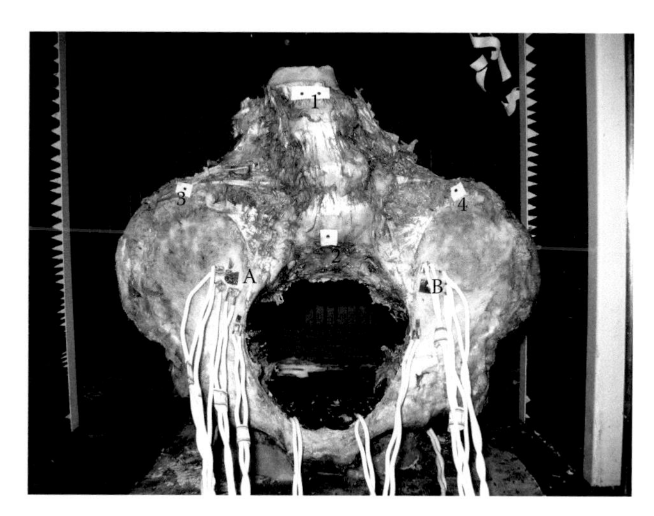


**FIGURE 11.7** In vitro experiment of the pelvis under vertical compression. 1–4, markers of vertical displacements; A/B, the strain rosettes.

To validate the aforementioned FE model of the pelvis, an in vitro experiment was carried out on a cadaveric pelvis. All soft tissues were detached from the pelvis prior to testing. Bilateral TIs were fixed with bone cement to simulate a sitting posture. The interrupted lumbar vertebrae were fixed with a bone cement stylolitic pallet, which was loaded by a material testing machine. The maximum loading of 500 N was achieved at a loading rate of 2 mm/min.

Four markers were installed on the sample (Figure 11.7). A charge-coupled device system positioned in front of the specimen was used to record images before and after loading. The system was connected to an image processing computer, in which a digital speckle correlation program calculated the displacement of each marker. At the same time, strain rosettes were used to measure the bone surface strain of the bilateral anterior fossa iliacas. Two strain rosettes were connected to the static strain indicator with a 1/4 bridge converter and common compensating gauge. The same method was used in previous studies (Niu et al. 2008).

The vertical displacements of markers 1 to 4 were, respectively, 5.12, 0.33, 0.01, and 0.02 mm. In the FE simulation, the displacements of the corresponding regions were 4.63, 0.12, −0.04, and 0.01 mm, respectively. In the in vitro experiment, the first principal strains were 152 and 162 με, and the third principal strains were −365 and −1181 με for the iliac fossas. The FE strain calculations at the corresponding regions were all within the first and third principal strains measured in the cadaveric specimen.

In vitro experiments are one of the most reliable methods for validating computational models, because cadaveric samples provide a somewhat physiological environment for testing (Liang et al. 2011). The widespread use of in vitro experiments in biomechanical research is limited by high cost and a limited supply of specimens. Only one sample was measured in the present study. It is too inconclusive to provide any measurement or conclusion with statistical significance, but the results are useful for a general validation of the FE simulation.

## 11.4 FINITE ELEMENT ANALYSIS OF THE RECONSTRUCTED MODELS

The distal femur and proximal tibia were also modeled in MIMICS and Geomagic software through the same process as the pelvis model. The models of the intact pelvis, femur, and tibia were reassembled to simulate hemipelvis reconstructions, as described in Section 11.2. As shown in Figure 11.8, two cancellous bone screws were used to fix the autograft bone and the sacrum at the SIJ location in the FCR and TPR models. At the symphysis pubica, the femur and tibia were fixed with titanium plates. Beam elements were used as mechanical equivalents of the titanium plate in the FE models.

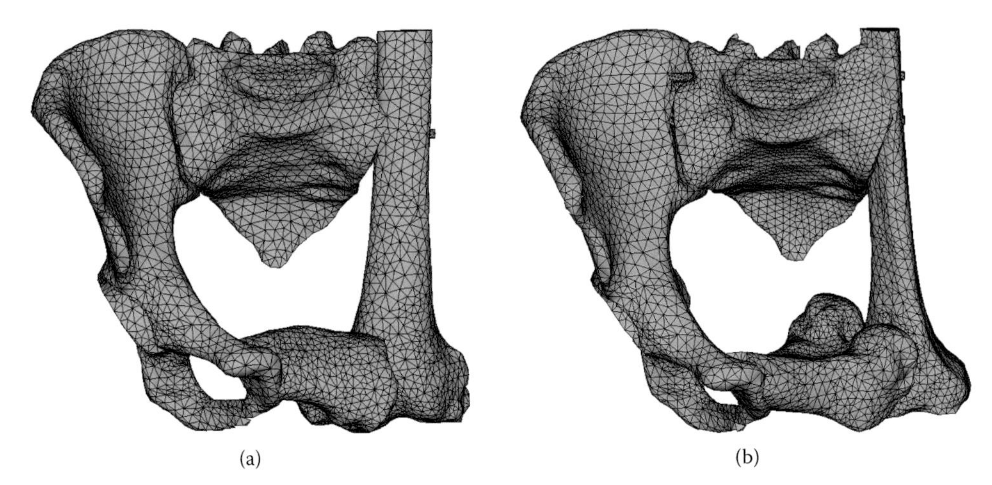


**FIGURE 11.8** Finite element models: (a) using the femoral condyle to substitute the tuberosity of the ischium after left ilium amputation; (b) using the tibial plateau to substitute the tuberosity of the ischium after left ilium amputation.

**TABLE 11.5**
**Maximum von Mises Stress of Various Parts in Three Models (MPa)**

| Model | Sacrum | Right Hemipelvis | Intact or Reconstructed Left Hemipelvis | Bridging Bone |
|---|---|---|---|---|
| Intact | 5.40 | 6.21 | 5.84 | N/A |
| FCR | 21.3 | 23.7 | 13.9 | 2.43 |
| TPR | 68.1 | 32.1 | 6.41 | 3.49 |

*Note:* FCR indicates the model using the femoral condyle to substitute the tuberosity of the ischium after left ilium amputation; TPR indicates the model using the tibial plateau to substitute the tuberosity of the ischium after left ilium amputation.

Both models were assigned the same boundary and loading conditions as described previously for the intact pelvic model. The peak von Mises stresses of the two reconstructed models and intact model are listed in Table 11.5. There was an obvious difference in results between the intact and FCR models. Stress concentrations in the FCR model occurred mainly around the connection between the femur and the left auricular surface (AS) of the sacrum, the right SIJ, and the left side of the screws. The peak von Mises stress of each region was larger in the FC-reconstructed model than in the intact model.

There was a general agreement of stress distributions between the two reconstructed models. The stress at the right ilium, sacrum, bridging bone, and screws in the TPR model was larger than that in the FCR model. As shown in Figure 11.9, the peak von Mises stress of the tibia in the TPR model was lower than that of the femur in the FCR model. However, the regions and extents of stress concentration differed greatly between the two models.

The surgical reconstruction was unable to completely recover the biomechanical function of the normal pelvis. The peak stresses in the sacrum in the two reconstructed models were both much larger than those in the intact model. The two long screws worked like shoulder poles, which ran through both sides of the sacrum and carried the load. The screws shielded stress from the SIJ. Therefore, the sacrum bore most of the load and was subjected to the greatest stress and strain. The peak stresses of the two screws were both distributed at the reconstructed side (Figure 11.10). The right iliac bone offered a wider contact area than the reconstructed femur or tibia, leading to a larger surface for loading and greater stress in the reconstructed side.

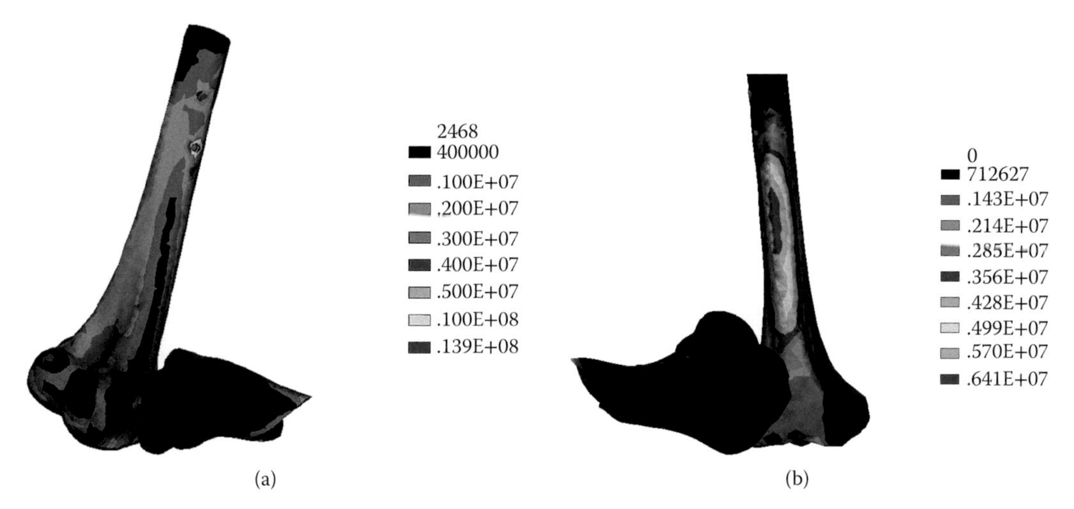


**FIGURE 11.9 (See color insert.)** The von Mises stress distribution of the autografts in the reconstruction models: (a) the femur in the FC-reconstructed model; (b) the tibia in the TP-reconstructed model.

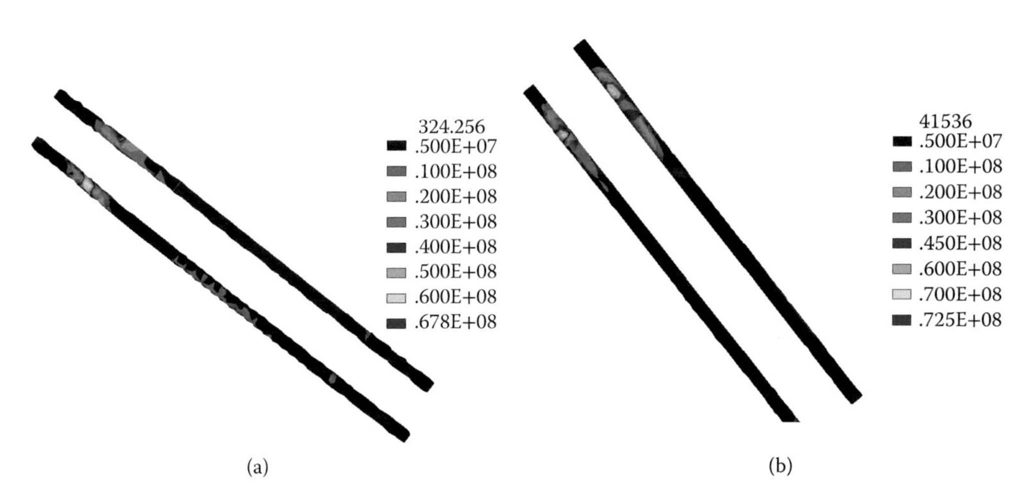


**FIGURE 11.10** The von Mises stress distributions of the screws in the (a) FC-reconstructed model and (b) TP-reconstructed model.

Because femoral strength is greater than that of the tibia, the femur in the FCR model shared more loading than the tibia in the TPR model. The stress peaks in the right ilium (35%), sacrum (219%), bridging bone (43%), and screws (6%) were all larger in the TPR model. The stress distributions were similar in the two reconstructed models, but the stress concentration in the TPR model occurred at the tibial shaft and had a relatively wider area, while the stress concentration in the FCR model was situated at the contact surface between the femur and screws and the area was narrower. This indicated that the TPR was prone to both local and complete fracture. Under a normal sitting posture, the stress in the bridging bone was very low. It had little biomechanical influence on the function of the reconstruction.

Therefore, from the perspective of biomechanics, priority should be given to FCR reconstruction due to its lower incidence of fracture. Additionally, the diameter of the screws should be appropriately enhanced, and their rigidity should be decreased in the healthy side to reduce the stress shielding.

## 11.5 LIMITATIONS AND FUTURE RESEARCH

In this chapter, we constructed one intact pelvic FE model and two postsurgical models, and used them to analyze hemipelvic reconstructive surgeries with an autografted distal femur and proximal tibia. Though we did CT scans of two reconstructed in vitro samples, as described in Section 11.5, we still modeled the intact and reconstructed pelvises from another cadaveric sample. The isogenic models were helpful for comparison with each other, because no individual variation needed to be considered in the analysis.

However, there were also some limitations in this study. To compare the FE computation and cadaveric experiment, only bone and joint structures were considered in the modeling. The ligaments and other soft tissues would influence the biomechanical properties and postsurgical outcome. How influential these tissues would be should be considered in clinical applications of this research. Additionally, only a sitting posture was analyzed in this study. Further clinical work should consider how to assemble an artificial limb on the reconstructed hemipelvis. It also raises another challenge for the biomechanical evaluation of the reconstructed structure, which has to support the upper body in more complicated conditions.

Of course, the FE model of the intact pelvis can also be used in many other fields to study injury protection, clinical operations, and implants. The model can also be changed or expanded to solve more biomechanically related questions in fields such as urology, obstetrics, gynecology, and the anorectal system. Further applications of the pelvic FE model must be combined with clinical practice and in vivo or in vitro experiments.

## ACKNOWLEDGMENTS

This study was supported by the Opening Project of the Shanghai Key Laboratory of Orthopaedic Implants (No. KFKT2013002) and the China Postdoctoral Science Foundation (No. 2013M530211).

## REFERENCES

Bodzay T., I. Flóris, and K. Váradi. 2011. Comparison of stability in the operative treatment of pelvic injuries in a finite element model. *Archives of Orthopaedic and Trauma Surgery* 131: 1427–33.

Böhme J., V. Shim, A. Höch, M. Mütze, C. Müller, and C. Josten. 2012. Clinical implementation of finite element models in pelvic ring surgery for prediction of implant behavior: A case report. *Clinical Biomechanics* 27: 872–8.

Bramer J. A., and A. H. Taminiau. 2005. Reconstruction of the pelvic ring with an autograft after hindquarter amputation: Improvement of sitting stability and prosthesis support. *Acta Orthopaedica* 76: 453–4.

Bréaud J., P. Baqué, J. Loeffler, F. Colomb, C. Brunet, and L. Thollon. 2012. Posterior urethral injuries associated with pelvic injuries in young adults: computerized finite element model creation and application to improve knowledge and prevention. *Surgical and Radiologic Anatomy* 34: 333–9.

Campanacci D., S. Chacon, N. Mondanelli et al. 2012. Pelvic massive allograft reconstruction after bone tumour resection. *International Orthopaedics* 36: 2529–36.

Cheng L., Y. Yu, R. Zhu et al. 2011. Structural stability of different reconstruction techniques following total sacrectomy: a biomechanical study. *Clinical Biomechanics* 26: 977–81.

Cheung, J. T., M. Zhang, A. K. Leung, and Y. B. Fan. 2005. Three-dimensional finite element analysis of the foot during standing a material sensitivity study. *Journal of Biomechanics* 38: 1045–54.

Eichenseer P. H., D. R. Sybert, and J. R. Cotton. 2011. A finite element analysis of sacroiliac joint ligaments in response to different loading conditions. *Spine* 36: E1446–52.

Enneking W. F., and W. K. Dunham. 1978. Resection and reconstruction for primary neoplasms involving the innominate bone. *The Journal of Bone and Joint Surgery. American Volume* 60: 731–46.

Gao Y. S., J. Mei, C. Q. Zhang, M. Ni, X. H. Wang, and B. Dou. 2011. To reconstruct the hemipelvis and acetabulum with homolateral proximal femur: a feasible way for hip reconstruction after tumorectomy involving the acetabulum. *European Journal of Orthopaedic Surgery and Traumatology* 21: 145–9.

Ghosh R., S. Gupta, A. Dickinson, and M. Browne. 2012. Experimental validation of finite element models of intact and implanted composite hemipelvises using digital image correlation. *Journal of Biomechanical Engineering* 134: 081003.

Ghosh R., S. Gupta, A. Dickinson, and M. Browne. 2013. Experimental validation of numerically predicted strain and micromotion in intact and implanted composite hemipelvises. *Proceedings of the Institution of Mechanical Engineers, Part H: Journal of Engineering in Medicine* 227: 162–74.

Grimer R. J., C. R. Chandrasekar, S. R. Carter, A. Abudu, R. M. Tillman, and L. Jeys. 2013. Hindquarter amputation: is it still needed and what are the outcomes? *The Bone and Joint Journal* 95-B: 127–31.

Hao Z., C. Wan, X. Gao, and T. Ji. 2011. The effect of boundary condition on the biomechanics of a human pelvic joint under an axial compressive load: a three-dimensional finite element model. *Journal of Biomechanical Engineering* 133: 101006.

Hao Z., C. Wan, X. Gao, T. Ji, and H. Wang. 2013. The effect of screw fixation type on a modular hemi-pelvic prosthesis: a 3-D finite element model. *Disability and Rehabilitation: Assistive Technology* 8: 125–8.

Hillmann A., C. Hoffmann, G. Gosheger, R. Rödl, W. Winkelmann, and T. Ozaki. 2003. Tumors of the pelvis: complications after reconstruction. *Archives of Orthopaedic and Trauma Surgery* 123: 340–4.

Hoffmann C., G. Gosheger, C. Gebert, H. Jürgens, and W. Winkelmann. 2006. Functional results and quality of life after treatment of pelvic sarcomas involving the acetabulum. *The Journal of Bone and Joint Surgery. American Volume* 88: 575–82.

Hosalkar H. S., and J. P. Dormans. 2007. Surgical management of pelvic sarcoma in children. *Journal of the American Academy of Orthopaedic Surgeons* 15: 408–24.

Hugate R. Jr., and F. H. Sim. 2006. Pelvic reconstruction techniques. *The Orthopedic Clinics of North America* 37: 85–97.

Huiskes R. 1987. Finite element analysis of acetabular reconstruction. Noncemented threaded cups. *Acta Orthopaedica Scandinavica* 58: 620–5.

Ivanov, A. A., A. Kiapour, N. A. Ebraheim, and V. Goel. 2009. Lumbar fusion leads to increases in angular motion and stress across sacroiliac joint: a finite element study. *Spine* 34: E162–9.

Ji T., W. Guo, X. D. Tang, and Y. Yang, 2010. Reconstruction of type II + III pelvic resection with a modular hemiplvic endoprosthesis: a finite element analysis study. *Orthopaedic Surgery* 2: 272–7.

Jia Y., L. Cheng, G. Yu et al. 2008. A finite element analysis of the pelvic reconstruction using fibular transplantation fixed with four different rod-screw systems after type I resection. *Chinese Medical Journal* 121: 321–6.

Kawahara N., H. Murakami, A. Yoshida, J. Sakamoto, J. Oda, and K. Tomita. 2003. Reconstruction after total sacrectomy using a new instrumentation technique: a biomechanical comparison. *Spine* 28: 1567–72.

Kiapour A., A. A. Abdelgaward, V. K. Goel, A. Souccar, T. Terai, and N. A. Ebraheim. 2012. Relationship between limb length discrepancy and load distribution across the sacroiliac joint: a finite element study. *Journal of Orthopaedic Research* 30: 1577–80.

Kunze M., A. Schaller, H. Steinke, R. Scholz, and C. Voigt. 2012. Combined multi-body and finite element investigation of the effect of the seat height on acetabular implane stability during the activity of getting up. *Computer Methods and Programs in Biomedicine* 105: 175–82.

Leung, A. S. O., L. M. Gordon, T. Skrinskas, T. Szwedowski, and C. M. Whyne. 2009. Effects of bone density alterations on strain patterns in the pelvis: application of a finite element model. *Proceedings of the Institution of Mechanical Engineers, Part H: Journal of Engineering in Medicine* 223: 965–79.

Liang J., Y. Yang, G. Yu, W. Niu, and Y. Wang. 2011. Deformation and stress distribution of the human foot after plantar ligaments release: a cadaveric study and finite element analysis. *Science China. Life Science* 54: 267–71.

Majumder S. A. Roychowdhury, and S. Pal. 2009. Effects of body configuration on pelvic injury in backward fall simulation using 3D finite element models of pelvis-femur-soft tissue complex. *Journal of Biomechanics* 42: 1475–82.

Mei J., M. Ni, Y. Chen et al. 2008. Anatomic and biomechanical study of the reconstructed pelvis with autograft after hindquarter amputation. *Chinese Journal of Orthopaedics* 28: 667–72.

Mindea S. A., S. Chinthakunta, M. Moldavsky, M. Gudipally, and S. Khalil. 2012. Biomechanical comparison of spinopelvic reconstruction techniques in the setting of total sacrectomy. *Spine* 37: E1622–7.

Niu W., Y. Fan, M. Ni, and J. Mei. 2009. Reconstructed pelvis with autogenous lower-extremity long bones after hindquarter amputation: biomechanical analysis in the sitting posture. *Science & Technology Review* 27 no. 20 (2009): 22–6.

Niu W., Y. Yang, Y. Fan, Z. Ding, and G. Yu. 2008 Experimental modeling and biomechanical measurement of flatfoot deformity. *The Seventh Asian-Pacific Conference on Medical and Biological Engineering,* 22–25 April 2008, Beijing, China, *IFMBE Proceedings* 19: 133–8.

Shim V. B., J. Böshme, P. Vaitl, C. Josten, and I. A. Anderson. 2011. An efficient and accurate prediction of the stability of percutaneous fixation of acetabular fractures with finite element simulation. *Journal of Biomechanical Engineering* 133: 094501.

Small S. R., M. E. Berend, L. A. Howard, D. Tunç, C. A. Buckley, and M. A. Ritter. 2013. Acetabular cup stiffness and implant orientation change acetabular loading patterns. *The Journal of Arthroplasty* 28: 359–67.

Sneppen O., T. Johansen, J. Heerfordt, I. Dissing, and O. Petersen O. 1978. Hemipelvectomy. Postoperative rehabilitation assessed on the basis of 41 cases. *Acta Orthopaedica Scandinavica* 49: 175–9.

Wang S., J. Xiong, C. Zhan et al. 2012. The anatomy of proximal femoral autografts for pelvic reconstruction: a cadaveric study. *Surgical and Radiologic Anatomy* 34: 305–9.

Yu B. S., X. M. Zhuang, Z. M. Li et al. 2010. Biomechanical effects of the extent of sacrectomy on the stability of lumbo-iliac reconstruction using iliac screw techniques: What level of sacrectomy requires the bilateral dual iliac screw technique? *Clinical Biomechanics* 25: 867–72.

Zhang Q. H., J. Y. Wang, C. Lupton et al. 2010. A subject-specific pelvic bone model and its application to cemented acetabular replacements. *Journal of Biomechanics* 43: 2722–7.

Zhou Y., L. Min, Y. Liu et al. 2013. Finite element analysis of the pelvis after modular hemipelvic endoprosthesis reconstruction. *International Orthopaedics* 37: 653–8.

Zhu R., L. Cheng, Y. Yu, T. Zander, B. Chen, and A. Rohlmann. 2012. Comparison of four reconstruction methods after total sacrectomy: a finite element study. *Clinical Biomechanics* 27: 771–6.

# *Section IV*

## *Lower Limb for Rehabilitation*

# 12 Foot–Ankle–Knee Model for Foot Orthosis

*Xuan Liu, Yubo Fan, and Ming Zhang*

## CONTENTS



## SUMMARY

Foot orthoses, for example laterally wedged orthoses, are often used to relieve knee pain and slow the progression of medial knee osteoarthritis. Investigations have been conducted to understand the effectiveness of foot orthoses for relief of the knee joint. However, the loading responses of knee internal structures are difficult to record from experimental studies. Thus, a three-dimensional finite element (FE) model of the human foot–ankle–knee complex was developed to simulate knee joint behavior under orthotic intervention. Subject-specific geometry of the lower extremity was reconstructed from magnetic resonance images. The geometry model generated was imported into ABAQUS for FE modeling. Material properties were obtained from the literature and previous FE work. Laterally wedged orthoses with wedge angles of 0, 5, and 10 degrees were fabricated. Gait analysis data and muscle forces were used as inputs to derive the established FE model. The model was validated through comparison with both in vivo and in vitro experimental data. After validation, the model was applied in a parametric study on the inclination angles of the laterally wedged orthoses. The predictions from the FE model suggested that laterally wedged orthoses with angles of 5 and 10 degrees could diminish the loading in the medial compartment of the knee. With further improvements to the model, together with experimental studies, this research will provide a useful platform for investigating orthotic intervention.

## 12.1 EXTENSION OF FOOT ORTHOTIC FUNCTION TO THE KNEE JOINT

Foot orthoses, external support devices around the foot, are widely used to relieve foot pain and facilitate locomotion; thus directly or indirectly modifying the structural and functional characteristics of the musculoskeletal system. Since Whitman developed a metal foot brace that could manipulate the foot into a proper position (Whitman, 1889), foot orthoses have been used extensively by clinicians for over a century. There is considerable research evidence that supports the therapeutic efficacy and significant mechanical effects of foot orthoses on daily activities such as standing, walking, and running. Foot orthoses are so prevalent in clinical practice because of their noninvasiveness, convenience of use, ease of fabrication, and affordability of materials. Although the most influential effects of foot orthosis focus on structural foot deformities, the use of a foot orthosis should be considered an adjunct to the treatment of lower extremity dysfunction related to poor alignment and faulty mechanics (Nawoczenski and Epler, 1997). There is great promise for an increased understanding and further development of foot orthoses as a valuable therapeutic tool in the treatment of mechanically based lower extremity diseases (Matthew, Werd and Knight, 2010).

Osteoarthritis (OA), as one of the most prevalent human musculoskeletal diseases (Pereira et al., 2011), imposes a significant economic burden on society. OA is a degenerative joint disease that is associated with a defective integrity of articular cartilage and related changes in the entire joint. OA can occur in almost all joints, and knee OA is the most commonly associated with complicated symptoms (pain, swelling, stiffness, etc.) that could further lead to disability of the lower extremity. Knee OA typically manifests as degradation of the knee joint, including erosion of the cartilage and meniscus, bone exposition, and development of bone spurs. OA, as a metabolically active disease, includes both destruction and repair mechanisms that may be triggered by biochemical and mechanical insults (Astephen et al., 2008). Since the knee joint plays a main role in weight-bearing activities, it is more susceptible to mechanical insults than other joints. The integral role of mechanical factors in the development and progression of knee OA is becoming widely acknowledged.

There are two branches of treatment options for knee OA: surgical and nonsurgical. Although surgery can significantly improve the knee joint environment for patients with severe OA, conservative treatment is a preferable noninvasive alternative for patients with mild OA. In this chapter, we introduce the extension of foot orthotic function to the knee joint, which could shed light on both the foot orthotic treatment for knee OA and orthosis design philosophy.

A wedge orthosis is simply a soft wedge inserted underneath a particular side of the foot. To attenuate the progression of medial knee OA, a laterally wedged orthosis, with full length elevation on the lateral side, has been suggested as an intervention strategy. The effects of wedge orthoses as a kind of conservative treatment of knee OA have been frequently studied by clinicians. Since direct measurement of knee joint loading is invasive, gait analysis has been used as an indirect method to quantify the mechanical loading on the lower extremity. In gait analysis, knee adduction moment (KAdM), the product of the frontal plane ground reaction force (GRF) and the moment arm, is used as a golden index of knee joint loading. However, results are inconsistent on the wedge orthotic effects on KAdM (Abdallah and Radwan, 2011; Kakihana et al., 2005; Kerrigan et al., 2002; Maly, Culham and Costigan, 2002). Besides the lingering controversy on the effects, experimental methods can only provide information on the resultant moment, leaving the distributed loading inside the joint unknown. Evaluation of the effectiveness of orthotic interventions and parametric analysis of the orthoses should be investigated further.

## 12.2 EXISTING FINITE ELEMENT ANALYSIS ON THE LOWER EXTREMITIES

Finite element analysis (FEA), as an advanced computer technique for structural analysis, was introduced to biomechanical research for evaluating stress in bones decades ago (Huiskes and Chao, 1983). Since then, many studies have aimed to assess the relationship between load-carrying functions and the morphology of tissues. The basic concept of FEA is the subdivision of a complicated structure into a finite number of discrete elements. Recently, computational methods, particularly FEA, have been used extensively to explore the biomechanical responses of joint internal structures.

FE models of the human knee with complex and irregular geometrical structures were first established in the late 1990s (Bendjaballah, Shirazi-Adl and Zukor, 1995). Over the past decade, a number of knee models have been developed. In industry, FE models of the lower extremity (Beillas et al., 2001; Yue, Shin and Untaroiu, 2011) focused on the response of bony structures to impact in automotive crashes. The loading and boundary conditions in these models were quite different from daily walking. FE models concerned with knee biomechanics (Beillas et al., 2004; Peña et al., 2006;

Shirazi and Shirazi-Adl, 2009; Adouni, Shirazi-Adl and Shirazi, 2012) have shown the potential of computational methods for investigating the behavior or response of knee structures with specified research interests. With FEA, we can predict the stress/strain of any elements in the structure and get an answer about which region sustains the greatest loading. The stress distribution predicted by FE models should be a more direct index of knee loading than the joint moment. In the case of orthotic intervention, existing FE models of the foot and ankle have contributed to the understanding of mechanical interactions between the foot and orthoses (Cheung and Zhang, 2008; Yu et al., 2008; Luo et al., 2011). Further, how the foot orthosis influences knee joint loading through thc foot, ankle, and shank is an intriguing topic.

To our knowledge, current FE models have not considered the interaction between lower extremity joints upon plantar loading. In addition, many FE studies applied simplified loading conditions to a single joint and have not taken into account subject-specific loading, for example plantar loading during walking. Thus, a comprehensive FE model of the foot–ankle–knee complex was developed together with a foot orthosis to simulate the walking stance phase (Liu and Zhang, 2013). The flexibility in simulating complicated loading situations comprising both plantar loadings through various foot orthoses and muscular loadings of the foot–ankle–knee complex using FEA is of great advantage for the quantitative evaluation of orthoses. An improved understanding of orthotic intervention through FEA can be helpful in delaying the progression of knee OA and improving orthotic designs.

## 12.3 STUDY ON LATERALLY WEDGED ORTHOSES USING THE FOOT–ANKLE–KNEE MODEL

### 12.3.1 Gait Analysis and Musculoskeletal Modeling

Inclination angles of laterally wedged orthoses were chosen as the intervention parameter. Orthoses with wedge angles of 0, 5, and 10 degrees were fabricated using a high-density EVA material (A. Algeo Ltd., Liverpool, UK) with a shore hardness of 65. The 5- and 10-degree orthoses were cut from wedged stripes, trimmed to fit the foot size of the subject, and inserted into the shoes. A normal male subject free from any knee joint disease and malalignment of the lower extremity participated in the study. The subject was 34 years old, with a height of 174 cm and a weight of 70 kg. Vicon and AMTI data were collected synchronously at frequencies of 200 and 1000 Hz, respectively. Thirty-nine reflective markers were attached to the skin according to anatomical landmarks. In the experiments, the subject went through several trials before data acquisition to get used to the foot supports and laboratory environment. Static posture trials were recorded at the beginning of the experiment for each pair of wedge orthoses for 5 seconds. The subject then performed three valid walking trials with each pair of orthoses. The subject was required to walk at a self-selected speed to screen natural motion.

The center of pressure (COP) and GRF obtained from gait analysis varied among the three wedge angles. Figure 12.1 illustrates the lateral-medial shift of COP underneath the foot support and GRF direction in the frontal plane with different angles. Use of the laterally wedged orthosis shifted the COP laterally on the foot, which made the GRF move closer to the knee joint center and finally reduce the lever arm of the GRF.

The kinematic data and GRF from gait analysis were inserted into to the musculoskeletal model to calculate muscle forces. OpenSim (Delp et al., 2007) was used for musculoskeletal modeling. Muscle-driven dynamic simulations could calculate variables, including muscle forces, that are not generally measurable in laboratory experiments. Three steps were performed in OpenSim. In step 1, the OpenSim model was scaled to match the height and mass of the specific subject. The motion data was then loaded into the OpenSim gait model. Due to experimental error and modeling assumptions, measured GRF and moments are often dynamically inconsistent with the model kinematics. In step 2, a residual reduction algorithm was applied to make the coordinates of the model more dynamically consistent with the measured GRF and moments. In step 3, the computed muscle control algorithm was used to compute muscle activations, which drive the generalized

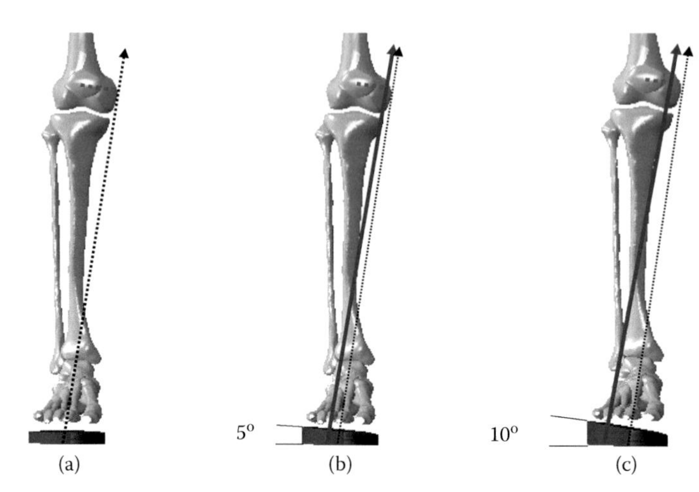


**FIGURE 12.1** The lateral-medial shift of the center of pressure and the ground reaction force (GRF) directions in the frontal plane with orthoses at (a) 0-degree, (b) 5-degree, and (c) 10-degree wedge angles. The blocks beneath the feet represent orthoses at three wedge angles. The directions of the GRF in wedge conditions are shown by arrows in the solid blue line. The GRF in the flat orthosis condition is the control, shown as a dotted arrow in (b) and (c). The dashed line represents the lever arm of the GRF, which passes through the center of the knee joint.

coordinates of a dynamic musculoskeletal model toward a desired kinematic trajectory. The calculated muscle forces were later input to the FE model together with gait analysis data to investigate the internal loadings of the knee joint under different orthotic interventions.

### 12.3.2 Finite Element Modeling of the Foot–Ankle–Knee Complex and Orthosis

The subject who was involved in the gait analysis also volunteered for magnetic resonance (MR) scanning of the right lower extremity. The MR images were scanned in a neutral unloaded position. The images were segmented using MIMICS v14 (Materialise, Inc., Leuven, Belgium) to obtain the boundaries of the skeleton and skin surface. After reconstruction from two-dimensional images, the three-dimensional objects generated were required to go through several steps before being imported to the FE software. Rapidform XOR3 (INUS Technology, Inc., Seoul, South Korea) was used to convert surface models to solid models. The models generated were then imported into ABAQUS v6.11 (Dassault Systems Simulia Corp., Providence, Rhode Island) for FE modeling and analysis.

The FE model of the human foot–ankle–knee complex (Figure 12.2) consisted of 30 bony segments, including the distal segment of the femur, patella, tibia, fibula, and 26 foot bones. All the bony and ligamentous structures in the foot and ankle region were embedded in a volume of soft tissue. In the knee joint, the menisci, articular cartilages, anterior cruciate ligament (ACL), posterior cruciate ligament (PCL), medial collateral ligament (MCL), and lateral collateral ligament (LCL) were built as soft structures. Foot and ankle ligaments and the plantar fascia were simulated as tension-only truss elements. The bones and other soft structures in the model were meshed with tetrahedral elements. Frictionless surface-to-surface contact behavior was defined between the contacting bony structures because of the lubricating nature of the articulating surfaces. It was assumed that the overall joint stiffness against shear loading was governed by the surrounding ligaments and muscles together with the contacting stiffness between the adjacent contoured articulating surfaces.

Geometric models of the orthoses for each wedge angle were created in Rapidform XOR3 based on the shape of the orthosis used in experiments and then assembled with the foot–ankle–knee complex in ABAQUS. The FE mesh of the foot supports was composed of an insole layer, a midsole layer, and an outsole layer. The three layers were tied together and the insole layer represented the wedge orthosis. A horizontal plate consisting of an upper concrete layer and a rigid bottom layer was

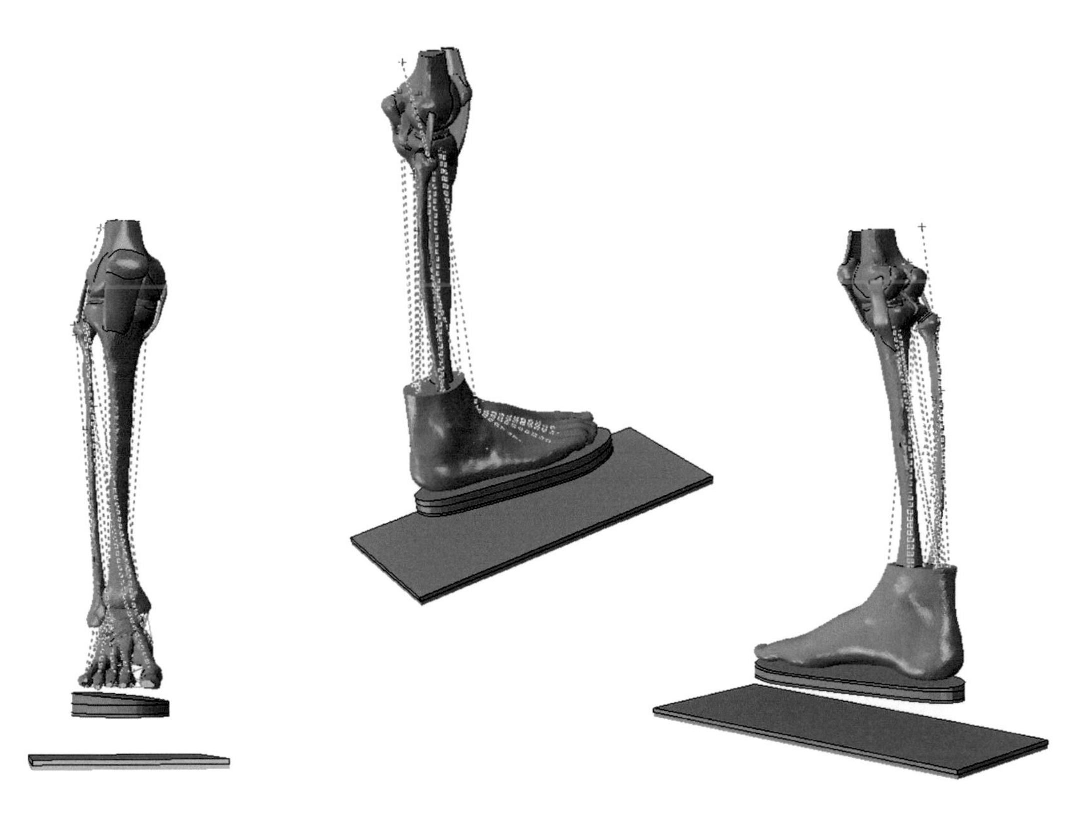

**FIGURE 12.2** Three-dimensional finite element model of the human foot–ankle–knee complex together with a laterally wedged orthosis. The blue dotted wires represent connector elements.

used to simulate the ground. The foot supports and ground were meshed with hexahedral elements. Surface-to-surface contact was also used to simulate the foot-insole and outsole-ground interfaces with the coefficient of friction ($\mu$) set at 0.6 (Zhang and Mak, 1999).

To reduce the complexity of the problem, linearly elastic materials were chosen to represent the mechanical properties of the overall bony structures, knee cartilages, menisci, foot ligaments, and the ground support of the FE model. The material properties of the foot structures, foot supports, and ground were in accordance with previous studies (Cheung et al., 2005; Cheung and Zhang, 2008). The knee ligaments were defined as hyperelastic materials. Stress-strain relationships were obtained from a model developed by Mesfar and Shirazi-Adl (2005). The Young's modulus and Poisson's ratio for the knee bony structures were assigned 16000 MPa and 0.3, respectively (Reilly and Burstein, 1975). The cartilage layers of the tibia, femur, and fibula were assigned a Young's modulus of 12 MPa and Poisson's ratio of 0.45 (Moglo and Shirazi-Adl, 2003). The menisci were assigned a Young's modulus of 59 MPa and Poisson's ratio of 0.49 (LeRoux and Setton, 2002).

The alignment of the model from the MR scan was modified to match the motion data through muscle control. The knee and ankle joint angles were adjusted according to recorded kinematic locations for each wedge. Foot–ground angles in three anatomical planes were altered directly in the model assembly by rotating the ground. GRF was applied at the location of COP underneath the ground support, which was allowed to translate and rotate in all degrees of freedom. As the boundary condition, the femur bone was cut approximately 10 cm above the femur condyles, and was fully constrained through a local rigid body constraint on the distal femur. Muscle forces were applied to ABAQUS connector elements, which model discrete physical connections between bodies. The insertion points of the ligaments and muscles were determined according to MR images together with anatomy software (Interactive Series, Primal Picture Limited, London, UK).

The FE model was validated through several comparative studies. An FE simulation of a walking mid-stance phase was conducted to compare tibiofemoral pressure with data from the cadaveric experiment. The specimen was from a middle-aged male donor with height of 170 cm and weight of 60 kg, free from any pathology and deformity of the lower extremity. A medial parapatellar approach to the knee was used to expose the tibiofemoral articular surfaces, which allowed sub-femoral insertion of the K-Scan pressure sensors (Tekscan Inc., Boston, Massachusetts). The tendons were separated for muscle loading. A material testing unit (Bose Inc., Eden Prairie, Minnesota) was used for compression tests. The FE model was also assigned plantar pressure boundary conditions recorded with an in vivo F-Scan (Tekscan Inc.). The FE-predicted pattern of plantar pressure in mid-stance was compared with the F-Scan data for validation.

### 12.3.3 Finite Element Predictions

Figure 12.3 depicts the tibiofemoral pressure and plantar pressure distribution obtained from cadaveric experiments and FE simulations using a flat insole in mid-stance. The K-Scan measurement of peak tibiofemoral contact pressure during mid-stance was 3.67 MPa (Figure 12.3a). The FE-predicted peak contact pressure was 3.89 MPa (Figure 12.3b), which was comparable with the cadaveric data. The contact area predicted by the FE model was approximately 477.0 $mm^2$, compared with 491.4 $mm^2$ from the K-Scan measurement. Loading transmitted through the menisci was obvious in the FE predictions at the exterior borders of both the medial and lateral compartments of the femoral cartilage but could not be observed in the K-Scan results because of the limited size

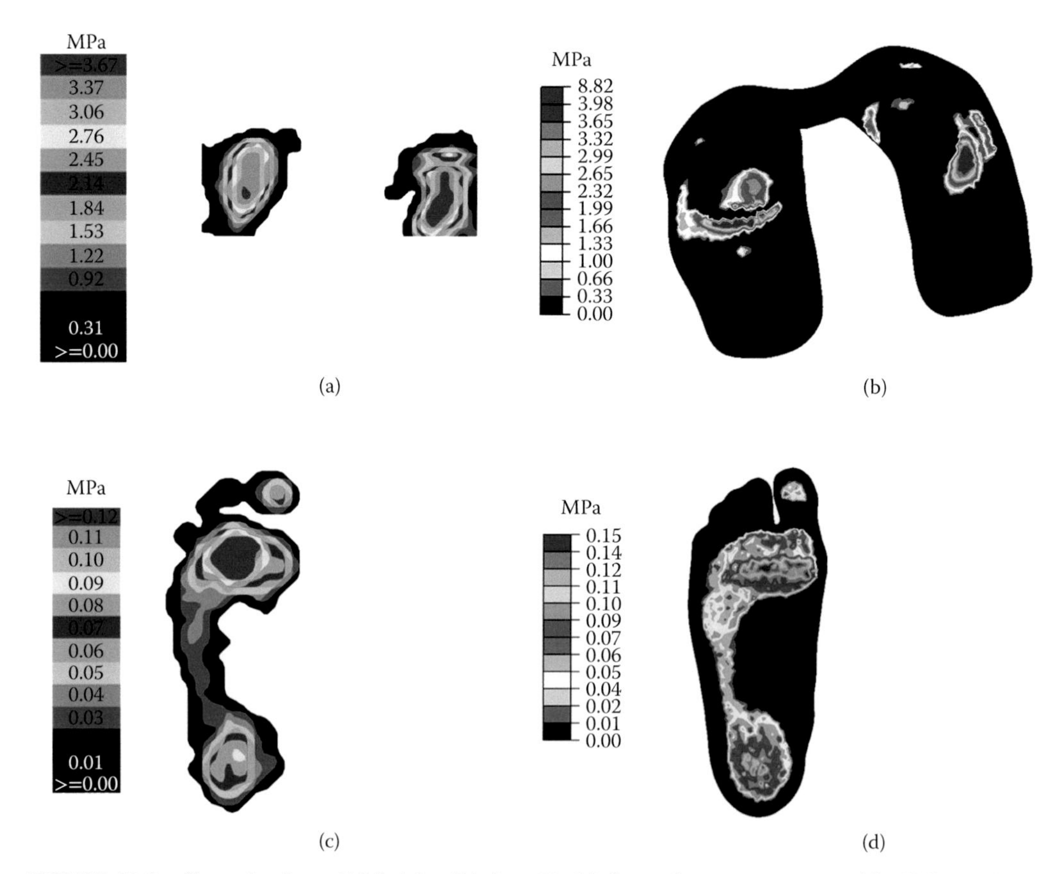


**FIGURE 12.3** **(See color insert.)** Model validation: (a) tibiofemoral pressure measured by K-Scan; (b) FE predicted tibiofemoral pressure; (c) plantar pressure measured by F-Scan; (d) FE predicted plantar pressure.

of the K-Scan sensor and the buckling phenomenon of the sensor edge during testing. The difference in pattern may also be attributed to the inconformity in knee geometries between the modeled subject and the cadaveric specimen. The peak plantar pressure patterns were comparable between FE predictions and F-Scan results when using a flat insole. The F-Scan measurement of peak plantar pressure during mid-stance was 0.11 MPa (Figure 12.3c). The FE-predicted peak plantar pressure was 0.15 MPa (Figure 12.3d). The contact area predicted by the FE model was approximately 68 $cm^2$ for a 0-degree wedge, compared with 70 $cm^2$ from F-Scan measurements. The difference between the plantar pressure data obtained from the F-Scan measurement and FE prediction was mainly due to differences in resolution.

The validated FE model was used to predict the strain, pressure, and shear stress in the tibiofemoral joint during mid-stance. The instant when the lowest GRF value appears was chosen as the cut-off point for mid-stance input data processing, at approximately 45% to 55% of the stance phase, close to heel rise. Figure 12.4 illustrates the minimum (compressive) principal strain distribution patterns in the femoral cartilage and menisci with 0-, 5-, and 10-degree laterally wedged orthoses during mid-stance. The peak minimum principal strain on the femoral cartilage and menisci among the three wedges were all predicted to occur in the medial side. The peak minimum principal strain at the medial femoral cartilage region decreased by approximately 1.3% using a 5-degree orthosis, and 10.8% using a 10-degree orthosis, compared with the 0-degree wedge model. Meanwhile, in comparison with the 0-degree wedge, the peak minimum principal strain at the medial meniscus decreased by approximately 7.7% using a 5-degree orthosis, and 25% using a 10-degree orthosis.

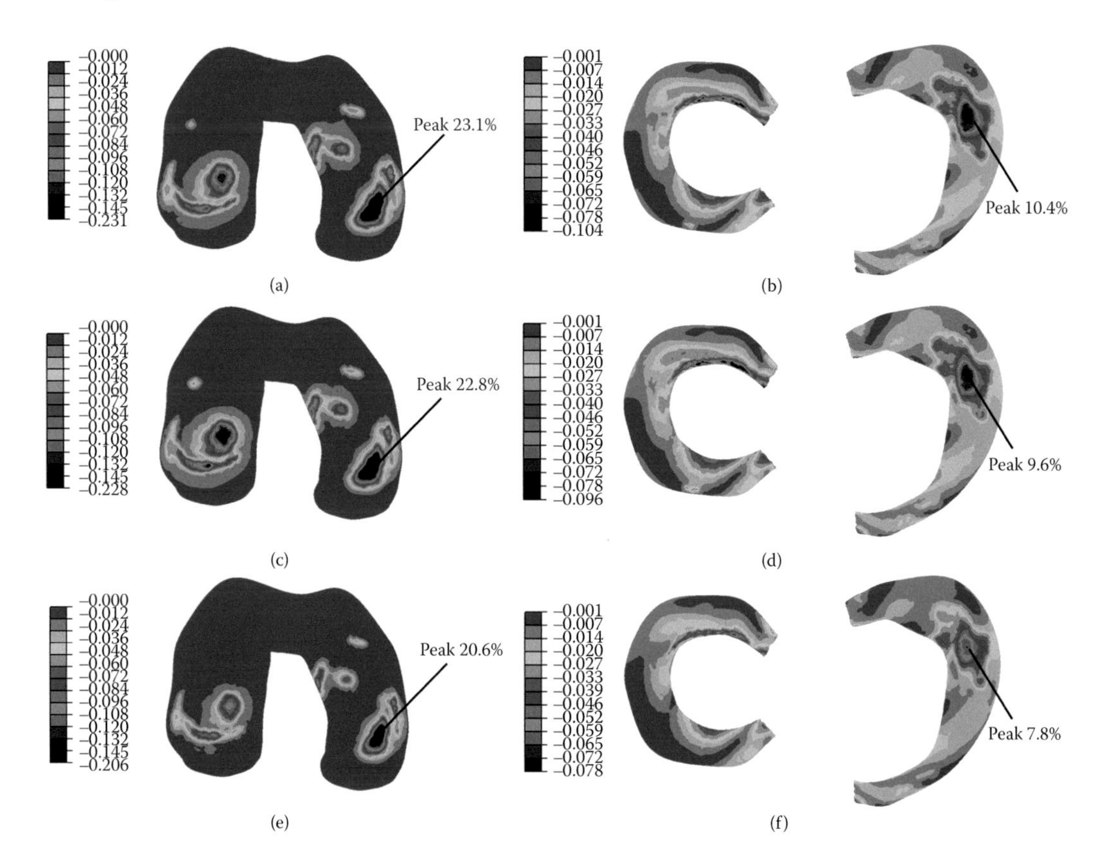


**FIGURE 12.4 (See color insert.)** Minimum (compressive) principal strain in the femur cartilage with (a) 0-, (c) 5-, and (e) 10-degree laterally wedged orthosis and minimum principal strain in the meniscus with (b) 0-, (d) 5-, and (f) 10-degree laterally wedged orthosis.

Figure 12.5 shows the contact pressure distribution patterns in the femoral cartilage and menisci with 0, 5-, and 10-degree laterally wedged orthoses during mid-stance. The peak contact pressure at the femur cartilage and menisci among the three wedge conditions were all predicted to occur in the medial side. In comparison to the 0-degree wedge, the peak contact pressure at the medial femoral cartilage region decreased by approximately 8.9% using a 5-degree orthosis, and 22.6% using a 10-degree orthosis. Meanwhile, the peak contact pressure at the medial meniscus decreased by approximately 10.4% using a 5-degree orthosis and 23.5% using a 10-degree orthosis when compared to the 0-degree model.

Figure 12.6 reveals the lateral-medial shear stress distribution patterns on the tibial plateau with 0-, 5-, and 10-degree laterally wedged orthoses during mid-stance. The peak shear stresses in the medial direction among the three wedge conditions were all predicted to occur on the medial plateau, while the peak shear stresses in the lateral direction all occurred on the lateral plateau. The lateral-medial shear stress in the tibial plateau appeared at the peripheral meniscal contact region, which indicated the tendency for the menisci to extrude (Figure 12.6). The peak shear stress in the medial side of the tibia plateau was much larger than the lateral side due to more prominent extrusion for the medial meniscus. In comparison to the 0-degree wedge, the peak shear stress in the medial direction at the tibial plateau decreased by approximately 13.1% using a 5-degree orthosis, and 13.9% using a 10-degree orthosis. Shear stress has been shown to be detrimental to cartilage health because it decreases the expression of cartilage matrix proteins (Lee et al., 2002). The finding

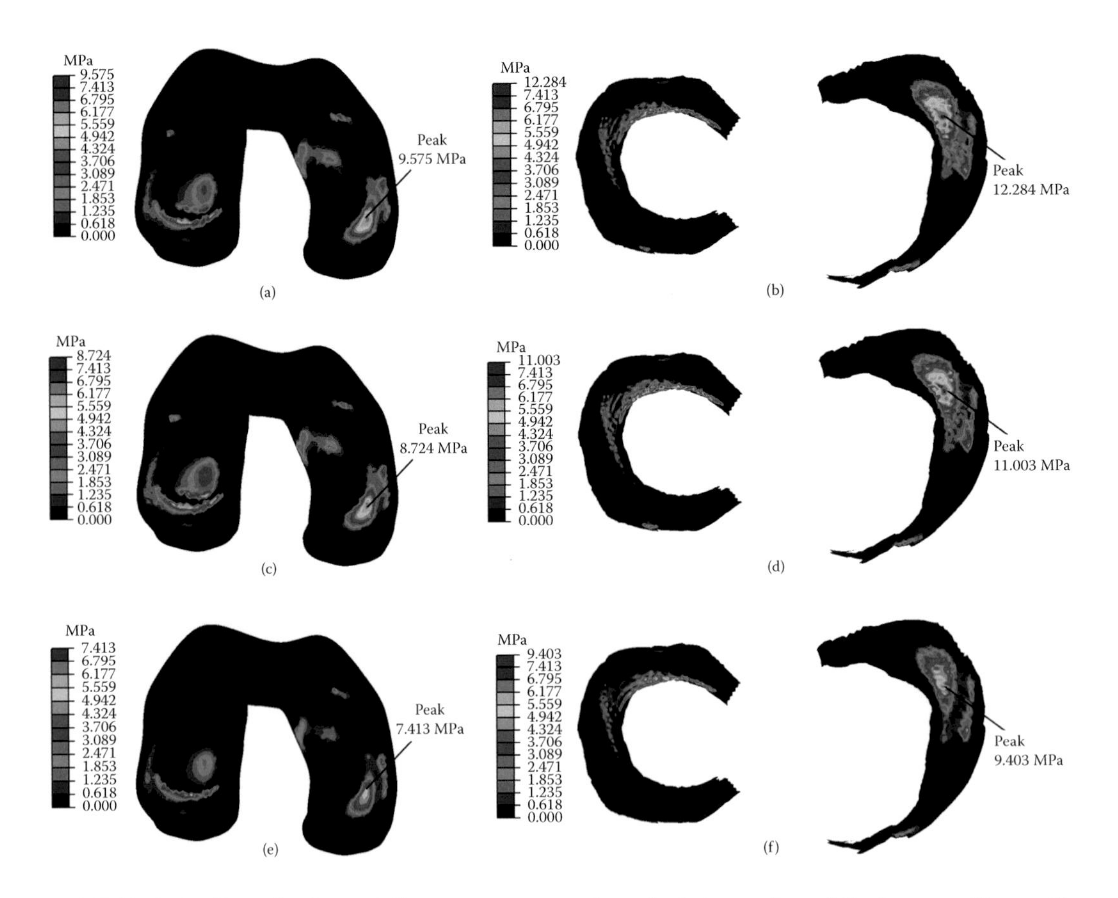


**FIGURE 12.5 (See color insert.)** Contact pressure on the articulation surface of the femur cartilage with (a) 0-, (c) 5-. and (e) 10-degree laterally wedged orthosis and contact pressure on articulation surfaces of the meniscus with (b) 0-, (d) 5-, and (f) 10-degree laterally wedged orthosis.

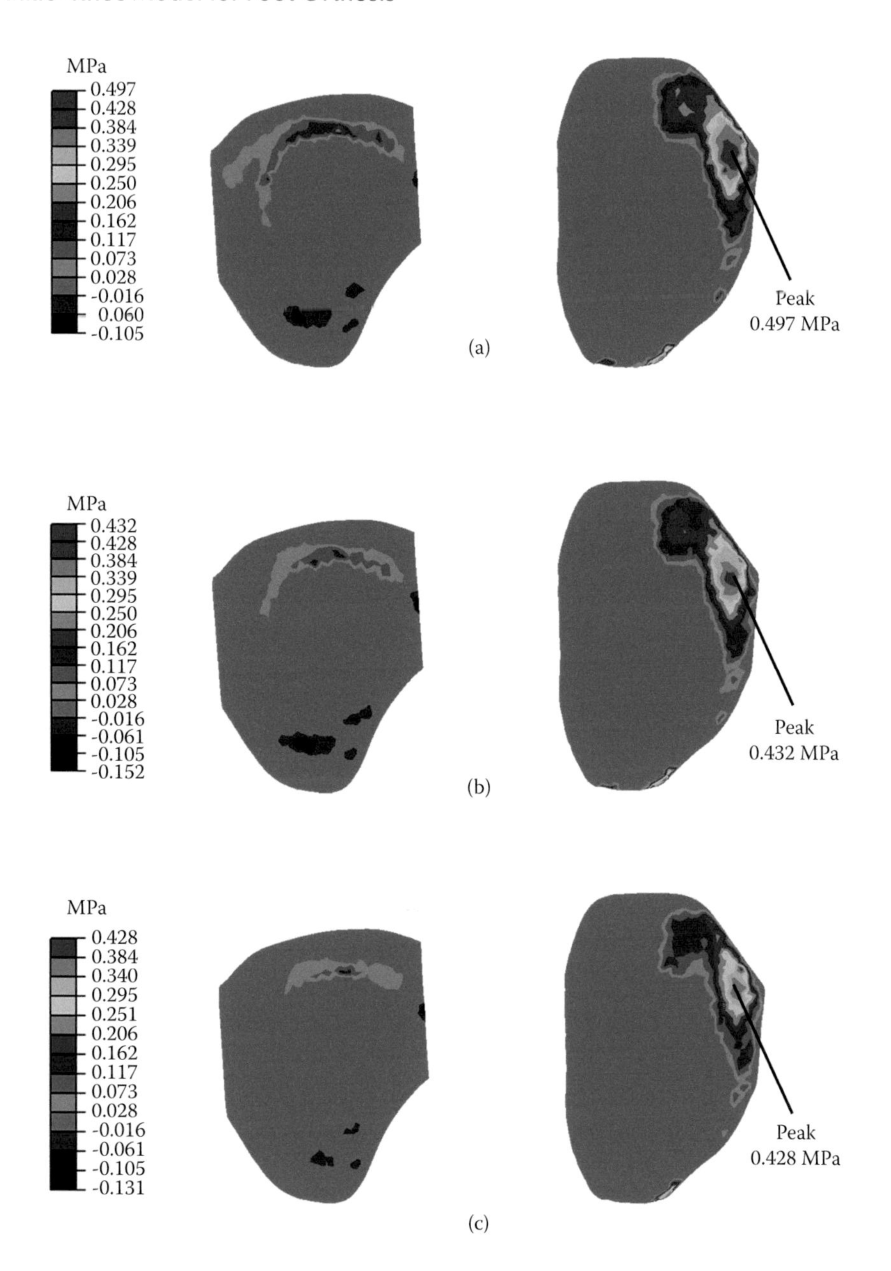


**FIGURE 12.6 (See color insert.)** Lateral-medial shear stress on the tibial plateau with (a) 0-, (b) 5-, and (c) 10-degree laterally wedged orthoses.

that a laterally wedged orthosis decreased the tibiofemoral shear stress in the medial direction indicated the effectiveness of this intervention.

The predicted minimal principal strain, contact pressure, and lateral-medial shear stresses in the tibiofemoral articulation during mid-stance were all revealed to be higher at the medial compartment (Figures 12.4, 12.5, and 12.6). This phenomenon was consistent with other FE studies (Shim et al., 2009; Yang et al., 2010) which also predicted the dominance of the medial component in the knee joint during mid-stance. The laterally wedged orthoses were predicted to release some of the loading on the medial joint, which proves the effectiveness of the orthosis. It should be noted that this FE model was based on a single subject, making it difficult to draw any general conclusions. Our FE simulation demonstrated the immediate effect of the laterally wedged insoles. The long-term therapeutic effects and clinical effectiveness of laterally wedged insoles could not be fully realized through this study. However, the predicted joint loading distribution suggested an influence on the morphology of the regional cartilage (Koo, Rylander, and Andriacchi, 2011), which has a strong correlation with joint degeneration.

## 12.4 CONCLUSIONS

Our FE predictions furnished necessary information on the distributions of joint internal loading, that could not be achieved from gait analysis techniques alone. The finding that laterally wedged orthoses redistributed the knee internal loadings may contribute to medial knee OA rehabilitation. Besides exploring joint biomechanical responses to interventions, designing foot orthoses that improve the joint loading environment should be a primary goal. With the implementation of parametric analyses, any features of the foot orthosis could be altered through FE simulation. Therefore, this foot–ankle–knee model will also provide scientific fundamentals for designing optimal foot orthoses.

## ACKNOWLEDGMENTS

This study is supported by grants from the National Natural Science Foundation of China (11272273, 11120101001) and the Research Grant Council of Hong Kong (PolyU 532611E).

## REFERENCES

Abdallah, A. A., and A. Y. Radwan. 2011. Biomechanical changes accompanying unilateral and bilateral use of laterally wedged insoles with medial arch supports in patients with medial knee osteoarthritis. *Clinical Biomechanics* 26:783–9.

Adouni, M., A. Shirazi-Adl, and R. Shirazi. 2012. Computational biodynamics of human knee joint in gait: from muscle forces to cartilage stresses. *Journal of Biomechanics* 45:2149–56.

Astephen, J. L., K. J. Deluzio, G. E. Caldwell, and M. J. Dunbar. 2008. Biomechanical changes at the hip, knee, and ankle joints during gait are associated with knee osteoarthritis severity. *Journal of Orthopaedic Research* 26:332–41.

Beillas, P., P. C. Begeman, K. H. Yang, A. I. King, P. J. Arnoux, H. S. Kang, K. Kayvantash, C. Brunet, C. Cavallero, and P. Prasad. 2001. Lower limb: advanced FE model and new experimental data. *Stapp Car Crash Journal* 45:469–94.

Beillas, P., G. Papaioannou, S. Tashman, and K. H. Yang. 2004. A new method to investigate in vivo knee behavior using a finite element model of the lower limb. *Journal of Biomechanics* 37:1019–30.

Bendjaballah, M. Z., A. Shirazi-Adl, and D. J. Zukor. 1995. Biomechanics of the human knee joint in compression: reconstruction, mesh generation and finite element analysis. *Knee* 2:69–79.

Cheung, J. T. M., and M. Zhang. 2008. Parametric design of pressure-relieving foot orthosis using statistics-based finite element method. *Medical Engineering & Physics* 30:269–77.

Cheung, J. T. M., M. Zhang, A. K., Leung, and Y. B. Fan. 2005. Three dimensional finite element analysis of the foot during standing: a material sensitivity study. *Journal of Biomechanics* 38:1045–54.

Delp, S. L., F. C. Anderson, A. S. Arnold, P. Loan, A. Habib, C. T. John, E. Guendelman and D. G. Thelen. 2007. OpenSim: open-source software to create and analyze dynamic simulations of movement. *IEEE Transactions on Bio-medical Engineering* 54: 1940–50.

Huiskes, R., and E. Y. Chao. 1983. A survey of finite element analysis in orthopedic biomechanics: the first decade. *Journal of Biomechanics* 16:385–409.

Kakihana, W., M. Akai, K. Nakazawa, T. Takashima, K. Naito, and S. Torii. 2005. Effects of laterally wedged insoles on knee and subtalar joint moments. *Archives of Physical Medicine and Rehabilitation* 86:1465–71.

Kerrigan, D. C., J. L. Lelas, J. Goggins, G. J. Merriman, R. J. Kaplan, and D. T. Felson. 2002. Effectiveness of a lateral-wedge insole on knee varus torque in patients with knee osteoarthritis. *Archives of Physical Medicine and Rehabilitation* 83:889–93.

Koo, S., J. H. Rylander, and T. P. Andriacchi. 2011. Knee joint kinematics during walking influences the spatial cartilage thickness distribution in the knee. *Journal of Biomechanics* 44:1405–9.

Lee, M. S., M. C. Trindade, T. Ikenoue, D. J. Schurman, S. B. Goodman, and R. L. Smith. 2002. Effects of shear stress on nitric oxide and matrix protein gene expression in human osteoarthritic chondrocytes in vitro. *Journal of Orthopaedic Research* 20:556–61.

LeRoux, M. A., and L. A. Setton. 2002. Experimental and biphasic FEM determinations of the material properties and hydraulic permeability of the meniscus in tension. *Journal of Biomechanical Engineering* 124:315–21.

Liu, X., and M. Zhang. 2013. Redistribution of knee stress using laterally wedged insole intervention: finite element analysis of knee–ankle–foot complex. *Clinical Biomechanics* 28:61–7.

Luo, G., V. L. Houston, M. A. Garbarini, A. C. Beattie, and C. Thongpop. 2011. Finite element analysis of heel pad with insoles. *Journal of Biomechanics* 44:1559–65.

Maly, M. R., E. G. Culham, and P. A. Costigan. 2002. Static and dynamic biomechanics of foot orthoses in people with medial compartment knee osteoarthritis. *Clinical Biomechanics* 17:603–10.

Matthew, B., E. Werd, and L. Knight. 2010. *Athletic Footwear and Orthotics in Sports Medicine*. New York: Springer.

Mesfar, W., and A. Shirazi-Adl. 2005. Biomechanics of the knee joint in flexion under various quadriceps forces. *Knee* 12: 424–34.

Moglo, K. E., and A. Shirazi-Adl. 2003. On the coupling between anterior and posterior cruciate ligaments, and knee joint response under anterior femoral drawer in flexion: a finite element study. *Clinical Biomechanics* 18:751–9.

Nawoczenski, D. A., and M. E. Epler. 1997. Orthotics in functional rehabilitation of the lower limb. Philadelphia, PA: W.B. Saunders.

Peña, E., B. Calvo, M. A. Martinez, and M. Doblare. 2006. A three-dimensional finite element analysis of the combined behavior of ligaments and menisci in the healthy human knee joint. *Journal of Biomechanics* 39:1686–701.

Pereira, D., B. Peleteiro, J. Araujo, J. Branco, R. A. Santos, and E. Ramos. 2011. The effect of osteoarthritis definition on prevalence and incidence estimates: a systematic review. *Osteoarthritis and Cartilage* 19:1270–85.

Reilly, D. T., and A. H. Burstein. 1975. The elastic and ultimate properties of compact bone tissue. *Journal of Biomechanics* 8:393–405.

Shim, V. B., P. Mithraratne, I. A. Anderson, and P. J. Hunter. 2009. Simulating in-vivo knee kinetics and kinematics of tibio-femoral articulation with a subject-specific finite element model, in *IFMBE Proceedings of World Congress on Medical Physics and Biomedical Engineering*, Munich, Germany.

Shirazi, R., and A. Shirazi-Adl. 2009. Computational biomechanics of articular cartilage of human knee joint: effect of osteochondral defects. *Journal of Biomechanics* 42:2458–65.

Whitman, R. 1889. Observations on seventy-five cases of flat foot, with particular reference to treatment. *The Journal of Bone & Joint Surgery* s1-1:122–37.

Yang, N. H., H. Nayeb-Hashemi, P. K. Canavan, and A. Vaziri. 2010. Effect of frontal plane tibiofemoral angle on the stress and strain at the knee cartilage during the stance phase of gait. *Journal of Orthopaedic Research* 28:1539–47.

Yu, J., J. T. Cheung, Y. Fan, Y. Zhang, A. K. Leung, and M. Zhang. 2008. Development of a finite element model of female foot for high-heeled shoe design. *Clinical Biomechanics* 23:S31–8.

Yue, N., J. Shin, and C. D. Untaroiu. 2011. *Development and validation of an occupant lower limb finite element model.* SAE World Congress and Exhibition, Detroit.

Zhang, M., and A. F. Mak. 1999. In vivo skin frictional properties. *Prosthetics and orthotics international* 23: 135–41.

# 13 Lower Residual Limb for Prosthetic Socket Design

*Winson C.C. Lee and Ming Zhang*

## CONTENTS



## SUMMARY

Comfort is among the most important issues when fitting a prosthesis. However, high stress applied to the residual limb, which is not particularly tolerant to loading, can cause discomfort, pain, and tissue breakdown. In an attempt to improve prosthesis fit, it is important to study the stress distribution at the residual limb-prosthetic socket interface. Computational finite element (FE) modeling allows for efficient parametric analysis and is a useful tool for investigating the load transfer mechanics at the limb-socket interface. Due to the complicated frictional and sliding actions at the interface, however, simulation of the mechanical interaction between the limb and socket is challenging. In addition, a prosthetic socket is usually shape-rectified so as to redistribute the load to load-tolerant regions of the residual limb. After donning the shape-rectified socket, some mechanical stresses known as pre-stresses are produced. Many previous models have incorporated some simplifying assumptions when simulating the friction-slip and pre-stresses. This chapter illustrates a technique that simulates the contact at the limb-socket interface, considering both the friction/slip and pre-stress conditions, by using an automated contact method.

## 13.1 INTRODUCTION

A lower-limb prosthesis can improve physical appearance and, more importantly, restore the lost functions of individuals with lower-limb amputations. To be effective, however, it should not produce any discomfort or pain. Prosthetic socket design is the most important factor in determining the comfort of using a prosthesis. To achieve a successful fit, it is important to understand the mechanical interaction between the prosthetic socket and the residual limb. FE modeling is a useful tool to understand the load transfer mechanics at the limb-socket interface. FE analyses have provided a better understanding of the effects on stress distribution over the residual limb of socket modifications (Zhang, 1995; Reynolds and Lord, 1992); material properties of the sockets (Quesada and Skinner, 1991) and liners (Simpson, Fisher, and Wright, 2001); alignment (Reynolds and Lord, 1992; Sanders and Daly, 1993); residual limb geometry (Zhang, 1995) and mechanical properties (Reynolds and Lord, 1992); and frictional properties at the interface (Zhang and Lord, 1995).

Simulation of the mechanical interaction between the limb and socket in FE modeling is challenging because of frictional and sliding actions at the interface. In addition, the residual limb is donned into a socket with a different shape from the naked residual limb surface. In some models, it was assumed that the residual limb and prosthetic socket were fully connected as one body assigned different mechanical properties (Reynolds and Lord, 1992; Sanders and Daly, 1993; Steege, Schnur, and Childress, 1987). This reduced the difficulties of modeling and computational time. However, it did not take into account slippage at the interface and large in-plane stresses that might develop on the surface of the limb. Another commonly adopted assumption is that the shape of the residual limb and the rectified socket are the same (Reynolds and Lord, 1992; Quesada and Skinner, 1991; Steege, Schnur, and Childress, 1987; Zachariah and Sanders, 2000). Under this assumption, modifications to the socket shape aiming to redistribute the load to load-tolerant regions cannot be implemented in the FE model. The stresses imposed on the residual limb after donning into the rectified socket, defined as pre-stresses in this chapter, were ignored under the above assumption.

Different methods have been used to account for friction/slip. Zhang et al. (1995) added interface elements between the socket and the limb and allowed sliding between the two bodies if the calculated shear stresses exceeded the frictional limit. However, if using interface elements the model must enforce point-to-point physical connections between the limb surface and the inner surface of the socket. An automated contact method was used by Zachariah and Sanders (2000). In this method, the residual limb and prosthetic socket were modeled as two deformable bodies. The FE software package simulated contact between the two bodies by automatically detecting any overlapping of interface nodes and imposing a nonpenetration condition constraint to the overlapped nodes. However, Zachariah and Sanders (2000) did not calculate pre-stresses produced by donning the residual limb into the shape-rectified socket.

In some models, simulation of donning the residual limb into a rectified socket has been implemented by applying radial displacements to the nodes of the unrectified socket to deform it into the rectified socket shape (Zhang et al., 1995; Silver-Thorn and Childress, 1997). This method requires a significantly longer modeling time because all the nodes at the areas that require shape modifications have to be identified and changes of coordinates imposed. It is also difficult to implement this with automeshing techniques in model development.

This chapter illustrates a technique that simulates the contact at the limb-socket interface, considering both the friction/slip and pre-stress conditions by using an automated contact method. Neither point-to-point connections to simulate the friction/slip condition nor radial nodal displacements to simulate pre-stress were required.

## 13.2 MODEL DEVELOPMENT

The model, also described in Lee et al. (2004), was based on the geometry of the residual limb of a 55-year-old, right-sided trans-tibial amputee. He was 158 cm tall and 80 kg in mass, with a residual limb length of 14.7 cm measured from the mid-patellar tendon to the distal end of the residual limb.

### 13.2.1 Geometries

The geometries of the residual limb surface and the internal bones were obtained by taking magnetic resonance imaging (MRI) scans of the residual limb in supine lying and knee extended positions, with axial cross-sectional images taken at 6-mm intervals. To reduce the distortion of the soft tissues at the posterior regions during the scanning procedure, the subject wore an unrectified

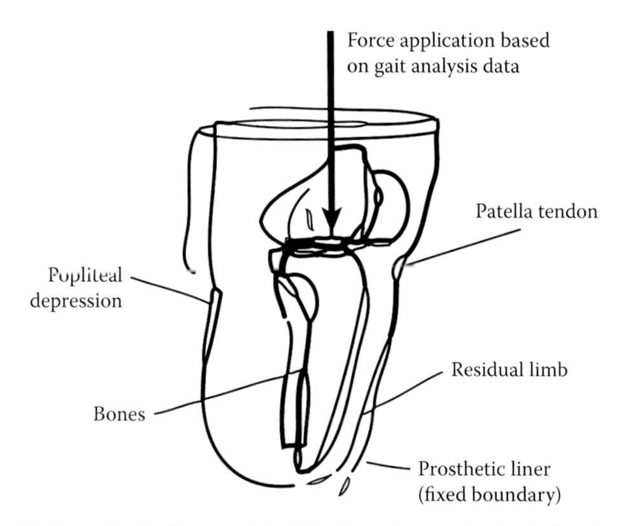


**FIGURE 13.1** Assembled prosthetic liner, residual limb, and bones in the finite element model. There are overlappings at the patellar tendon and popliteal depression regions.

socket, based on a loose plaster cast of the limb surface. The bones and the limb surfaces were identified and segmented using MIMICS software (Materialise, Inc., Leuven, Belgium).

The unrectified cast, representing the residual limb surface, was digitized using the BioSculptor™ system and exported to the computer-aided design (CAD) software ShapeMaker™ (Seattle Limb System). This shape was modified into a soft liner placed within a hard patellar tendon-bearing (PTB) socket. A rectification template (Radcliff and Foort, 1961) was applied onto this shape by adding buildups at pressure-sensitive areas and undercuts at pressure-tolerant areas to prepare the inner surface of the prosthetic liner.

The surfaces of the bones, residual limb, and prosthetic liner were imported to SolidWorks 2001 (SolidWorks Corporation, Massachusetts). Assembling of the liner, limb, and bones was performed based on the relative positions among the three structures. Positions of the bones relative to the residual limb surface were based on the same set of MRI scans. The position of the shape-modified prosthetic liner relative to the residual limb surface was based on the outputs from Shapemaker™, in which the prosthetic liner was produced by adding buildups and undercuts directly to appropriate regions of the residual limb model. The assembled liner, limb, and bones are shown in Figure 13.1. It can be seen in the figure that there were some overlapping areas at the interface where socket undercuts were made.

The surfaces were then converted into solid models using SolidWorks. The soft tissue model was generated by geometrically subtracting the bones from the solid limb. The prosthetic liner was given a thickness of 4 mm. The solid models representing bones, soft tissues, and liner were then imported to the finite element package ABAQUS (Hibbitt, Karlsson & Sorensen, Inc., Pawtucket, Rhode Island). An FE mesh with a total of 22,301 three-dimensional tetrahedral elements was built using ABAQUS automeshing techniques. The meshed geometries of the residual limb, prosthetic liner, and bones are shown in Figure 13.2.

### 13.2.2 Material Properties

The material properties were assumed to be linearly elastic, isotropic, and homogeneous. Young's modulus was 200 kPa for soft tissues and 10 GPa for bones. Poisson's ratio was assumed to be 0.49 for soft tissues and 0.3 for bones (Zhang and Lord, 1995). The prosthetic liner was assigned a Young's modulus of 380 kMPa and Poisson's ratio of 0.39, resembling the mechanical properties of pelite.

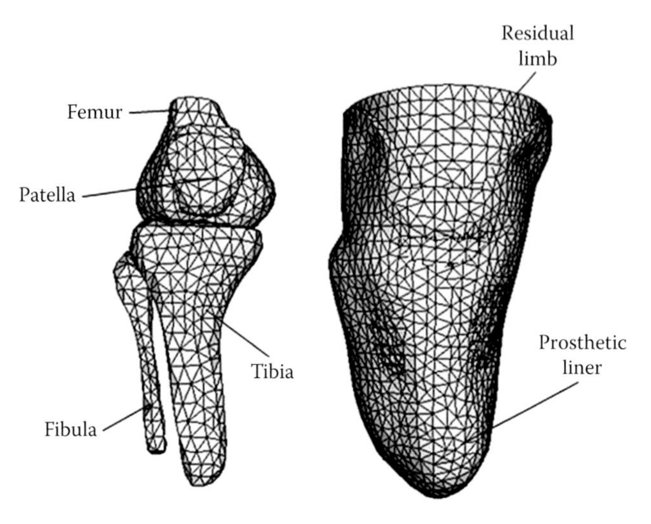


**FIGURE 13.2** Finite element mesh of the residual limb, prosthetic liner, and bones.

### 13.2.3 Boundary Conditions and Pre-Stress Simulations

The prosthetic liner was rigidly fixed, with the assumption that the hard socket would offer rigid support. The bones and soft tissues were tied together with different mechanical properties. The residual limb and liner were modeled as two separate structures, and their interaction was simulated using automated contact methods. The inner surface of the liner and the residual limb surface were defined as master and slave surfaces, respectively. In simulating contact using ABAQUS (Figure 13.3), normal vectors, for example, $N_2$, are computed for all nodes on the master surface by averaging the normal vectors of the outward edges (1-2 and 2-3 segments) that make up the master surfaces. Additional normal vectors, for example $N_{S/2}$, are computed at the middle of each segment. These normal vectors are used to define a set of smooth varying normal vectors on the whole master surface. An "anchor" point on the master surface was calculated for each node on the slave surface (slave node) so that the vector formed by the slave node and the anchor point coincided with one of the smooth varying normal vectors ($N_a$) of the master surface. A tangent plane was determined at every "anchor" point that was perpendicular to the normal vector. Under the master-slave contact algorithm in ABAQUS, the corresponding slave nodes (node 5) are automatically constrained not to penetrate into the tangent planes on the master surface when two surfaces come into contact.

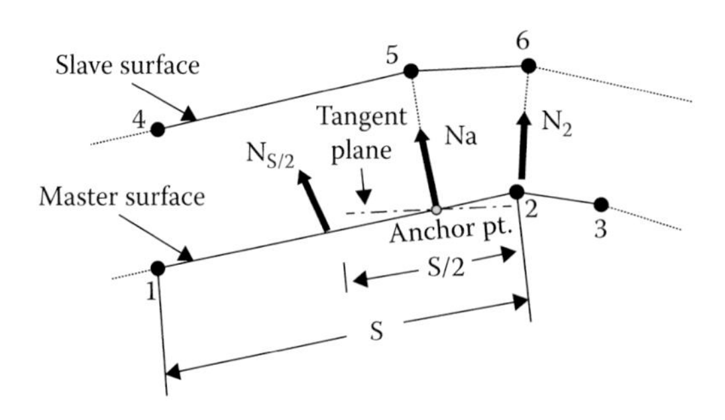


**FIGURE 13.3** Master-slave contact algorithm. An anchor point and a tangent plane are computed for every slave node based on the computed normal vectors ($N_2$, $N_{S/2}$, $N_a$). Each slave node (e.g., node 5) is constrained not to penetrate its tangent planes.

To establish the pre-stress condition from donning the limb into the rectified socket, an axial force of 50 N was applied at the center of the knee. Initially, some nodes on the slave surface (limb surface) penetrated into the master surface (inner surface of the liner) because of the socket rectifications. Under the master-slave contact algorithm, the solver in ABAQUS moved the penetrating slave nodes onto their corresponding tangent planes of the master surface. Stresses (pre-stresses) were developed at both the master and slave surfaces over the overlapping regions.

### 13.2.4 Analysis of the Effects of Pre-Stress on Full Weight Bearing

The calculated pre-stresses and the deformations at the pre-stress stage were retained, and a full body weight (W) of 800 N was added at the tibial plateau (Figure 13.1). There were no artificial constraints imposed between the master and slave surfaces when they were separated as no interface elements were defined at the interface. When the nodes on the slave surface contacted their corresponding tangent planes of the liner, the solver constrained those nodes not to penetrate into the tangent planes and stress was developed at both master and slave surfaces. A coefficient of friction ($\mu$) of 0.5 was assigned for the liner-limb interface (Zhang and Mak, 1999). During the contact phase, sliding was allowed only when the shear stress exceeded the critical shear stress value $\tau > \tau_{crit} = \mu p$, where $p$ is the value of normal stress. During the sliding phase, if the shear stress was reduced and dropped lower than the critical shear stress value, sliding stopped. It was assumed that the static and kinetic coefficients of friction were the same in this model.

To understand how the pre-stresses influenced the predictions, a second model was built for comparison, which was the same as the first model except that the initial geometry of the residual limb copied the shape of the liner and no pre-stress was applied onto the residual limb at the first analysis step. The change in shape of the residual limb after socket donning was simulated in the second model; however, no pre-stress existed as there was no overlapping region between the limb and the liner at initial configuration.

### 13.2.5 Prediction of Interface Stress throughout a Gait Cycle

Loading was simulated over an entire gait cycle. The loads were calculated according to the kinematic data of the lower limb and the prosthesis, and the ground reaction forces applied to the prosthetic foot measured by a Vicon Motion Analysis System (Oxford Metrics, UK) and a force platform (AMTI, USA) (Jia, Zhang, and Lee, 2004). The equivalent forces and moments applied at the knee joint during walking were calculated using inverse dynamics. To simplify the problem during the calculation of the joint loads, assumptions were made that there was no relative movement between the residual limb and socket during walking and only inertial effects in the sagittal plane were considered.

## 13.3 MODEL FINDINGS

### 13.3.1 Pre-Stress Effects on Full Weight Bearing

When the limb was donned into the socket, high normal stress was produced at the regions where socket undercuts were made, including the areas around the patellar tendon, popliteal depression, anteromedial tibial, and anterolateral tibial. Figure 13.4a displays the normal stress distribution when a full body-weight loading (800 N) was applied with prior pre-stressing. The normal stresses over regions where socket undercuts were made increased further, up to a maximum of 228 kPa over the mid-patellar tendon region.

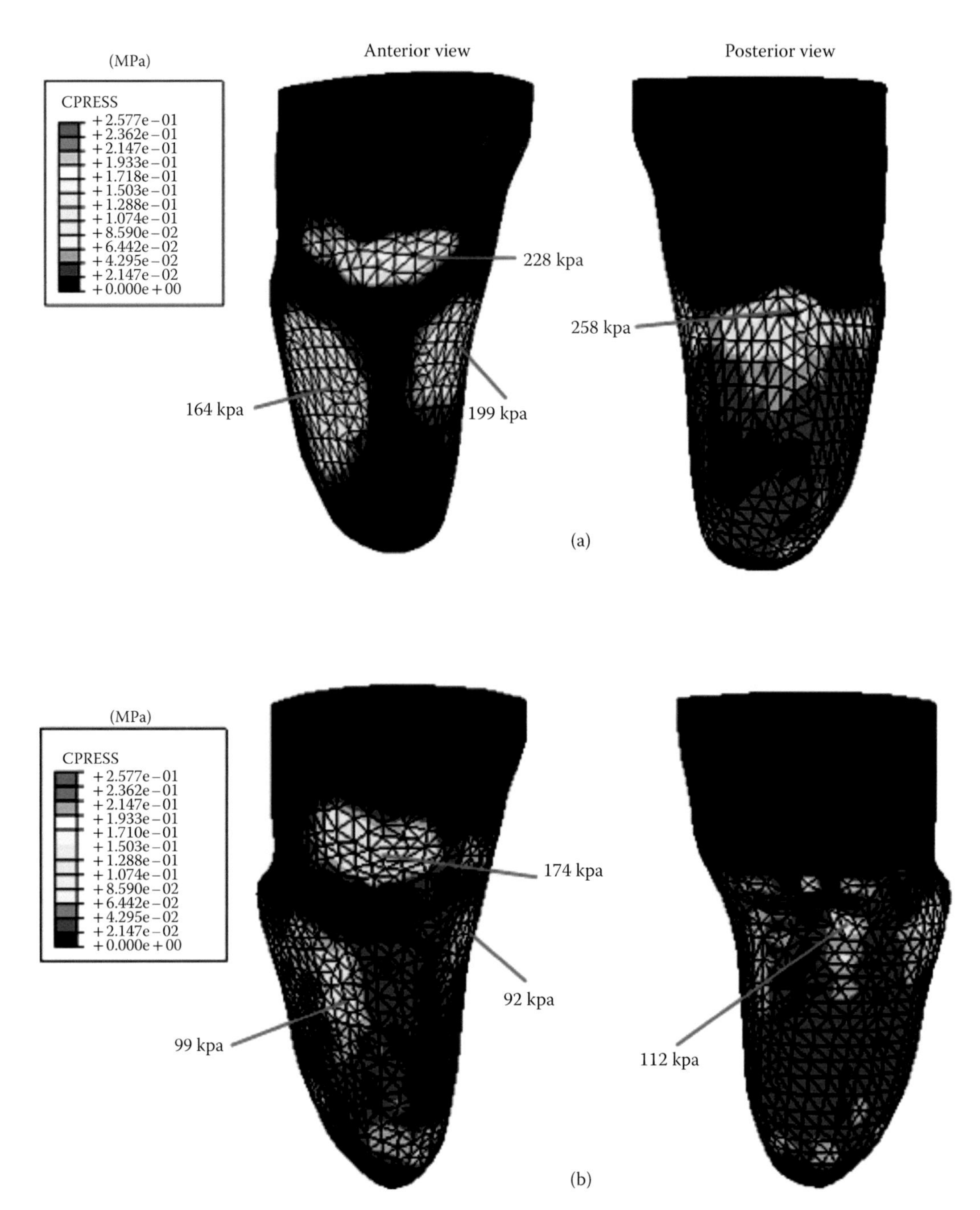


**FIGURE 13.4** **(See color insert.)** Contact normal stress distribution (a) with pre-stress considered and (b) with pre-stress ignored upon application of 800 N.

Figure 13.5a shows the resultant shear stress distribution at the limb-liner interface when 800 N was applied with pre-stress considered. The resultant shear stress is a combination of longitudinal and circumferential components of shear stresses in the plane of the contact interface. High resultant shear stresses were predicted at the four critical regions where socket undercuts were made. The maximum value was 77 kPa over the anteromedial tibial region.

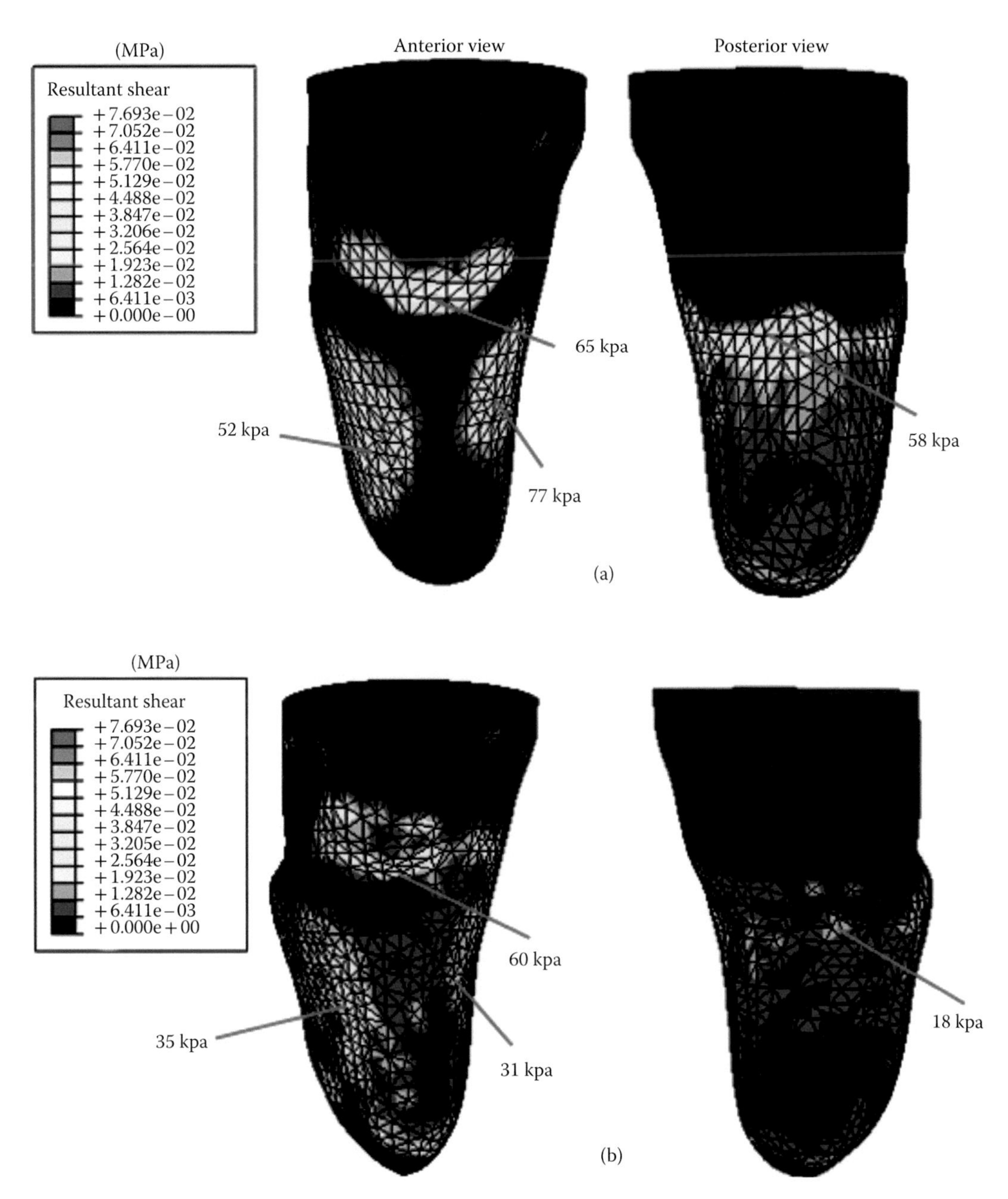


**FIGURE 13.5** **(See color insert.)** The anterior and posterior views of resultant shear stress distribution (a) with pre-stress considered and (b) with pre-stress ignored upon application of 800 N.

Stress distribution patterns over the residual limb and peak stress values over the four pressure-tolerant regions were different in the second model, which ignored pre-stress and incorporated a simplifying assumption that the shape of the limb and rectified socket were the same. Figure 13.4b shows the normal stress distribution and Figure 13.5b shows the resultant shear stress distribution over the limb upon application of 800 N to the second model. Stresses were more evenly distributed in the second model. Peak normal and resultant shear stresses were lowered over the four pressure-tolerant regions where socket undercuts were made, but higher stresses fell on regions that were not pressure tolerant.

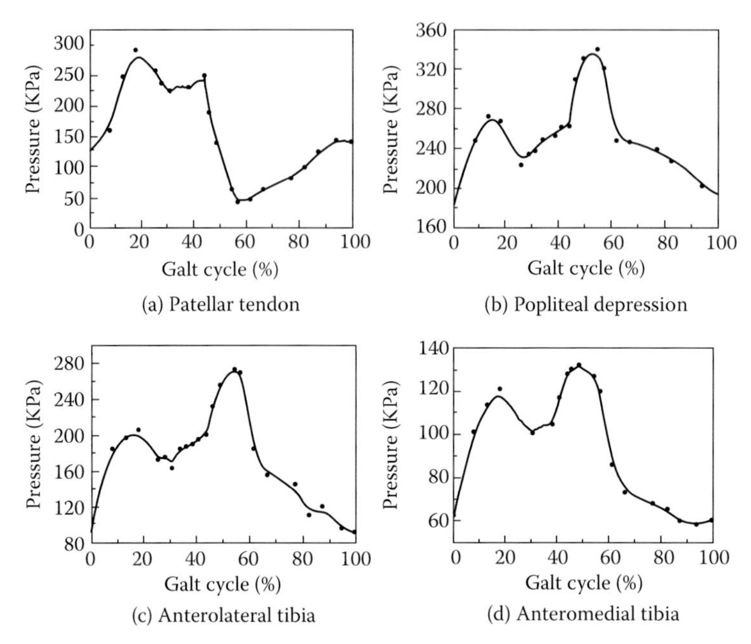


**FIGURE 13.6** Peak pressures on residual during the whole gait cycle (the finite element model considered the pre-stresses after donning the prosthetic socket).

### 13.3.2 Interface Stress throughout a Gait Cycle

Figures 13.6 and 13.7 display the peak pressures and resultant stresses over the patella tendon, popliteal depression, anterolateral tibia, and anteromedial tibia regions during a gait cycle. Generally speaking, all the stress curves had a double-peaked shape, which was similar to the resultant ground reaction force (GRF). Around the first peak, the GRF produced a moment to extend the limb and increased the pressure over the anterior-proximal and posterior-distal sides, and decreased pressures over the anterior-distal and posterior-proximal sides.

At the initiation of the swing phase, even though the GRFs disappeared, angular acceleration was positive. The inertial forces explained the non-zero normal and shear stresses at the prosthetic socket-residual limb interface during the swing phase.

## 13.4 APPLICATIONS

This study used an automated contact method to simulate the contact between the prosthetic socket and the residual limb, while considering pre-stresses. Due to the difference in limb and socket shape, there were some regions of the limb that penetrated into the socket at the initial configuration. The penetrated limb surface was automatically deformed such that it just contacted the inner surface of the socket, thus producing a pre-stress condition. At subsequent stages when loading was applied, the limb surface was automatically constrained not to penetrate the inner surface of the socket. The limb was allowed to slide if the shear stress exceeded the frictional limit.

The difference in shape between the residual limb and socket imposes challenges in contact simulation at the limb-socket interface. Simplifications are usually made at this stage, such as assuming that the limb and socket have the same shape. However, ignoring this pre-stress could lead to inaccuracies in the model. In this investigation, it was shown that peak normal and shear stresses over the patellar tendon, anterolateral and anteromedial tibia, and popliteal depression would be noticeably decreased if such simplifying assumptions were incorporated into the model.

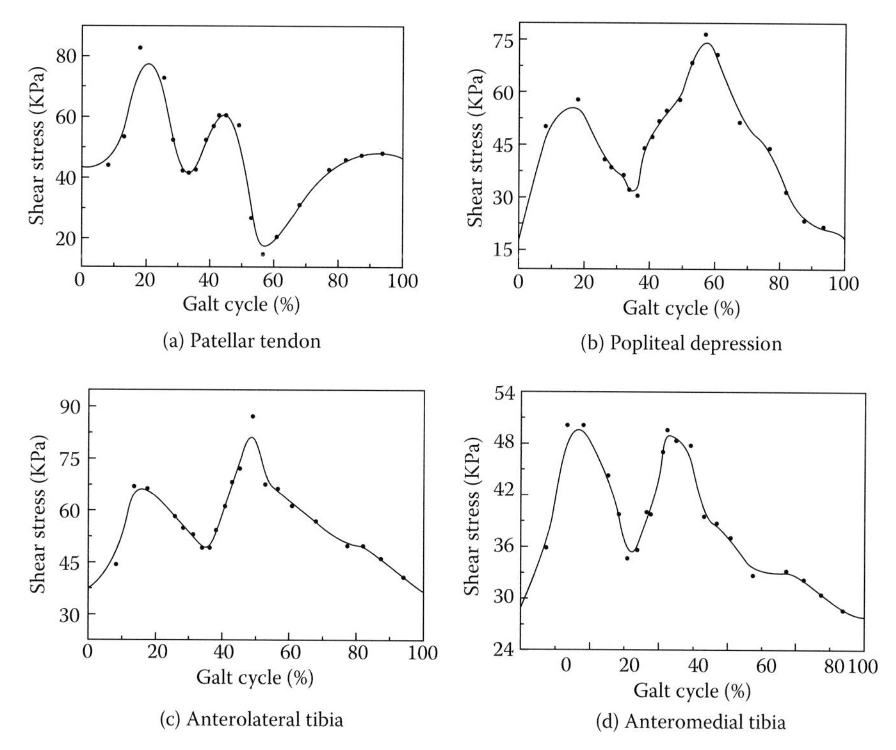


**FIGURE 13.7** Peak resultant shear stresses on residual during the whole gait cycle (the finite element model considered the pre-stresses after donning the prosthetic socket).

The real procedure of socket donning was not simulated in this investigation. Instead, the final deformation state of the limb immediately after donning into a socket was studied. Simulation of socket donning is challenging because donning involves wiggling the limb into the socket and these large relative motions are difficult to define. If the wiggling motions are not simulated and the limb is forced straight into the socket, severe distortions of the limb could happen due to the complicated shape of the socket. In addition, large sliding actions between the limb and the socket require large computational resources.

Three main factors, the geometry of the models, material properties, and loading/boundary conditions, determine the accuracy of an FE model. In this model, the accuracy of the geometric representation was assumed within a tolerable level as the geometries of the residual limb (with bones) and the prosthetic socket were obtained from MRI images and a prosthetic digitizer, respectively. However, the temporal variation in the residual limb volume and shape was not considered. Previous studies have shown that when the prosthesis is not worn for 30 minutes or more the measured interface stress can change significantly (Zachariah and Sanders, 1996; Silver-Thorn and Childress, 1997). This suggested drastic changes in shape and volume may occur in the residual limb, particularly when unloaded.

The residual limb was assumed to be linearly elastic, nonviscoelastic, homogeneous, and isotropic. Better characterization of the material properties of different locations of the residual limb, considering time dependency and loading direction as well as nonlinear material properties of the soft tissue, is desired. In addition, it was assumed that the soft tissue was a passive structure. In reality, the muscles, particularly at the posterior cuff of the residual limb, would have some degree of contraction during walking. Muscle contractions, leading to stiffness changes at different regions of the limb, could alter the stress distribution at the limb-socket interface. Little is known about the effect of muscle contractions on interface stresses. The difference in interface stress between a passive soft tissue structure and a soft tissue with muscle contraction deserves further investigation.

Assumptions were made in loading and boundary conditions. In calculating the equivalent forces and moments applied at the knee joint using inverse dynamics, the residual limb together with the prosthesis was assumed to be one rigid structure. It was also assumed that slippage did not occur at the socket wall-liner interface and at the soft tissue-bone interface, which possibly influenced the predicted residual limb-prosthetic socket interface stress.

Although the FE models have limitations that need further refinements, the models have practical applications in parametric analysis. With these models, we can observe the changes in stress distribution at the interface between the prosthetic socket and the residual limb under various design parameters, such as modifications to the socket shape and stiffness of the liner and socket. If the models are further expanded to include the limb and the whole prosthesis incorporating the prosthetic foot, shank, and socket, the effects of shank flexibility, prosthetic foot designs, alignment, and the mass distribution of different components on the interface stress can be studied.

## REFERENCES

Jia, X.H., M. Zhang, and W.C.C. Lee. 2004. Load transfer mechanics between trans-tibial prosthetic socket and residual limb–dynamic effects. *J Biomech* 37(9):1371–7.

Lee, W.C.C., M. Zhang, X.H. Jia, and J.T.M. Cheung. 2004. Finite element modeling of contact interface between trans-tibial residual limb and prosthetic socket, *Med Eng Phys* 26:655–62.

Quesada, P., and H.B. Skinner. 1991. Analysis of a below-knee patellar tendon-bearing prosthesis: a finite element study. *J Rehab Res Dev* 28:1–12.

Radcliff, C.W., and J. Foort. 1961. *The Patellar-Tendon-Bearing Below-Knee Prosthesis*. Berkeley, CA: Biomechanics Laboratory, University of California.

Reynolds, D.P., and M. Lord. 1992. Interface load analysis for computer-aided design of below-knee prosthetic sockets. *Med Biol Eng Comput* 30:419–26.

Sanders, J.E., and C.H. Daly. 1993. Normal and shear stresses on a residual limb in a prosthetic socket during ambulation: comparison of finite element results with experimental measurements. *J Rehab Res Dev* 30:191–204.

Silver-Thorn, M.B., and D.S. Childress. 1997. Generic, geometric finite element analysis of the trans-tibial residual limb and prosthetic socket. *J Rehab Res Dev* 34:171–86.

Simpson, G., C. Fisher, and D.K. Wright. 2001. Modeling the interactions between a prosthetic socket, polyurethane liners and the residual limb in trans-tibial amputees using non-linear finite element analysis. *Biomed Sci Instrum* 37:343–47.

Steege, J.W., D.S. Schnur, and D.S. Childress. 1987. Prediction of pressure in the below-knee socket interface by finite element analysis. In *ASME Symposium on the Biomechanics of Normal and Pathological Gait*, 39–44.

Zachariah, S.G., and J.E. Sanders. 1996. Interface mechanics in lower-limb external prosthetics: A review of finite element models. *IEEE Trans Rehabil Eng* 4(4):288–301.

Zachariah, S.G., and J.E. Sanders. 2000. Finite element estimates of interface stress in the trans-tibial prosthesis using gap elements are different from those using automated contact. *J Biomech* 33:895–9.

Zhang, M. 1995. Biomechanics of the residual limb and prosthetic socket interface in below-knee amputees. PhD thesis, University of London.

Zhang, M., M. Lord, A.R. Turnersmith, and V.C. Roberts. 1995. Development of a non-linear finite element modeling of the below-knee prosthetic socket interface. *Med Eng Phys* 17:559–66.

Zhang, M., and A.F.T. Mak. 1999. In *vivo* friction properties of human skin. *Prosthet & Orthot* 23:135–41.

# 14 Residual Limb Model for Osteointegration

*Winson C.C. Lee*

## CONTENTS



## SUMMARY

Direct anchoring of a lower-limb prosthesis to the bone (osseointegration) has been shown to solve some common problems associated with conventional socket prostheses. During the rehabilitation phase, amputees fitted with osseointegrated implants apply static loading against the abutment on a weighing scale to prepare the bone to tolerate the forces likely to be experienced during walking. However, the weighing scale measures only the vertical force. Moments and other directions of forces, which can affect the bone-implant interface stresses and the rehabilitation outcome, are not measured. When the amputee starts to walk, in addition, there is a risk of bone mechanical failure. This chapter illustrates the development of a finite element (FE) model to study the stresses in the bone and at the bone-implant interface.

Bone-implant interface stresses were compared under three loading conditions: (1) vertical force corresponding to the load clinically prescribed in a weight-bearing exercise; (2) loads applied on the three axes, corresponding to the "true" load measured by a triaxial load transducer during the same exercise; and (3) loads experienced during independent level walking. An additional parametric analysis was performed to predict the bone structural integrity under different conditions. The model revealed that the weighing scale in fact applied much greater and less uniform stresses on the bone than expected. During walking, high stress occurred at a location different from that experienced during the weight-bearing exercise. In addition, with a bone loss of 40% and the application of a push-off loading at four times its normal force, bone failure would occur. These findings imply that triaxial loading should be monitored during weight-bearing exercises and carefully prescribed.

It is also important to assess the bone quality of an amputee when considering the use of an osseointegrated prosthesis.

## 14.1 INTRODUCTION

A lower-limb prosthesis is conventionally attached to the residual limb by a prosthetic socket. Due to high mechanical stresses applied from the prosthetic socket to the residual limb, residual limb pain and soft tissue breakdown commonly occur (Gallagher, Allen, and MacLachlan, 2001; Hagberg and Brånemark, 2001; Mak, Zhang, and Boone, 2001). In addition, fitting problems occur due to insufficient length or volume fluctuations of the residual limb. In an attempt to solve these problems, a surgical method that connects a prosthesis directly to the femur has been developed (osseointegration; Aschoff and Grundei, 2004; Brånemark et al., 2001). A threaded titanium implant is inserted into the bone. A coupling device (abutment), which penetrates through the soft tissues and skin, connects the external prosthetic components to the titanium implant (Brånemark et al., 2001).

In addition to alleviating skin problems and residual limb pain due to the absence of the prosthetic socket, amputees fitted with transfemoral osseointegrated prostheses can enjoy a greater hip range of motion and better sitting comfort compared to patients using socket-type prostheses (Hagberg et al., 2005). Amputees also experience increased proprioception of the limb and its contact with the ground because of the direct contact with the bone (Brånemark et al., 2001). Patients can walk further and be more active than conventional prosthesis users (Brånemark et al., 2001; Robinson, Brånemark, and Ward, 2004; Sullivan et al., 2003). Osseointegration is primarily recommended for transfemoral amputees who experience complications in using a conventional socket-type prosthesis (Brånemark et al., 2001; Sullivan et al., 2003).

Osseointegration in lower-limb amputees requires two stages of surgery. The first stage is to insert a threaded titanium implant into the bone. After the surgery, relative motion between the implant and the bone is prevented as much as possible, as micro-motion could potentially disrupt interfacial bone formation (Simmons and Pilliar, 2000; Soballe et al., 1992). During the second surgery, which is performed approximately six months later, the previous surgical scar is reopened. An abutment, penetrating through the soft tissue and skin, is connected to the implant using an abutment-retaining bolt.

After the second surgery, amputees have to undergo some rehabilitation exercises. One important exercise involves the application of a prescribed static load to the abutment. This exercise prepares the bone to tolerate the forces likely to be incurred during walking. It also strengthens the bone, which is weakened due to disuse atrophy. Initially a smaller amount of load is applied, and the load is increased incrementally over a period of months until full standing weight can be borne safely and without pain (Hagberg et al., 2005). If pain is perceived at any stage, the prescribed load is decreased to a non-painful level (Hagberg et al., 2005). Full weight bearing is achieved at a minimum of three months after the second surgery.

Currently, the load applied during weight-bearing exercises is monitored using a conventional weighing scale that measures only the vertical force. If the limb is loaded at an angle in the frontal and transverse planes, however, the nonvertical forces and the moments generated are not measured. The effects of those additional loads on the bone-implant interface stresses are of interest as they impact on the process of rehabilitation.

Due to disuse atrophy, in addition, significant bone loss can occur after some period of time using socket prostheses (Meerkin, 2000). The reduced bone density together with the high loads experienced during walking put lower-limb amputees at risk of bone cracking when they switch to an osseointegrated prosthesis. Bone fracture might be possible when the osseointegrated prostheses are subjected to accidentally high loadings.

This chapter illustrates the development of an FE model that analyzes the mechanical stresses in the residual bone and at the bone-implant interface. Loading applied to the model was based on actual experimental data from a triaxial load transducer (Frossard et al., 2003a, 2003b; Lee et al., 2007).

The model compares the bone stresses adjacent to the implant for three loading cases: (1) loads applied on the long axis of the limb only, which corresponds to the load clinically prescribed during the weight-bearing exercise; (2) loads applied on the three axes corresponding to the "true" load measured simultaneously by a triaxial load transducer during the same exercise; and (3) three-dimensional loads during independent walking. A comparison of these loadings could help establish guidelines for refining the training protocol during rehabilitation. Also, an additional parametric analysis is performed to predict the bone structural integrity under different conditions.

## 14.2 MODEL DEVELOPMENT

A quasi-static implicit FE analysis was performed using ABAQUS (ABAQUS Inc., Rhode Island) to predict stresses at the bone-implant interface as well as in the entire residual femur bone. Details of the models, which are also described in the literature (Lee et al., 2008), are described below.

### 14.2.1 Geometry

A generic model (Figure 14.1) based on the dimensions of a commercial implant-abutment fixation was created in SolidWorks (SolidWorks Corporation, Massachusetts). The implant had a diameter of 20 mm, thread pitch of 1.75 mm, and length of 100 mm. The implant and the abutment were tied together because this study focused on the bone and bone-implant interface rather than the internal behavior of the fixation itself. The femur geometry was obtained from the BEL Repository, which was based on the third-generation standardized femur (Size #3306, Pacific Research Laboratories Inc., Virginia) with an intramedullary canal. The femur model was sectioned so that the residual bone had a length of 230 mm as measured from the greater trochanter to the distal cut end. Threads on the femur were generated by geometrically subtracting the bone from the threaded implant. To simplify the model, parallel circumferential grooves were used instead of helical threads on both the implant and the bone.

### 14.2.2 Mesh

An FE mesh with three-dimensional tetrahedral elements was built using ABAQUS automeshing techniques. The optimal mesh density for the bone adjacent to the bone-implant interface was determined by performing a mesh refinement test such that the convergence of peak stress was within 2% of the previous coarser mesh. A coarser mesh density was assigned to the rest of the model. In total, 211,117 elements were assigned to the entire model.

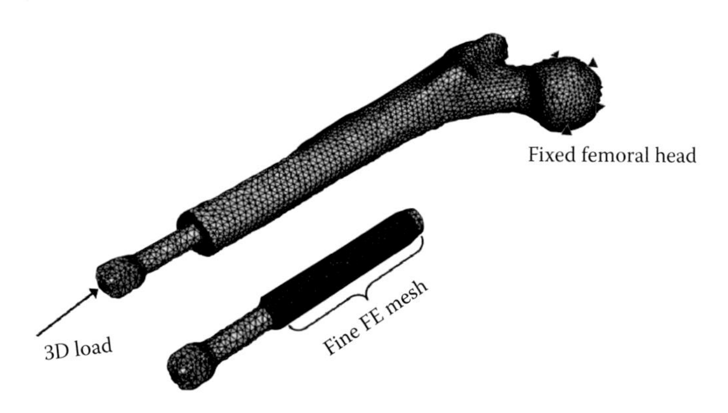


**FIGURE 14.1** **(See color insert.)** The finite element model of the bone and the fixator (implant and abutment).

### 14.2.3 Material Properties and Contact Simulation

The femur and implant were assumed to be linearly elastic, homogeneous, and isotropic. The femur (Reilly and Burstein, 1975; Zhang, Dong, and Fan, 2005) and the implant (Xu, Crocombe, and Hughes, 2000; Xu, Xu, and Crocombe, 2006) were assigned elastic moduli of 15 GPa and 115 GPa, respectively. A Poisson's ratio of 0.3 was used for the two materials.

Previous animal studies investigating the interface properties between the bone and the titanium implant suggested that bone is closely apposite to the titanium implant, such that the bone surrounding the implant has to plastically deform to bring about relative motion (Brånemark et al., 1998). To simulate the close apposition between implant and bone, the threaded implant and the intramedullary canal of the femur were tied together to prevent any relative motions between the nodes at the interface.

### 14.2.4 Load Applications: Weight-Bearing Exercise and Level Walking

Loads exerted on the abutment during simulated weight-bearing exercises and independent walking of a 92.6-kg male unilateral transfemoral amputee patient using an osseointegrated prosthesis were measured (Cairns et al., 2006; Frossard et al., 2003a, 2003b; Lee et al., 2007). During the weight-bearing exercises, the patient was asked to transfer body weight through the abutment against a weighing scale at magnitudes of 200 N, 400 N, and 900 N for about 10 seconds at each load. The three-dimensional forces and moments were measured simultaneously using a triaxial commercial load transducer (Model 45E15A, JR3 Inc., Woodland, California). The loads applied on the abutment during straight, level walking were also measured by the same commercial transducer, which was mounted between the abutment and the prosthetic knee.

In the FE model, three loading cases were applied at the distal end of the abutment. Loading A comprised a vertical force with magnitudes of 200 N, 400 N, and 900 N. Loading B was the actual forces and moments on the three axes measured by the triaxial load transducer when the patient transferred body weight against a weighing scale at magnitudes of 200 N, 400 N, and 900 N. A comparison of loading A and loading B provides information regarding the "true" stresses at the bone-implant interface compared to the apparent stresses when moments and nonvertical component forces are neglected. Loading C comprised a series of quasi-static loads (with 35 data points) encountered during straight, level walking. The femoral head was fully fixed.

The FE model computed the stress at the threaded region of the bone under the three different loading cases. The stress ratio, von Mises stress (kPa) divided by body weight (kg), was used to describe the stress magnitude.

### 14.2.5 Parametric Analysis Investigating the Possibility of Bone Mechanical Failure

To simulate accidentally high-impact forces, the axial component of the loads experienced during the heel-off phase of the gait was multiplied up to four times. The material properties and the bone length were also changed to simulate various conditions. The bone length (from greater trochanter to bone distal end) varied in the range of 180 mm to 220 mm. The bone's Young's modulus (E) was changed from 3,000 to 15,000 MPa. A Young's modulus of 15,000 MPa resembled a normal human bone. A Young's modulus of 3000 MPa resembled a bone density drop of 40% (Meerkin, 2000), and was calculated based on a previous finding that the Young's modulus of the bone was proportional to its density to the power of three (Carter and Hayes, 1977).

Bone failure was predicted by assessing whether the bone strain was above the yield strain (1.2) of human femoral bone (Helgason et al., 2008). This is based on previous findings that suggested the yield strain of the bone is relatively independent of changes in bone density.

## 14.3 MODEL FINDINGS

### 14.3.1 Loading A

When only vertical forces were applied, the stresses on the bone adjacent to the implant were distributed quite evenly (Figure 14.2). The stresses over the entire implanted region of the bone were in the ranges of 0.20 to 6.31 kPa/kg, 0.40 to 12.5 kPa/kg, and 0.86 to 27.7 kPa/kg, when a 200 N, 400 N, and 900 N axial force, respectively, was applied.

### 14.3.2 Loading B

The stress distribution on the implanted region of the bone was less uniform compared to loading A, when the "true" forces and moments on the three axes, measured simultaneously during the same exercises, were applied (Figure 14.3). The peak stresses in loading B in the implanted region

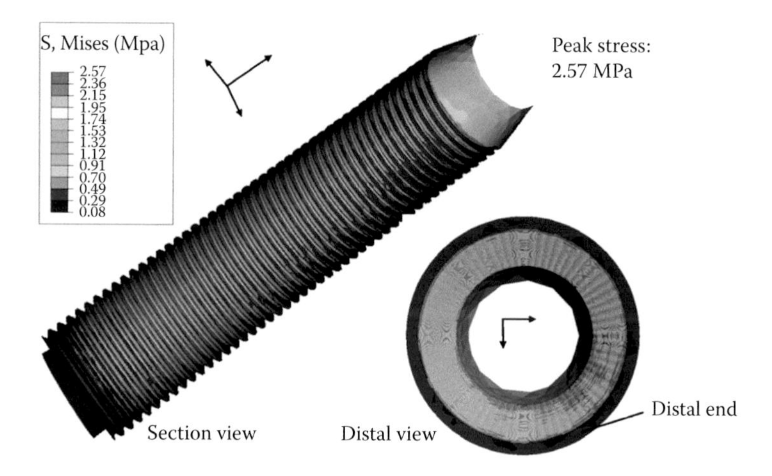


**FIGURE 14.2** **(See color insert.)** Stress distribution at the bone adjacent to the implant at loading condition A.

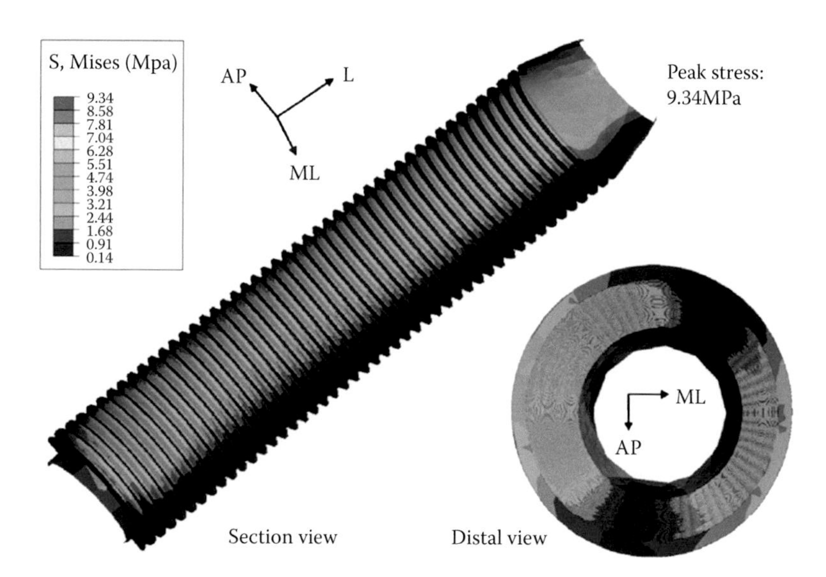


**FIGURE 14.3** **(See color insert.)** Stress distribution at the bone adjacent to the implant at loading condition B.

were approximately three times higher than in loading A, with ranges of 0.26 to 15.8 kPa/kg, 0.52 to 38.3 kPa/kg, and 1.55 to 101 kPa/kg when the amputee applied 200 N, 400 N, and 900 N loads, respectively, against the weighing scale. At all three loading magnitudes, the highest stresses were found at the lateral aspect of the bone.

### 14.3.3 Loading C

Stresses peaked at the push-off phase of the gait (approximately 55% of the gait cycle), reaching 92.0 kPa/kg (Figure 14.4). At the late stance phase of the gait cycle, stresses at the posterolateral and anteromedial regions of the bone-implant interface were significantly lower than the stresses along the posteromedial and anterolateral regions. Across the gait cycle similar patterns of stress were seen, although the location of the greater nodal stress regions varied considerably (Table 14.1).

### 14.3.4 Effects of Various Parameters

At push off, high strain was found at the neck of the femur. The maximum principal strains (MPS) increased when the bone density decreased. The MPSs were still below the bone yield strain of 1.2 (Table 14.2). With a bone loss of 40% (simulated by assigning a Young's modulus of 3000 MPa) and axial loads multiplied four times, however, the MPS was greater than 1.2, which indicated possible bone failure. The MPS increased slightly when bone length increased.

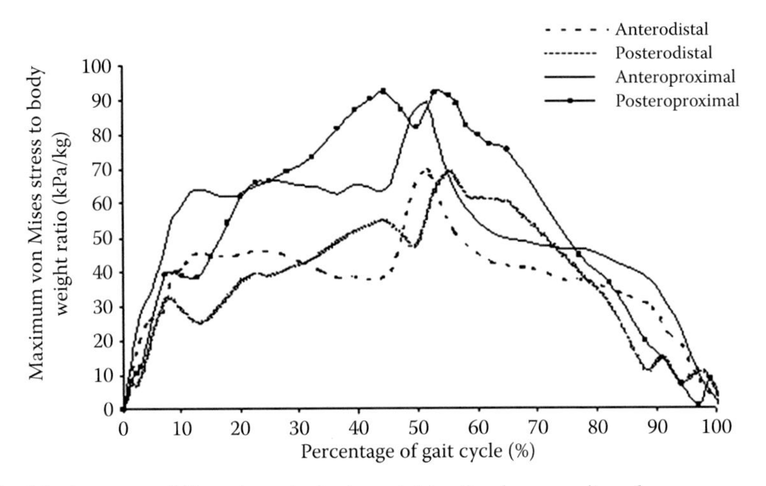


**FIGURE 14.4** Maximum von Mises stress-to-body-weight ratio along a gait cycle.

**TABLE 14.1**
**Locations of High Stresses at Different Phases of the Gait Cycle**

| Percentage of Gait Cycle | Location of High Stress at the Bone-Implant Interface |
|---|---|
| 0–15 | Posterolateral |
| 15–40 | Anterolateral |
| 45–50 | Anterolateral |
| 50–100 | Posteromedial |

**TABLE 14.2**
**Effects of Changing Bone's Young's Modulus (E) and Length and Multiplying Load on Maximum Principal Strains (MPS) at Push-Off Phase of the Gait**

| I: Change of E While Normal Walking Load Was Applied | | II: Change of Axial Loads, While Keeping E = 15,000 MPa | | III: Change of Axial Loads, While Keeping E = 3,000 MPa | | IV: Change of Bone Length While Normal Walking Load Was Applied and E = 3,000 MPa | |
|---|---|---|---|---|---|---|---|
| E (MPa) | MPS | Load multiple | MPS | Load multiple | MPS | Bone length (mm) | MPS |
| 15000 | 0.09 | 1 | 0.09 | 1 | 0.38 | 180 | 0.37 |
| 9000 | 0.13% | 2 | 0.18 | 2 | 0.79 | 200 | 0.38 |
| 3000 | 0.38 | 4 | 0.36 | 4 | **1.56** | 220 | 0.39 |

## 14.4 APPLICATIONS

In this study, FE modeling was used to compare the effects of different loading cases, bone densities, and limb lengths on stresses in the bone implanted with a transfemoral osseointegrated prosthesis. Loading B, corresponding to the actual forces and moments measured by a triaxial load transducer, produced less uniform stress distribution than loading A, with peak stresses being up to three times higher than in loading A. This indicates that the rehabilitation program controlled by visual feedback from the weighing scale may in fact be producing much greater stresses than expected. These higher stresses may cause intolerable pain, which typically delays the weight-bearing exercise and rehabilitation program. Loading C represents a series of loads applied on the abutment during walking. Although the magnitudes of the peak stresses predicted in loading C were comparable to those occurring when the subject was asked to transfer nearly the full body weight in loading B, high stress occurred at different locations of the implanted region.

The lengthy rehabilitation program has been perceived by amputees and the clinical team as a shortcoming. The results of this study imply that one possible way to refine the weight-bearing exercise is to incorporate a triaxial load transducer, which provides feedback on three-dimensional forces and moments to the patients and clinicians. At the early phase of the rehabilitation, the transducers can be used to increase the likelihood of forces being applied predominantly along the bone axis. This distributes the stress evenly and reduces the magnitude of peak stresses. The uniformly distributed stress allows the quality of the bone to be improved along the entire interface and the lower stress magnitude will likely reduce the sensation of pain. Three-dimensional loads can then be applied progressively in proportion to the loads that would be expected during walking and other activities of daily living.

The model also suggests the importance of assessing the bone quality of amputees who are to receive implanted prostheses prior to the implantation procedure. With a bone loss of 40% and the application of a load that corresponds to four times the push-off loading, bone failure will occur. This model simulates a realistic scenario. Previous studies have indicated that the density of the bone of a transfemoral amputee drops by 40% after 6 months of using a socket prosthesis (Meerkin, 2000). The bone loss is due to disuse atrophy as the residual bone experiences lesser forces. An increased force of up to four times normal loading would likely occur during an accident, such as a fall. In daily activities like sit-to-stand, stair climbing, or squatting from a standing position, the compressive loading at the knee joint can reach 2.5 times body weight.

The FE model has certain limitations. The geometry of a generic femur was used. Use of a standardized femur has the advantage of allowing precise control over the implant-bone interface, mesh density, and applied loading. The limitation of using a standardized femur is that it is not patient-specific; however, the modeling methodology developed in this study allows comparative assessment of different loading regimens on a well-defined femur-implant geometry.

Another limitation of the FE model was the use of linear and homogeneous bone material properties. In reality, there is likely to be a combination of damaged bone, new woven bone transformed from hematoma, and some fibrous tissue at the interface after the surgical implantation. Future studies with patient-specific femur geometry derived from CT scans may also include patient-specific bone properties; however, the focus of this study was a comparison of relative stress levels between different loading cases. The tied contact interface between the implant and the bone did not permit any relative motion between the implant and the bone, thus simulating the limiting case of complete osseointegration. This assumption is consistent with the observation of close apposition between the bone and the implant during the six months from the initial implantation surgery to the commencement of loading after the second surgery. However, there is much potential for a better understanding of the mechanical properties of the osseointegrated bone-implant interface through histological studies and mechanical testing.

## REFERENCES


Aschoff, H., and Grundei, H., 2004. The Endo- Exo-Femurprosthesis: A new concept of prosthetic rehabilitation engineering following thigh-amputation: Some cases and early results. *Proceedings of the International Society for Prosthetics and Orthotics 11th World Congress*, Hong Kong SAR.

Brånemark, R., Brånemark, P.-I., Rydevik, B.L., and Myers, R.R., 2001. Osseointegration in skeletal reconstruction and rehabilitation: A review. *Journal of Rehabilitation Research and Development* 38(2):175–81.

Brånemark, R., Ohrnell, L.-O., Skalak, R., Carlsson, L., and Brånemark, P.-I., 1998. Biomechaical characterization of osseointegration: An experimental in vivo investigation in the beagle dog. *Journal of the Orthopaedic Research Society* 16(1):61–9.

Cairns, N.J., Frossard, L. A., Hagberg, K., and Branemark, R., 2006. Static load bearing during early rehabilitation of transfemoral amputees using osseointegrated fixation. *Proceedings of the 30th Annual Scientific Meeting of the International Society for Prosthetics and Orthotics*, pp. 51–52, Perth, Australia.

Carter, D.R., and Hayes, W.C., 1977. The compressive behavior of bone as a two-phase porous structure. *The Journal of Bone and Joint Surgery. American Volume* 59(7):954–62.

Frossard, L., Beck, J., Dillon, M., Chappell, M., and Evans, J.H., 2003a. Development and preliminary testing of a device for the direct measurement of forces and moments in the prosthetic limb of transfemoral amputees during activities of daily living. *Journal of Prosthetics and Orthotics* 15(4):135–42.

Frossard, L., Lee Gow, D., Contoyannis, B., Nunn, A., and Branemark, R., 2003b. Load applied on the abutment of transfemoral amputees fitted with an osseointegrated implant during load bearing exercises using a long pylon. *Proceedings International Society for Prosthetics and Orthotics-Australia*, pp. 55–56, Melbourne, Australia.

Gallagher, P., Allen, D., and MacLachlan, M.I., 2001. Phantom limb pain and residual limb pain following lower limb amputation: A descriptive analysis. *Disability and Rehabilitation* 26:522–30.

Hagberg, K., and Brånemark, R., 2001. Consequences of non-vascular trans-femoral amputation: A survey of quality of life, prosthetic use and problems. *Prosthetics and Orthotics International* 25:186–94.

Hagberg, K., Häggström, E., Uden, M., and Brånemark, R., 2005. Socket versus bone-anchored trans-femoral prostheses: Hip range of motion and sitting comfort. *Prosthetics and Orthotics International* 29(2): 153–63.

Helgason, B., Viceconti, M., Rúnarsson, T.P., and Brynjólfsson, S., 2008. On the mechanical stability of porous coated press fit titanium implants: A finite element study of a pushout test. *Journal of Biomechanics* 41(8):1675–81.

Lee, W.C.C., Doocey, J.M., Frossard, L., Branemark, R., Pearcy, M.J., and Evans, J.H., 2008. FE stress analysis of the interface between the bone and an osseointegrated implant for amputees: Implications to refine the rehabilitation program. *Clinical Biomechanics* 23(10):1243–50.

Lee, W.C.C., Frossard, L., Hagberg, K., Haggstrom, E., and Brånemark, R., 2007. Kinetics analysis of transfemoral amputees fitted with osseointegrated fixation performing common activities of daily living. *Clinical Biomechanics* 22(6):665–73.

Mak, A., Zhang, M., and Boone, D., 2001. State-of-the-art research in lower-limb prosthetic biomechanics-socket interface: A review. *Journal of Rehabilitation Research & Development* 38(2):161–74.

Meerkin, J.D., 2000. Musculo-skeletal adaptation and altered loading environments : an amputee model. PhD thesis, Queensland University of Technology.

Reilly, D.T., and Burstein, A.H., 1975. The elastic and ultimate properties of compact bone tissue. *Journal of Biomechanics* 8(6):393–405.

Robinson, K.P., Brånemark, R., and Ward, D.A., 2004. Future developments: Osseointegration in transfemoral amputees. In *Atlas of Amputations and Limb Deficiencies: Surgical, Prosthetic and Rehabilitation Principles*, edited by D.G. Smith, J.W. Michael, and J.H. Bowker. American Academy of Orthopaedic Surgeons in US.

Simmons, C.A., and Pilliar, R.M., 2000. A biomechanical study of early tissue formation around bone-interfacing implants. In *Bone Engineering*, edited by J.E. Davies. Toronto, Canada: Em Squared.

Soballe, K., Brockstedt-Rasmussen, H., Hansen, E.S., and Bunger, C., 1992. Hydroxyapatite coating modifies implant membrane formation. Controlled micromotion studied in dogs. *Acta Orthopaedica Scandinavica* 63(2):128–40.

Sullivan, J., Uden, M., Robinson, K.P., and Sooriakumaran, S., 2003. Rehabilitation of the trans-femoral amputee with an osseointegrated prosthesis: The United Kingdom experience. *Prosthetics and Orthotics International* 27:114–20.

Xu, W., Crocombe, A.D., and Hughes, S.C., 2000. Finite element analysis of bone stress and strain around a distal osseointegrated implant for prosthetic limb attachment. Proceedings of the Institution of Mechanical Engineers Part H. *Journal of Engineering in Medicine* 214:595–602.

Xu, W., Xu, D.H., and Crocombe, A.D., 2006. Three-dimensional finite element stress and strain analysis of a transfemoral osseointegration implant. Proceedings of the Institution of Mechanical Engineers Part H. *Journal of Engineering in Medicine* 220:661–70.

Zhang, M., Dong, X., and Fan, Y., 2005. Stress analysis of osseointegrated transfemoral prosthesis: A finite element model. *Proceedings of the 2005 IEEE Engineering in Medicine and Biology 27th Annual Conference*, Shanghai, China.

# Section V

## Spine

# 15 Spine Model for Vibration Analysis

*Lixin Guo, Ming Zhang, and Ee-Chon Teo*

## CONTENTS



## SUMMARY

Epidemiological investigations suggest that whole body vibration (WBV) contributes significantly to injuries and functional disorders of the skeleton and joints, including the spine. Although a large number of investigations have drawn attention to the risks of WBV on the human spine, many dynamic characteristics of the spine and injury mechanisms under vibration loading are not clear. In this study, a detailed three-dimensional finite element (FE) model of the spine T12-Pelvis segment was developed for investigating its biomechanical characteristics. Validation was conducted for static and dynamic conditions, respectively. This study analyzed the frequency characteristics and modal modes of the intact and injured spine, analyzed the transient response characteristics of the intact and injured spine, and performed a material sensitivity analysis of the spine FE models. The findings of this study may be helpful for further understanding the dynamic characteristics of the human spine and its mechanism of injury under vibration loading, and provide a useful reference for WBV-related injury treatment in clinics and product development in industry.

## 15.1 INTRODUCTION

Surveys have shown that as many as 85% of adults experience lower back pain that interferes with their work or recreational activities and up to 25% of people between the ages of 30 and 50 years report symptoms of lower back pain. Of all patients suffering lower back pain, 90% recover within six weeks regardless of the type of treatment received. The remaining 10% who continue to have problems after three months or longer account for 80% of disability costs related to the spine. Despite increasing public attention to cumulative trauma disorders of the upper extremities, occupational lower back disorders account for the most significant proportion of industrial-related musculoskeletal disorders (Shirazi-Adl and Parnianpour 2001).

Lower back disorders arising from industrial labor are a complex multifactorial problem. A complete understanding can only be gained by considering personal and environmental risk factors, which include both biomechanical and psychosocial factors. The results of epidemiological studies have associated some occupational conditions with low back pain symptoms, such as physically heavy work, static work postures, repetitive work, exposure to vibration, and so on (Shirazi-Adl and Parnianpour 2001).

Epidemiological investigations have suggested that WBV contributes significantly to injuries and functional disorders of the skeleton and joints (Frymoyer et al. 1980). Long-term WBV has been found to pose health risks for the lumbar spine, especially for the lower lumbar motion segment L3-L5 (Frymoyer et al. 1980; Pankoke, Hofmann, and Wolfel 2001; Cheung, Zhang, and Chow 2003). It has also been shown that dynamic loads pose a greater risk and could induce a twofold (Kasra, Shirazi-Adl, and Drouin 1992; Keller, Colloca, and Beliveau 2002) and threefold (Lee, Esler, and Midren 1993) increase in stress, strain, and joint force in comparison with static conditions. Many studies have explored the link between lower back pain and WBV in special working populations (Fairley 1995; Chen, Chang, and Shih 2003; Cann, Salmoni, and Eger 2004). Accumulated loads were found to be risk factors for occupational lower back pain during prolonged dynamic asymmetric tasks (Mientjes et al. 1999; Kumar 2001).

In view of the harmfulness of vibration to the human body, many studies have reported on the problem of WBV (Pope and Novotny 1993; Goel, Park, and Kong 1994; Pankoke, Hofmann, and Wolfel 2001; Kong and Goel 2003; Guo and Teo 2005) to provide insights into spinal injury and degeneration and to investigate therapeutic methods for the treatment of diseases of the human spine. Pankoke, Hofmann, and Wolfel (2001) built a nonlinear FE reduced model of a seated man with considerations for body stature, body mass, and postures for investigating WBV. Guo and Teo (2005) developed a detailed FE model of the spine T12-Pelvis segment, obtained the resonant frequencies of different spinal segments, and also predicted the dynamic trend of the spine T12-Pelvis segment in the sagittal plane under WBV. Kong and Goel (2003) built a head-to-sacrum FE model to predict the frequency response properties of the human upper body. Goel, Park, and Kong (1994) developed a three-dimensional FE model of the ligamentous L4-S1 segment with an upper-body mass of 40 kg and analyzed the dynamic response of the spine under loading with different frequencies.

Considering injury or degeneration of the human spine, the effects of loss of intradiscal pressure have been analyzed under static conditions using FE analysis (Kim et al. 1991; Goto et al. 2002) and experimental studies (Frei et al. 2001). In addition, Goel and Kim (1989) predicted that total denucleation might induce laminae separation of intervertebral discs. Shirazi-Adl and colleagues (Shirazi-Adl, Shrivastava, and Ahmed 1984; Shirazi-Adl and Drouin 1987) demonstrated that degenerated discs caused high loads to be distributed around the edges of the disc. Sharma, Langrana, and Rodriguez (1995, 1998) investigated the roles of facets and ligaments in spinal stability by addressing the facet load transmission. Goto et al. (2002), using FE methods, also indicated that decreased intradiscal pressure might increase the burden on facet joints and cause deformation of these joints, and concluded that disc degeneration occurs before facet joint osteoarthritis (Adams and Hutton 1980; Butler et al. 1990).

This chapter introduces FE modeling, model validation, modal analysis, response analysis, and material sensitivity analysis of the human spine, which can be used to further understand the biodynamic characteristics of the human spine and provide a reference for the prevention and treatment of WBV-related spine diseases.

## 15.2 FINITE ELEMENT MODELING OF THE HUMAN SPINE

A detailed and validated FE model of the human spine is a key prerequisite for analyzing the mechanical characteristics of the spine. To date, the spine has been modeled numerous times for different motion segments, such as the cervical spine, thoracic spine, and lumbar spine. In this study, a detailed ligamentous FE model of the spinal segment T12-Pelvis is developed and used to analyze biomechanical characteristics of the spine, including static and dynamic characteristics, as well as material property sensitivity and spinal injuries.

A three-dimensional FE model of the spinal segment L3-L5 was developed based on actual geometrical data of embalmed vertebrae of the lumbar spine using a flexible touch-point digitizer (Guo et al. 2005). Next, an FE model of the motion segment T12-Pelvis was developed using both a flexible touch-point digitizer and laser scanning. The T12-Pelvis model includes lumbar spine segment L1-L5, thoraco-lumbar joint T12-L1, lumbo-saracic joint L5-S1, and sacro-ilium joints. This model can represent the mechanical characteristics of a human lumbar spine and its adjunctive joints in both static and dynamic conditions (Guo and Teo 2005; Guo, Zhang, and Zhang 2011).

The geometry of each model was measured from cadaveric vertebrae, which were obtained from an adult male spinal specimen with no physical abnormalities or traumas. Using a digitizing technique, the geometrical data of T12 and L1 to L5 vertebrae were generated. Then, significant points that can describe the main shape of the vertebrae were selected as key points to build the vertebral configuration. Geometrical data of the sacrum was measured by laser scanning. The L3-L5 model was developed based on actual geometrical data of the embalmed vertebrae of the lumbar spine measured using a flexible touch-point digitizer, as shown in Figure 15.1.

After establishing the L3-L5 model, the T12-Pelvis FE model was developed based on actual geometrical data of the embalmed vertebrae of the lumbar spine, using a flexible touch-point digitizer and scanning technique for different vertebrae. The T12-Pelvis model was constructed in the commercial finite element analysis software ANSYS and included seven vertebrae from T12 to S1 and six intervertebral discs, as shown in Figure 15.2. A sitting/standing posture was assumed for modeling the lumbar spine.

The T12-Pelvis FE model was composed of vertebrae, intervertebral discs, ligaments, and endplates. The vertebral body consisted of cortical bone, cancellous bone, and a posterior bony structure, and the intervertebral disc consisted of a nucleus pulposus, annulus ground substance, and annulus fibers. Annulus fibrosus was assumed to consist of three consecutive laminar layers. In each layer, annulus fibers were oriented on average at ±30° to the endplates. The total volume of annulus fibers was assumed as 19% of annulus volume. Different kinds of ligaments were incorporated into the FE model using cable elements (tension only) (Guo, Zhang, and Teo 2007). The vertebrae, endplate disc annulus, and disc nucleus were created by isotropic three-dimensional block elements. The spinal ligaments and the intervertebral disc configurations were built based on data available in the literature.

Because the pelvis was regarded as a hard structure and all the translational degrees of freedom of joint areas near the pelvis or ilium were fixed during the FE modal analyses, only the contact areas of the sacroiliac joint near the ilium were modeled. For the consideration of the small motions of the sacroiliac joint (Skalak and Shu, 1987), imaginary cartilage was assumed to mimic the cartilage, joint gap, synovia, and soft tissue between the sacrum and the ilium using isotropic brick elements. The anterior sacroiliac ligament and the posterior sacroiliac ligament

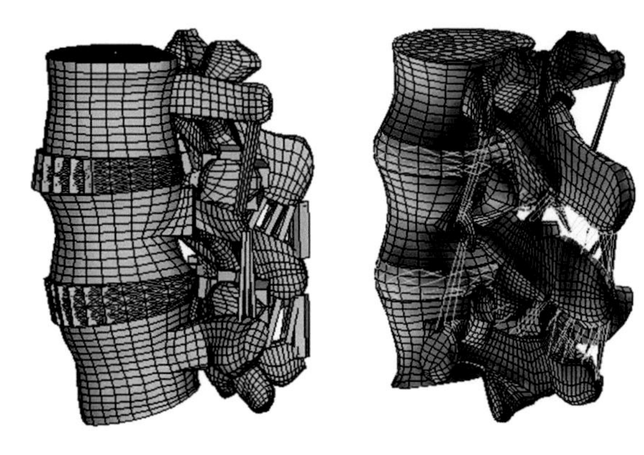

**FIGURE 15.1** The ligamentous finite element model of the L3-L5 segment.

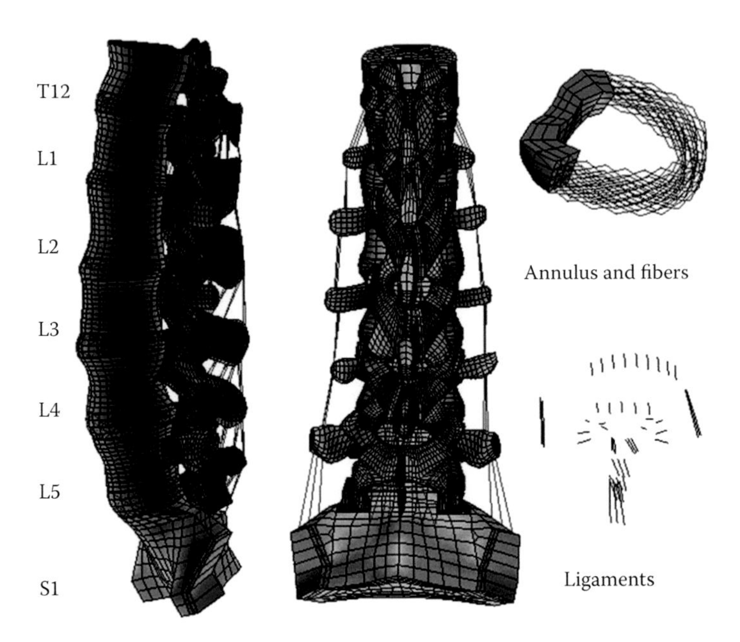


**FIGURE 15.2** The finite element model of the T12-Pelvis spine segment in lateral view and back view.

were modeled using three-dimensional cable elements. The interosseous sacroiliac ligament was also modeled using three-dimensional cable elements and embedded in the imaginary cartilage between the sacroiliac joint-like fibers. Because the ilium was fixed in the model, which might provide high stability for the sacroiliac joint, the other ligaments relevant to the sacroiliac joint were neglected (Guo and Teo 2005).

For modeling, the procedures, methods, and material properties were based on the literature (Goel, Park, and Kong 1994; Kim et al, 1991; Sharma, Langrana, and Rodriguez 1995, 1998). The nonlinear stress-strain behaviors of the ligaments were also obtained from the literature (Sharma, Langrana, and Rodriguez 1995). The nucleus pulposus has a high Poisson's ratio and low Young's modulus (incompressible, like water). More details about modeling of the human spine have been given elsewhere (Guo et al. 2005; Guo and Teo 2005). The material properties and element types of all the spinal components are shown in Table 15.1. The three-dimensional sliding surface-contact conditions of the facet contact articulation surfaces were built, supporting large deformation and making it suitable for simulating facet articulation (Guo, Zhang, and Zhang 2011). For all the analyses, the translation degrees of freedom of the inferior surfaces were fixed.

In modal analyses, the upper body mass of the human body is an important factor for the frequency characteristics of FE models. In this study, the distributed mass of 40 kg was assigned to the T12-Pelvis FE model, as well as the L1-L5 and L1-S1 finite element models. Eighty percent of upper body weight was assigned to the top vertebra of the model and the other mass was evenly assigned to the vertebrae of the L1-S1 segments. The vertical line of the lumped mass was assigned 1.0 cm anterior to the center of the L3-L4 motion segment, which assumes a normal sedentary posture (Guo et al. 2008).

For FE models with one segment and two segments, only a lumped mass of 40 kg was imposed on the top of the superior vertebra. Besides consideration of the distribution of upper body weight, mass changes of the upper body were also studied, from 30 kg to 60 kg in some special studies, for the assumption of thin or fat people during the modal analysis.

All modeling and simulations were carried out in the commercially available finite element analysis software ANSYS. After the modal analyses, the natural frequencies and vibration modes of the FE models were extracted, which could be used for further special analyses.

**TABLE 15.1**
**Element Types and Material Properties of Spinal Components**

| Components | Element Type | Young's Modulus (MPa) | Poisson's Ratio | Cross-Sectional Area (mm²) | Density (kg/mm³) |
|---|---|---|---|---|---|
| Cortical bone | 8-node solid | 1,200 | 0.30 | | $1.7 \times 10^{-6}$ |
| Cancellous bone | | 100 | 0.2 | | $1.1 \times 10^{-6}$ |
| Bony posterior element | | 3,500 | 0.25 | | $1.4 \times 10^{-6}$ |
| Annulus | | 4.2 | 0.45 | | $1.05 \times 10^{-6}$ |
| End plate | | 500 | 0.25 | | $1.2 \times 10^{-6}$ |
| Nucleus pulpous | | 1 | 0.49 | | $1.02 \times 10^{-6}$ |
| Imaginary cartilage | | 10 | 0.4 | | $1.0 \times 10^{-6}$ |
| Annulus fiber | 3D-cable | 500 | | | $1.0 \times 10^{-6}$ |
| Capsular ligaments | | | | 30 | $1.0 \times 10^{-6}$ |
| Intertransverse ligaments | | | | 1.8 | $1.0 \times 10^{-6}$ |
| Supraspinous ligaments | | | | 30 | $1.0 \times 10^{-6}$ |
| Interspinous ligaments | | | | 40 | $1.0 \times 10^{-6}$ |
| Ligamentum flavum | | | | 40 | $1.0 \times 10^{-6}$ |
| Anterior longitudinal ligaments | | | | 63.7 | $1.0 \times 10^{-6}$ |
| Posterior longitudinal ligaments | | | | 20 | $1.0 \times 10^{-6}$ |
| Anterior sacroiliac ligaments | | 20 | | 160 | $1.0 \times 10^{-6}$ |
| Posterior sacroiliac ligaments | | 20 | | 300 | $1.0 \times 10^{-6}$ |
| Interosseous sacroiliac ligament (regarded as fiber) | | 500 | | | $1.0 \times 10^{-6}$ |

## 15.3 VALIDATION OF SPINE FINITE ELEMENT MODELS

To validate the L3-L5 segment model, a static FE analysis was conducted (Berkson, Nachemson, and Schultz 1979; Shirazi-Adl and Drouin 1987; Kim et al. 1991; Kasra, Shirazi-Adl, and Drouin 1992; Goel, Park, and Kong 1994; Pankoke, Hofmann, and Wolfel 2001). For effective comparison with other published results, the same loads and boundary restrictions were applied. Figure 15.3a shows that the compressive displacements of the L3-L4 segment and L4-L5 segment fall in the range of the experimental deviation (Berkson, Nachemson, and Schultz 1979) under a 400-N compressive force. Figure 15.3b shows the comparison of force-displacement results of the L3-L5 segment between this study and the results of Kim et al. (1991), and indicates that the two results are in agreement within the compressive displacement of 1.5 mm. After the nucleus was removed from the L4-L5 spinal segment, the denucleated FE model was also validated against the published results of Frei et al. (2001). Figure 15.3c shows that the displacements of the intact model and the denucleated model of the L4-L5 segment all fall in the range of the experimental results (Frei et al. 2001) under 1000-N compressive forces.

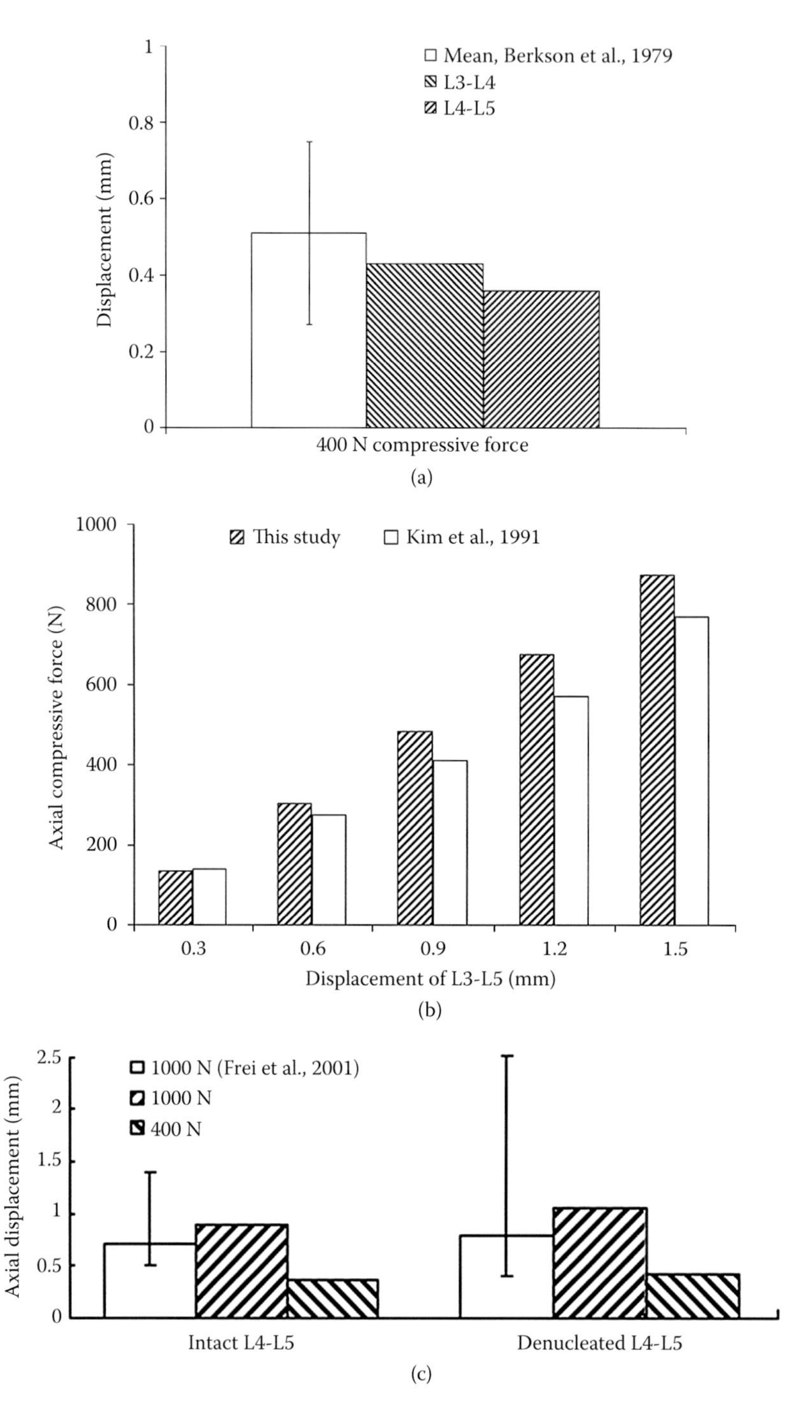


**FIGURE 15.3** Validation of the finite element model against published results: (a) against an experimental study; (b) against a finite element analysis of the L3-L5 segment; (c) against an experimental study for the intact model and the denucleated model of the L4-L5 segment under 1000 N and 400 N compressive forces.

Considering the frequency characteristics of the finite element models, Goel, Park, and Kong (1994) reported that the first-order natural frequencies of their ligamentous L3-L5 FE model were 3.8 Hz in flexion-extension and 17.5 Hz in axial mode. The first-order resonant frequencies of the current FE model were 3.96 Hz in flexion-extension and19.6 Hz in axial mode. Kasra, Shirazi-Adl, and Drouin (1992) performed an experimental frequency response analysis on a specimen of

T12-L1 carrying a mass of 40 kg, and reported a first-order resonant frequency of 23.5 Hz in axial mode under a 30-N preload. In addition, the static facet contact force in the sagittal plane was 2.1% of the 400-N compressive load for the same model. These results agree with other published results, for example 1%~5% from Shirazi-Adl and Drouin (1987).

In order to investigate the resonant frequency characteristics of the human spine, different numbers of spinal segments were analyzed using FE modal analysis. Table 15.2 lists the resonant frequencies of FE models with one and two motion segments, L1-L5, L1-S1, and T12-Pelvis, and other published results (Pope et al. 1987; Panjabi et al. 1986; Sandover and Dupuis 1987; Kong and Goel 2003; Kasra, Shirazi-Adl, and Drouin 1992; Goel, Park, and Kong 1994).

The resonant frequencies of the one motion segment FE models of T12-L1 and L2-L3 in the vertical direction, carrying a mass of 40 kg, are 25.7 and 26.7 Hz, respectively. The frequency of the two segment L3-L5 model is 19.6 Hz. The frequencies of the entire lumbar spine of L1-L5 and L1-S1 in the vertical direction, carrying a distributed mass of 40 kg, are 11.5 and 9.12 Hz. The resonant

**TABLE 15.2**
**Resonant Frequencies of Finite Element Models in the Vertical Direction for Different Spinal Motion Segments**

| Spinal Motion Segments | | Resonant Frequency (Hz) | References | Notes |
|---|---|---|---|---|
| One motion segment | T12-L1 | 23.5 | Kasra, Shirazi-Adl, and Drouin (1992) | Experimental study |
| | T12-L1 | 25.7 | | 40 kg upper body |
| | L2-L3 | 26.0 | Kasra, Shirazi-Adl, and Drouin (1992) | |
| | L2-L3 | 26.7 | | 40 kg upper body |
| | L3-L4 | 27.7 | Kong and Goel (2003) | |
| Two motion segments | L4-S1 | 17.5 | Goel, Park, and Kong (1994) | 40 kg upper body |
| | L4-S1 | 16.4 | | 40 kg upper body |
| | L3-L5 | 18.0 | Kong and Goel (2003) | |
| | L3-L5 | 19.6 | | 40 kg upper body |
| Lumbar spine | L1-L5 | 12.2 | Kong and Goel (2003) | |
| | L1-L5 | 11.5 | | 40 kg upper body |
| Lumbosacrum | L1-S1 | 10.6 | Kong and Goel (2003) | |
| | L1-S1 | 9.12 | | 40 kg upper body |
| Whole thoraco-lumbar spine | T1-S1 | 8.40 | Kong and Goel (2003) | With rib cage |
| Head to sacrum | H-S1 | 8.32 | Kong and Goel (2003) | With rib cage |
| | H-S1 | 8.91 | Kong and Goel (2003) | With muscle and rib cage |
| | H-S1 | 6.82 | Kong and Goel (2003) | With preload and rib cage |
| Lower thorax to sacrum | T12-S1 | 7.68 | | 40 kg upper body |
| Lower thorax to pelvis | T12-Pelvis | 8.23 | | 30 kg upper body |
| | T12-Pelvis | 7.17 | | 40 kg upper body |
| | T12-Pelvis | 5.84 | | 60 kg upper body |
| Human body | | 4~6 | Pope et al. (1987); Panjabi et al. (1986); Sandover and Dupuis (1987) | Experimental study |
| | | 4~8 | Kong and Goel (2003) | |

frequency of the FE model of T12-Pelvis, carrying a distributed mass of 40 kg, in the vertical direction is 7.17 Hz (Guo et al. 2005; Guo and Teo 2005; Guo, Zhang, and Teo 2007; Guo et al. 2011).

In addition, considering the modeling complexity, the current ligamentous FE model of the human spine does not include muscles attached on the spine, the mass distribution of the human upper body, or the deformation between ilia and pelvis. These are limitations of this study design, but may not influence the analytical results. Other investigations have indicated that the muscles of the lumbar spine produce only a small influence on the amplitude of the vertical driving point impedance within 15 Hz. Our study mainly investigates the vertical vibration modes of the lumbar spine, so it may diminish the influence of the mass distribution of the upper body to the minimum.

## 15.4 DYNAMIC CHARACTERISTICS ANALYSIS OF THE INJURED SPINE

In this study, we investigated how injury to the spine (including denucleation and facetectomy) may affect the vibrational trend of the whole lumbar spine, for example, the motion relation of spine motion segments. At the same time, this study also aimed to investigate how injury to a motion segment may influence adjacent segments (Guo et al. 2009b, 2009c). The hypothesis of this study is that injuries to the spine might change the dynamic characteristics of the whole human lumbar spine and adversely affect adjacent motion segments under WBV. To complete these studies, an FE model of a whole lumbar spine, including its adjacent motion segments, was developed and used to explore the dynamic characteristics of the human spine using FE modal analyses.

Three injury conditions at the L4-L5 segment were assumed, that is, case 1 simulates denucleation at the L4-L5 intervertebral disc, case 2 simulates removal of facet joints and their capsular ligaments at the L4-L5 segment, and case 3 simulates denucleation and removal of facet articulations and their capsular ligaments at the L4-L5 segment.

After FE modal analysis, the resonant frequencies and modal modes of the spine model were obtained. The first-order vertical resonant frequency (FOVRF) of the intact T12-Pelvis model was 7.21 Hz. For the injured spines, the FOVRFs of the T12-Pelvis model were 6.89 Hz, 7.17 Hz, and 6.83 Hz for case 1, case 2, and case 3, respectively. These results indicate that all the simulated injuries may decrease the resonant frequency of the whole spine system. The FOVRF decreases by 5.3% if the spine suffers a denucleation and a facetectomy with capsular ligament removal. In addition, for validation and to gain a thorough understanding of the dynamic characteristics of the human spine, additional analyses were also conducted on several shorter segments of the T12-Pelvis model. For a one motion segment, the FOVRF of T12-L1 was 25.7 Hz. For two motion segments, the FOVRF of L4-S1 was 16.4 Hz. The FOVRFs of L1-L5 and L1-S1 were 11.5 and 9.12 Hz, respectively.

Figure 15.4 shows the anteroposterior (A-P) deformation (for the four model conditions) of all the vertebrae from T12 to S1 for the first-order vertical vibration mode (FOVVM) of the model. Figure 15.5

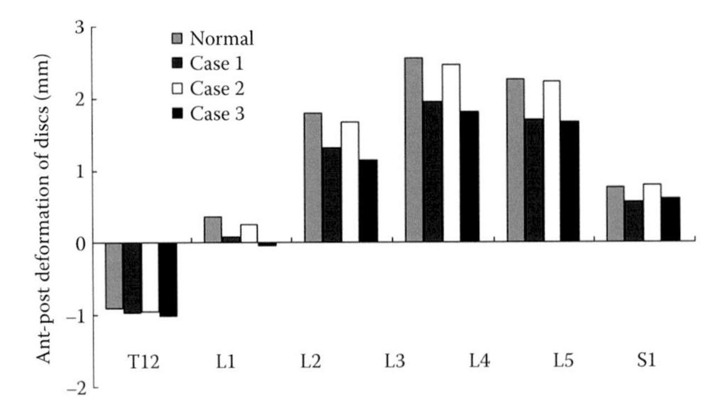


**FIGURE 15.4** Deformation of vertebrae in the anteroposterior direction for the FOVVM. Normal: for the intact spine; case 1: for denucleation at the L4-L5 intervertebral disc; case 2: for removal of facet joints and their capsular ligaments at the L4-L5 segment; case 3: for denucleation and removal of facet joints and their capsular ligaments at the L4-L5 segment.

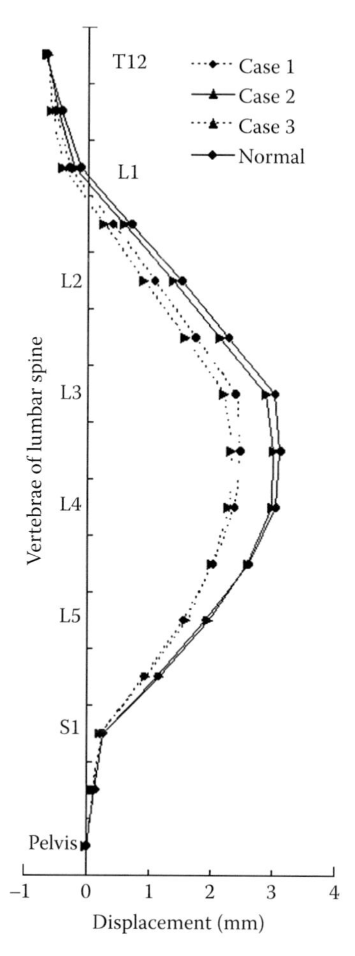


**FIGURE 15.5** Curved plot of spine deformation in the sagittal plane for the FOVVM. The points between every two vertebrae represent intervertebral discs.

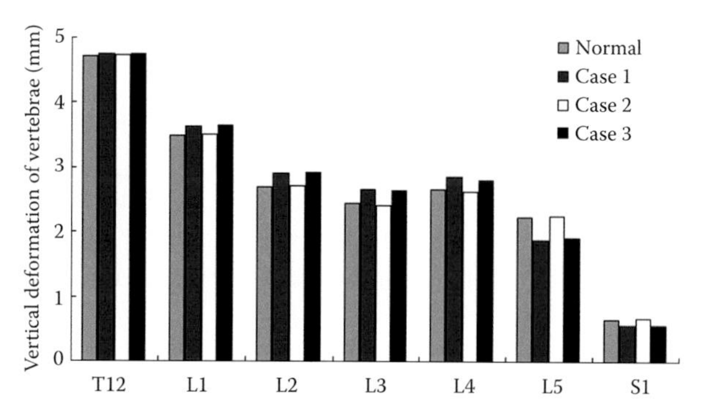


**FIGURE 15.6** The vertical deformation of the vertebrae for the FOVVM.

shows a curved plot of spine deformation in the sagittal plane for the FOVVM and the marked points indicate the vertebrae and intervertebral discs. Figure 15.6 exhibits the deformations in the vertical direction. From Figures 15.4 and 15.6, it can be seen that the human upper body primarily undergoes vertical vibration with little motion in the A-P direction during WBV. Figure 15.7 indicates the rotational angles, in the sagittal plane, of each vertebra for the FOVVM. From the figures, it can be seen that the lumbar spine segments not only show vertical vibration but also exhibit A-P motions

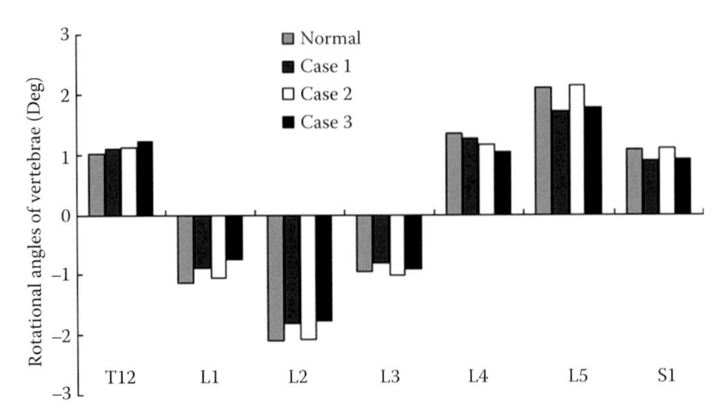


**FIGURE 15.7** Rotational angles of the vertebrae in the sagittal plane for the FOVVM.

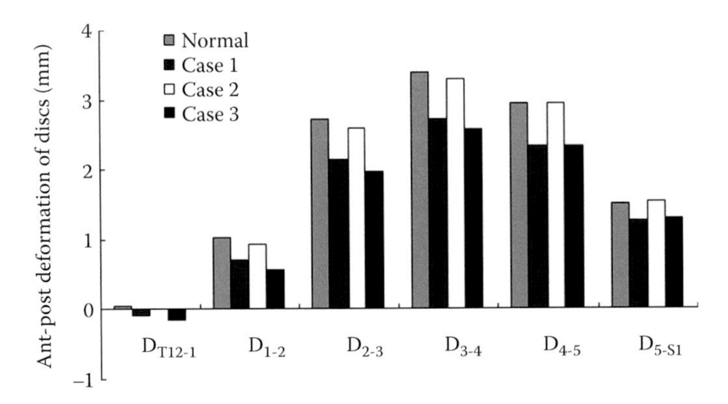


**FIGURE 15.8** The anteroposterior deformation of the anterior points of the disc annulus in the sagittal plane for the FOVVM.

(including sagittal plane translation and flexion-extension rotation) during WBV. Figure 15.8 shows the A-P displacements of the anterior points of each disc annulus in the sagittal plane for the FOVVM. In Figure 15.8, the symbols $D_{T12\text{-}1}$, $D_{1\text{-}2}$, $D_{2\text{-}3}$, $D_{3\text{-}4}$, $D_{4\text{-}5}$, and $D_{5\text{-}S1}$ represent the disc between every two vertebrae, for example, $D_{1\text{-}2}$ means the disc between L1 and L2.

From the above simulations, it can also be seen that the vertebrae L3 and L4 and the intervertebral disc between them exhibit the maximum deformation in the A-P direction (Figures 15.8 and 15.4). In addition, although the facetectomy with removal of capsular ligaments has a slight influence on the lumbar spine, denucleation produces a more obvious influence (variations in vibration amplitudes) on the adjacent discs. This phenomenon indicates that the intervertebral disc plays an important role in the dynamic characteristics of the lumbar spine during WBV.

The biomechanical behavior of the human spine is complex. FE models with few motion segments aren't sufficient for studying the dynamic biomechanical characteristics of the whole lumbar spine. Therefore, a more complex T12-Pelvis model was used in this study. To grasp the findings of this study, some limitations should be stated. The T12-Pelvis FE model did not include the muscles, cervical spine, thoracic spine, and rib cage of the upper body. The addition of the cervical spine and thoracic spine may decrease the resonant frequency of the T12-Pelvis FE model and the addition of muscles and a rib cage may increase the resonant frequency of the model. In addition, the potential problem of anteroposterior oscillation of the human body during WBV may involve the active-regulation effect of muscles and the nonlinear influence of large muscle deformation. Therefore, the current model also needs further improvement in the modeling of muscles and distributing the mass of the upper body in terms of both anatomic and mechanical aspects, so that the model can simulate the dynamic and impact conditions of flexion-extension, lateral bending, and torsion of the human spine. However, the current study mainly dealt with vertical vibration in the sagittal plane, so this might decrease the influence of the aforementioned limitations to the minimum. The lower thorax to pelvis model can be used to predict the biomechanical behavior of the lumbar spine under vibration and impact loads. The findings might be useful as guidelines for clinical treatment and product development in industry.

## 15.5 TRANSIENT RESPONSE ANALYSIS OF THE HUMAN SPINE

Epidemiological investigations have suggested that WBV contributes significantly to injuries and functional disorders of the skeleton and joints. Many clinical observations and investigations have explored the link between low back pain and WBV in special working populations, such as drivers of trucks, tanks, helicopters, and so on. Although a large number of investigations have drawn attention to the harmful effects of WBV on the human spine and the influence of injury and degeneration of intervertebral discs on facet joints, the influence of WBV on facet joints is not clearly known, especially for damaged intervertebral discs.

In this study, the biomechanical response of the human spine under vibration loading was analyzed using transient FE analysis. The stress and deformation of intervertebral discs and facet contact forces were explored under WBV. At the same time, the effects of denucleation of intervertebaral discs, and different loads, frequencies, and damping on facet contact pairs were analyzed and compared with a normal spine.

A ligamentous FE model of a spinal L3-L5 segment was used. To investigate the effects of injured discs, and changes in frequency, load, and damping on facet contact forces of the lumbar spine under WBV, various conditions were considered for the transient dynamic analyses. For example, case 1: Intact/5Hz/no damp/40N, means that, for case 1, transient dynamic analysis was carried out on the intact model without damping under a 5-Hz cyclic axial load of 40 N. Case 7: Denucleated/5Hz/damp0.08/40N, means that, for case 7, the dynamic analysis was carried out on the denucleated model with a damping ratio of 0.08 under the 5-Hz cyclic axial force of 40 N. For all the dynamic analyses, the preload of the upper body mass of 40 kg was imposed on the top of vertebra L3. The total loads were 400 ± 40 N for all the cases except case 3, and 400 ± 20 N for case 3.

Because the upper body was preloaded in this study and its influence is very large when compared with the sinusoidal compressive force, the influence of preload must be separated from the results in order to better evaluate the effects of dynamic loads (Guo, Zhang, and Teo 2007). Therefore, a few formulations are given.

To evaluate the effect of dynamic loads on FE models by comparing with static loads, the dynamic-static amplitude ratio $R_{DS}$ was defined as follows:

$$R_{DS} = \frac{V^i_{\max} - V^i_{\min}}{aV^i_{static}} \qquad (i = 1, \ \ldots, \ 8) \tag{15.1}$$

$$a = \frac{2F_{dynamic}}{F_{preload}}$$

To compare the effects of frequency, load, and damping on intact models and denucleated models, the case-to-case amplitude ratio $R_{CC}$ was defined as follows:

$$R_{CC} = \frac{V^i_{\max} - V^i_{\min}}{V^j_{\max} - V^j_{\min}} \qquad (i, j = 1, \ \ldots, \ 8) \tag{15.2}$$

where $V^i_{\max}$ and $V^i_{\min}$ are the maximum and minimum values of dynamic curves in the first vibration period, which were obtained by the cyclic axial load and preload for case $i$. $V^i_{static}$ is the value of different models under the static preload. $i$, $j$ are the serial numbers of cases in Table 15.3. $V^i_{\max}$, $V^i_{\min}$, and $V^i_{static}$ correspond with the values $D_{max}$, $D_{min}$, and $S_{400}$ in Table 15.3, respectively. $a$ is

**TABLE 15.3**
**Dynamic Maximum and Minimum Facet Contact Forces for Different Cases and Static Facet Contact Force under a 400-N Static Compressive Load**

| Case* | Contact Forces | | Case | Contact Forces | | Case | Contact Forces | |
|---|---|---|---|---|---|---|---|---|
| | Items | Values (N) | | Items | Values (N) | | Items | Values (N) |
| | $D_{max}$ | 11.8 | | $D_{max}$ | 15.6 | | $D_{max}$ | 9.73 |
| Case 1 | $D_{min}$ | 3.39 | Case 2 | $D_{min}$ | 2.57 | Case 3 | $D_{min}$ | 6.35 |
| | $S_{400}$ | 8.41 | | $S_{400}$ | 8.41 | | $S_{400}$ | 8.41 |
| | $D_{max}$ | 11.2 | | $D_{max}$ | 10.7 | | $D_{max}$ | 31.6 |
| Case 4 | $D_{min}$ | 4.93 | Case 5 | $D_{min}$ | 6.43 | Case 6 | $D_{min}$ | 12.8 |
| | $S_{400}$ | 8.41 | | $S_{400}$ | 8.41 | | $S_{400}$ | 18.3 |
| | $D_{max}$ | 31.7 | | $D_{max}$ | 31.5 | | | |
| Case 7 | $D_{min}$ | 14.8 | Case 8 | $D_{min}$ | 14.7 | | | |
| | $S_{400}$ | 18.3 | | $S_{400}$ | 18.3 | | | |

*Note:* $D_{max}$ and $D_{min}$ are the maximum and minimum values of facet contact forces during the first vibration period. The static facet contact force $S_{400}$ is the predicted facet contact force under 400-N static compressive loads.
*Instruction of different cases in this study.
*Case 1:* Intact/5Hz/no damp/40N
*Case 2:* Intact/10Hz/no damp/40N
*Case 3:* Intact/5Hz/no damp/20N
*Case 4:* Intact/5Hz/damp0.08/40N
*Case 5:* Intact/5Hz/damp0.15/40N
*Case 6:* Denucleated/5Hz/no damp/40N
*Case 7:* Denucleated/5Hz/damp0.08/40N
*Case 8:* Denucleated/5Hz/damp0.15/40N

the load coefficient of the dynamic load against the preload. In this study, the dynamic load force $F_{dynamic}$ is 40 N, and the preload force $F_{preload}$ is 400 N.

After the transient dynamic analyses, the values of facet contact forces in the sagittal plane were recorded. The plots of predicted results as a function of time revealed a cyclic response with some fluctuations and distortions, as shown in Figures 15.9 and 15.10. For the sake of comparison, the mechanical responses of the static axial loads of 400 N were also computed. Table 15.3 exhibits the dynamic maximum and minimum of the facet contact forces of different cases in the first vibration period.

For example, the dynamic maximum and minimum facet contact forces are 11.8 and 3.39 N for the intact model without damping under a 5-Hz cyclic axial load of 40 N and upper body preload of 40 kg (case 1), and the corresponding static facet contact force is 8.41 N under a 400-N static compressive force. The higher frequency cyclic compressive loads will increase the facet contact force. The results in Table 15.3 and Figure 15.9a show that the dynamic maximum and minimum facet contact forces are 15.6 and 2.57 N, respectively, using a 10-Hz cyclic load of 40 N for the intact model without damping (case 2). In addition, when comparing different magnitudes of external load (Figure 15.9b), although the external load doubled (from 20 N—case 3—to 40 N), the amplitude of the contact force increased by 1.49 (Table 15.4). This implies that the facet force contact force exhibits nonlinear behavior under different magnitudes of external vibration loads.

The dynamic maximum and minimum facet contact forces of the denucleated model (case 6) were 31.6 and 12.8 N (see Table 15.3 and Figure 15.9a). The static facet force of the denucleated model (case 6) under a 400-N compressive load was 18.3 N. Compared with the intact

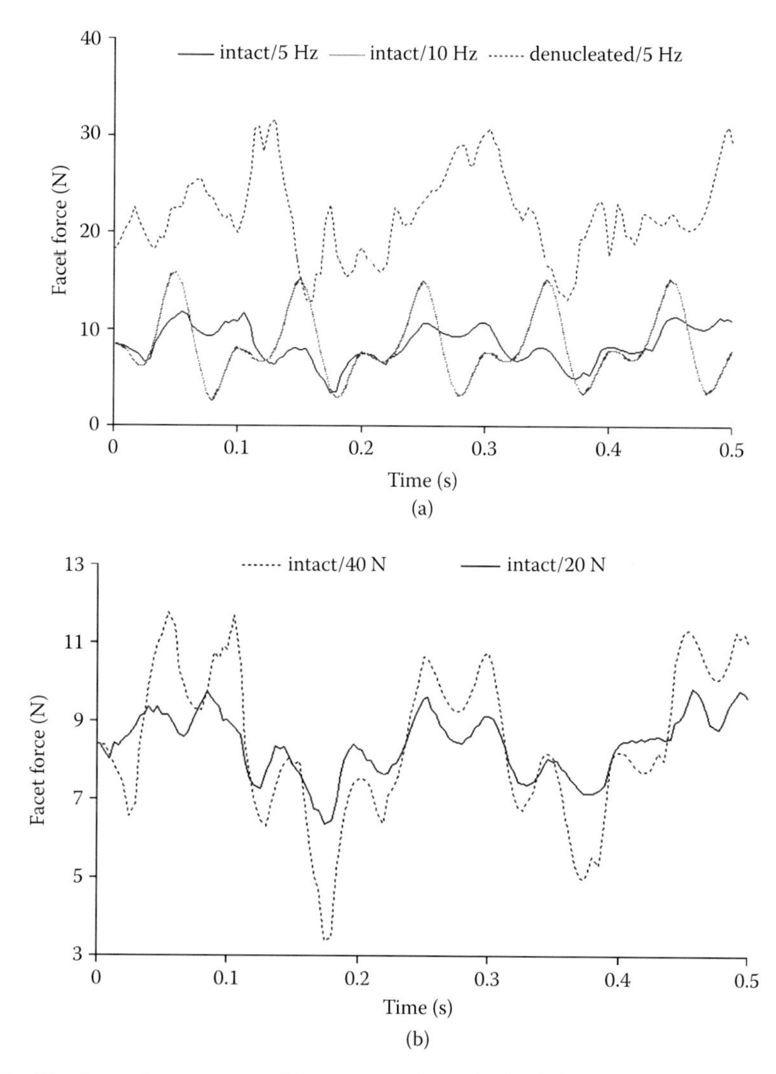


**FIGURE 15.9** The dynamic responses of facet contact force in the finite element models without damping under axial cyclic loads for different frequencies and loads and degenerated discs: (a) for case 1, case 2, and case 6; (b) for case 1 and case 3.

## TABLE 15.4
## Case-to-Case Amplitude Ratio $R_{CC}$ of Facet Contact Forces for Different Cases

| Cases | About | Intact Model | | | | | Denucleated Model | | |
|---|---|---|---|---|---|---|---|---|---|
| | $R_{CC}$ | Case 1 | Case 2 | Case 3 | Case 4 | Case 5 | Case 6 | Case 7 | Case 8 |
| Versus | $R_{CC}$ | — | 1.55 | 0.402 | 0.746 | 0.506 | | | |
| case 1 | $1/R_{CC}$ | | 0.645 | 2.49 | 1.34 | 1.98 | | | |
| Versus | $R_{CC}$ | | | | | | — | 0.901 | 0.895 |
| case 6 | $1/R_{CC}$ | | | | | | | 1.11 | 1.12 |

*Note:* The ratios were obtained based on Equations 15.1 and 15.2.

model, the maximum facet contact force of the denucleated model was 2.18 times greater (case 1). Obviously, the removal of the nucleus leads to a greater static facet contact force (Guo and Teo 2006).

The dynamic maximum facet contact forces of the intact model with damping were 11.2 N for a damping ratio of 0.08 (case 4) and 10.7 N for a damping ratio of 0.15 (case 5). Damping may decrease the amplitude of vibration and alleviate the extent of vibration of the lumbar spine. The maximum facet contact force of the intact model with a 0.08 damping ratio (case 4) reached 0.746 times that of case 1, and the maximum force of the intact model with a 0.15 damping ratio (case 5) reached 0.506 times that of case 1 (Figure 15.10a and Table 15.4).

However, from Figure 15.10b, it can be found that the damping did not have the same effects on the denucleated models; the peak values are similar across all denucleated models.

Data about displacement and stress were acquired from the dynamic simulations for further analysis. The plots of predicted stress and bulging of the L4-L5 disc as a function of time revealed a cyclic response, with some fluctuations and distortions, as shown in Figures 15.11 and 15.12.

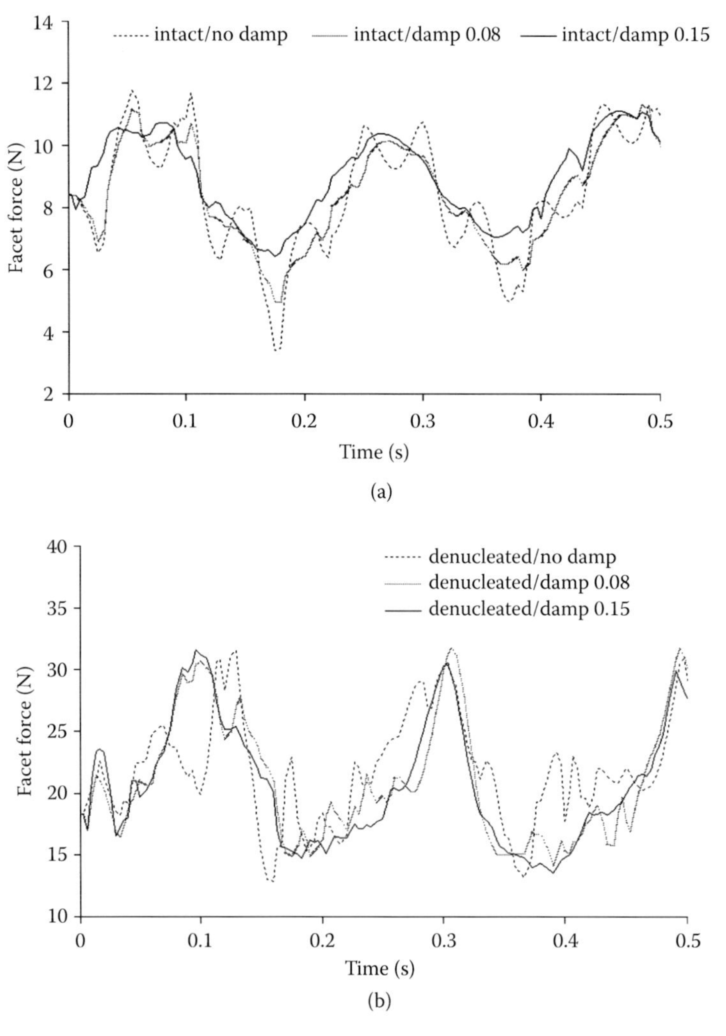


**FIGURE 15.10** The dynamic responses of facet contact force in the finite element models with damping under 5-Hz axial cyclic loads of 40 N for intact and denucleated models: (a) for intact models with different damping ratios of 0.0, 0.08, and 0.15; (b) for denucleated models with different damping ratios of 0.0, 0.08. and 0.15.

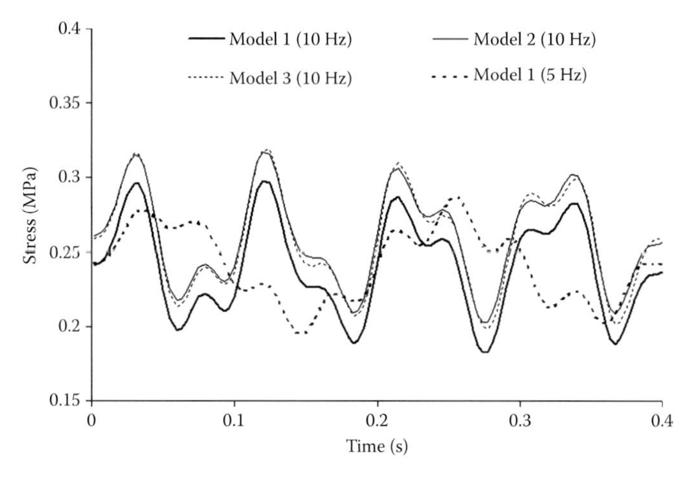


**FIGURE 15.11** The stress at the posterolateral region of the L4-L5 disc annulus under 5-Hz and 10-Hz axial loads for model 1, model 2, and model 3.

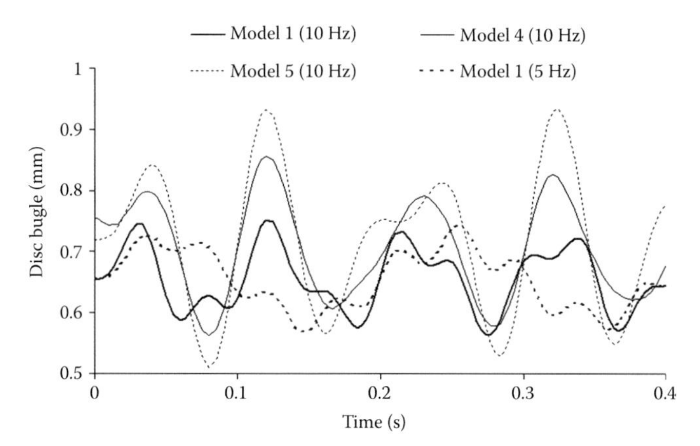


**FIGURE 15.12** The anteroposterior disc bulge of the L4-L5 disc annulus under 5-Hz and 10-Hz axial loads for model 1, model 4, and model 5.

For comparison against dynamic results, the mechanical responses of the –360-N and –440-N static axial loads were also computed. Table 15.3 summarizes the biomechanical responses of the annulus and nucleus from the intact and injured models for L4-L5 under vibration and static conditions. Figure 15.13 shows the deformation rotation angles of the L4-L5 disc under 10-Hz axial loads for all intact and injured conditions.

For example, for the facet-injured model (model 2), the maximum von Mises stress around the annulus circumference reached 0.263 MPa at the anterior region, 0.279 MPa at the posterior region, and 0.317 MPa at the posterolateral region when applying both the vibration load and the upper-body preload. For comparison, the corresponding maximum values of the intact model (model 1) are 0.265 MPa at the anterior region, 0.235 MPa at the posterior region, and 0.297 MPa at the posterolateral region (shown in Table 15.3). Removal of the facet also increases the bulging of the annulus and the nucleus stress. The maximum values are 0.938 mm for anteroposterior disc bulging and 0.103 MPa for stress within the nucleus. The corresponding values of the intact model are 0.751 mm for anteroposterior disc bulging and 0.097 MPa for nucleus stress. From Table 15.4, for the intact model, WBV increases the vibration amplitude by 34.6% at the anterior region of the annulus circumference, and by 97% and 106.6% at the posterior and posterolateral regions,

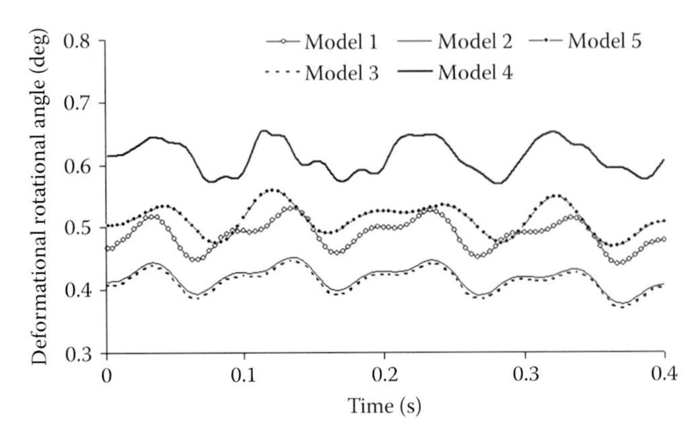


**FIGURE 15.13** The deformation rotation angles of the L4-L5 disc in flexion under 10-Hz axial loads for the normal and injured conditions.

respectively. Comparing the static stress values of the annulus from model 2 and model 1, removal of the facet decreases the stress at the anterior region, but increases the stress at the posterior and posterolateral regions. Denucleation (model 4 and model 5) increases the stress of the annulus. During WBV at 10 Hz, comparing model 3 with model 1 and model 2, ligaments show only a minor influence on the intervertebral disc. Figure 15.13 shows that the facets play an important role in the deformation rotation angles of the denucleated motion segment.

In addition, to investigate the effect of sinusoidal compression forces, different magnitudes of sinusoidal forces were imposed and analyzed for the intact model with the upper body: 20 N, 30 N, 40 N, 50 N, and 60 N. The results show that the stress variation at the annulus circumference is almost linear.

## 15.6 MATERIAL SENSITIVITY ANALYSIS OF SPINE FINITE ELEMENT MODELS

The resonant frequency of structural systems is directly related to their geometrical configuration, mass distribution, and material property. For the human spine, the shape of articulating bony vertebrae and associated tissues, and mass distribution of all the spinal components, are relatively constant. However, due to aging or disease, the material characteristics of spinal components might change (Diamant, Shahar, and Gefen 2005). Thus, it is believed the material properties of spinal components are a main factor affecting the frequency characteristics of FE models of the human spine. The material sensitivity of single intervertebral discs (Shirazi-Adl, Shrivastava, and Ahmed 1984; Spilker, Jakobs, and Schultz 1986), single motion segments (Rao and Dumas 1991), and coupled motion segments (Kumaresan, Yoganandan, and Pintar 1999; Ng, Teo, and Lee 2004) have been analyzed by comparing the mechanical responses of FE models under static conditions. It is unknown how this material property sensitivity will affect the dynamic characteristics of the human spine.

Accordingly, this study will principally analyze the influence of material property sensitivity of spinal components on frequency characteristics of the human spine (Guo et al. 2009a). The results may help us to understand how changes in material properties may influence the resonant frequency of FE models of the human spine.

Changes in the material properties of spinal components may arise due to age, gender, or component degeneration (Diamant, Shahar, and Gefen 2005), which might influence dynamic characteristics. Therefore, a sensitivity analysis was carried out to determine how material parameters (Young's modulus of vertebral body, endplate, ligament, and intervertebral disc) may influence the predicted frequency characteristics of the spine.

Table 15.5 shows the material property (for the basic model parameters) of the T12-Pelvis FE model and various ranges of Young's modulus of the spinal components reported (Goel et al. 1995; Shirazi-Adl, Ahmed, and Shrivatava 1986; Sharma, Langrana, and Rodriguez 1995; Lavaste et al. 1992; Goel et al. 1993; Kumaresan, Yoganandan, and Pintar 1999; Kumaresan et al. 1999; Lee et al. 2000; Kim 2001). For example, the Young's modulus of the nucleus pulpous varies from 0.13 to 4.0 MPa (Table 15.5). The Young's modulus of annulus ground substance varies from 0.8 to 4.2 MPa, a change of over 500%. The analyzed spinal components include a cortical shell, cancellous bone, bony posterior element, nucleus pulposus, annulus ground substance, annulus fiber, and ligaments. The ligaments were partitioned into two groups. Ligament group 1 consisted of anterior longitudinal ligaments and posterior longitudinal ligaments, and ligament group 2 consisted of the remaining ligaments. Several spinal component groups were also assigned, for example, the vertebra group includes cortical bone, cancellous bone, and a bony posterior element; the ligament group includes all ligaments; and the intervertebral disc group includes a nucleus pulposus, annulus ground substance, and annulus fiber. To ease comparison, the elastic modulus of each spinal component was assumed to vary by ±30% against the basic model in this study. In addition, ±80% variations are assigned to the Young's modulus of the disc annulus based on the parameters of the basic model, making the Young's modulus vary from 0.84 to 7.56 MPa (Guo et al. 2009a).

In this study, FE modal analyses were run 29 times to extract the resonant frequencies of the T12-Pelvis model under the high and low values of material properties. Also, a further 17 runs were specially performed to analyze how variations in the Young's modulus of the disc annulus may influence the resonant frequency of the T12-Pelvis model. An additional two runs were also conducted to examine variations in the Poisson's ratio of the nucleus pulposus.

Figure 15.14 shows the first-order vertical resonant frequencies of the T12-Pelvis models obtained after ±30% variations of the Young's moduli of the component groups of vertebral bodies VT1 to VT5. It can be seen that bony posterior elements (VT3) have a more obvious influence on the resonant frequencies than cortical bone, cancellous bone, and endplates. The reason is that the lumbar spine can only flex through very small angles in spite of the human upper body conducting a vertical vibration. Figure 15.15 exhibits the effect of variations in the elastic modulus on the

**TABLE 15.5**
**Analyzed Component Groups of the Human Spine in This Study**

| Component Groups | | Description |
|---|---|---|
| Vertebra tissue | VT1 | Cortical bone |
| | VT2 | Cancellous bone |
| | VT3 | Bony posterior element |
| | VT4 | Endplate |
| | VT5 | Vertebra (including endplate, cortical bone, cancellous bone, posterior element and endplate) |
| Disc tissue | DT1 | Nucleus pulposus |
| | DT2 | Annulus ground substance |
| | DT3 | Annulus fiber |
| | DT4 | Intervertebral disc (including nucleus pulposus and disc annulus and annulus fiber) |
| Ligament tissue | LT1 | Anterior longitudinal ligaments and posterior longitudinal ligaments |
| | LT2 | All ligaments except LT1 |
| | LT3 | All ligaments |
| All spinal components | All | Including all spinal components |

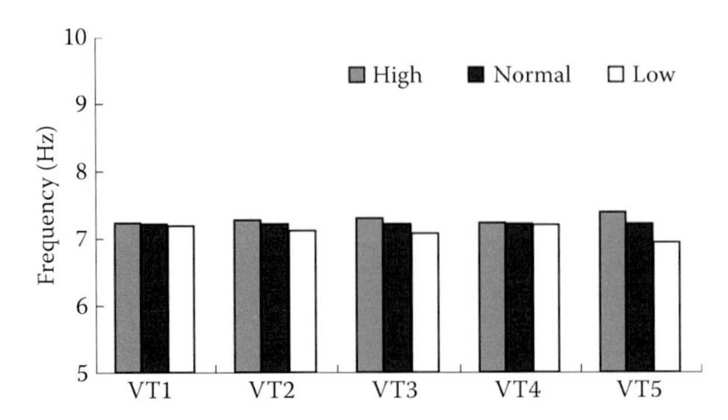


**FIGURE 15.14** The resonant frequencies of the finite element models due to variations of the material property for the component groups VT1, VT2, VT3, VT4, and VT5.

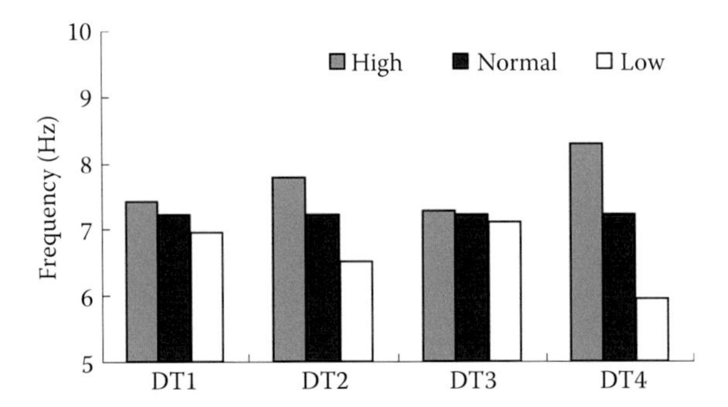


**FIGURE 15.15** The resonant frequencies of the finite element models due to variations of the material property for the component groups DT1, DT2, DT3, and DT4.

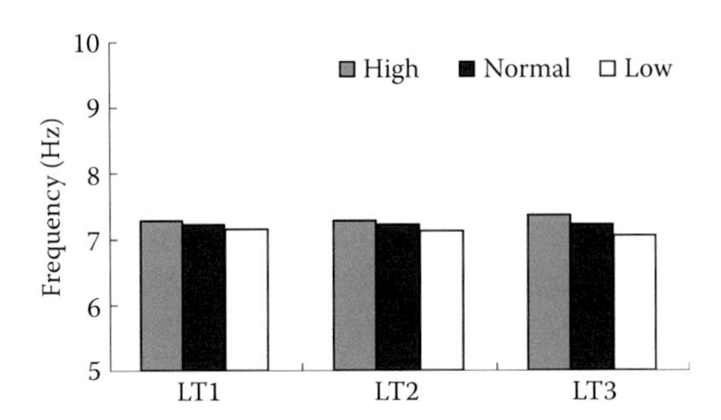


**FIGURE 15.16** The resonant frequencies of the finite element models due to variations of the material property for the component groups LT1, LT2, and LT3.

resonant frequencies of the model. The results indicate that variations in the elastic modulus of the disc annulus have a greater influence than other disc components. Figure 15.16 exhibits the effect of the elastic modulus variations of ligaments on the resonant frequencies of the model. The results show that the influence of the elastic modulus variations in LT2 is slightly greater than in LT1. Figures 15.17 and 15.18 show the resonant frequencies and their relative percentage changes (against the basic model) for the spinal component units: endplate, vertebra, intervertebral disc, ligament, and all spinal components due to ±30% variations of the material properties of the T12-Pelvis FE model, respectively. The results (Figures 15.17 and 15.18) indicate that the intervertebral disc plays

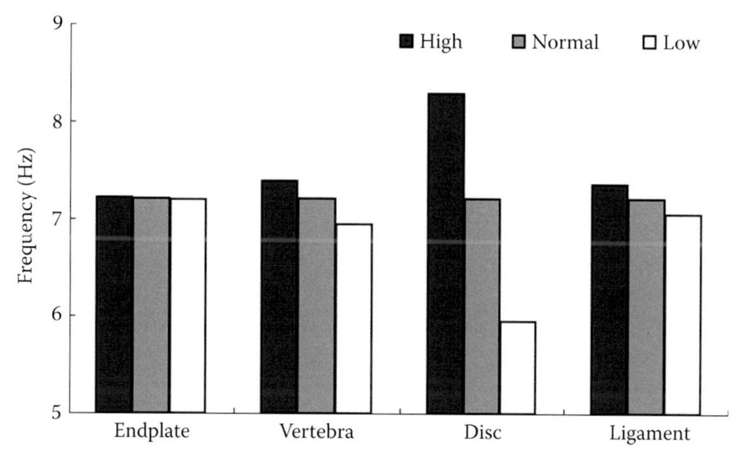


**FIGURE 15.17** Comparing the effects of variations of the material properties of the main spinal component groups on the resonant frequencies of the spine.

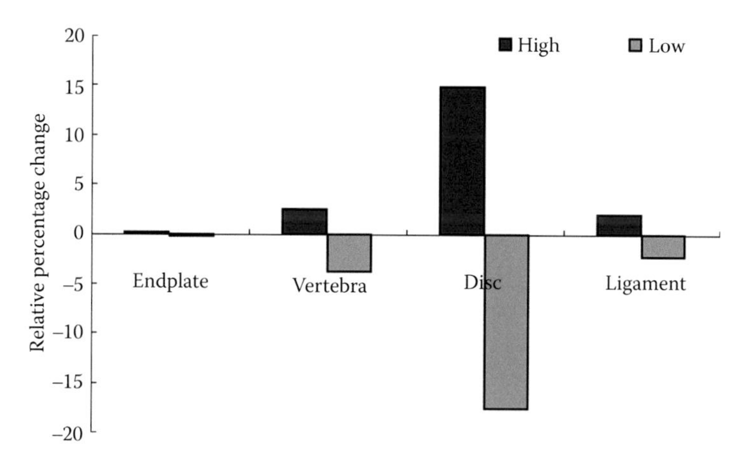


**FIGURE 15.18** The relative percentage changes of resonant frequencies of the T12-Pelvis finite element model for different spinal component groups due to the variation of material properties against the basic model.

a greater role in determining the resonant frequency of the spine. It can be found that the first-order vertical resonant frequency of the model decreases by 19.2% if the Young's moduli of all the spinal components are decreased by 30%, based on the material properties of the basic model.

Of all the spinal components, the disc annulus contributed most prominently to the resonant frequency of the human spine (Figures 15.17 and 15.18). Therefore, additional analyses were carried out to understand how variations in the Young's modulus of the disc annulus influence the resonant frequency of the human spine.

The resonant frequency and percentage changes in the frequency of the model due to the variations (against the basic model) of the Young's modulus of the disc annulus were also analyzed (Guo et al. 2009a). The resonant frequencies of the model varied from −29.0% to 15.5% of the intact model, while the Young's modulus of the disc annulus varied from 0.84 MPa to 7.56 MPa. The results also show that the resonant frequency of the human spine model will decrease remarkably when the Young's modulus of the disc annulus is less than 2.0 MPa. In addition, the effect of variations in the Poisson's ratio of the nucleus pulpous on the resonant frequency was also analyzed. The results show that the resonant frequency of the human spine might decrease by 18% and 16% if the Poisson's ratio of the nucleus pulposus decreases by 30% and 20%, respectively.

Changes in the material properties of multiple spinal components due to age, gender, and component degeneration may occur concurrently in vivo, and how these variations may affect the biomechanical behavior of the human spine is unknown. In this study, a detailed three-dimensional FE model of the whole lumbar spine segment T12-Pelvis was developed, including T12-L1, L5-S1, and sacroiliac joints. Experimental studies have demonstrated that the nucleus pulposus of intervertebral discs has a solid-like behavior under dynamic loading and with degeneration and aging (Iatridis et al. 1997). In addition, although the viscoelastic/poroelastic nature of the disc exists in the spine under time-durative compressive loading, an FE modal analysis does not necessarily need to include the time parameter. Therefore, the nucleus pulposus used in this study was assigned solid-like properties. To understand the findings of this study, it should be noted that the T12-Pelvis FE model did not include the muscle, cervical spine, thoracic spine, and rib cage of human upper body. Addition of the cervical spine and thoracic spine may decrease the resonant frequency of the T12-Pelvis FE model and the addition of muscles and a rib cage may increase the resonant frequency of the T12-Pelvis FE model.

The bony posterior elements might suffer more from a vibration environment due to their greater influence on vertical vibration than cortical bone, cancellous bone, and endplates. Under compressive cyclic loading over a long time, the trabecular bone might undergo physiological changes. Keller et al. (1989) reported that the compressive strength and stiffness of the trabecular bone of the spine bony posterior elements increased with increasing bone density. Further studies (Keller et al. 1990; Polikeit, Nolte, and Ferguson 2004) indicated that variations in the material properties of the vertebral body might result in spine disease. From the simulation results of different material parameters defined for vertebrae, intervertebral discs, and ligaments, as shown in Figures 15.14 through 15.16, the intervertebral disc was shown to be the main component influencing the resonant frequency of the human spine. In addition, in other material property sensitivity studies (Kumaresan, Yoganandan, and Pintar 1999; Ng, Teo, and Lee 2004) under static compression loading, similar findings were also achieved. Kumaresan, Yoganandan, and Pintar (1999) reported that variations in the material properties of intervertebral discs resulted in significant changes in angular rotation and disc stress. Of all the components in the intervertebral disc, the annulus ground substance produces the greatest effect on the vertical resonant frequencies of the spine (Figures 15.15 and 15.17). The annulus ground substance also plays a significant role in the static biomechanical behavior of the spine (Ng, Teo, and Lee 2004). Rao and Dumas (1991) and Kumaresan, Yoganandan, and Pintar (1999) also underscored the importance of the material property of soft tissue structures in their FE studies of the human spine.

As shown in Figure 15.15, the nucleus pulposus (DT1) is less sensitive to resonant frequency variation than the annulus (DT2), though it had a lower elastic modulus than the annulus ground substance in this study. The reason is that the nucleus pulposus is enclosed within the annulus and two vertebrae. Generally, the Young's modulus of the nucleus pulposus will increase with aging and degeneration of the spine (Keller et al. 1990). However, as the elastic modulus of the nucleus increases, its Poisson's ratio decreases (Goel et al. 1995; Kumaresan, Yoganandan, and Pintar 1999). The simulation results in this study show that the vertical resonant frequency will decrease if the Poisson's ratio of the nucleus pulposus decreases. For situations of serious degeneration with cavitation, this phenomenon will reduce the vertical resonant frequency of the spine. However, the loss of disc height (Kumaresan, Yoganandan, and Pintar 1999) may also increase the resonant frequency.

Of all the components in the intervertebral disc, the disc annulus contributes most prominently to the resonant frequency of the human spine (Figures 15.17 and 15.18). Therefore, it is necessary to further understand how the Young's modulus of the annulus ground substance affects the resonant frequency. From this study, it can be found that the percentage change of the resonant frequency relative to the basic condition is greater than 20% if the Young's modulus of the disc annulus is lower than 1.5 MPa, and the relative percentage changes of the resonant frequencies of the model are nonlinear. The degressive rate of the relative percentage change is faster due to the decrease of the elastic modulus of the disc annulus than the ascensive rate due to the increase of elastic modulus of the disc annulus. This implies that a young individual might have a lower resonant frequency in comparison to an elderly person with a similar truck mass. With an increase in age, the rigidity or elastic modulus of the annulus fibrous will increase (Keller et al. 1989), and may increase the resonant frequency of the spine in cases of severe degeneration.

Although there were many early investigations about the material property sensitivity of the human spine, no investigations have reported on how changes in material properties influence the dynamics of the spine. This study, in certain aspects, may provide insights into the material property sensitivity resulting from aging and spine degeneration. At the same time, by understanding the material property sensitivity of the spine, it may be helpful to adopt correct material property parameters in FE modeling.

## ACKNOWLEDGMENTS

This work was supported by the National Natural Science Foundation of China (51275082, 508175041), the Program for New Century Excellent Talents in University (NCET-08-0103), the Research Fund for the Doctoral Program of Higher Education (20100042110013), and the Fundamental Research Fund of Central Universities (N130403009).

## REFERENCES

Adams, M.A., and Hutton, W.C. 1980. The effect of posture on the role of the apophysial joints in resisting intervertebral compressive forces. *Journal of Bone and Joint Surgery (British Volume)* 62:358–362.

Berkson, M.H., Nachemson, A., and Schultz, A.B. 1979. Mechanical properties of human lumbar spine motion segments—Part II: Responses in compression and shear influence of grass morphology. *Journal of Biomechanical Engineering* 101:53–57.

Butler, D., Trafimow, J.H., Andersson, G.B., McNeill, T.W., and Huckman, M.S. 1990. Discs degenerate before facets. *Spine* 15:111–113.

Cann, A.P., Salmoni, A.W., and Eger, T.R. 2004. Predictors of whole-body vibration exposure experienced by highway transport truck operators. *Ergonomics* 47:1432–1453.

Chen, J.C., Chang, W.R., and Shih, T.S. 2003. Predictors of whole-body vibration levels among urban taxi drivers. *Ergonomics* 46:1075–1090.

Cheung, J.T.M., Zhang, M., and Chow, D.H.K. 2003. Biomechanical responses of the intervertebral joints to static and vibrational loading: a finite element study. *Clinical Biomechanics* 18:790–799.

Diamant, I., Shahar, R., and Gefen, A. 2005. How to select the elastic modulus for cancellous bone in patient-specific continuum models of the spine. *Medical & Biological Engineering & Computing* 43:465–472.

Fairley, T.E. 1995. Predicting the discomfort caused by tractor vibration. *Ergonomics* 38:2091–2106.

Frei, H., Oxland, T.R., Rathonyi, G.C., and Nilte, L.P. 2001. The effect of nucleotomy on lumbar spine mechanics in compression and shear loading. *Spine* 26:2080–2089.

Frymoyer, J.W., Pope, M.H., Costanza, M.C., Rosen, J.C., Goggin, J.E., and Wilder, D.G. 1980. Epidemiologic studies of low-back pain. *Spine* 5:419–423.

Goel, V.K., and Kim, Y.E. 1989. Effects of injury on the spinal motion segment mechanics in the axial compression mode. *Clinic Biomechanics* 4:161–167.

Goel, V.K., Kong, W.Z., Han, J.S., Weinstein, J.N., and Gilbertson, L.G. 1993. A combined finite element and optimization investigation of lumbar spine mechanics with and without muscles. *Spine* 18:1531–1541.

Goel, V.K., Monroe, B.T., Gilbertson, L.G., and Brinkmann, P. 1995. Interlaminar shear stresses and laminae separation in a disc: Finite element analysis of the L3-L4 motion segment subjected to axial compressive loads. *Spine* 20:689–698.

Goel, V.K., Park, H., and Kong, W. 1994. Investigation of vibration characteristics of the ligamentous lumbar spine using the finite element approach. *Journal of Biomechanical Engineering* 116:377–383.

Goto, K., Tajima, N., Chosa, E., Totoribe, K., Kuroki, H., Arizumi, Y., and Arai, T. 2002. Mechanical analysis of the lumbar vertebrae in a three-dimensional finite element method model in which intradiscal pressure in the nucleus pulposus was used to establish the model. *Journal of Orthopaedic Science* 7:243–246.

Guo, L.X., and Teo, E.C. 2005. Prediction of the modal characteristics of the human spine at resonant frequency using finite element models. Proceedings of the Institution of Mechanical Engineers Part H: *Journal of Engineering in Medicine* 219:277–284.
Guo, L.X., and Teo, E.C. 2006. Influence of injury and vibration on adjacent components of spine using finite element methods. *Journal of Spinal Disorders & Techniques* 19:118–124.
Guo, L.X., Teo, E.C., Lee, K.K., and Zhang, Q.H. 2005. Vibration characteristics of human spine under axial cyclic loads: Effect of frequency and damping. *Spine* 30:631–637.
Guo, L.X., Wang, Z.W., Zhang, Y.M. et al. 2009a. Material property sensitivity analysis on resonant frequency characteristics of the human spine. *Journal of Applied Biomechanics* 25:64–72.
Guo, L.X., Zhang, M., and Teo, E.C. 2007. Influences of denucleation on contact force of facet joints under whole body vibration. *Ergonomics* 50:967–978
Guo, L.X., Zhang, M., Wang, Z.W., Zhang, Y.M., Wen, B.C., and Li, J.L. 2008. Influence of anteroposterior shifting of trunk mass centroid on vibrational configuration of human spine. *Computers in Biology and Medicine* 38:146–151.
Guo, L.X., Zhang, M., Zhang, Y.M., and Teo, E.C. 2009b. Vibration modes of injured spine at resonant frequencies under vertical vibration. *Spine* 34:E682–E688.
Guo, L.X., Zhang, M., Zhang, Y.M. et al. 2009c. Influence prediction of tissue injury on frequency variations of the lumbar spine under vibration. *OMICS: A Journal of Integrative Biology* 13:521–526.
Guo, L.X., Zhang, Y.M., and Zhang, M. 2011. Finite element modeling and modal analysis of the human spine vibration configuration. *IEEE Transactions on Biomedical Engineering* 58:2987–2990.
Iatridis, J.C., Setton, L.A., Weidenbaum, M., and Mow, V.C. 1997. Alterations in the mechanical behavior of the human lumbar nucleus pulposus with degeneration and aging. *Journal of Orthopaedic Research* 15: 318–322.
Kasra, M., Shirazi-Adl, A., and Drouin, G. 1992. Dynamics of human lumbar intervertebral joints. Experimental and finite-element investigations. *Spine* 17:3–102.
Keller, T.S., Colloca, C.J., and Beliveau, J.G. 2002. Force-deformation response of the lumbar spine: A sagittal plane model of posteroanterior manipulation and mobilization. *Clinical Biomechanics* 17:185–196.
Keller, T.S., Hansson, T.H., Abram, A.C., Spengler, D.M., and Panjabi, M.M. 1989. Regional variations in the compressive properties of lumbar vertebral trabeculae. Effects of disc degeneration. *Spine* 14:1012–1019.
Keller, T.S., Holm, S.H., Hansson, T.H., and Spengler, D.M. 1990. The dependence of intervertebral disc mechanical properties on physiologic conditions. *Spine* 15:751–761.
Kim, Y. 2001. Prediction of mechanical behaviors at interfaces between bone and two interbody cages of lumbar spine segments. *Spine* 26:1437–1442.
Kim, Y.E., Goel, V.K., Weninstein, J.N., and Lim, T.H. 1991. Effect of disc degeneration at one level on the adjacent level in axial mode. *Spine* 16:331–335.
Kong, W.Z., and Goel, V.K. 2003. Ability of the finite element models to predict response of the human spine to sinusoidal vertical vibration. *Spine* 28:1961–1967.
Kumar, S. 2001. Theories of musculoskeletal injury causation. *Ergonomics* 44:17–47.
Kumaresan, S., Yoganandan, N., and Pintar, F.A. 1999. Finite element analysis of the cervical spine: A material property sensitivity study. *Clinical Biomechanics* 14:41–53.
Kumaresan, S., Yoganandan, N., Pintar, F.A., and Maiman, D.J. 1999. Finite element modeling of the cervical spine: Role of intervertebral disc under axial and eccentric loads. *Medical Engineering & Physics* 21: 689–700.
Lavaste, F., Skalli, W., Robin, S., Roy-Camille, R., and Mazel, C. 1992. Three-dimensional geometrical and mechanical modeling of the lumbar spine. *Journal of Biomechanics* 25:1153–1164.
Lee, C.K., Kim, Y.E., Lee, C.S., Hong, Y.M., Jung, J.M., and Goel, V.K. 2000. Impact response of the intervertebral disc in finite-element model. *Spine* 25:2431–2439.
Lee, M., Esler, M.A., and Midren, J. 1993. Effect of extensor muscle activation on the response to lumbar posteroanterior force. *Clinical Biomechanics* 8:115–119.
Mientjes, M.I.V., Norman, R.W., Wells, R.P., and McGill, S.M. 1999. Assessment of an EMG-based method for continuous estimates of low back compression during asymmetrical occupational tasks. *Ergonomics* 42:868–879.
Ng, H.W., Teo, E.C., and Lee, V.S. 2004. Statistical factorial analysis on the material property sensitivity of the mechanical responses of the C4-C6 under compression, anterior and posterior shear. *Journal of Biomechanics* 37:771–777.
Panjabi, M.M., Andersson, G.B., Jorneus, L., Hult, E., and Mattsson, L. 1986. In vivo measurements of spinal column vibrations. *Journal of Bone and Joint Surgeries of America* 68:695–702.

Pankoke, S., Hofmann, J., and Wolfel, H.P. 2001. Determination of vibration-related spinal loads by numerical simulation. *Clinic Biomechanics* 16:S45–56.

Polikeit, A., Nolte, L.P., and Ferguson, S.J. 2004. Simulated influence of osteoporosis and disc degeneration on the load transfer in a lumbar functional spinal unit. *Journal of Biomechanics* 37:1061–1069.

Pope, M.H., and Novotny, J.E. 1993. Spinal biomechanics. *Transactions of the ASME Journal of Biomechanical Engineering* 115:569–574.

Pope, M.H., Wilder, D.G., Jorneus, L., Broman, H., Svensson, M., and Andersson, G. 1987. The response of the seated human to sinusoidal vibration and impact. *Journal of Biomechanics in Engineering* 109(4): 279–284.

Rao, A.A., and Dumas, G.A. 1991. Influence of material properties on the mechanical behaviour of the L5-S1 intervertebral disc in compression: A nonlinear finite element study. *ASME Journal of Biomechanical Engineering* 13:139–151.

Sandover, J., and Dupuis, H. 1987. A reanalysis of spinal motion during vibration. *Ergonomics* 30:975–985.

Sharma, M., Langrana, N.A., and Rodriguez, J. 1995. Role of ligaments and facets in lumbar spinal stability. *Spine* 20:887–900.

Sharma, M., Langrana, N.A., and Rodriguez, J. 1998. Modeling of facet articulation as a nonlinear moving contact problem: Sensitivity study on lumbar facet response. *Journal of Biomechanics in Engineering* 120:118–125.

Shirazi-Adl, A., Ahmed, M., and Shrivatava, S.C. 1986. Mechanical response of a lumbar motion segment in axial torque alone and combined with compression. *Spine* 11:914–927.

Shirazi-Adl, A., and Drouin, G. 1987. Load-bearing role of facets in a lumbar segment under sagittal plane loadings. *Journal of Biomechanics* 20:601–613.

Shirazi-Adl, A., and Parnianpour, M. 2001. Finite element model studies in lumbar spine biomechanics. In *Biomechanical Systems: Techniques and Applications, Volume 1: Computer Techniques and Computational Methods in Biomechanics*, edited by C.T. Leondes, 1–36. Boca Raton, FL: CRC Press.

Shirazi-Adl, S.A., Shrivastava, S.C., and Ahmed, A.M. 1984. Stress analysis of the lumbar disc-body unit in compression. A three-dimensional nonlinear finite element study. *Spine* 9:120–134.

Skalak, R., and Shu, C.E. 1987. *Handbook of Bioengineering*. New York: McGraw-Hill.

Spilker, R.L., Jakobs, D.M., and Schultz, A.B. 1986. Material constants for a finite element model of the intervertebral disk with a fiber composite annulus. *ASME Journal of Biomechanical Engineering* 108:1–11.

# 16 Cervical Spinal Fusion and Total Disc Replacement

*Zhongjun Mo, Lizhen Wang, Ming Zhang, and Yubo Fan*

## CONTENTS



## SUMMARY

Cervical intervertebral disc degeneration is a leading source of neck and arm pain. The traditional treatment for cervical disc disease was spinal fusion, which was regarded as the gold standard method of treatment. However, a high rate of complications, including adjacent segment degeneration, was often associated with such fusion. In an effort to avoid adjacent segment disease, non-fusion techniques, including total disc replacement and nucleus replacement, were developed as alternatives. Unfortunately, clinical research has indicated that cervical disc replacement did not significantly reduce the postoperative rate of adjacent segment degeneration and adjacent segment disease. Biomechanical factors play an important role in the degeneration or rehabilitation progress of the musculoskeletal system. In the present study, the biomechanical effect of spinal fusion and total disc replacement on the surrounding tissues was analyzed based on the finite element method. The results showed that adjacent intervertebral disc pressure was not apparently different from the intact healthy condition after either spinal fusion or total disc replacement. It also showed that total disc replacement would increase the facet joint contact force and capsular ligament tension. The increase in facet joint loading is a potential source of facet joint degeneration.

## 16.1 INTRODUCTION

### 16.1.1 Intervertebral Disc Disease

Alteration in the biomechanical properties of a disc is a typical characteristic of intervertebral disc degeneration. The exact pathogenesis of the degenerative process is still unknown (An, Thonar, and Masuda 2003). It has been suggested that disc degeneration might be predetermined genetically, occurring during the aging process, or may be caused by overloading or underloading (Goel et al. 2006). Loss of nutrition to the disc may be an important factor in its degeneration. It is likely that almost any abnormal loading condition can lead to adaptive changes that may in turn lead to the degeneration of the disc (Eck et al. 2002).

As the source of disc degeneration, knowledge of the biomechanical load is needed to understand the degenerative progress. In theory, once degeneration sets in, the intervertebral disc goes through a cascade of degenerative changes that result in biomechanical alteration of load transfer through the disc, in turn causing changes in the mechanical properties and composition of the tissue. Structural disorganization of the intervertebral disc along with loss in proteoglycans stemming from degeneration causes the hydrostatic mechanism to fail (Eck et al. 2002). There are different ways a degenerated disc can lead to low back pain or neck pain. Depending on whether the degeneration occurs in the nucleus pulposus or the annulus, the degenerated disc may be classified into several modalities (Figure 16.1).

A degenerated annulus could have fissures that consist of microscopic fragmentation of individual fibers. Annular tears at the corners of the vertebral body separating the annulus from the end plates are also seen. Disc bulging may occur due to a decrease in the radial tensile strength of the annulus. Degeneration of the nucleus occurs following loss of water content leading to collagenation of the nucleus, which results in an increase of the elastic modulus of the nucleus.

Nucleus degeneration combined with annular degeneration may cause disc herniation into the spinal canal, causing low back pain due to nerve pinching. Thinning of the disc and loss of disc height are also commonly seen. This loss of disc height combined with gradual ossification of the end plate and protrusion of the disc tissue causes stenosis, which again leads to back pain.

### 16.1.2 Gold Standard for the Treatment of Intervertebral Disc Disease: Spinal Fusion

Currently, spinal fusion is regarded as the gold standard for the treatment of intervertebral disc disease. Spinal fusion was first described for the treatment of Pott disease and spinal deformity. Cervical spine fusion was used initially as a treatment for degenerative spondylotic conditions. These early authors suggested that spinal fusion offered the surgeon an opportunity to remove the pathological process (infection, arthritis, or deformity), to eliminate painful motion, and to allow decompression of the neural elements with subsequent stabilization of the index segment.

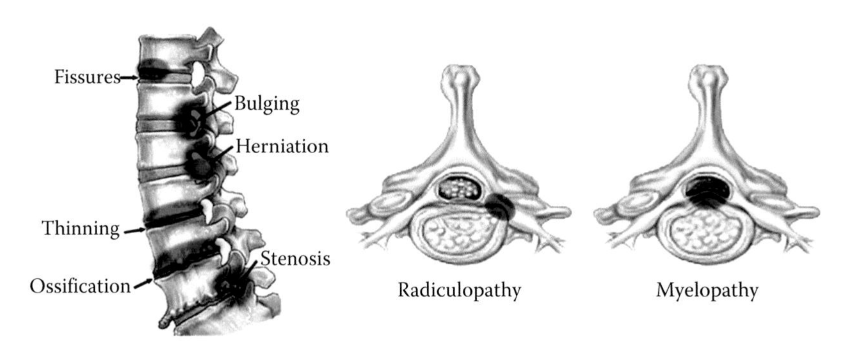


**FIGURE 16.1** The modalities of spinal disc degeneration.

Fusion techniques for the spinal column were based on the concept that instability causes pain and hence restriction of the mobility would offer relief. Clinical trials have shown that spinal fusion offered favorable outcomes by stabilizing the spine column and providing pain relief and consistent improvement of neurologic symptoms. The overall rates of improvement in neurological status were 83.6%, 83.2%, and 88.9% at 24, 36, and 60 months after cervical spinal fusion (Burkus et al. 2010). A prospective study reported a four-year overall success rate of 72.5% for the fusion surgery (Sasso et al. 2011).

However, approximately 3% of patients with spinal fusion annually suffer from adjacent segment degeneration (ASD) (Hilibrand and Robbins 2004). Baba et al. (1993) observed that 25% of their 100 patients subsequently developed new spinal canal stenosis above the previously fused segments. Gore and Sepic (1984) observed new spondylosis in 25% of 121 patients and progression of preexisting spondylosis in another 25% of 121 patients who had previously undergone anterior cervical fusion. Bohlman et al. (1993) found that 9% of their 122 patients went on to develop adjacent segment diseases requiring additional surgery. In addition, Williams, Allen, and Harkess (1968) found that 17% of their 60 patients undergoing anterior cervical decompression and fusion developed adjacent segment disease and needed additional surgery.

Previous research attributed this complication to fusion-induced abnormal kinematics, hypermobility, and an increase in intradiscal pressure (IDP) at the adjacent level. These symptoms are directly associated with adjacent disc degeneration (Eck et al. 2002; Dmitriev et al. 2005). Finite-element models of the cervical spine have been used to investigate the effect of cervical spine fusion on adjacent levels (Hong-Wan, Ee-Chon, and Qing-Hang 2004). It was reported that increasing the stiffness of intervertebral graft materials is associated with increased internal stress at adjacent levels.

### 16.1.3 Promising Alternative to Spinal Fusion: Total Disc Replacement

Non-fusion treatments, including nucleus replacement and total disc replacement, were developed for preserving movement functions and decreasing physical strain and stress on the adjacent segments. As an alternative to spinal fusion, total disc replacement aimed to restore segment mobility and load-carrying capacity and was expected to avoid inducing adjacent segment degeneration. Total disc replacement surgery involves removing the pain-causing disc and replacing it with a mechanical device that mimics normal spine kinematics. In the lumbar region, disc replacement is indicated for patients with disc-related back pain, whereas in the cervical region, wider indications could be addressed, including myelopathy and radiculopathy.

Clinical studies have shown that total disc replacement offers satisfactory clinical results for the treatment of disc degeneration diseases. A prospective randomized controlled clinical trial of the Prestige replacement disc showed that the overall rates of maintenance or improvement in neurological status in the total disc replacement group were 91.6%, 92.8%, and 95.0%, respectively, at 24, 36, and 60 months (Burkus et al. 2010). Another clinical trial demonstrated a 93.3% success for Bryan replacement disc arthroplasty with a neck disability index (NDI) score of 51 preoperatively and 16.7 at 48 months (Garrido, Taha, and Sasso 2010).

In theory, cervical disc replacement should decrease the development of adjacent segment degeneration by maintaining normal disc kinematics. In cadaveric studies, cervical arthroplasty has shown the ability to maintain motion and mechanics within physiologic ranges at the operative segment and decrease stresses on adjacent segments (Dmitriev et al. 2005). Although few clinical studies have specifically aimed to evaluate adjacent segment degeneration after cervical disc replacement, some reports have shown that total disc arthroplasty did not influence the incidence of adjacent segment disease in the cervical spine (Yang et al. 2012; Jawahar et al. 2010). Meta-analysis indicated that there were fewer incidences of adjacent segment disease and adjacent segment surgery in comparison to cervical disc replacement with anterior cervical discectomy and fusion, but the difference was not statistically significant (Yang et al. 2012). Jawahar et al. (2010) used three devices to evaluate the equivalent efficiency of total disc replacement and anterior discectomy and fusion, including Kineflex-C (SpinalMotion Inc., Mountain View, California), Mobi-C (LDR Spine, Austin, Texas), and Advent Cervical Disc (Blackstone Inc., Parsippany, New Jersey). The author concluded that total disc arthroplasty is equivalent to anterior cervical discectomy and fusion for providing relief from symptoms in the treatment of one- and two-level disc degeneration diseases of the cervical spine. However, the risk of developing adjacent segment degeneration is equivalent after both procedures.

Mechanical intervention in normal disc motion may adversely affect the disc and lead to accelerated rates of degeneration by interfering with the normal nutritional supply (Eck et al. 2002). Intervertebral discs lack a true blood supply. Instead, they are dependent on nutrients diffusing through the extracellular matrix from peripheral blood vessels and vertebral endplates. Increased pressure in the disc acts to alter the diffusion characteristics of nutrients from the periphery and leads to an accumulation of waste products in the disc. Failure to remove waste products adequately from the disc can lead to increased lactate levels and decreased pH, which can impair metabolism and lead to cell death (Buckwalter 1995).

Currently, more than 20 types of artificial cervical discs have been used clinically and have exhibited satisfactory performance. Various artificial discs have been developed for cervical disc replacement, including the Prestige (Medtronic Sofamor Danek, Memphis, Tennessee), Bryan (Medtronic Sofamor Danek, Memphis, Tennessee), PCM (Cervitech, Inc., Rockaway, New Jersey), ProDisc-C (Synthes Spine Solutions, Westchester, Pennsylvania), and Mobi-C (LDR, France). The majority of these artificial discs adopt polymer-on-metal or metal-on-metal designs to form a sliding articulation. Ball-in-socket design is the most commonly used of these prostheses. A primary difference between ProDisc-C and Mobi-C is that the former has a fixed core (three degrees constraint) and the latter has a mobile core (one degree constraint and two degrees controlled). The designs exhibit different biomechanical functions.

### 16.1.4 Biomechanical Study of Spinal Fusion and Total Disc Replacement

Distinguishing different artificial discs and detailing spinal fusion from the biomechanical viewpoint may assist in optimizing cervical spine surgery. Various approaches, including isolated device testing, in vitro experiment, in vivo measurement, and finite element modeling (FEM), were implemented to evaluate the biomechanical properties of multiple total disc replacement (Cunningham et al. 2003; McAfee et al. 2006; Phillips et al. 2009; Shikinami et al. 2004; Cakir et al. 2005; Chang et al. 2007a, 2007b; Jirkova and Horak 2009). As early as 1973, FEM has been applied to the human spine and has played an increasingly important role in improving the understanding of the fundamental biomechanics of the spine and advancing the performance of medical implants (Bowden 2006).

Specifically, the region C3-C7, where most spinal fusion procedures are performed, is bordered by a highly mobile upper cervical region, which accommodates approximately half of all cervical motion. The kinematics of the cervical spine have never been assessed in vitro in a model that includes the upper cervical region and therefore most closely approximates the in vivo potential for motion transfer to the upper cervical spine after cervical fusion.

This study was conducted to investigate the biomechanical performance of spinal fusion and total disc replacement with two different artificial discs: the ProDisc-C, with a ball-in-socket design, and the Mobi-C, with a ball-in-socket design, adopting the ball-in-socket design with controlled mobility. At first, a three-dimensional finite element model of the three-level ligamentous cervical segment (C3-C6) was built, and subsequently these distinct designs were implanted at C4/C5. The spinal kinematics of the index and adjacent segments, the adjacent intervertebral disc pressure, the facet joint contact force, and the capsular ligament tension were calculated.

## 16.2 MODEL DEVELOPMENT

### 16.2.1 Finite Element Modeling of an Intact Healthy Cervical Spine

A healthy male subject (32 years old, 65 kg, and 172 cm) without any radiographic evidence of degeneration participated in the study. The subject was briefed on the research procedure and signed a form of consent. The study plan was approved by the ethical committee of the corresponding institute. The geometry of the four vertebrae was reconstructed based on computer tomography (CT) scan images taken from the subject. The CT images included the C3-C6 vertebrae and were obtained at 0.5-mm intervals and at a resolution of 0.6 mm (Brilliance iCT, Philips, Netherlands). Commercial software (MIMICS 10.1, Materialise, Leuven, Belgium) was used to transform the planar CT scans into solid vertebral volumes and exported as STL format files. The solid volume was then imported into an inverse engineering software (Rapidform 2006, INUS Technology Inc., Korea) to locally smooth the surface and convert into a Nonuniform rational basis spline (NURBS) surface geometry structure. The vertebrae were imported into an FE package (ABAQUS 6.9.1, Simulia Inc, USA) and meshed with tetrahedron elements. A lay of triangle shell elements were generated from the external surface of the trabecular bone to represent the cortical bone and endplates with a thickness of 0.4 mm (Keaveny and Buckley 2006). The nodes on the shell element were shared with the external tetrahedron element surface of the cancellous bone (Figure 16.2a).

The intervertebral disc was represented as a continuous solid structure occupying the intervertebral space and partitioned into an annulus and nucleus pulposus and meshed with hexahedron elements (Figure 16.2b). The nucleus pulposus occupied 43% of the total disc volume and was located slightly posterior to the center of the disc, which was consistent with anatomic measurements (Denozière and Ku 2006), and the Poisson's ratio was set as 0.49, which represents an incompressible material property. The modeled annulus fibrosus consisted of a composite material based on clinical observations (Drake et al. 2008). A layer of annulus fiber was embedded in the matrix of annulus ground substance with an inclination of between 15 and 30 degrees with respect to the transverse plane. The content of the annulus fiber was approximately 19%, similar to the natural collagen content of the annulus (Denozière and Ku 2006). The fibers were meshed into truss elements.

Spinal ligaments play an important role in spinal biomechanics and stability. In the present study, the main cervical spinal intervertebral ligaments were incorporated into the model (Narayan, Srirangam, and Frank 2001), including the anterior longitudinal ligament (ALL), posterior longitudinal ligament (PLL), flaval ligament (FL), and interspinous ligament (ISL). Their insertion points were chosen to mimic anatomic observations as closely as possible (Drake et al. 2008).

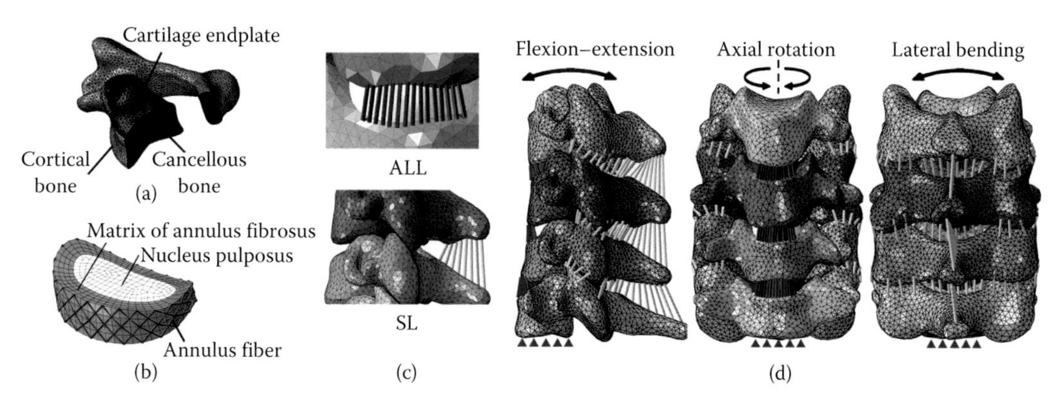


**FIGURE 16.2 (See color insert.)** Details of the finite element cervical model: (a) vertebra, (b) disc, (c) ligament insertion point, (d) entire cervical segment and illustration of spinal motion.

All the incorporated ligaments were modeled as incompressible discrete axial connectors with nonlinear load-deformation behavior, as exhibited by natural ligaments (Narayan, Srirangam, and Frank 2001). The material properties of the intact healthy cervical spine are presented in Table 16.1.

During assembly of the cervical components, the interacting surfaces of the intervertebral discs and the vertebral endplates were tied together to prevent sliding. The tips of the ligaments were also tied to the corresponding anatomic insertion position on the bony vertebrae (Figure 16.2c). A frictionless sliding contact property was applied to the inferior articular process and the concave articular surface of the facet joint and of the uncovertebral joint.

The kinematic motions of the spine are quite complex when performing routine daily activities. However, motions can be simplified into the three anatomic planes: flexion (bending anteriorly) and extension (bending posteriorly), lateral bending, and axial torsion (Figure 16.2d). In the present

**TABLE 16.1**
**Material Properties of Cervical Spine Components**

| Component | Young's Modulus (MPa) | Poisson's Ratio | Description |
|---|---|---|---|
| Cortical bone | 12,000 | 0.30 | 0.5 mm thickness |
| Cancellous bone | 100 | 0.25 | |
| Endplate cartilage | 1,200 | 0.30 | 0.5 mm thickness |
| Matrix of annulus fibrosus | 3.4 | 0.45 | |
| Nucleus pulposus | 1 | 0.49 | 43% volume of disc |
| Annulus fiber | 450 | 0.45 | 19% volume of annulus |
| Ligament | Nonlinear tension only discrete axial connectors | | |

**C3-C5 (load-deflection relationship)**

| ALL | | PLL | | SL & ISL | | LF | | CL | |
|---|---|---|---|---|---|---|---|---|---|
| D(mm) | F(N) | D(mm) | F(N) | D(mm) | F(N) | D(mm) | F(N) | D(mm) | F(N) |
| 0 | 0 | 0 | 0 | 0 | 0 | 0 | 0 | 0 | 0 |
| 1 | 28 | 1 | 25 | 1 | 7 | 2 | 38 | 2 | 55 |
| 2 | 52 | 2 | 44 | 2 | 12.5 | 4 | 60 | 4 | 130 |
| 3 | 72 | 3 | 62 | 3 | 18 | 6 | 80 | 6 | 180 |
| 4 | 89 | 4 | 78 | 4 | 22.5 | 8 | 108 | 8 | 210 |
| 5 | 102 | 5 | 89 | 5 | 26 | | | 10 | 230 |
| 6 | 115 | | | 6 | 30 | | | | |
| | | | | 7 | 32.5 | | | | |

**C5-C6 (load-deflection relationship)**

| ALL | | PLL | | SL & ISL | | FL | | CL | |
|---|---|---|---|---|---|---|---|---|---|
| D(mm) | F(N) | D(mm) | F(N) | D(mm) | F(N) | D(mm) | F(N) | D(mm) | F(N) |
| 0 | 0 | 0 | 0 | 0 | 0 | 0 | 0 | 0 | 0 |
| 1 | 20 | 1 | 20 | 1 | 8 | 2 | 30 | 2 | 75 |
| 2 | 40 | 2 | 40 | 2 | 14 | 4 | 68 | 4 | 145 |
| 3 | 58 | 3 | 60 | 3 | 20 | 6 | 102 | 6 | 204 |
| 4 | 78 | 4 | 78 | 4 | 25 | 8 | 130 | 8 | 250 |
| 5 | 98 | 5 | 92 | 5 | 29 | 10 | 145 | 10 | 265 |
| 6 | 112 | | | 6 | 32.5 | | | | |
| 7 | 120 | | | 7 | 35 | | | | |

*Note:* ALL, anterior longitudinal ligament; CL, capsular ligament; LF, flaval ligament; ISL, interspinous ligament; PLL, posterior longitudinal ligament; D(mm), deflection in millimeters; and F(N), force in Newtons.

study, the inferior endplate of the C6 vertebrae was constrained by fixing all degrees of freedom in the intact healthy model (Panjabi et al. 2001). An axial precompression of 74 N was applied at the center of the superior endplate of the C3 vertebrae. This precompression simulated the in vivo loading due to head weight and the reaction of cervical muscles, as a simplified "follower load." A 1.8-Nm moment was imposed on the superior endplate surface of C3 vertebrae in the three anatomical planes to simulate flexion, extension, left-right lateral bending, and left-right axial rotation (Panjabi et al. 2001).

### 16.2.2 Validation of the Intact Healthy Cervical Spine

In general, a validated model is one that matches previously observed experimental behavior. There are many ways to validate a cervical spine model. The most important model parameters to validate are those that will be studied later during simulation (Bowden 2006). For example, in the present study, the model was used to predict spinal kinematics and intradiscal pressure; then the spinal kinematics and intradiscal pressure were validated, but validation of cortical stress was not necessary (Griffin 2001). An important consideration during the validation step is directly related to the subject-specific nature of biological modeling. However, the large variation in material and geometric considerations between individuals warrant such considerations. Most cervical models were validated against published experimental data from different subjects, since it is difficult to perform subject-specific validation experiments.

In the present study, the rotational motions in the intact healthy cervical spine imposing 1.8 Nm in the three anatomic planes (Figure 16.3) fell into the standard deviation of in vitro experimental data (Panjabi et al. 2001; Finn et al. 2009). In detail, the range of motion (ROM) was greater than the average value of Panjabi's in vitro testing at the C3-C4 level and less at C4-C5 and C5-C6 during

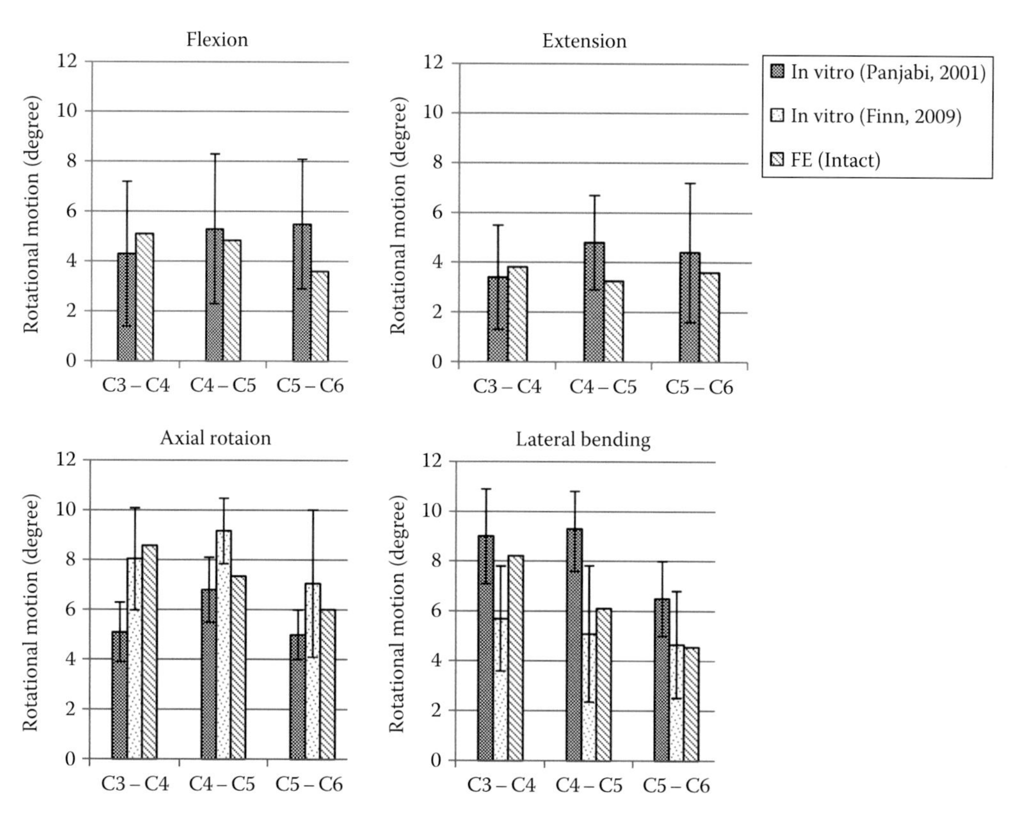


**FIGURE 16.3** Model validation: comparison of spinal motion with experimental data.

flexion and extension. At all the levels considered, the ROM was greater than the average value of Panjabi et al.'s (2001) study but less than the values of Finn et al.'s (2009) study during axial rotation. An inverse phenomenon was observed during lateral bending. Although these values were different than the average value of in vitro studies, they fell into the standard deviation of the experimental data. Typically, the standard error measurements associated with different studies are high, due to the inter-subject differences, making validation of gross kinematics difficult. As such, more rigorous and difficult validation is necessary by comparing results that depend on both kinematic and material considerations such as intervertebral disc pressures.

Intradiscal pressure is considered the main biomechanical factor associated with disc degeneration and has been widely studied in previous in vitro studies. In the study by McNally et al. (1996), the intradiscal pressure profile was more consistent through the disc cross-section in the posterior lateral to anterior-posterior direction, with a value of about 1.8 MPa. However, the degenerative disc had multiple regions of pressure concentrations. In the present study, the maximum intradiscal pressure was observed during flexion or lateral bending with values of 2.0 MPa, 1.83 MPa, and 1.83 MPa, at C3-C4, C4-C5, and C5-C6, respectively. Although these values are slightly greater than McNally's in vitro experimental data, the difference is minute and can be used for comparison. Another important consideration was that finite element methods give more discretized field results than current experimental methods. Intervertebral disc pressure results are generally measured experimentally using a single needle-type pressure sensor inserted into the nucleus pulposus. Comparing the pressure results from the current experimental measurements, the finite element model could provide more discretized field results. The global maximum pressures may potentially lie outside the region measured by the sensor.

## 16.3 APPLICATION 1: A SIMULATION OF CERVICAL SPINAL ARTHRODESIS

### 16.3.1 Model Development

Anterior cervical decompression and fusion is one of the gold standard treatments for degenerative disc diseases. In the present study, a simplified anterior screw/plate fixture system with an allograft of trabecular bone was used to simulate the spinal fusion surgery. The main dimension (width × height) of the anterior plate was 12 mm × 25 mm with a thickness of 2 mm, and the natural intervertebral disc was replaced by allografted cancellous bone (Figure 16.4a).

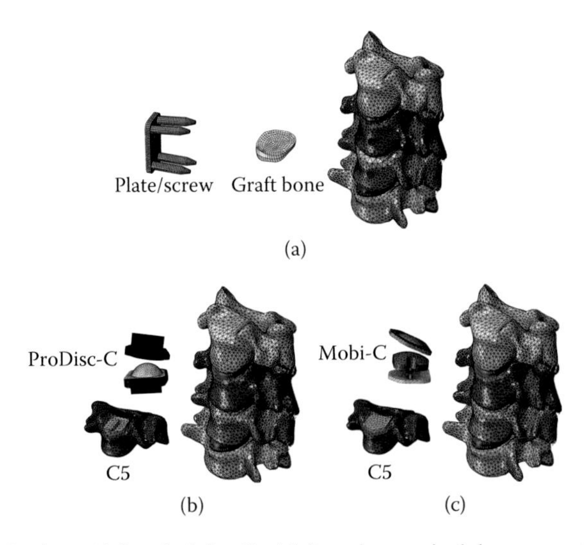


**FIGURE 16.4 (See color insert.)** Surgical details. (a) Anterior cervical decompression and fusion. (b) Total disc replacement with ProDisc-C. (c) Total disc replacement with Mobi-C.

The fixation components, including plate and screw, were made of titanium alloy (Ti6Al4V). The Young's modulus and Poisson's ratio of Ti6Al4V were 114 GPa and 0.3, respectively.

During this simulation, the anterior longitudinal ligament and the entire disc were completely removed, and then the intervertebral space was replaced with the allograft bone and was considered totally fused to the adjacent vertebrae. The bone-screw interface was considered completely integrated and constrained. The loading and boundary conditions were the same as the intact healthy cervical spine.

### 16.3.2 Cervical Spinal Motion after Anterior Fusion

After spinal fusion, the ability for motion was almost fully lost at the operative level (C4-C5) for all the spinal motions (Figure 16.5). The rotation motions imposing the moment load were 0.1°, 0.1°, 0.3°, 0.3° during flexion, extension, axial rotation, and lateral bending, respectively. At the adjacent level, the motions only increased by 3% during lateral bending at C5-C6, and only decreased by 6% during flexion at the common site.

The results showed that spinal fusion will restrain motion at the operative segment under moment loading, but has a limited effect at the adjacent segment. Since there was complete loss of motion at the operative level and limited compensatory motion at the adjacent level, motions of the entire segment (C3-C6) decreased by between 30% and 36% during all spinal motions.

### 16.3.3 Influence of Spinal Fusion on Facet Joint

The adjacent discs transfer load through the vertebral body, and the facet joints and posterior ligaments transfer load through the posterior elements. The facet joints and posterior ligaments, particularly the capsular ligament, have been shown to play a significant role in torsion, transverse shear, and extension.

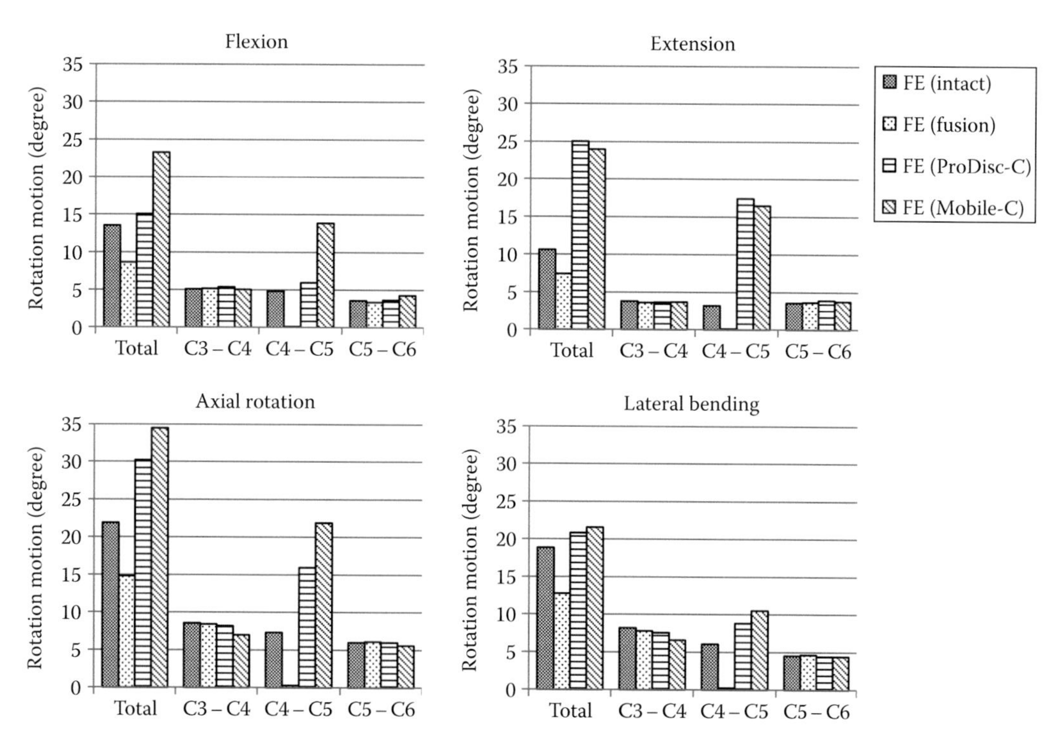


**FIGURE 16.5** Spinal motions after different surgical reconstructions.

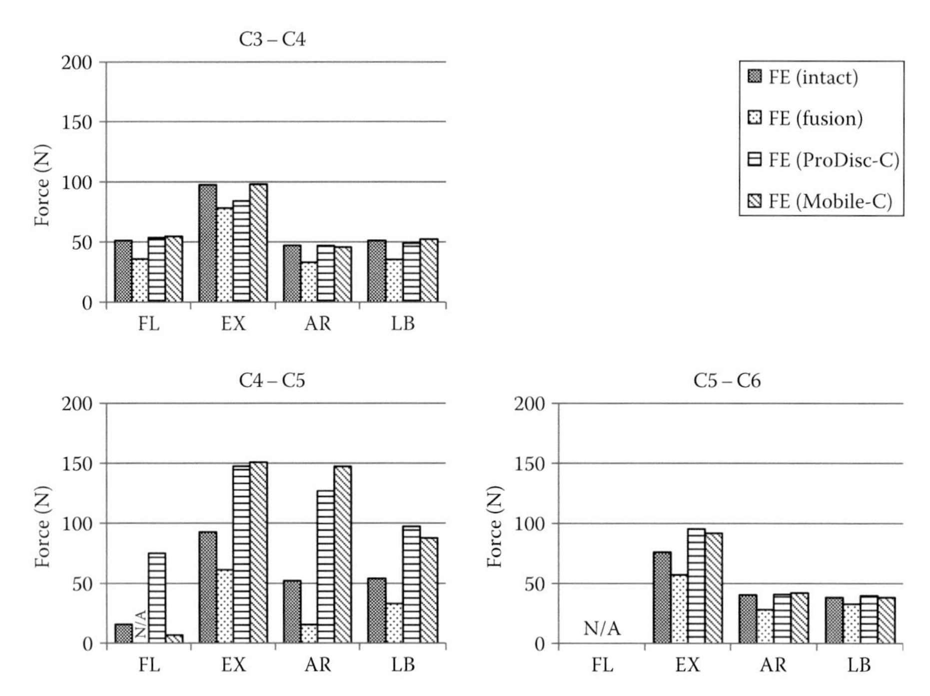


**FIGURE 16.6** The force transmitted through the facet joint. FL: flexion, EX: extension, AR: axial rotation, LB: lateral bending.

In the present study, the maximum facet contact force was observed during extension (Figure 16.6), meaning that the facet joint plays an important role in restricting extension. In theory, the loss of motion at the index segment after spinal fusion would cause an enormous decrease in facet joint contact force. In fact, the maximum facet contact force only decreased by 34% at the operative segment after spinal fusion, which is much lower than the 98% decrease in range of motion. Also, although the range of motion changed little, the maximum facet joint contact force decreased by 20% and 25% at the superior adjacent segment (C3-C4) and the inferior adjacent segment (C5-C6), respectively.

The maximum capsular ligament tension in the intact healthy cervical spine was observed during lateral bending, and a relatively lower tension was observed during flexion and axial rotation, with extension producing the lowest ligament tension (Figure 16.7). This means that the capsular ligaments have a significant effect in restricting lateral bending, flexion, and torsion, but limited effect in restricting extension. Unlike the facet joint contact force, the maximum ligament tension decreased by 81% at the operative segment after spinal fusion, which was consistent with the decrease of spine motion. At the adjacent segments, the ligament tension decreased by 4% and 26% at C3-C4 and C5-C6, respectively.

The results showed that the facet joint contact force and ligament tension were more sensitive than the spinal kinematics at the adjacent segment.

### 16.3.4 Adjacent Segment Intradiscal Pressure after Spinal Fusion

Although the posterior elements, particularly the facet joints, have been shown to play a significant role in torsion, transverse shear, and extension, it has been well established that the disc carries the majority of the load for compression loading, which appears mostly in flexion and lateral bending.

In the present study, the maximum intervertebral disc pressure was observed during lateral bending and flexion (Figure 16.8). Under the moment loading, the maximum intervertebral disc pressure changed little at both the superior and inferior adjacent segments.

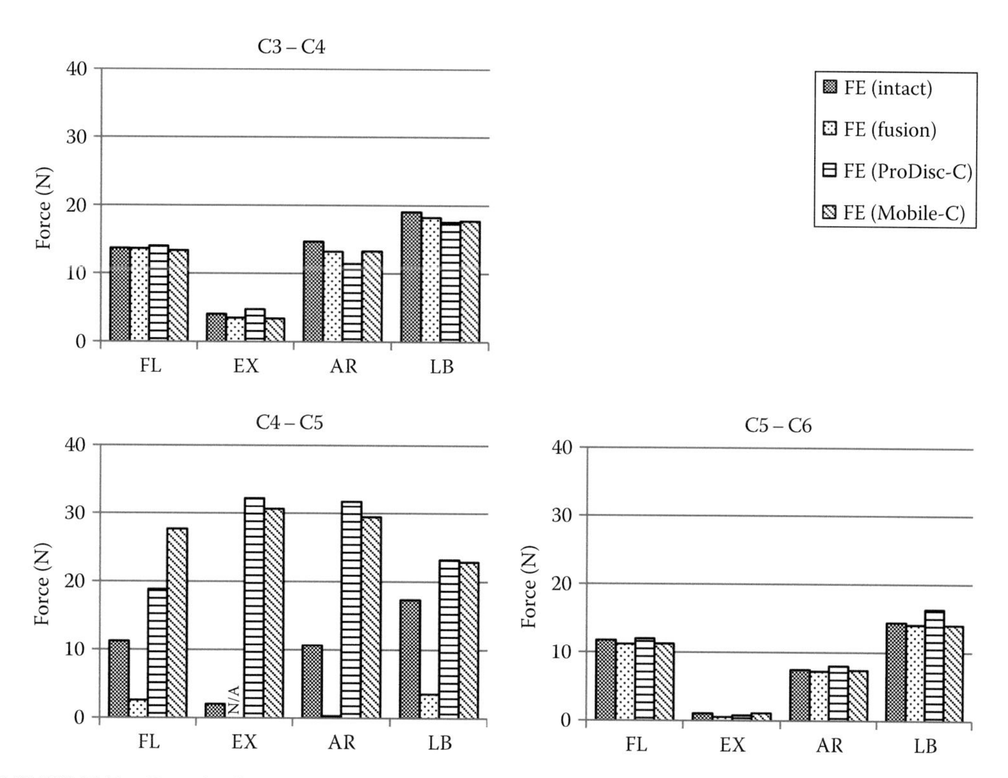


**FIGURE 16.7** Capsular ligament tension. FL: flexion, EX: extension, AR: axial rotation, LB: lateral bending.

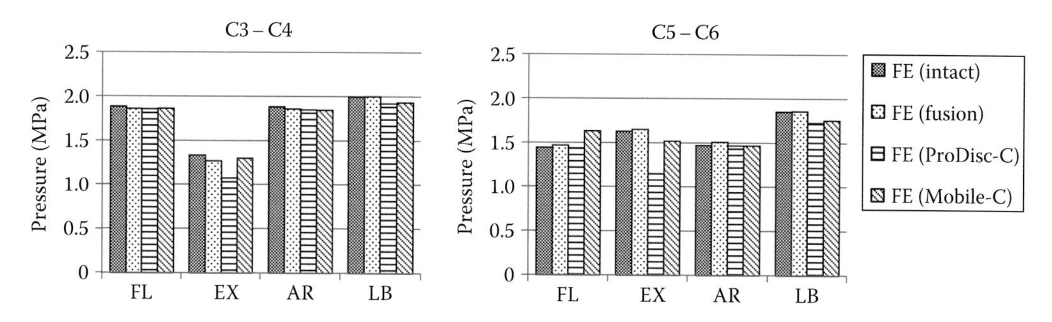


**FIGURE 16.8** Intervertebral disc pressure in the adjacent motion segment. FL: flexion, EX: extension, AR: axial rotation, LB: lateral bending.

## 16.4 APPLICATION 2: SIMULATION OF CERVICAL SPINAL ARTHROPLASTY

### 16.4.1 Model Development

Both the ProDisc-C and Mobi-C, which use a ball-in-socket design, are typical semi-constrained artificial discs for total disc replacement. Both artificial discs are composed of three components: two metal endplates and an ultra-high-molecular-weight polyethylene (UHMWPE) inlay to form a metal-on-UHMWPE articulation. The primary difference between these two implants is that the ProDisc-C has a fixed core, which restricts the absolute translation against the inferior endplate, whereas the Mobi-C has a mobile core, which encourages restoration of the instantaneous axis of rotation and reduces stress at the bone-implant interface. The mobile core was expected to reduce the stress on the posterior facet joints. In summary, the constraints of ProDisc-C are more rigorous than those of Mobi-C.

In the present study, these two artificial discs were sized so as to restore the intervertebral height to 5 mm at the index segment. The primary dimensions of both artificial discs were 14 mm × 12 mm (width × length). The metal endplates in both the ProDisc-C and Mobi-C were cobalt alloy (CoCrMo). The Young's modulus and Poisson's ratio of the CoCrMo were 114 GPa and 0.3, respectively, and the corresponding material properties of the UHMWPE were 1 GPa and 0.49, respectively.

The anterior longitudinal ligament and the whole intervertebral disc were completely removed, and the intervertebral space was replaced with ProDisc-C (Figure 16.4b) or Mobi-C (Figure 16.4c). Since the endplate is coated with a porous layer, the implant endplate was considered to be totally integrated with the vertebral bone. A frictionless finite sliding contact was applied between the interaction surface of the UHMWPE inlay and the CoCrMo alloy endplates in both ProDisc-C and Mobi-C. The loading and boundary conditions were the same as the intact healthy cervical spine.

### 16.4.2 Spinal Motion after Total Disc Replacement

The rotational motions increased at the operative level (C4-C5) after total disc replacement with either ProDisc-C or Mobi-C (Figure 16.5). With the ProDisc-C, the ROM increased by between 24% and 438% during flexion, extension, axial rotation, and lateral bending. With the Mobi-C, the corresponding value increased by between 72% and 408%. The ROMs of ProDisc-C were lower than Mobi-C during all spinal movements, except in extension.

As shown in Figure 16.5, both ProDisc-C and Mobi-C models allowed for up to five times greater extension than the fusion model. The anterior longitudinal ligament and the disc play an important role in constraining extension. During distraction, the disc supported 70% of the load. After excision of the anterior longitudinal ligament and intervertebral disc, the index site lost its ability to constrain extension. Although the metal endplate in both ProDisc-C and Mobi-C was integrated with the bony structure, the artificial disc component was separated and not constrained in the axial direction.

At the adjacent segment, the ROM increased by 10% at C5-C6 during extension with ProDisc-C, and increased by 20% during flexion with Mobi-C. Compared to the operative level, the ROM at the adjacent level remained almost unaffected with both ProDisc-C and Mobi-C. Due to the increased motion at the operative level and the limited compensatory motion at the adjacent segment, the entire cervical segmental motion (C3-C6) increased by between 11% and 135% after the total disc replacement surgery. During flexion, motion of the ProDisc-C more closely resembled the intact healthy cervical spine than Mobi-C.

### 16.4.3 Influence of Total Cervical Disc Replacement on Facet Joint

Similar to the intact healthy cervical spine, the maximum facet joint contact force was seen during extension (Figure 16.6). At the operative segment (C4-C5), the facet joint force in extension increased by 60% and 63% in ProDisc-C and Mobi-C, respectively. In the ProDisc-C model, the values increased by 20% at C5-C6 but decreased by 14% at C3-C4 in extension. In the Mobi-C model, the forces increased by 20% at C5-C6 but only 1% at C3-C4. Compared to the increase at the operative level, the change in contact force at the adjacent segment was relatively small.

The maximum tension in the capsular ligament was observed during lateral bending in the intact healthy cervical spine (Figure 16.7). However, the maximum ligament tension was observed during extension and axial rotation after both total cervical disc replacement surgeries. Because of the excision of the anterior longitudinal ligament and intervertebral disc, the cervical spine lost its ability to restrict the transfer motion in the transverse plane, which would lead to relative sliding between the superior and inferior vertebrae. In extension, the cervical spine displayed hypermobility (more than five times that of the healthy cervical spine). The superior metal endplate separated from contact in both the ProDisc-C and Mobi-C. This left the posterior ligaments, particularly the capsular ligament, to restrict motion in the transverse plane. For the same reason, the capsular ligament also experienced high tension during axial rotation after total disc replacement. Although there were no reports of facet joint degeneration after total disc replacement, the enormous increase of facet joint contact force and capsular ligament tension would be a potential source for such complications.

### 16.4.4 Intradiscal Pressure at Adjacent Segment after Total Disc Replacement

The maximum intradiscal pressure at both the superior and inferior adjacent segments of both total disc replacements was similar to the intact healthy vertebrae (Figure 16.8). Specifically, the disc pressure decreased by 4% and 7% with the ProDisc-C, and 3% and 5% with the Mobi-C.

## ACKNOWLEDGMENTS

This project was supported by the National Natural Science Foundation of China (Nos. 10925208, 11120101001, 11202017, 11272273, and 81271666), The Hong Kong Polytechnic University Research Grant (No. G-UA40), Beijing Natural Science Foundation (7133245), Young Scholars for the Doctoral Program of the Ministry of Education of China (20121102120039), the National Science and Technology Pillar Program (No. 2012BAI18B05,2012BAI18B07), and the 111 Project (No. B13003).

## REFERENCES

An, H.S., Thonar, E.J.A., and Masuda, K. 2003. Biological repair of intervertebral disc. *Spine* 28: S86–92.

Baba, H., Furusawa, N., Imura, S., Kawahara, N., Tsuchiya, H., and Tomita, K. 1993. Late radiographic findings after anterior cervical fusion for spondylotic myeloradiculopathy. *Spine* 18: 2167–73.

Bohlman, H.H., Emery, S.E., Goodfellow, D.B., and Jones, P.K. 1993. Robinson anterior cervical discectomy and arthrodesis for cervical radiculopathy. Long-term follow-up of one hundred and twenty-two patients. *J Bone Joint Surg Am* 75: 1298–1307.

Bowden, A. 2006. Finite element modeling of the spine, in *Spine Technology Handbook*, ed. S.M. Kurtz and A.A. Edidin, pp. 443–471. Burlington: Elsevier Inc.

Buckwalter, J.A. 1995. Aging and degeneration of the human intervertebral disc. *Spine* 20: 1307–1314.

Burkus, J.K., Haid, R.W., Traynelis, V.C., and Mummaneni, P.V. 2010. Long-term clinical and radiographic outcomes of cervical disc replacement with the Prestige disc: Results from a prospective randomized controlled clinical trial. *J Neurosurg Spine* 13: 308–318.

Cakir, B., Richter, M., Käfer, W., Puhl, W., and Schmidt, R. 2005. The impact of total lumbar disc replacement on segmental and total lumbar lordosis. *Clin Biomech* 20: 357–364.

Chang, U., Kim, D.H., Lee, M.C., Willenberg, R., Kim, S., and Lim, J. 2007a. Range of motion change after cervical arthroplasty with ProDisc-C and Prestige artificial discs compared with anterior cervical discectomy and fusion. *J Neurosurg Spine* 7: 40–46.

Chang, U.K., Kim, D.H., Lee, M.C., Willenberg, R., Kim, S.H., and Lim, J. 2007b. Changes in adjacent-level disc pressure and facet joint force after cervical arthroplasty compared with cervical discectomy and fusion. *J Neurosurg Spine* 7: 33–39.

Cunningham, B.W., Gordon, J.D., Dmitriev, A.E., Hu, N.B., and McAfee, P.C. 2003. Biomechanical evaluation of total disc replacement arthroplasty: An in vitro human cadaveric model. *Spine* 28 (Suppl): S110–S117.

Denozière, G., and Ku, D.N. 2006. Biomechanical comparison between fusion of two vertebrae and implantation of an artificial intervertebral disc. *J Biomech* 39: 766–775.

Dmitriev, A.E., Cunningham, B.W., Hu, N., Sell, G., Vigna, F., and McAfee, P.C. 2005. Adjacent level intradiscal pressure and segmental kinematics following a cervical total disc arthroplasty: An in vitro human cadaveric model. *Spine* 30: 1165–1172.

Drake, R.L., Vogl, A.W., Mitchell, A.W.M., Tibbitts, R., and Richardson, P. 2008. *Gray's Atlas of Anatomy*. Philadelphia, PA: Churchill Livingstone Elsevier.

Eck, J.C., Humphreys, S.C., Lim, T., Jeong, S.T., Kim, J.G., Hodges, S.D., and An, H.S. 2002. Biomechanical study on the effect of cervical spine fusion on adjacent-level intradiscal pressure and segmental motion. *Spine* 27: 2431–2434.

Finn, M.A., Brodke, D.S., Daubs, M., Patel, A., and Bachus, K.N. 2009. Local and global subaxial cervical spine biomechanics after single-level fusion or cervical arthroplasty. *Eur Spine J* 18: 1520–1527. doi: 10.1007/s00586-009-1085-7.

Garrido, B.J., Taha, T.A., and Sasso, R.C. 2010. Clinical outcomes of Bryan cervical disc arthroplasty: A prospective, randomized, controlled, single site trial with 48-month follow-up. *J Spinal Disord Tech* 23: 367–371. doi: 10.1097/BSD.0b013e3181bb8568.

Goel, V.K., Sairyo, K., Vishnubhotla, S.L., Biyani, A., and Ebraheim, N. 2006. Spine disorders: implications for bioengineers, in *Spine Technology Handbook*, ed. S.M. Kurtz and A.A. Edidin, pp. 145–182. Burlington: Elsevier Inc.

Gore, D.R., and Sepic, S.B. 1984. Anterior cervical fusion for degenerated or protruded discs. A review of one hundred forty-six patients. *Spine* 9: 667–671.

Griffin, M.J. 2001. The validation of biodynamic models. *Clin Biomech (Bristol, Avon)* 16 (Suppl 1): S81–S92.

Hilibrand, A.S., and Robbins, M. 2004. Adjacent segment degeneration and adjacent segment disease: The consequences of spinal fusion? *Spine J* 4: S190–S194.

Hong-Wan, N., Ee-Chon, T., and Qing-Hang, Z. 2004. Biomechanical effects of C2–C7 intersegmental stability due to laminectomy with unilateral and bilateral facetectomy. *Spine* 29: 1737–1745.

Jawahar, A., Cavanaugh, D.A., Kerr, III, E.J., Birdsong, E.M., and Nunley, P.D. 2010. Total disc arthroplasty does not affect the incidence of adjacent segment degeneration in cervical spine: Results of 93 patients in three prospective randomized clinical trials. *Spine J* 10: 1043–1048.

Jirkova, L., and Horak, Z. 2009. Analysis of influence location of intervertebral implant on the lower cervical spine loading and stability. In *13th International Conference on Biomedical Engineering*, Vols 1–3, 1724–1727.

Keaveny, T.M., and Buckley, J.M. 2006. Biomechanics of vertebral bone, in *Spine Technology Handbook*, ed. S.M. Kurtz and A.A. Edidin, pp. 63–98. Burlington: Elsevier Inc.

McAfee, P.C., Cunningham, B.W., Hayes, V., Sidiqi, F., Dabbah, M., Sefter, J.C., Hu, N.B., and Beatson, H. 2006. Biomechanical analysis of rotational motions after disc arthroplasty: Implications for patients with adult deformities. *Spine* 31 (Suppl): S152–S160.

McNally, D.S., Shackleford, I.M., Goodship, A.E., and Mulholland, R.C. 1996. In vivo stress measurement can predict pain on discography. *Spine* 21: 2580–2587.

Narayan, Y., Srirangam, K., and Frank, A.P. 2001. Biomechanics of the cervical spine, Part 2. Cervical spine soft tissue responses and biomechanical modeling. *Clin Biomech (Bristol, Avon)* 16: 1–27.

Panjabi, M.M., Crisco, J.J., Vasavada, A., Oda, T., Cholewicki, J., Nibu, K., and Shin, E. 2001. Mechanical properties of the human cervical spine as shown by three-dimensional load-displacement curves. *Spine* 26: 2692–2700.

Phillips, F.M., Allen, T.R., Regan, J.J., Albert, T.J., Cappuccino, A., Devine, J.G., Ahrens, J.E., Hipp, J.A., and McAfee, P.C. 2009. Cervical disc replacement in patients with and without previous adjacent level fusion surgery: A prospective study. *Spine* 34: 556–565. doi: 10.1097/BRS.0b013e31819b061c.

Sasso, R.C., Anderson, P.A., Riew, K.D., and Heller, J.G. 2011. Results of cervical arthroplasty compared with anterior discectomy and fusion: Four-year clinical outcomes in a prospective, randomized controlled trial. *J Bone Joint Surg Am* 93: 1684–1692. doi: 10.2106/JBJS.J.00476.

Shikinami, Y., Kotani, Y., Cunningham, B.W., Abumi, K., and Kaneda, K. 2004. A biomimetic artificial disc with improved mechanical properties compared to biological intervertebral discs. *Adv Funct Mate*r 14: 1039–1046. doi: 10.1002/adfm.200305038.

Williams, J., Allen, M., and Harkess, J. 1968. Late results of cervical discectomy and interbody fusion: Some factors influencing the results. *J Bone Joint Surg* 50: 277–286.

Yang, B., Li, H., Zhang, T., He, X., and Xu, S. 2012. The incidence of adjacent segment degeneration after cervical disc arthroplasty (CDA): A meta analysis of randomized controlled trials. *PloS one* 7: e35032.

# 17 Spine Model for Disc Replacement

*Qi Li, Lizhen Wang, Zhongjun Mo, and Yubo Fan*

## CONTENTS



## SUMMARY

Total disc replacement (TDR), a new surgical technique for treating degenerative disc disease, has resulted in many reliable clinical outcomes. The biomechanical effects of TDR need to be investigated to evaluate the long-term impact of using such devices. The finite element method (FEM) can simulate the various geometries and materials of the prosthesis and predict the range of motion (ROM) and many other biomechanical parameters before and after operation, while reducing both research time and expense in comparison to cadaver and animal studies. Therefore, we developed computational models that can be used to understand cervical spine biomechanics and evaluate the long-term effectiveness of disc prostheses. A three-dimensional, nonlinear FEM of intact human C3-C7 segments was developed from a reconstruction of computer tomography (CT) images. Specifically, the C3-C7 segments were chosen because of the large amount of clinical and experimental data available for these joints. The geometrically accurate spinal FE model consisted of five separate vertebrae, four intervertebral discs, and five kinds of ligaments. The intact model was validated by comparing the ROM predicted by FEM to in vitro experimental data. An artificial disc prosthesis (ProDisc-C, Synthes) was then integrated into the C3-C7 model in ABAQUS to make up a complete spine model for disc replacement. This validated FE model can be used to investigate the biomechanical effects of total disc replacement, such as facet force and tension force in ligaments.

## 17.1 INTRODUCTION

The spine supports and stabilizes the human body (Figure 17.1) while performing activities such as walking and bending. The spine is made up of 24 vertebrae that are stacked on top of each other to form a column. The spinal column houses and protects the spinal cord in its spinal canal.

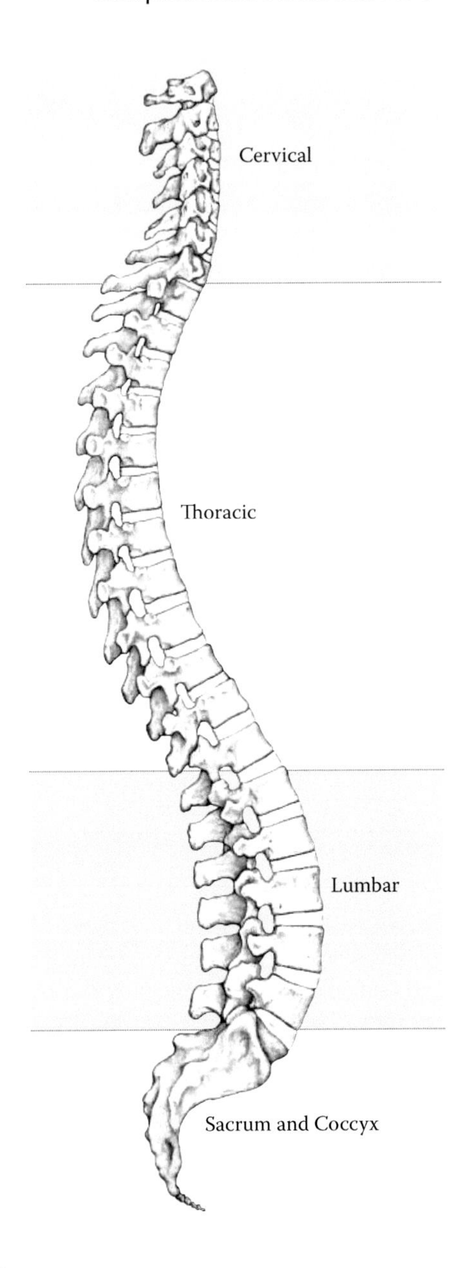


**FIGURE 17.1** Human spine.

The vertebrae are separated by intervertebral discs, which maintain an appropriate space to support motion and allow nerves to pass.

Intervertebral disc degeneration due to natural aging can often cause severe lower back pain (Luoma et al. 2000). Anterior cervical discectomy fusion (ACDF) is an effective surgical treatment for cervical degenerative disc disease (DDD) and has been reported with a good rate of success and early recovery (Clements and O'Leary 1990; Cauthen et al. 1998; Yue, Brodner, and Highland 2005). However, degeneration of adjacent levels associated with immobility of the fused levels after ACDF continues to raise concerns (Hilibrand and Robbins 2004; Ishihara et al. 2004; Chang et al. 2007). As an alternative to ACDF, cervical total disc replacement (CTDR) is designed to decrease the rate of adjacent-level disease by restoring disc height, preserving operative level motion, and reasonably alleviating the excessive stresses on adjacent levels. Considering the advantages, some surgeons advocate arthroplasty instead of arthrodesis, even though long-term follow-up results haven't been satisfactorily reported.

In 1966, Fernstrom implanted stainless steel balls into the intervertebral space after discectomy of cervical and lumber vertebrae (Fernstrom 1966). This is considered to be the first generation of spinal arthroplasty. The clinical result of Fernstrom's innovation was not quite satisfactory, primarily because of settlement. The stainless steel balls fell into the endplate and failed to maintain the intervertebral height of patients within months after implantation. Today, great advances have been made in the biomechanical and kinematic characterization of disc prostheses. Roughly 20 kinds of spinal arthroplasty devices are currently in use worldwide, with each product displaying its own unique properties. ProDisc, Bryan, and PCM (Porous Coated Motion) are the most popular of these devices.

The ProDisc prosthesis (Figure 17.2) uses a cobalt chromium (CoCrMo) alloy for its upper and lower plates and an ultra-high-molecular-weight polyethylene (UHMWPE) core. It is a semiconstrained prosthesis. The keels on the surface anchor it into the vertebral endplates with a fixed center of rotation.

The Bryan prosthesis (Figure 17.3) is composed of two titanium-alloy shells and a polyurethane core. It is an unconstrained device that theoretically allows range of motion in all planes. A flexible membrane with inner lubricating liquid surrounds the artificial core.

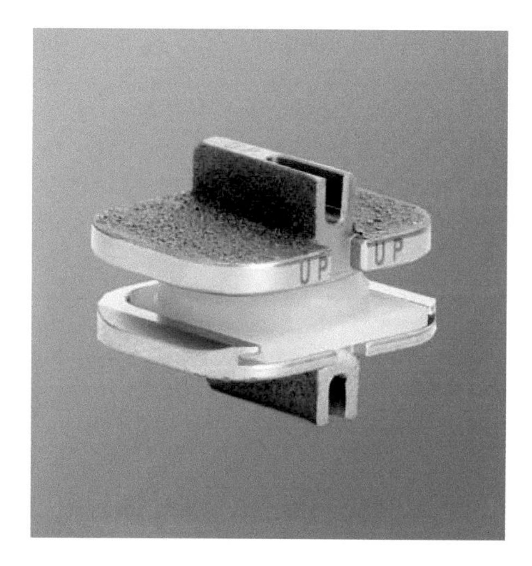


**FIGURE 17.2** ProDisc prosthesis.

**FIGURE 17.3** Bryan prosthesis.

The PCM prosthesis (Figure 17.4) has two porous-surfaced Co-Cr endplates and a polyethylene bearing surface attached to the endplate. The articulating surface of this device creates a larger radius of articulation and increased translation.

Various cadaver and FEM studies to date have investigated biomechanical changes after cervical TDR procedures. A simplified C3-C6 FE model (Sung 2006) with an elastomer-type artificial disc was shown to reasonably restore the mobility of a spinal segment. It has been reported that the ROM of a prosthesis implanted at C4-C5 was reduced by about 50% to 70% during flexion, extension, and right lateral bending, whereas the ROM increased by about 18% during axial rotation (Jirkova and Horak 2009). A recent in vitro experiment found that the ROM was not statistically different between an implanted cervical spine and an intact one (Barrey et al. 2009). It has been concluded that use of artificial disc prostheses did not alter the motion patterns at either the instrumented level or adjacent segments compared with the harvested condition, except in extension (DiAngelo et al. 2004).

Previous in vitro cervical spine studies analyzed the range of motion after one- and two-level cervical arthroplasty versus arthrodesis and hybrid constructs (Cho et al. 2010). Cho's results demonstrated that two-level arthrodesis decreases the entire ROM, whereas two-level arthroplasty increases the entire ROM, and two-level arthroplasty does not significantly change the ROM at the adjacent level. It has been highlighted that the operative-level ROM was preserved with one- and two-level arthroplasty, whereas the ROM at the distal adjacent level increased by the greatest amount after two-level arthrodesis (Cunningham et al. 2003).

Most previous biomechanical studies investigated the operative- and adjacent-level kinematic or mechanical responses to single-level cervical disc arthroplasty. Few studies have qualified or analyzed fundamental biomechanical changes after multi-level cervical arthroplasty, which has seen a recent surge in popularity.

Noticeable limitations remain with current models of spine arthroplasty. Improvements could be made by incorporating (1) anisotropy of the bony structure; (2) proper prestress in muscles and ligaments; (3) accurate properties for annulus fibrosus and nucleus pulposus; and (4) a spinal cord and nerves.

**FIGURE 17.4** PCM prosthesis.

## 17.2 DEVELOPMENT OF AN IMPLANTED CERVICAL MODEL

### 17.2.1 Model of an Intact Spine

CT images taken at 1-mm intervals were obtained from C3-C7 of a 35-year-old healthy male volunteer with a height of 175 cm and weight of 69 kg (Figure 17.5). The subject did not have any prior history of cervical disc disease nor radiographic changes in the cervical segments. Commercial software MIMICS 10.1 (Materialise, Inc., Leuven, Belgium) was then used to develop an initial three-dimensional geometrical cervical model from the CT data.

Because of inherent irregularities, all parts of the initial geometrical model were loaded into the data processing software RapidForm 2011 (INUS, Korea) for smoothening. Then the smoothed parts were converted into IGES format and imported into FE analysis software ABAQUS 6.11 (Simulia, USA). The FE meshes were then generated.

The cancellous bones were represented by tetrahedron elements. Triangle shell elements were generated from the external surfaces of the tetrahedron elements to represent the cortical shell and endplates with a thickness of 0.4 mm (Kurtz and Edidin 2006). The facet joints were modeled with frictionless sliding contacts between the relevant bony surfaces.

The intervertebral disc consists of an annulus fibrosus and a nucleus pulposus. The fibers in the annulus fibrosus were modeled using three-dimensional linear contact elements that only sustain tension forces. The fiber content in the annulus fibrosus was assumed to be approximately 20% of the matrix volume (Guerin and Elliott 2006). The nucleus pulposus and the ground substance matrix for the annulus fibrosus were modeled using 20-noded solid elements, as shown in Table 17.1 (Guerin and Elliott 2006; Denozière and Ku 2006).

Five major ligaments, the anterior longitudinal ligament (ALL), posterior longitudinal ligament (PLL), flaval ligament (FL), capsular ligament (CL), and spinous ligament (SL), were considered and modeled as nonlinear tension-only elements since they sustain only tension forces. The insertion points were determined according to anatomic observations (Yoganandan, Kumaresan, and Pintar 2000; Panjabi, Oxland, and Parks 1991). As a result, an intact FE C3-C7 spine model was developed (Figure 17.6). The mechanical properties of the components of the FE model are shown in Table 17.1.

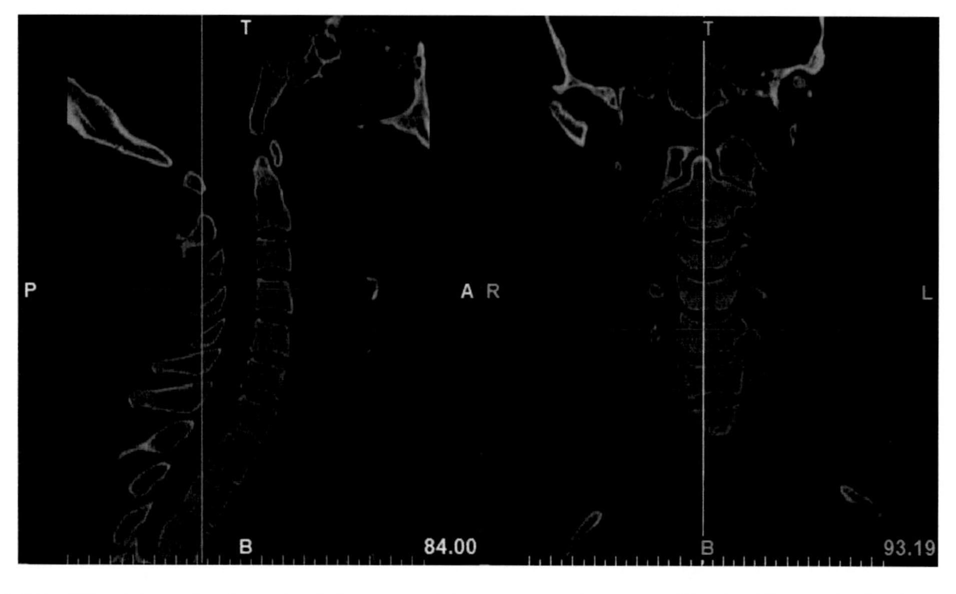


**FIGURE 17.5 (See color insert.)** CT images at 1-mm intervals on C3-C7 of a 35-year-old healthy male volunteer.

**TABLE 17.1**
**Material Properties of the Cervical Segment Component and Artificial Device in the Finite Element Model**

| Component | Element Type | Young's Modulus *E* (MPa) | Poisson's Ratio *V* |
|---|---|---|---|
| Cortical shell | Shell | 12,000 | 0.29 |
| Cancellous bone | Tetrahedral | 100 | 0.25 |
| Endplate | Shell | 12,000 | 0.29 |
| Annulus | Hexahedral | 3.4 | 0.40 |
| Nucleus | Hexahedral | 1 | 0.49 |
| Annulus fiber | Truss | 450 | 0.45 |
| CoCrMo | Tetrahedral | 220,000 | 0.32 |
| UHMWPE | Tetrahedral | 1,000 | 0.49 |
| Titanium | Hexahedral | 114,000 | 0.35 |
| Allograft | Hexahedral | 100 | 0.25 |

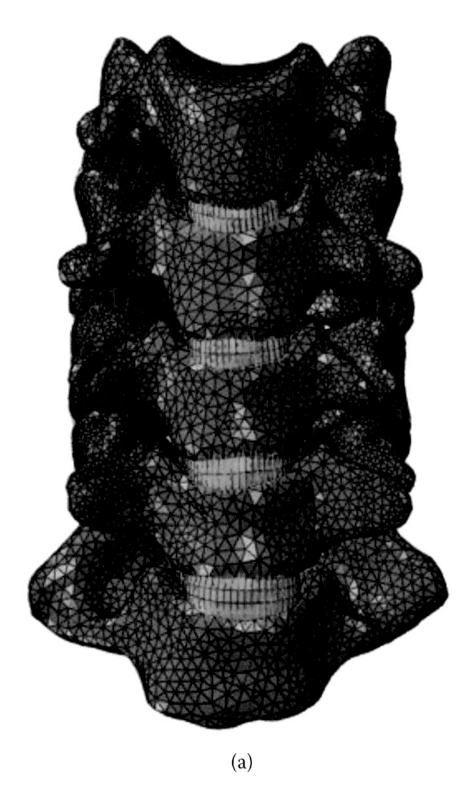

(a)

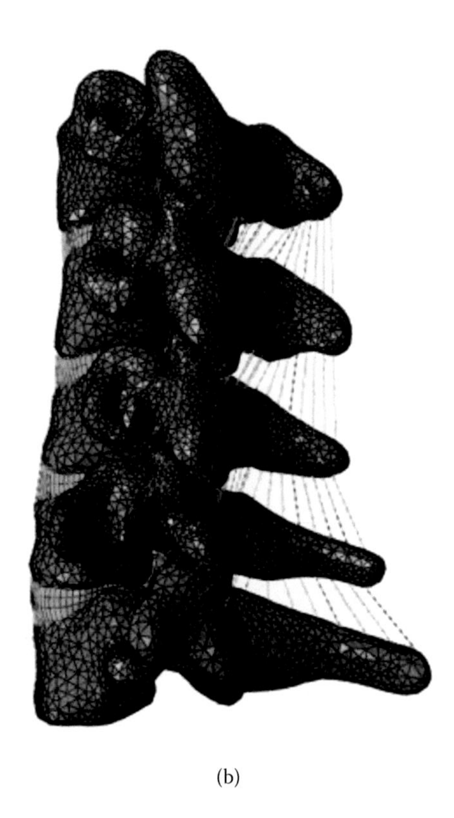

(b)

**FIGURE 17.6 (See color insert.)** Finite element model of intact C3-C7: (a) anterior view and (b) lateral view.

### 17.2.2 Model of Total Disc Replacement

The ProDisc-C (Synthes Spine USA Products LLC, Westchester, Pennsylvania) artificial disc was selected for the TDR model. An appropriate prosthesis size was chosen by analyzing the CT data and the developed intact cervical model. The three-dimensional model (IGES format) of the prosthesis was developed using CAD (computer-aided design) engineering software Pro/Engineer, and

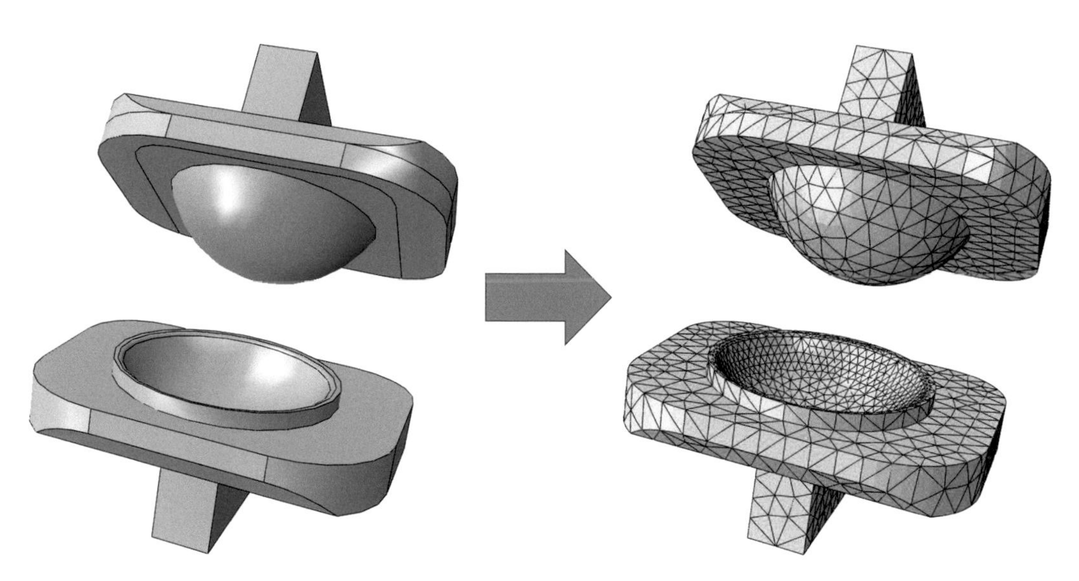

**FIGURE 17.7** Finite element model of ProDisc-C prosthesis.

then imported to ABAQUS for meshing (Figure 17.7). The material properties of the CoCrMo alloy and UHMWPE were obtained from previous studies (see Table 1.1).

According to the standard TDR procedure, the ALL and intervertebral disc were totally excised at the C4-C5 and C5-C6 levels. Next, the operative level was implanted with a ProDisc-C prosthesis. Referencing kinematic features, the polyethylene core was fixed to the CoCrMo lower plate and three-dimensional nonlinear contact conditions were added between the upper plate and the polyethylene core.

### 17.2.3 Loading and Boundary Conditions

The inferior surface of C7 was constrained in all degrees of freedom. Flexion, extension, axial rotation (AR), and lateral bending (LB) were simulated by applying a moment of 1.8 Nm, combined with a 73.6-N follower load imposed on a reference node on the superior surface of the C3 vertebra.

### 17.2.4 Model Validation

The intact model was validated by comparing the ROM predicted by FEM to in vitro experimental data with an applied moment of 1.8 Nm. The motion predicted by FE simulations for flexion-extension, lateral bending, and axial rotation across four motion segments of the intact cervical spine (C3/C4,C4/C5,C5/C6, and C6/C7) were in good agreement with experimental results (Moroney et al. 1988; Panjabi et al. 2001; Finn et al. 2009) (Figure 17.8). Therefore, the present model was able to predict any biomechanical changes induced on the surgically altered cervical spine segment.

## 17.3 APPLICATION OF THE MODEL

### 17.3.1 Effect of Two-Level Cervical Arthroplasty

The first two-level cervical arthroplasty was reported in 2004 and has since become a viable method for managing multilevel non-fusion cervical cord compression. However, even though multilevel cervical arthroplasty is increasing in popularity, limited biomechanical data is available. We studied the main parameters of a two-level implanted cervical spine: ROM, and facet force and tension force in ligaments.

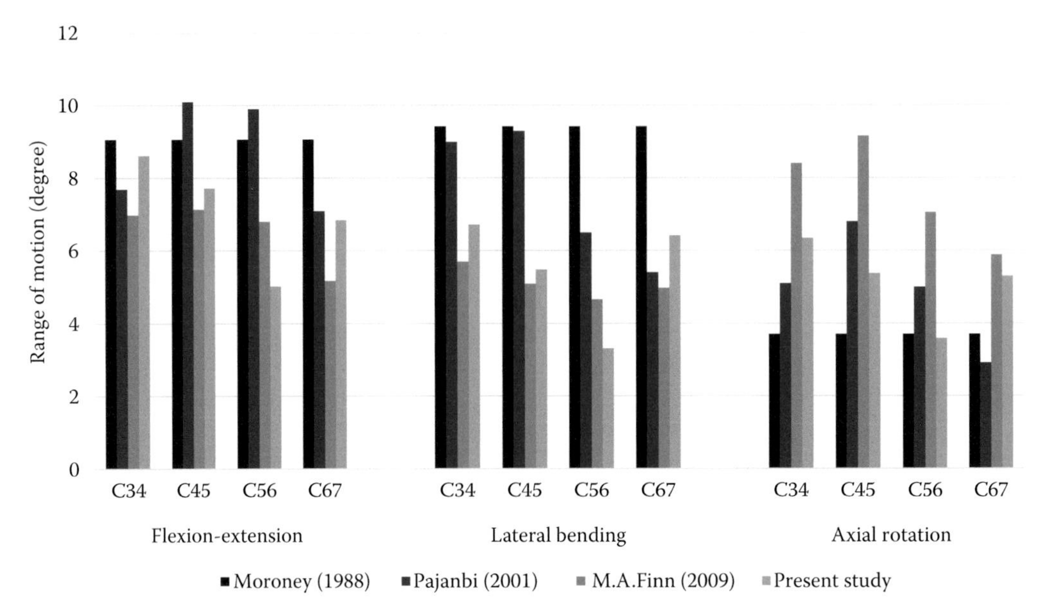


**FIGURE 17.8** Comparison of range of motion for C3-C7 under 1.8-Nm moment loading in flexion-extension, lateral bending, and axial rotation.

#### 17.3.1.1 Comparison of Range of Motion

Figure 17.9 shows a comparison of ROM for intact and TDR models. After two-level TDR, ROM increased by 9% in flexion to 32% in axial rotation in comparison to the intact model at the implanted level. ROM at the adjacent segments did not show significant differences owing to a compensatory mechanism at these levels.

#### 17.3.1.2 Ligament Force

Ligaments have an important role in spinal biomechanics and stability. In the present study, cervical spinal ligament force increased by 27.4%–78.2% in the two-level arthroplasty segments (Figure 17.10). In this acute instability-type model, with resection of the structures required to insert the cervical prosthesis, the dramatically increased tension force in the ligaments will increase the risk of degeneration in these ligaments.

#### 17.3.1.3 Facet Joint Force

In the present study, the maximum force transmitted through the facet joint is shown in Figure 17.11. The facet joint force generally increased at the arthroplasty operative levels in comparison to the intact model. During extension, the greatest increase (57.9%) was seen at the C4-C5 level in the two-level arthroplasty spine.

The findings in the present investigation highlight the important trends of ROM, the force transmitted through the facet joint, and the tension force of the cervical spinal ligaments after two-level cervical arthroplasty. Our results indicated that the changes of motion at the surgical and adjacent levels following TDR were favorable; the entire ROM did not show significant differences since a compensatory decrease was seen at the adjacent segments. This result is consistent with trends observed in previous studies. A relatively large increase of ROM after TDR (36%) was observed in axial rotation at the surgical level. This may imply that resection of the whole annulus had a more significant effect when compared to resection of the unilateral facet joint and part of the posterior annulus.

The increased stress on the facet joint after artificial disc replacement (ADR) in the lumbar spine has been cited as a reason for degenerative changes in the implanted segments and poor clinical

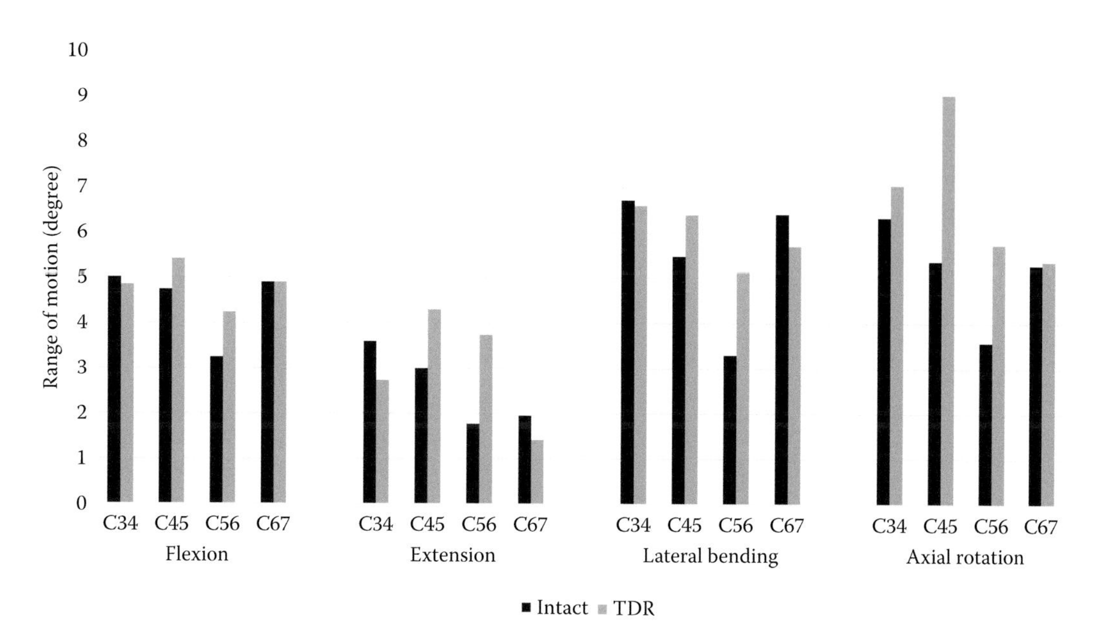


**FIGURE 17.9** Range of motion at C3-C7 segments in the intact and total disc replacement models.

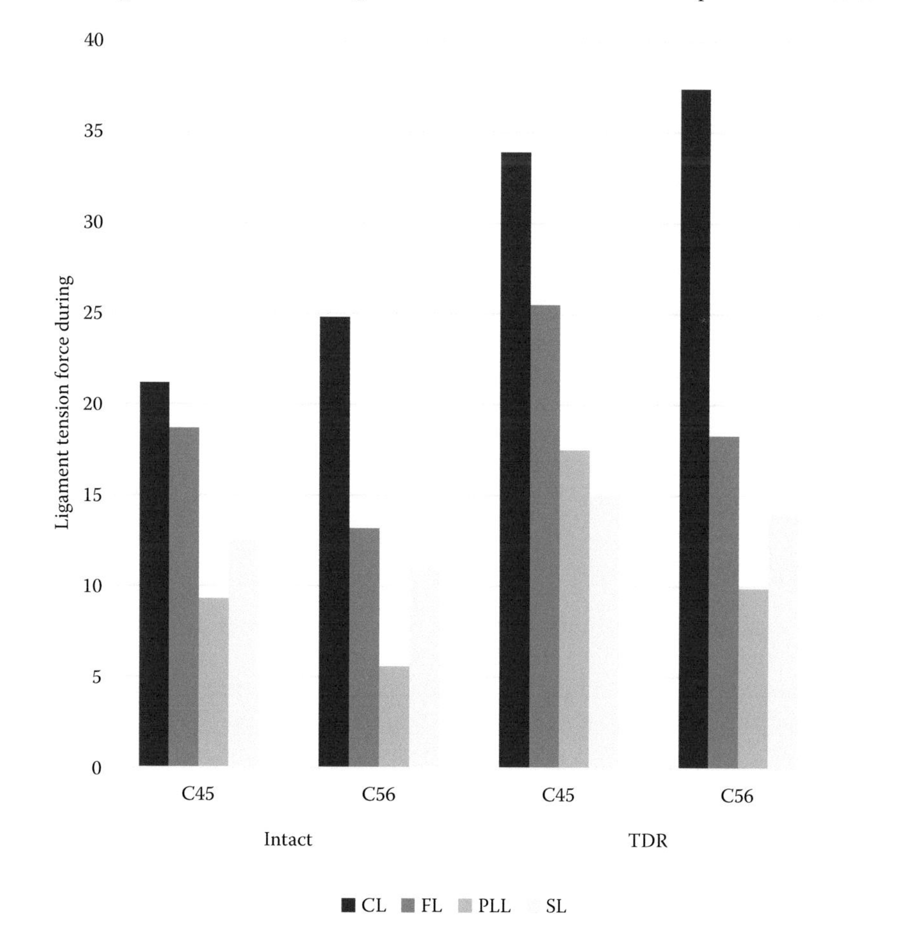


**FIGURE 17.10** Ligament tension force during flexion in C4/C5, C5/C6 of the intact and total disc replacement models.

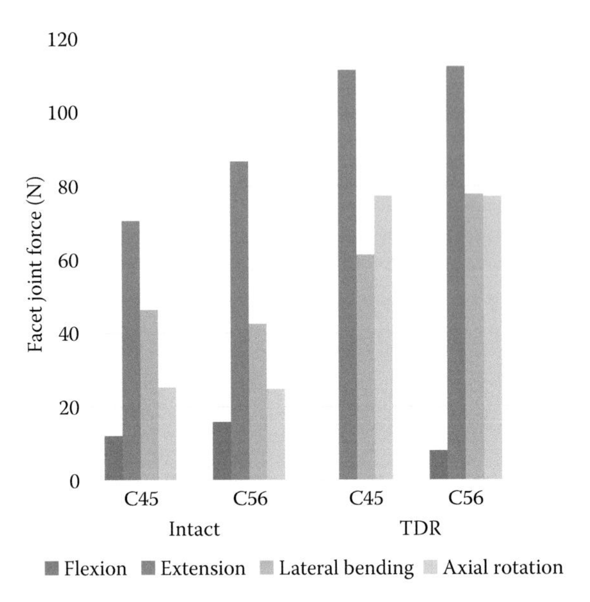


**FIGURE 17.11** Facet joint force in C4/C5 and C5/C6 of the intact and total disc replacement models.

results, but there is not sufficient biomechanical or clinical evidence available to draw definitive conclusions. Previous studies have shown that the facet joint force may increase by 8%–113% (Galbusera et al. 2006; Chang et al. 2007). In the present study, the maximum force transmitted through the facet joint was seen during extension and increased at all arthroplasty levels. The greatest increase (57.9%) was seen at the C4-C5 level in the two-level arthroplasty spine. This study concluded that facet joint force increased after arthroplasty, and, compared to earlier single-level TDR studies, the increase of facet joint force in two-level TDR was greater. This corresponds to the results of previous studies.

In summary, this study has described a three-dimensional, geometrically and materially nonlinear FE model of the lower cervical multilevel units C3-C7. The model was validated with previously published experimental data. Then the model was applied to investigate the change in biomechanical parameters after two-level total disc replacement. The analyses suggest that two-level TDR (using ProDisc-C cervical prosthesis) can restore the mobility of an intact intervertebral disc reasonably well. The FE model developed for this study may be a helpful tool in the evaluation and design of cervical spinal procedures.

## ACKNOWLEDGMENTS

This work was supported by the National Natural Science Foundation of China (Nos. 10925208, 11120101001, 11202017), the Beijing Natural Science Foundation (7133245), Young Scholars for the Doctoral Program of the Ministry of Education of China (20121102120039), the National Science and Technology Pillar Program (Nos. 2012BAI18B05, 2012BAI18B07), and the 111 Project (No. B13003).

## REFERENCES

Barrey, C., T. Mosnier, J. Jund, G. Perrin, and W. Skalli. 2009. In vitro evaluation of a ball-and-socket cervical disc prosthesis with cranial geometric center. *J Neurosurg Spine* 11:538–46.

Cauthen, J. C., R. E. Kinard, J. B. Vogler, D. E. Jackson, O. B. DePaz, O. L. Hunter, L. B. Wasserburger, and V. M. Williams. 1998. Outcome Analysis of Noninstrumented Anterior Cervical Discectomy and Interbody Fusion in 348 Patients. *Spine* 23:188–92.

Chang, U. K., D. H. Kim, M. C. Lee, R. Willenberg, S. H. Kim, and J. Lim. 2007. Changes in adjacent-level disc pressure and facet joint force after cervical arthroplasty compared with cervical discectomy and fusion. *J Neurosurg Spine* 7:33–9.

Cho, B. Y., J. Lim, H. B. Sim, and J. Park. 2010. Biomechanical analysis of the range of motion after placement of a two-level cervical ProDisc-C versus hybrid construct. *Spine (Phila Pa 1976)* 35:1769–76.

Clements, D. H., and P. F. O'Leary. 1990. Anterior cervical discectomy and fusion. *Spine (Phila Pa 1976)* 15:1023–5.

Cunningham, B. W., J. D. Gordon, A. E. Dmitriev, N. Hu, and P. C. McAfee. 2003. Biomechanical evaluation of total disc replacement arthroplasty: An in vitro human cadaveric model. *Spine (Phila Pa 1976)* 28:S110–7.

Denozière, G., and D. N. Ku. 2006. Biomechanical comparison between fusion of two vertebrae and implantation of an artificial intervertebral disc. *J Biomech* 39:766–75.

DiAngelo, D. J., J. T. Robertston, N. H. Metcalf, B. J. McVay, R. C. Davis, and K. T. Foley. 2003. Biomedical testing of an artificial cervical joint and an anterior cervical plate. *J Spinal Disorders Techn* 16:314–23.

Fernstrom, U. 1966. Arthroplasty with intercorporal endoprothesis in herniated disc and in painful disc. *Acta Chir Scand Suppl* 357:154–9.

Finn, M. A., D. S. Brodke, M. Daubs, A. Patel, and K. N. Bachus. 2009. Local and global subaxial cervical spine biomechanics after single-level fusion or cervical arthroplasty. *Eur Spine J* 18:1520–7.

Galbusera, F., A. Fantigrossi, M. T. Raimondi, M. Sassi, M. Fornari, and R. Assietti. 2006. Biomechanics of the C5-C6 spinal unit before and after placement of a disc prosthesis. *Biomech Modeling Mechanobiol* 5:253–61.

Guerin, H. A., and D. M. Elliott. 2006. *Structure and Properties of Soft Tissues in the Spine.* Burlington, VT: Academic Press.

Hilibrand, A. S., and M. Robbins. 2004. Adjacent segment degeneration and adjacent segment disease: The consequences of spinal fusion? *Spine J* 4:190S–94S.

Ishihara, H., M. Kanamori, Y. Kawaguchi, H. Nakamura, and T. Kimura. 2004. Adjacent segment disease after anterior cervical interbody fusion. *Spine J* 4:624–8.

Jirkova, L, and Z. Horak. 2009. Analysis of influence of location of intervertebral implant on the lower cervical spine loading and stability. *13th International Conference on Biomedical Engineering* 23:1724–7.

Kurtz, S. M., and A. Edidin. 2006. *Spine Technology Handbook.* Waltham, MA: Academic Press.

Luoma, K., H. Riihimaki, R. Luukkonen, R. Raininko, E. Viikari-Juntura, and A. Lamminen. 2000. Low back pain in relation to lumbar disc degeneration. *Spine (Phila Pa 1976)* 25:487–92.

Moroney, S. P., A. B. Schultz, J. A. Miller, and G. B. Andersson. 1988. Load-displacement properties of lower cervical spine motion segments. *J Biomech* 21:769–79.

Panjabi, M. M., J. J. Crisco, A. Vasavada, T. Oda, J. Cholewicki, K. Nibu, and E. Shin. 2001. Mechanical properties of the human cervical spine as shown by three-dimensional load–displacement curves. *Spine* 26:2692–700.

Panjabi, M. M., T. R. Oxland, and E. H. Parks. 1991. Quantitative anatomy of cervical spine ligaments. Part I. Upper cervical spine. *J Spinal Disord* 4:270–6.

Sung, K. H. 2006. Finite Element Modeling of Multi-Level Cervical Spinal Segments (C3-C6) and Biomechanical Analysis of an Elastomer-Type Prosthetic Disc. *Medical Engineering & Physics* 28:534–41.

Yoganandan, N., S. Kumaresan, and F. A. Pintar. 2000. Geometric and mechanical properties of human cervical spine ligaments. *J Biomech Eng* 122:623–9.

Yue, W. M., W. Brodner, and T. R. Highland. 2005. Long-term results after anterior cervical discectomy and fusion with allograft and plating: A 5- to 11-year radiologic and clinical follow-up study. *Spine (Phila Pa 1976)* 30:2138–44.

# 18 Spine Model for Applications in Aviation Protection

*Cheng-fei Du, Lizhen Wang, Ya-wei Wang, and Yubo Fan*

## CONTENTS



## SUMMARY

Military aircrew may be more prone to developing musculoskeletal disabilities involving the spine due to the anatomical characteristics of the spine and the complex accelerative environment. In some uncommon situations (such as ejection or flight maneuver), the impact load may be severe enough to injure the pilot. The spine, especially the cervical spine and thoracolumbar junction, are particularly susceptible to fracture in high-stress conditions. Many factors that affect the degree of injury to the pilot have been investigated through physical testing and simulation. Compared with physical experiments, computer simulations provide more detailed information at lower costs and offer better repeatability. The aim of this chapter is to introduce the development of a finite element model of the spine and its application in investigating how to protect the spine from injury under accelerative impact conditions often faced by aircrew.

## 18.1 INTRODUCTION

The human spine consists of 33 bony vertebral elements, which are connected with each other by fibrocartilaginous intervertebral discs. Vertebral joint capsules and various ligaments are also included in the spinal structure to stabilize successive vertebrae. The inferior nine vertebrae are fused into the sacrum and coccyx. The remaining 24 vertebrae are divided into three regions: cervical (7 vertebrae), thoracic (12 vertebrae), and lumbar (5 vertebrae), according to the regions they occupy (Figure 18.1). The anatomy of the spine makes it particularly adept at absorbing energy and protecting the body from impact loading. However, injury will still occur if the impact energy surpasses the tolerance of the spine.

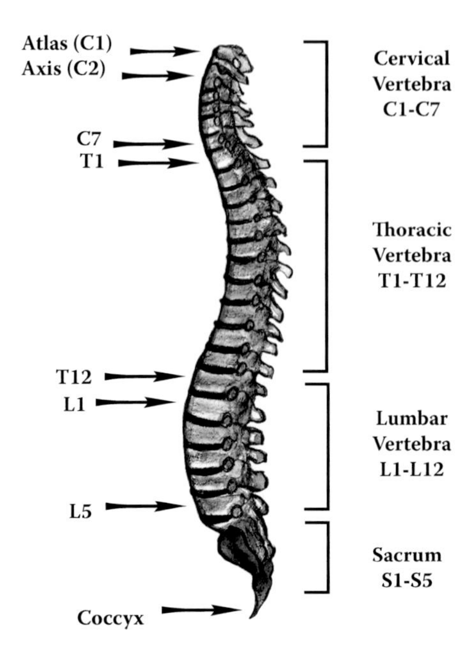


**FIGURE 18.1** **(See color insert.)** Anatomical structure of the human spine.

Aircraft occupants, especially military aircrew, subjected to complex dynamic impacts are prone to developing spinal complications or disabilities. For example, the firing of the ejection gun applies an accelerative impact of more than 15 G on the aviator along the long axis of the spine. The tremendous thrust force can lead to large flexion compression of vertebrae and thus induce a high risk of vertebral fracture or disc rupture. Although the ejection seat and restraint system have been in use for many years, the incidence of ejection-associated spinal injury remains high. According to an investigation by Lewis (2006), the overall percentage of ejection survivals who sustained spinal fractures was 29.4% and the majority of fractures were located around the thoracolumbar region.

It is worth mentioning that landing on the ground after ejection is also an important event that may cause spinal injury or aggravate the level of lesions produced during ejection (Sturgeon, 1987). Sturgeon also compared the distribution pattern of vertebral fractures occurring during ejection with landing-induced vertebral fracture and found that ejection vertebral fractures tended to be broadly distributed between T3 and L1, while landing fractures were more commonly located below L1.

Despite advancements in the crashworthiness of helicopters, both the rate and severity of injuries sustained in helicopter accidents remain relatively high. An analysis of medical records showed that 86.5% of aircrew involved in helicopter accidents suffered injury (Sasirajan, Narinder, and Dahiya, 2007). Of these, 43.4% were spinal lesions, the leading cause of injury, followed by injury to the head and face (22.6%). Similar to ejection injuries, the thoracolumbar junction was the most common site of injury and compression fractures represented the majority of injury patterns.

Developments in engine performance and aerodynamics have given modern fighter pilots increased agility in the air, which can potentially lead to cervical spine injury. Such agile fight maneuvers are capable of exposing pilots to multiaxial forces and the unrestrained head-neck complex may bear the brunt of these forces. In such instances, several types of injury may occur, such as compression fractures, ligament tearing (Schall, 1989), and muscle pain (Green, 2003; Lecompte et al., 2008; De Loose et al., 2009). The weight of the helmet and oxygen mask worn by the pilot will accentuate the strain on the neck muscles and induce further flexion compression to the cervical spine (Hamalainen, 1993). In addition, the increasing number of display and sighting systems adds to the complexity of the acceleration environment and the potential for injury (Newman, 2006).

In general, the rate and severity of spinal injury is influenced by multiple factors: age and size of the pilot, body posture and level of muscle tension just before impact, design of the cockpit and seat, performance of restraint systems, and features of the accelerative load (such as onset rate, peak magnitude, and duration).

Many researchers have experimentally investigated the biomechanical mechanism that acts to protect the spine from injury produced by exposure to impact accelerations. By means of in vitro biomechanical testing, much research has been undertaken to investigate which of the two components, vertebral body or endplate, fractures first (Brown, Hansen, and Yorra, 1957; Roaf, 1960; Holdsworth, 1970; Willen et al., 1984; Karlsson et al., 1998) in order to understand the mechanism of spinal fracture under compressive dynamic loading. Some investigators were interested in ascertaining the percent of total body weight supported by the individual vertebrae (Ruff, 1950; Stech and Payne, 1963; Pearsall, Reid, and Livingston, 1996). By conducting subject tests, the dynamic response of various regions of the body influenced by the acceleration environment can be measured and evaluated directly (Ewing et al., 1976; Leupp, 1983; Hearon and Brinkley, 1986; Frazier et al., 1988). Mannequins also have been widely used for obtaining dynamic information in high-acceleration impacts, with the aim of validating the effectiveness of protection devices or devising countermeasures to injury induced by impact loading (Leupp, 1983; Buhrman, 1991; Erin and John, 2006).

Experimental work is often expensive, requires considerable preparation, and provides only limited information as to the response of the body during impact and the possibility of injury. Furthermore, sometimes it is difficult to confirm whether the test environment is comparable to the real situation. Thus, many studies have been undertaken on the mechanism of spinal injury produced by impact loading or design of protection devices using numerical simulations. The most famous of these is the Dynamic Response Index Model (DRI), which is essentially a one-degree-of-freedom lumped mass, spring model of the head-spine-torso (Payne, 1961). It has been extensively correlated with injury data and provides a useful criterion for evaluating safety in an axial ($G_z$) acceleration environment.

The multibody dynamics method may be another effective tool to investigate the response of the body in an accelerative environment. This type of model allows the body to be described as a set of rigid segments, coupled at joints with appropriate mechanical properties. By comparing the displacement, acceleration of the body, or tension in the belt during different ejection conditions, the influence of several factors can be investigated, such as rate of onset, ejection angle, preejection alignment, and initial state of the belt (Belytschko, Schwer, and Privitzer, 1978; Obergefell and Kaleps, 1988).

Compared to the multibody method, finite element models may offer more detailed information, such as stress and strain within a region of interest. Early investigations on the dynamic properties of the spine using the finite element method were limited to single vertebrae or a functional segment (Lee et al., 2000; Wilcox et al., 2004; Qiu et al., 2006). Later, these methods increased in popularity and were used to study the dynamic response of multi-segments to impact loading. Based on the actual geometry of an embalmed human cadaver, Teo et al. (2004) constructed a detailed head-neck (C0-C7) finite element model and estimated the effect of muscle on avoidance of severe cervical injury from the ejection process. Panzer and Cronin (2009) developed a detailed cervical spine model with accurate representations of soft tissues, which was used to evaluate global kinematics and the tissue-level response of a human head and neck in a frontal crash (Panzer, Fice, and Cronin, 2011); this model was also used to investigate ligament injuries in whiplash injuries during traffic accidents (Fice and Cronin, 2012). These models were all developed for kinematic analysis of a human head-neck after an auto crash, which differs from impact loading in aviation.

Few multisegment finite element models of the thorax or lumbar spine were found to investigate dynamic responses or vibration. Wayne developed a finite element model of the upper body from the head to sacrum to predict the vibration response of the human spine (Wayne and Goel, 2003). All parts of the upper body skeleton, except L2-L5, were simplified as beams. By using a comprehensive nonlinear finite element model of a spine multi-segment (T12-L5), Zhang, Li, and Tan (2008) demonstrated that differences between a sitting posture and napping posture have little effect on the response of the thoracolumbar spine to a mine blast-induced impact.

It is important to comprehensively and systematically analyze the mechanism of spine injury caused by impact loading in various aviation environments. In this study, detailed finite element models of the head-neck complex and thoracolumbar-pelvis complex (T9-S1) were developed and the processes to validate them were introduced. The response of the thoracolumbar spine during the ejection scenario, a typical example of impact loading in an aviation environment, was investigated. Finally, the application of the two finite element models of the spine is discussed briefly.

## 18.2 FINITE ELEMENT MODEL OF THE HEAD-NECK COMPLEX

### 18.2.1 Development of the Model

The geometry of the full head-neck model was based on a computed tomography (CT) image of a 42-year-old Chinese man. The linear distance between the C7/T1 joint and the head-neck joint was 120.1 mm in the model, compared to 118.8 mm reported by Robbins (1983) for a seated mid-size male. Cancellous bone was modeled using three-dimensional elements as orthotropic elastic materials to account for the increase in stiffness in the superior-inferior direction, rather than in the transverse direction, due to the trabecular structure (Mosekilde and Danielsen, 1987).

Cortical bone and bony endplates were modeled using two-dimensional quadrilateral elements assigned as isotropic materials. The thickness of the cortical bone and bony endplates were 0.5 and 0.6 mm, respectively (Panjabi et al. 2001). Intervertebral discs usually respond to multiple load vectors.

Although many single-entity definitions that incorporate Young's modulus in compression and tension have been used for intervertebral disc modeling, a more complex and realistic modeling approach was also developed for obtaining detailed internal information. The nucleus pulposus comprised approximately 50% of the total disc in the cross-sectional area (Pooni et al., 1986), and was modeled as a fluid. With regard to the annulus component, fibers were simulated using incompressible truss elements and ground substance was modeled as elastomeric foam using a strain energy function.

Like the intervertebral discs, zygapophysial joints respond to multiple load vectors. A majority of finite element simulations of the zygapophysical joints include facet bones, capsular ligaments, and an air gap between the two cartilages. Advanced modeling approaches incorporated the simulation of articular cartilage using solid elements, synovial fluid using fluid elements, and synovial membrane using membrane elements (Yoganandan, Kumaresan, and Pintar, 2001).

The ligaments were modeled using a set of one-dimensional tension-only elements distributed to match the width of the ligament and attached to their respective vertebra at the origin and insertion locations described in various anatomical studies (Panjabi et al., 1991; Yoganandan, Kumaresan, and Pintar, 2000). An exponential curve was used for the nonlinear and subtraumatic regions to ensure force and stiffness continuity. The muscle groups were usually incorporated using the origin-insertion location details, and Hill's muscle model was used to define muscle reaction for both passive and active responses (Panzer, Fice, and Cronin, 2011). Figure 18.2 shows a finite element model of a human head-neck established to investigate pilot cervical spine injuries. This model contains a total of 43,384 nodes and 53,560 elements.

### 18.2.2 Model Validation

To reduce computation time, head and vertebrae are simplified as a rigid body in the validation. The kinematic response of the whole cervical spine model is validated using frontal impact studies on human volunteers (Ewing, 1972). The cervical spine model was evaluated by prescribing the average measured experimental forward acceleration of the T1 vertebra (Figure 18.3). The global response of the model head center of gravity (CG) was compared to the experimental response

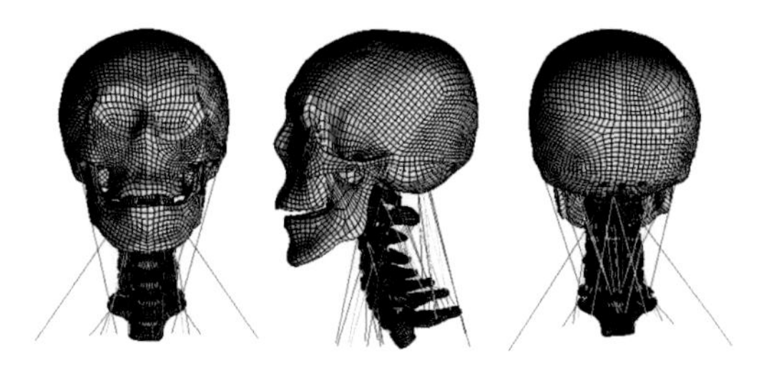

**FIGURE 18.2** Detailed finite element model of the head-neck complex.

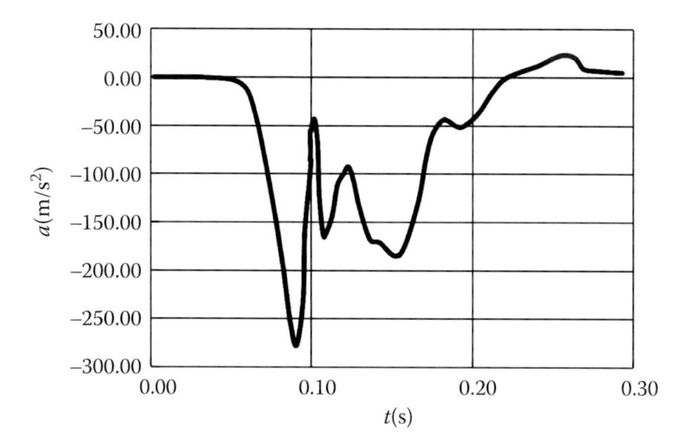


**FIGURE 18.3** $+x$ acceleration of T1 as boundary condition.

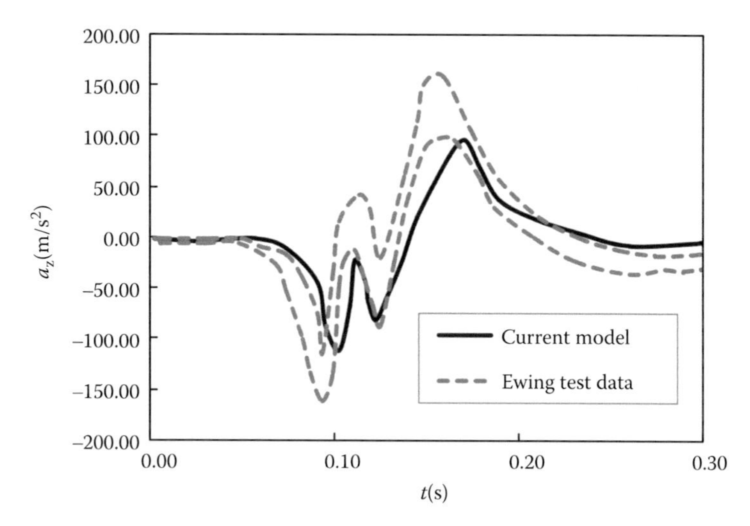


**FIGURE 18.4** $+z$ acceleration of the head center of gravity.

corridors as shown in Figures 18.4 and 18.5. From these two figures, it can be seen that the model can generate reasonable results under frontal impact loads, but not well enough, which may be caused by using a passive muscle model and too many simplifications of the developed model in this validation. Further improvement and detailed validation need to be carried out in future work.

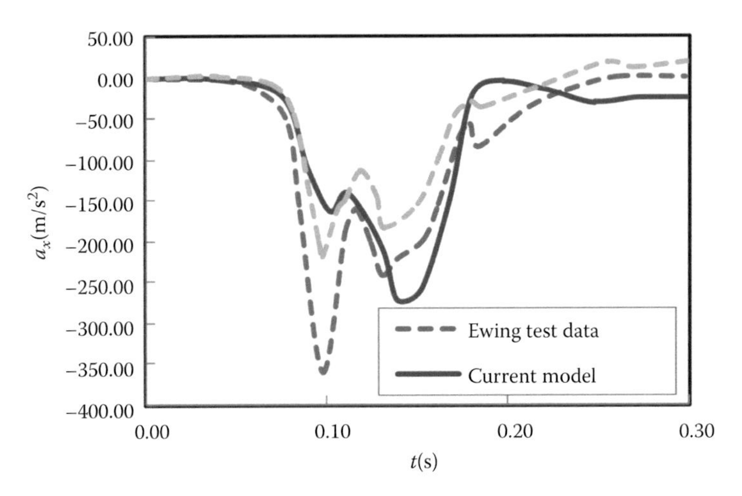


**FIGURE 18.5** +$x$ acceleration of the head center of gravity.

### 18.2.3 Discussion and Applications

An understanding of the mechanics of the cervical spine is essential for training and protection of aircraft pilots wearing equipment. Although physical models and rigid body models have provided significant insight into the response of the cervical spine to external load, advanced finite element models can address many limitations associated with these methods, including prediction of localized stress and strain within the tissues. The development of the head-neck complex finite element model is a complicated task and full of challenges because of geometrical characteristics, different types of tissues, and nonlinear tissue materials. Accurate geometry and material properties, representative loading conditions, and proper verification and validation using experimental studies must be included in biofidelic human model development. With the development of visualization techniques such as CT scans and magnetic resonance imaging, accurate geometry that is anatomically correct and numerically representative can be easily obtained. The tissue materials are relatively difficult to characterize since tissues are typically anisotropic with nonlinear, viscoelastic behavior. Experimental mechanical data are usually used to describe such behavior. However, the activation of muscle controlled by the nerve system affects the dynamic response of the head-neck complex, which is hard to quantify because the activation varies with persons and their psychic status.

When the detailed finite element model of the head-neck complex is applied to actual injury analysis, simplifications are usually made to save computation time. For example, the head and vertebrae are usually simplified as a rigid body when soft tissue injuries are the focus of the analysis and attention is not paid to fractures (Panzer, Fice, and Cronin, 2011). The efficiency of calculating the head-neck complex finite element model is mainly influenced by the nonlinearity of the material properties of soft tissue and the nonlinearity of contact behavior. Explicit finite element dynamic algorithms are usually used for cervical injury analysis under impact loadings, such as in auto crashes and ejection of pilots. These types of loadings have a large peak value and short duration (no more than 1 second). There is another type of loading in aviation, usually called sustained loading. This types of loading occurs when the airplane undertakes maneuvers, most of which continue for more than 10 seconds. As mentioned above, soft tissue injuries, such as ligament tearing and muscle pain, often happen during such maneuvers. Applying explicit finite element dynamic algorithms to such analysis has a large effect on the efficiency of the calculations because the time step is decided by the size of the minimum element.

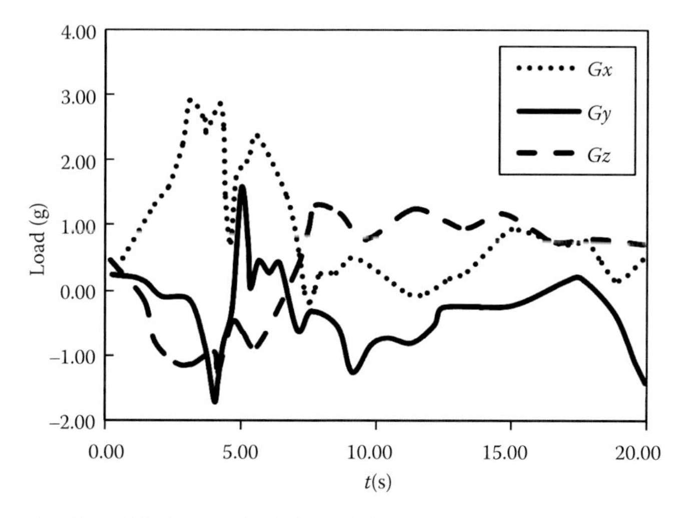


**FIGURE 18.6** Acceleration of the human body in a Herbst maneuver.

The model introduced above was developed for analyzing pilot cervical spine injuries under multiaxial sustained loading during flight maneuvers. Figure 18.6 shows the flight simulation result of pilot load under Herbst maneuver, the period of which is much longer (20 s) than the validation case (0.3 s). Thus, calculating efficiency was an important factor for using this model under sustained loading, and then a high-quality mesh with fewer elements was used. Moreover, the calculation algorithm should be optimized for reducing the computation time in future work. This model can also be used in other applications, such as in injury analysis of a pilot's head and neck during ejection and landing.

## 18.3 FINITE ELEMENT MODEL OF THE THORACOLUMBAR-PELVIS COMPLEX

### 18.3.1 Development of the Model

To establish the finite element model, the geometry of the vertebrae and pelvis was obtained from reconstruction of CT scans of a healthy male subject of height 174 cm and weight 75 kg. The vertebrae were made up of post elements and the vertebral body consisted of a bony endplate, cortical wall, and cancellous bone. The thickness of the endplate and cortex of the bone were assumed to be 0.35 mm (Silva et al., 1994). All these components were meshed in the finite element pre-processor (Hypermesh 11.0; Altair Engineering Corp, Michigan, USA). The endplate and cortical bone could be generated easily using element offset from a layer of shell elements. The post elements were divided into blocks and a source surface was chosen to first mesh into shell elements, then the solid map tool was used to drag the shell elements from block to block for building solid elements. All the components of the vertebrae were meshed as eight-node brick elements. A detailed model of the vertebrae (L2) is shown in Figure 18.7.

As in many previous investigations of the spine using finite element methods, the intervertebral disc comprised the nucleus pulposus and fiber-reinforced annulus ground substance with a proportion according to histological findings (44% nucleus, 56% annulus) (Schmidt et al., 2007). The nucleus pulposus and ground substance used solid elements, as with the vertebrae, and the fibers were modeled as three-dimensional cable elements that sustained only tension. Eight crisscross fibrous layers were defined in the radial direction and were oriented at an average angle of ±30° to the endplates. The disc structure is shown in Figure 18.7.

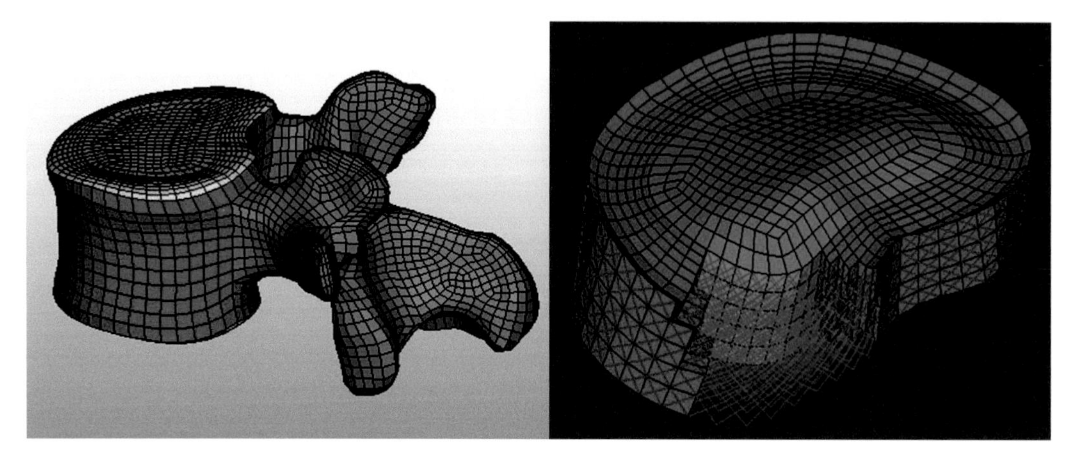

**FIGURE 18.7** **(See color insert.)** Model of L2 vertebra (left) and disc of L2-L3 (right).

Seven ligaments, the anterior (ALL) and posterior longitudinal ligaments (PLL) and the intertransverse (ITL), flavum (FL), supraspinous (SSL), interspinous (ISL), and capsular ligaments (FC), were represented by three-dimensional tension-only cable elements. Their attachment points to the vertebrae were in accordance with anatomical observations proposed by Agur (1999) and the cross-sectional area of each ligament was obtained from the literature (Goel et al., 1995).

In most cases, the pilot is in the seat when subjected to the acceleration environment. Therefore, models of the pelvis and seat should be developed. By utilizing Hypermesh, the pelvis and seat could be meshed using four-node tetrahedral solid and four-node shell elements, respectively. The backward declination angle of the seat was assumed as 17° and the angle between the seat pan and horizontal plane was 7°. The whole model, including the thoracolumbar spine, pelvis, and seat, is shown in Figure 18.8.

### 18.3.2 Material Properties

The vertebrae and pelvis were all taken as isotropic homogeneous elastic materials. An isotropic, incompressible, hyperelastic Mooney–Rivlin material model was used to simulate the fluid-like behavior of the nucleus pulposus and the hyperelastic properties of the annulus ground substance. The stress-strain behavior of the annulus collagen fibers were described by a nonlinear function (Shirazi-Adl, Ahmed, and Shrivastava, 1986). Since external lamellae are stiffer than internal lamellae, the fibers in different annulus layers were weighted (outermost layers 1–2, 1.0; layers 3–4, 0.9; layers 5–6, 0.75; innermost layers 7–8, 0.65) (Brickley-Parsons and Glimcher, 1984). The force-deflection curve of the ligaments was derived from related research by Schmidt, Heuer, and Drumm (2007). The seat was assumed to be a rigid body since its deformation was negligible compared to the spine. The material properties of the components are tabulated in Table 18.1.

### 18.3.3 Loading and Boundary Conditions

The facet joint had a gap less than 0.5 mm and could transmit only compressive forces. Simulation of quasi-static conditions for the purpose of validating the current model was performed in a commercial finite element package (ABAQUS 6.11; Simula Corporation, Pennsylvania, USA) and the dynamic analysis was undertaken using explicit dynamic software (LS-DYNA 971 R5; Livermore Software Technology Corporation, California, USA). The interaction between facet joints was assumed as frictionless surface-to-surface contact. During dynamic analysis, the hourglass coefficient of 0.01 was used to avoid the emergence of an hourglass scenario.

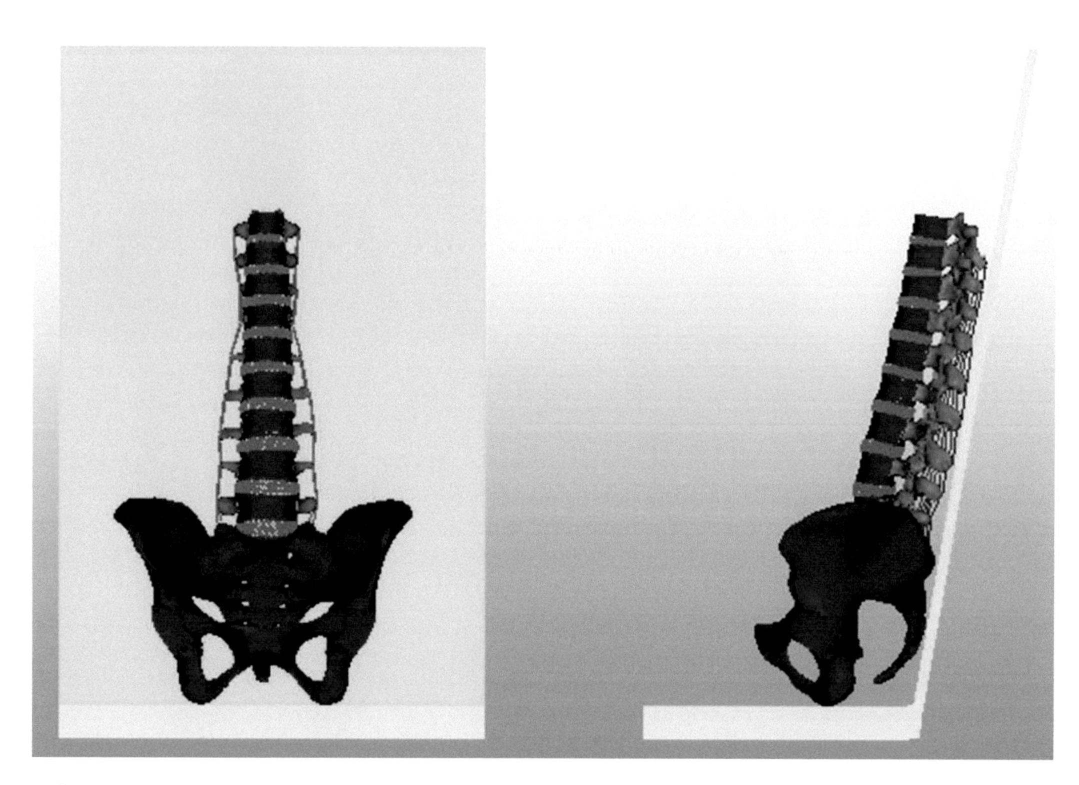

**FIGURE 18.8** **(See color insert.)** Model of the thoracolumbar-pelvis complex along with the seat.

## TABLE 18.1
## Material Properties Used in the Finite Element Model of the Thoracolumbar-Pelvis Complex

| Component | Young's Modulus (MPa) | Poisson's Ratio | Density (kg/mm²) | Cross-Section Area (mm²) |
|---|---|---|---|---|
| Cortical bone | 14,000 | 0.30 | 1.83E-06 | |
| Cancellous bone | 100 | 0.2 | 0.17E-06 | |
| Posterior elements | 3,500 | 0.25 | 1.83E-06 | |
| Endplate | 10,000 | 0.25 | 1.06E-06 | |
| Pelvis | 5,000 | 0.2 | 1.83E-06 | |
| Annulus | Mooney–Rivlin c = 0.18, c = 0.045 | | 1.0E-06 | |
| Nucleus pulpous | Mooney–Rivlin c1 = 0.12, c2 = 0.03 | | 1.2E-06 | |
| Collagenous fiber | Calibrated stress-strain curves | | | |
| Ligament | | | | |
| ALL | Calibrated deflection-force curves | | 1.0E-06 | 63.7 |
| PLL | Calibrated force-deflection curves | | 1.0E-06 | 20 |
| FL | | | 1.0E-06 | 40 |
| ITL | | | 1.0E-06 | 3.6 |
| ISL | | | 1.0E-06 | 40 |
| SSL | | | 1.0E-06 | 30 |
| CL | | | 1.0E-06 | 60 |

*Note:* ALL, anterior longitudinal ligament; PLL, posterior longitudinal ligament; FL, ligamentum flavum; ITL, intertransverse ligament; ISL, interspinous ligament; SSL, supraspinous ligament; CL, capsular ligament.

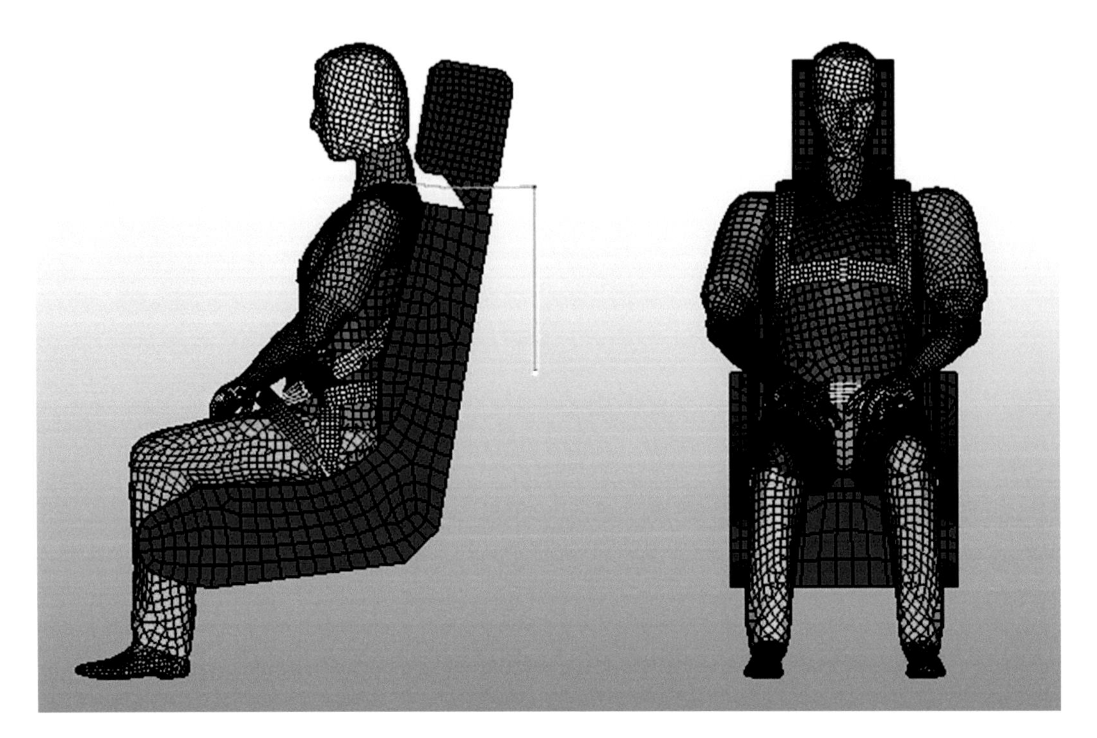

**FIGURE 18.9** **(See color insert.)** Side and front views of a multibody dynamics model, including dummy, restraint systems, and seat.

The effect of the restraint harness on the pilot can be derived from multibody dynamics. A rigid body dynamics model of the pilot was developed, having the same weight and height as the subject, providing the source data to the previous finite element model. This model consists of 16 interconnected rigid segments: head, neck, upper trunk, buttocks, thighs, legs, foots, upper arms, forearms, and hands. These segments were connected through joints and the properties of each joint, such as range of motion, stiffness, friction, and dampening, were included. Given the compression effect of the intervertebral disc under axial loading, the joint connecting the upper and lower torso was replaced by a spring. The harness restraint was modeled directly in Hypermesh using a belt routing tool, which makes it convenient to design a safety belt with specific thickness and width along the surface of the dummy model. The geometric information and material properties of the harness restraint were derived from military specifications or experimental measurement. The model of the rigid dummy was adjusted to the required posture according to the purpose of the specific study and the constraint and boundary conditions were all set correspondingly. The model of a restrained dummy with pre-ejection posture is shown in Figure 18.9.

The hip joint and thorax can be chosen as sites to evaluate the dynamic response of a restrained dummy during various impact loadings. The response of these regions can be recorded as the input data to the finite element model. For the finite element model of the thoracolumbar-pelvis complex, the imposed loading has the same properties as that applied on the multibody dynamics model.

### 18.3.4 Model Validation

Before the finite element model can be applied to the study of the response of the spine under various impact loadings, the validity of the model should be verified. The lumbar segment was chosen for validation due to the lack of experimental data relating to the thoracolumbar spine. The calibration process of the current model was conducted prior to validation. A calibration factor was introduced to weight the stress-strain curve of the collagen fibers. The calibration factor was varied to obtain the optimal value that could permit the range of motion predicted by the model to match well with the in vitro experimental results under a bending moment in different directions (Schmidt et al., 2006). The calibration factors of different ligaments were acquired in a similar way (Schmidt et al., 2007). The calibration results of the collagen fibers are shown in Figures 18.10 and 18.11.

Validation was then undertaken by comparing the data predicted by the current model with results from other studies in the literature. The range of motion of each segment under different moments in three principal planes and the disc compression under a follower load of 1200 N were calculated for comparison with the experimental and simulated data reported by Renner, Natarajan, and Patwardhan (2007). The magnitudes of the bending moments were 6 NM for extension, 8 NM for flexion, 4 NM for torsion, and 6 NM for lateral bending. The loading was exposed on the superior surface of L1 and the boundary condition was simulated by fixing S1 in the same way as the in vitro experiment. The range of motion of each vertebra was determined by taking the sum of two motions in each plane, and the degree of disc compression was evaluated by directly measuring the change in thickness. The validation results are shown in Figure 18.12.

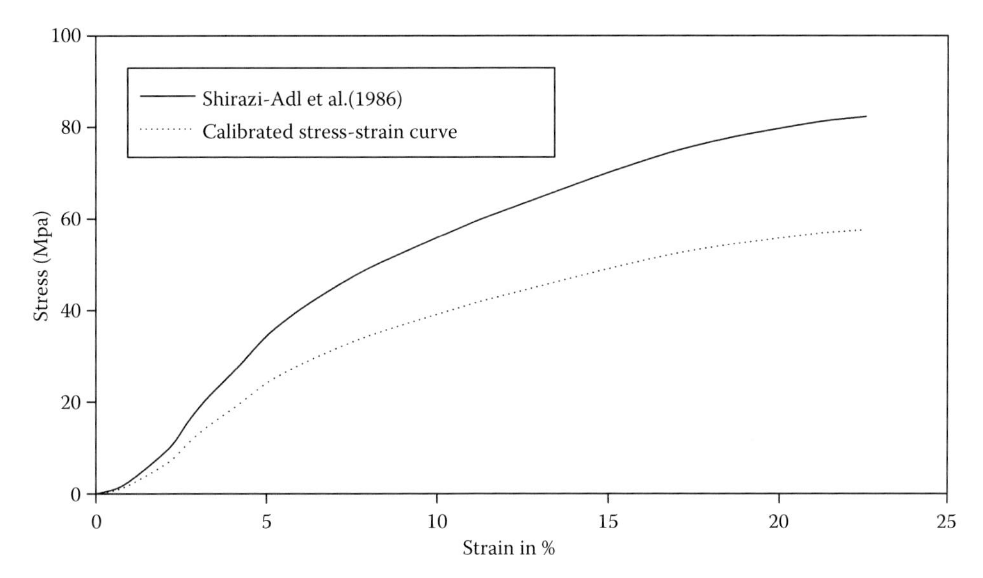


**FIGURE 18.10** Calibrated stress-strain curve of collagenous fiber.

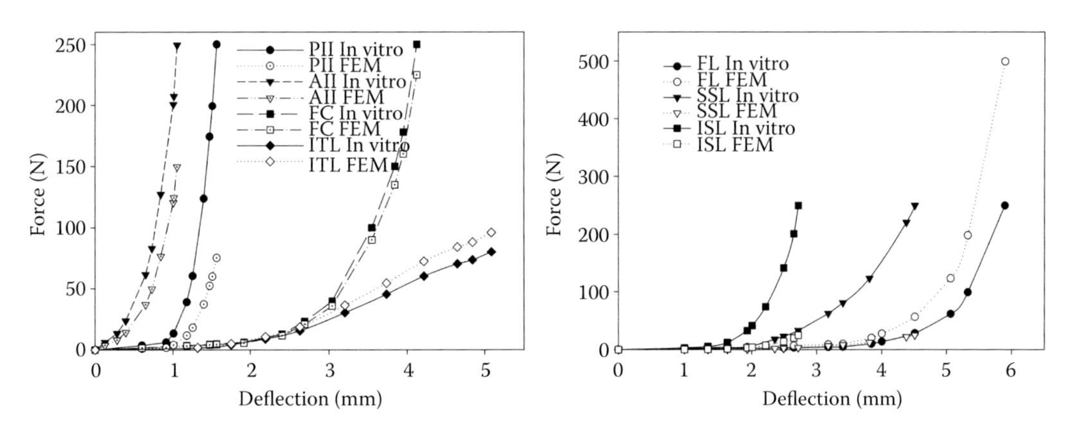


**FIGURE 18.11** Calibrated defection-force curves of ligaments.

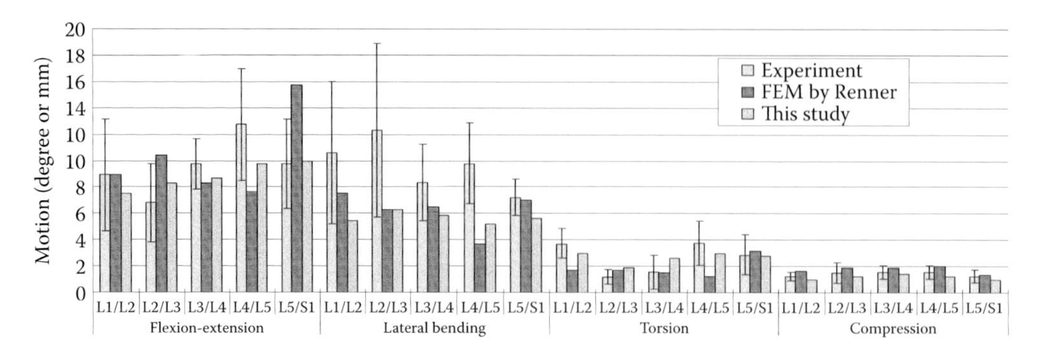


**FIGURE 18.12** Comparison of results predicted by the finite element model and experimental data.

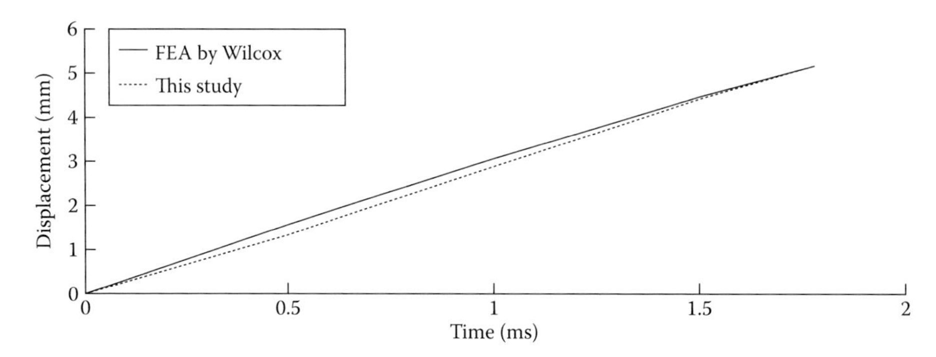


**FIGURE 18.13** Dynamic validation results of the model.

The T12-L1 segment was chosen to conduct dynamic validation. The dynamic response of the superior endplate of the upper vertebrae predicted by this model was compared with experiment results reported by Wilcox and coworkers (2004). Since the stiffness of the intervertebral disc increases dramatically at higher loading rates, the material parameters of the annulus ground substance and nucleus pulposus were all increased by 10 times according to the investigation by Race, Broom, and Robertson (2000) before dynamic analysis was performed.

The results predicted by the model were in close agreement with the published experimental data (Figure 18.13). Accordingly, this model can be further used for study of the dynamic response of the thoracolumbar spine under various impacts.

During impact loading, distribution of the mass of the trunk to each vertebra was also taken into consideration. The density of each vertebra was amplified to fulfill the different weight of it account for the mass of upper body (Pearsall, 1996), and the vertebral body and post element share 80% and 20%, respectively. A rigid beam was constructed by two nodal points: one was connected with the upper endplate of the T8 vertebra, and another point representing the weight of the rest parts of the upper body was in the sagittal plane, 10 mm anterior to the disc of L2-L3, 300 mm superior to the upper endplate of L1.

### 18.3.5 Application of the Finite Element Model

The validated finite element model can be used to study the dynamic response of the thoracolumbar spine during various impacts in a typical aviation environment. The ejection scenario is discussed in the following paragraphs as an example.

An accelerative load with peak value of 15 G, duration of 0.2 s, and onset rate of 150 G/s was exerted on the seat pan at an angle of 5° to the back of the seat. In the whole process of simulation, all the components in the finite element model were constrained to move only in the sagittal plane of the pilot and the pelvis and T9 were allowed to rotate only around the reference axis of the seat.

The results of simulation show that the harness plays a vital role in limiting the movement of the pilot. Figure 18.14 shows a comparison of spinal deformations with and without the use of a harness. Obviously, without protection from a harness, the upper body would move forward quickly and result in excessive flexion over a very short time.

With the multibody dynamic model, the effect of restraint systems on the aircraft occupant can be analyzed quantitatively. Figure 18.15 shows the dynamic response of the hip and thorax (near T9)

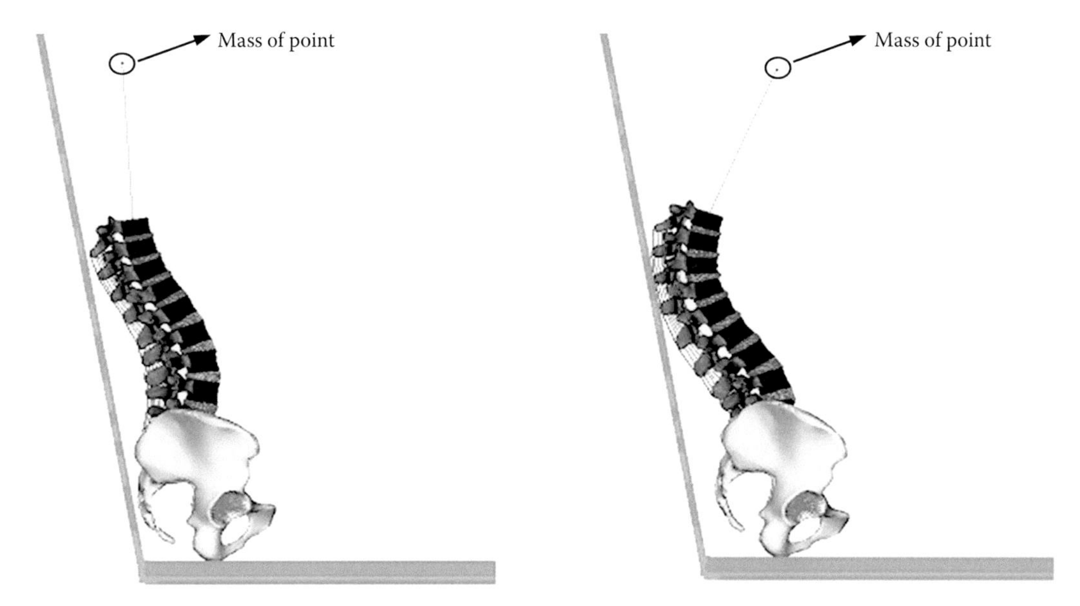


**FIGURE 18.14** **(See color insert.)** Response of the thoracolumbar spine with (left) and without (right) the harness.

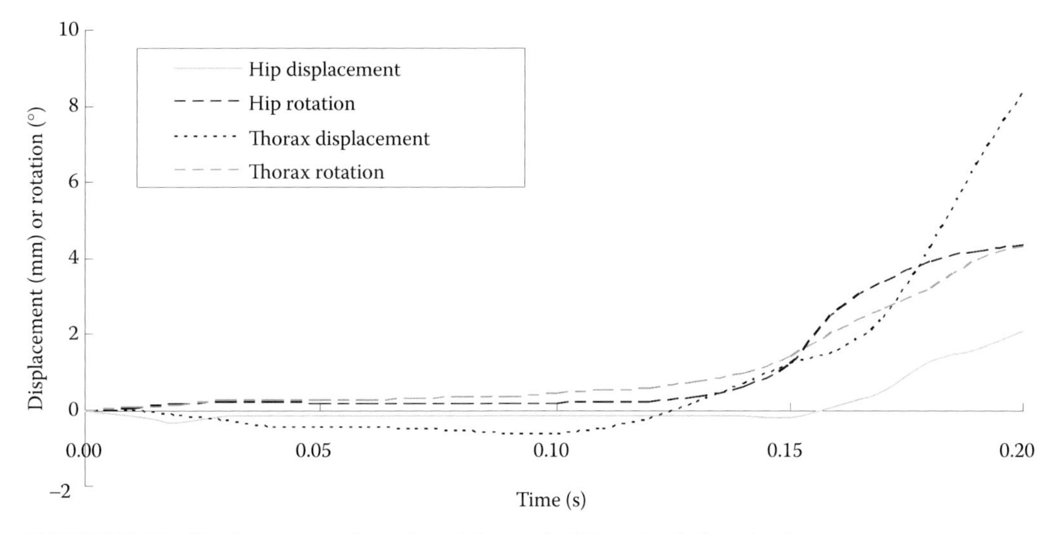


**FIGURE 18.15** Displacement and rotation of the trunk of the pilot during ejection.

of a pilot in the condition of ejection occurring. The displacement of the thorax in the direction vertical to the seat back and the displacement of the hip joint along with the pan of the seat were evaluated. The hip has a similar rotational response to the thorax, while the thorax region translated far more than the hip. The possible reason is that the compression of the spring caused slack in the belt, which produced more space for the trunk of the pilot to move.

With the boundary condition from restraint systems, more accurate mechanical information on the spine can be acquired. The von Mises stress distribution on the thoracolumbar spine at 0.12 s (Figure 18.16) indicates that high stresses were mainly concentrated on the anterior parts of T12, L1, and the region close to the pedicle bases of the lower lumbar spine. A similar stress distribution was reported by Qiu et al. (2006) and can be used to explain why the thoracolumbar spine, especially the region of T12 and L1, is the most common site where fractures appear. In this model, the maximum stress on the cortical bone reached 103 Mpa, which is close to the yield stress of cortical bone (110 Mpa).

With the present model, a parametric study can be undertaken to analyze how factors such as the features of the accelerative load, body posture, and configuration of the restraint harness affect the dynamic response of the thoracolumbar spine to ejection and can be used to inform the design of protective devices for ejection-induced spinal injury. Figure 18.17 is an example of the effect of the onset rate of impact loading on the peak values of stress on the thoracolumbar spine. Both the slope of the stress historical curve and the peak value throughout the duration of impact increased with the increase of the onset rate of loading.

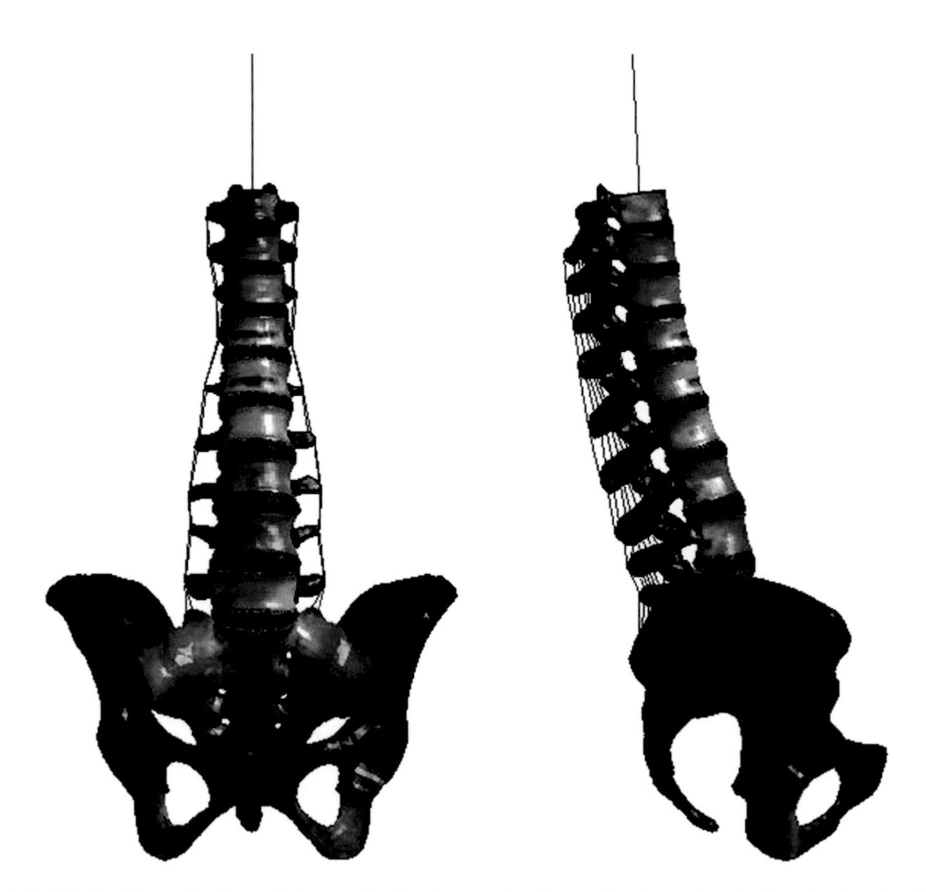

**FIGURE 18.16 (See color insert.)** Stress distribution on the thoracolumbar spine during ejection ($t = 0.12$ s).

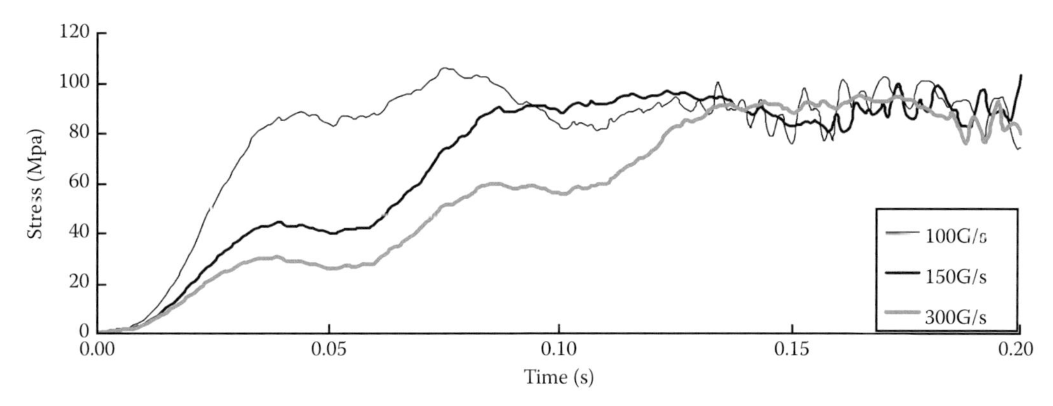


**FIGURE 18.17** Effect of various onset rates of impact loading on the maximum stress on the thoracolumbar spine.

The effect of the angle (θ) between the direction of ejection and the back of the seat on the dynamic response of the thoracolumbar spine can also be investigated. When the ejection line is parallel to the seat back, θ is 0°. If the direction of ejection causes the spine to flex, θ will be positive, whereas θ will be negative if the direction of ejection leads the spine to a state of extension. From the simulation results, it was found that a negative value of θ resulted in a greater degree of lumbar lordosis and generated high pressure in the nucleus pulpous in each segment of the spine.

Two popular approaches to investigate the effect of a dynamic environment on pilots are multibody techniques and finite element techniques. Both approaches offer specific advantages and disadvantages. Multibody techniques are attractive for simulating dummy segment motions and complex joint behavior with low computational expense. Finite element techniques allow the calculation of local deformations in dummy segments and more detailed and more accurate mechanical information can be obtained. However, detailed finite element models require excessive computing power. Moreover, in constructing these models, the choice of materials, mesh of the medium, and technique for the numerical calculations tends to be rather complex. Hence, constructing models and running analyses using finite element techniques is a time-consuming process.

With the advent of super agile aircraft having significant maneuverability, the overall G environment experienced by the pilot during flight and ejection is becoming ever more severe. A comprehensive study on the dynamic response of pilots in this complicated environment may be essential and beneficial for understanding how to protect the body from injury. With rapid advances in recent years, numerical simulation has become widely used for analyzing the dynamic response of the human body in impact situations. Information about displacement, velocity, and acceleration of the body can be obtained, and the joint reaction force and restraint tension can be predicted using multibody dynamics. Multibody techniques are more suitable for simulating dynamic situations in which loading is applied over a prolonged period (such as several seconds) since this approach requires less computing power. However, by using multibody dynamics methods, it is difficult to know where and when the injury will take place under the impact conditions because detailed internal mechanical information on the body cannot be obtained. Furthermore, finite element models can be used to investigate injury to soft tissue, which is usually accompanied by bone fracture during impact.

Since both multibody dynamics and finite element methods have different advantages and disadvantages when dealing with dynamic problems, a combination of these two methods will be helpful for studying the spine in aviation applications. As used in this study, the integration of the influence of the belt on the pilot evaluated by a multibody dynamics model to a finite element model can permit more accurate analyses of the dynamic response of the spine following impact loading.

## ACKNOWLEDGMENTS

This work was supported by the Beijing Natural Science Foundation (7133245), Young Scholars for the Doctoral Program of the Ministry of Education of China (20121102120039), the National Natural Science Foundation of China (Nos. 10925208, 11120101001), and the 111 Project (No. B13003).

## REFERENCES

Agur, A. M. R. 1999. *Grant's Atlas of Anatomy*, 11th ed. Baltimore, MD: Williams and Wilkins.

Belytschko, T., L. Schwer, and E. Privitzer. 1987. Theory and application of a three-dimensional model of the human spine. *Aviation, Space, and Environmental Medicine* 49(1 Pt. 2):158.

Brickley-Parsons, D., and M. J. Glimcher. 1984. Is the chemistry of collagen in intervertebral discs an expression of Wolff's law? *Spine* 9:148–163.

Brown, T., R.J. Hansen, and A.J. Yorra. 1957. Some mechanical tests on the lumbosacral spine with particular reference to the intervertebral discs: A preliminary report. *The Journal of Bone & Joint Surgery* 39:1135–64.

Buhrman, J.R. 1991. Vertical impact tests of humans and anthropomorphic manikins. AL-TR-1991-0129. Wright-Patterson AFB, OH: Armstrong Laboratory.

De Loose, V., M. Van den Oord, F. Burnotte, D. Van Tiggelen, V. Stevens, B. Cagnie, L. Danneels, and E. Witvrouw. 2009. Functional assessment of the cervical spine in F-16 pilots with and without neck pain. *Aviation, Space and Environmental Medicine* 80(5):477–481.

Erin, C., and P. John. 2006. The characterization of spinal compression in various-sized human and mannequin subjects during +Gz impact. *Air Force Research Lab Wright-Patterson AFB OH: Human Effectiveness Directorate* No. AFRL-WS-05-2259.

Ewing, C. L., and D. J. Thomas. 1972. Human head and neck response to impact acceleration. Monograph 21 USAARL 73-1. Naval Aerospace and Regional Medical Centre.

Ewing, C. L., D. J. Thomas, L. Lustick, W. H. L. Muzzy, G. Willems, and P. L. Majewski. 1976. The effect of duration, rate of onset and peak sled acceleration on the dynamic response of the human head and neck. *Proceedings of the 20th Stapp Car Crash Conference* 20:3–41.

Fice, J. B., and D. S. Cronin. 2012. Investigation of whiplash injuries in the upper cervical spine using a detailed neck model. *Journal of Biomechanics* 45(6):1098–1102.

Frazier, J. W., J. W. Mcdaniel, V. D. Skowpronski, and M. A. Nilss. 1988. Body displacement measured during sustained +GZ, -GZ, and ±GY acceleration using a stereoscopic photographic system. *Wright-Patterson AFB, OH: Aerospace Medical Research Laboratory* AAMRL-TR-88-015l.

Green, N. D. 2003. Acute soft tissue neck injury from unexpected acceleration. *Aviation, Space and Environmental Medicine* 74(10):1085–1090.

Goel, V. K., B. T. Monroe, L. G. Gilbertson, and P. Brinckmann. 1995. Interlaminar shear stresses and laminae separation in a disc: Finite element analysis of the L3–L4 motion segment subjected to axial compressive loads. *Spine* 20:689–698.

Hamalainen, O. 1993. Flight helmet weight, +Gz forces, and neck muscle strain. *Aviation, Space, and Environmental Medicine* 64:55–57.

Hearon, B. F., and J. W. Brinkley. 1986. Comparison of human impact response in restraint systems with and without a negative G strap. *Aviation, Space, and Environmental Medicine* 57:301–312.

Holdsworth, F. 1970. Fractures, dislocations, and fracture-dislocations of the spine. *Journal of Bone & Joint Surgery* 52A:1534–1551.

Karlsson, L., Lundin, O., Ekstrom, L., Hansson, T., and Sward, L. 1998. Injuries in adolescent spine exposed to compressive loads: An experimental cadaveric study. *Journal of Spinal Disorders* 11:501–507.

Lecompte, J., O. Maisetti, A. Guillaume, W. Skalli, and P. Portero. 2008. Neck strength and EMG activity in fighter pilots with episodic neck pain. *Aviation, Space and Environmental Medicine* 79(10):947–952.

Lee, C. K., Y. E. Kim, C. S. Lee, Y. M. Hong, J. M. Jung, and V. K. Goel. 2000. Impact response of the intervertebral disc in a finite-element model. *Spine* 25:2431–2439.

Leupp, D. G. 1983. ACES II negative Gz restraint investigation. Wright-Patterson AFB, OH: Aerospace Medical Research Laboratory, AFAMRL-TR-83-049.

Lewis, M. E. 2006. Survivability and injuries from use of rocket-assisted ejection seats: Analysis of 232 cases. *Aviation, Space, and Environmental Medicine* 77(9):936–943.

Mosekilde, L., and C. C. Danielsen. 1987. Biomechanical competence of vertebral trabecular bone in relation to ash density and age in normal individuals. *Bone* 1987:79–85.

Newman, D. G. 2006. Multi-sensor integration systems for the tactical combat pilot. *Aviation, Space and Environmental Medicine* 77:85–88.
Newman, D. G., and D. Ostler. 2011. The geometry of high angle of attack maneuvers and the implications for $G_y$–induced neck injuries. *Aviation, Space and Environmental Medicine* 82:819–824.
Obergefell, L. A., and I. Kaleps. 1988. Simulation of body motion during aircraft ejection. *Mathematical and Computer Modelling* 11: 436–439.
Panjabi, M. M., N. C. Chen, E. K. Shin, and J. L. Wang. 2001. The cortical shell architecture of human cervical vertebral bodies. *Spine* 26(24):2478–2484.
Panjabi, M. M., T. R.Oxland, and E. H. Parks. 1991. Quantitative anatomy of cervical spine ligaments part 2: middle and lower cervical spine. *Journal of Spinal Disorders* 4: 277–85.
Panzer, M. B., and D. S. Cronin. 2009. C4-C5 segment finite element model development, validation, and load-sharing investigation. *Journal of Biomechanics* 42(4):480–490.
Panzer, M. B., J. B. Fice, and D. S. Cronin. 2011. Cervical spine response in frontal crash. *Medical Engineering and Physics* 33(9):1147–1159.
Payne, P. R. 1961. The dynamics of human restraint systems impact acceleration stress. *National Academy of Sciences, National Research Council*, Publication No. 977, Washington, D.C.
Pooni, J. S., D. W. L. Hukins, P. F. Harris, R. C. Hilton, and K. E. Davies. 1986. Comparison of the structure of human intervertebral discs in the cervical, thoracic and lumbar region of the spine. *Surgical and Radiologic Anatomy* 8:175–182.
Pearsall, D. J., J. G. Reid, and L. A. Livingston. 1996. Segmental inertial parameters of the human trunk as determined from computed tomography. *Annals of Biomedical Engineering* 24:198–210.
Qiu, T. X., K. W. Tan, V. S. Lee, and E. C. Teo. 2006. Investigation of thoracolumbar T12–L1 burst fracture mechanism using finite element method. *Medical Engineering Physics* 28:656–664.
Race, A., N.D. Broom, and P. Robertson. 2000. Effect of loading rate and hydration on the mechanical properties of the disc. *Spine* 25(6): 662–669.
Renner, S. M., R. N. Natarajan, and A. G. Patwardhan. 2007. Novel model to analyze the effect of a large compressive follower pre-load on range of motions in a lumbar spine. *Journal of Biomechanics* 40(6):1326–1332.
Roaf, R. 1960. A study of the mechanics of spinal injuries. *Journal of Bone & Joint Surgery* 42B:810–823.
Robbins, D. H. 1983. Anthropometric specifications for mid-sized male dummy. *University of Michigan Transportation Research Institute UMTRI-83-53-2*.
Ruff, S. 1950. Brief acceleration: Less than one second. *German Aviation Medicine, World War II* 1:584–597.
Teo, E. C., Q. H. Zhang, K. W. Tan, and V. S. Lee. 2004. Effect on muscles activation on head-neck complex under simulated ejection. *Journal of Musculoskeletal Research* 8(4):155–165.
Sasirajan, J., T. Narinder, and Y. S. Dahiya. 2007. A retrospective analysis of injuries among aircrew involved in helicopter accidents. *Journal of Indian Aerospace Medicine* 51:48–53.
Schall, D. G. 1989. Non-ejection cervical spine injuries due to +Gz in high performance aircraft. *Aviation, Space and Environmental Medicine* 60(5):445–456.
Schmidt, H., F. Heuer, J. Drumm, Z. Klezl, L. Claes, and H. J. Wilke. 2007. Application of a calibration method provides more realistic results for a finite element model of a lumbar spinal segment. *Clinical Biomechanics* 22(4):377–384.
Schmidt, H., F. Heuer, U. Simon, A. Kettler, A. Rohlmann, L. Claes, and H.J. Wilke. 2006. Application of a new calibration method for a three-dimensional finite element model of a human lumbar annulus fibrosus. *Clinical Biomechanics* 21(4): 337–344.
Shirazi-Adl, A., A. M. Ahmed, and S. C. Shrivastava. 1986. Mechanical response of a lumbar motion segment in axial torque alone and combined with compression. *Spine* 11(9):914–927.
Silva, M. J., C. Wang, T. M. Keaveny, and W. C. Hayes. 1994. Direct and computed tomography thickness measurements of the human, lumbar vertebral shell and endplate. *Bone* 15:409–414.
Stech, E. L., and P. R Payne. 1963. The Effect of Age on Vertebral Breaking Strength, Spinal Freuency and Tolerance to Acceleration in Human Beings. Frost Engineering Development Corp. Report 122-101.
Sturgeon, W. R. 1987. Canadian forces aircrew ejection, descent, and landing injuries 1 January 1975-31 December. AD-A206116
Wayne, Z. K., and V. K. Goel. 2003. Ability of the finite element models to predict response of the human spine to sinusoidal vertical vibration. *Spine* 28:1961–1967.
Werner, U. 1999. Ejection-associated injuries within the German Air Force from 1981–1997. *Aviation, Space, and Environmental Medicine* 70(12):1230–1234.

Wilcox R. K., D. J. Allen, R. M. Hall, D. Limb, D. C. Barton, and R. A. Dickson. 2004. A dynamic investigation of the burst fracture process using a combined experimental and finite element approach. *European Spine Journal* 13:481–488.

Willen, J., S. Lindahl, L. Irstam, B. Aldman, and A. Nordwall. 1984. The thoracolumbar crush fracture. An experimental study on instant axial dynamic loading: the resulting fracture type and its stability. *Spine* 9:624–631.

Yoganandan, N., S. Kumaresan, and F. A. Pintar. 2000. Geometric and mechanical properties of human cervical spine ligaments. *Journal of Biomechanical Engineering* 122:623–629.

Yoganandan, N., S. Kumaresan, and F. A. Pintar. 2001. Biomechanics of the cervical spine. Part 2. Cervical spine soft tissue response and biomechanical modeling. *Clinical Biomechanics* 16(1):1–27.

Zhang, Q. H., J. Z. Li., and H. N. S. Tan. 2008. A finite element study of the response of thoracolumbar junction to accidental mine blast scenario. *7th Asian-Pacific Conference on Medical and Biological Engineering* 2008:129–132.

# *Section VI*

## *Head and Hand*

# 19 Head Model for Protection

*Lizhen Wang, Peng Xu, Xiaoyu Liu, Zhongjun Mo, Ming Zhang, and Yubo Fan*

## CONTENTS



## SUMMARY

Head injury is a leading cause of morbidity and death in both industrialized and developing countries. It is estimated that brain injuries account for 15% of the burden of fatalities and disabilities, and represent the leading cause of death in young adults. Brain injury may be caused by an impact or a sudden change in the linear and/or angular velocity of the head. However, the woodpecker does not experience any head injury at the high speed of 6–7 m/s with a deceleration of 1000 g when it drums a tree trunk. It is still not known how woodpeckers protect their brain from impact injury. In order to investigate this, finite element (FE) models of human and woodpecker heads were established to study the dynamic intracranial responses based on micro-CT and CT images, respectively. The mechanical properties in the woodpecker's head were investigated using a mechanical testing system. It was shown that the macro/micro morphology of the cranial bone and beak can be recognized as a major contributor to nonimpact injuries. This biomechanical analysis makes it possible to visualize events during woodpecker pecking and may inspire new approaches to prevention and treatment of human head injury.

## 19.1 INTRODUCTION

Head injury remains an acute problem in road transport safety, especially for pedestrians and motorcyclists around the world (McIntosh, 2005; Martin et al., 2008). Considering the competitive team sports at the 2004 Olympic Games, it was shown that 24% of all the injuries reported were head injuries (Junge et al., 2006). According to the European Brain Injury Consortium (EBIC) survey, 51% of head injuries were from car accidents or sports related to falling (Finfer and Cohen, 2001; Yang et al., 2006). Holbourn (1943) was the first to cite angular acceleration as an important mechanism in head injury. In 2003, King et al. presented a review of brain injury mechanisms with a distinction between the effects of linear and angular acceleration on the mechanisms of concussion. An intriguing example from nature is the case of woodpeckers, who drum tree trunks at a speed of 6–7 m/s with a linear acceleration of approximately 1000 g and an angular acceleration of 297 krad/s with no head injuries (May et al., 1979; Wang, Cheung et al., 2011). Indeed, the woodpecker drums about 10–20 bouts continuously, and every bout takes about 50 milliseconds. It drums about 12,000 times per day on average (Spring, 1965). Woodpeckers perform rhythmic drumming with their beaks on surfaces such as dead tree limbs to catch and feed themselves with worms, or to attract mates and announce their territorial boundaries (Spring, 1965). The kinematics of the woodpecker's head has been collected in previous studies (Wang et al., 2011).

## 19.2 DEVELOPMENT OF A FINITE ELEMENT MODEL OF HUMAN AND WOODPECKER HEADS

Due to the fact that in real-world collisions, head injury occurs from a combination of translational and rotational acceleration, neither of which needs to be extremely high for the current measurement devices, sophisticated computer models of the head can provide useful information in the investigation of human injury due to impact such as intracranial response under real world head impact conditions. It was necessary to understand the head anatomy for constructing its geometrical model.

### 19.2.1 Human and Woodpecker Head Anatomy

The human head consists of different tissue layers such as the scalp, skull bone, and dural, arachnoidal, and pial membranes, as well as cerebrospinal fluid (CSF), that cover the brain (Figure 19.1). The skull bone can be viewed as a three-layered sandwich structure with an inner and outer table of cortical bone and spongy bone sandwiched between them. The brain, with its covering membranes and CSF, is connected to the spinal cord through the foramen magnum. The inferior part of the skull base is attached to the neck by articulation through occipital condyles, ligaments, and muscles.

The woodpecker's brain is tightly packed by relatively dense yet spongy bone, especially evident at the occiput, in the contre-coup position from the beak. The woodpecker has a very narrow subdural space and relatively little cerebrospinal fluid, which might reduce fluid transmission of shock waves. In addition, it has powerful protractor quadrati and protractor pterygoidei muscles, which could form a muscular shock absorber and distributor, holding the beak in "resilient rigidity." The last but most important feature is that the woodpecker's skull is encircled by musculotendinous bands that extend posteriorly from the floor of the mouth, around, up, and over the back of head, and then forward to the right nostril—a curious sling-like structure (May et al., 1976; Wang et al., 2011).

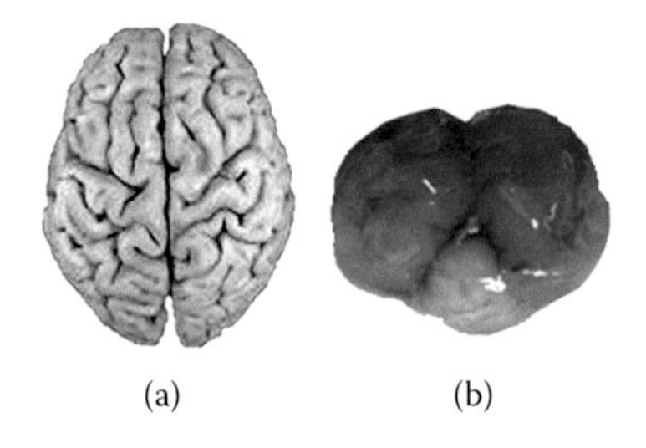


**FIGURE 19.1** Anatomy of the brain: (a) human and (b) woodpecker.

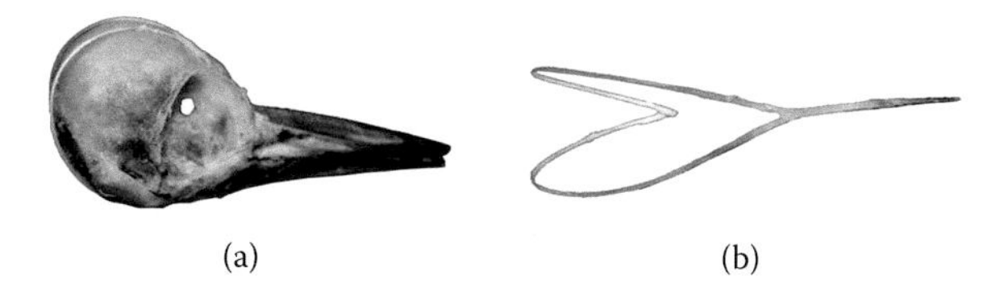

(a) (b)

**FIGURE 19.2** Great spotted woodpecker's (a) head and (b) hyoid bone.

Apart from the possible neurophysiological implications of this interesting arrangement, this muscular sling might function as an isometric shock absorber and distributor by limiting movement around the anterior-posterior axis (Figure 19.2).

### 19.2.2 Geometry of the Model

The geometry of the human head was developed based on CT scan images of a female corpse (37 years old, dead from a car accident without head injury). The images were transverse sections scanned at 1-mm intervals along the vertical axis of the body. The head model simulated all essential anatomical features of the human head, including skull, brain, maxillary, and mandibular bone (Figure 19.3a). Due to the irregularity and complexity of the skull's bony structures, the geometry of some of the bones, such as the orbital and nasal bones, were simplified and smoothed, but the structural features of the head were preserved. The architecture of the skull bone resembles a sandwich structure containing cortical and cancellous bone layers. The cortical layer of the cranial bone is generally thinner. Thus, the thin layer is modeled with shell elements with a thickness of 1 mm while the cancellous bone is modeled as solid elements of varying thicknesses at different regions. The temporo-mandibular articulation was included by ligaments.

The great spotted woodpecker (*Dendrocopos major*) was selected for its wide distribution in northern China. It was fed with yellow mealworm (*Tenebriomolitor L.*) in separate metal cages. Then, the morphology of the woodpecker's cranial bone was obtained based on image processing of microcomputed tomography (micro-CT, Skyscan1076, Skyscan, Belgium). To investigate the dynamic response of the woodpecker's head, a geometrically accurate three-dimensional FE model of the woodpecker's head, including the upper/lower beak, skull, brain, and hyoid bone, was developed based on the actual geometry and anatomic detail from micro-CT scanning (Figure 19.3b).

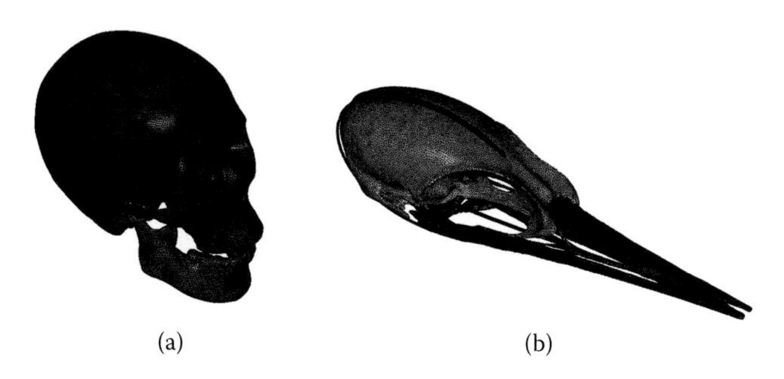

(a) (b)

**FIGURE 19.3** The finite element models of (a) human head and (b) woodpecker head.

### 19.2.3 Material Properties

It is well known that most biological materials display both elastic and viscous properties. Brain tissue is a hydrated soft tissue consisting of about 78% water. It is linked to a soft gel when defined as a general engineering material (Holbourn, 1943). Experimental characterization of brain materials from a variety of species suggests that brain tissue exhibits incompressible viscoelastic behavior (Stalnaker, 1969; Shuck and Advani, 1972). The bulk modulus of brain tissue was considered to be similar to that of water with a value of approximately 2 GPa, according to Stalnaker (1969). Traditionally, the material behavior of the brain is approximated by a viscoelastic model that is a combination of linear springs and dashpots. The behavior of the material is characterized as viscoelastic in shear with the deviatoric stress rate dependent on the shear relaxation modulus, while the hydrostatic behavior of the brain is considered elastic. The shear modulus of a viscoelastic brain could be expressed by

$$G(t) = G_\infty + (G_0 + G_\infty)e^{(-\beta t)}$$

where $G_0$ is the short-term shear modulus, $G_\infty$ is the long-term shear modulus, $\beta$ is a decay constant, and $t$ is the duration.

An elastic-plastic material model was used for the cortical and cancellous bones of the human head. Damage elements were introduced to predict bone fracture. The element elimination option was used. This option removes any element with a strain that exceeds a preset ultimate strain magnitude in every time step. Since the model was meshed very finely, it was thought that elimination of some elements would only have a minimal effect on the balance of energy transfer. As listed in Tables 19.1 and 19.2, a Young's modulus of 560 MPa was assigned for the cancellous bone (Carter and Hayes, 1977; Carter et al., 1980; Ding and Hvid, 2000; Ding et al., 2002; Ding et al., 2006). A Young's modulus of 6 GPa was assigned for the cortical bone (Carter and Hayes, 1977; Carter et al., 1987).

For the woodpecker, the material properties of the skull, beak, and hyoid were derived from the data of the mechanical test in our previous study (Wang et al., 2011). A homogenous density and linearly viscoelastic material model in combination with a large deformation theory was chosen to model the brain tissue (Stalnaker, 1969; Hosey, 1982; Ruan et al., 1993; Ruan, 1994). The behavior of this material was characterized as viscoelastic in shear with a deviatoric stress rate dependent on the shear relaxation modulus, while the compressive behavior of the brain was considered elastic, as in Table 19.2.

### 19.2.4 Model Validation and Boundary Condition

The mesh of the model was developed using the preprocessor Hypermesh Version 14.0 (Altair Engineering, Inc., Troy, Michgan) (Figure 19.3). The numerical simulation was performed with

**TABLE 19.1**
**Mechanical Properties of Human and Woodpecker Cranial Bones**

| Structures | Density (Kg/m$^{-3}$) | Young's Modulus (GPa) | Possion's Ratio |
|---|---|---|---|
| Cortical (human) | 2100 | 6 | 0.25 |
| Cancellous (human) | 1000 | 0.56 | 0.30 |
| Cranial bone (woodpecker) | — | 0.31 | 0.4 |
| Beak (woodpecker) | — | 1 | 0.3 |
| Hyoid bone (woodpecker) | — | 1.13 | 0.2 |

**TABLE 19.2**
**Mechanical Properties of Human and Woodpecker Brains**

| Structures | Density (Kg/m$^{-3}$) | Decay Constant (s$^{-1}$) | Shear Modulus (GPa) | | Bulk Modulus (GPa) |
|---|---|---|---|---|---|
| | | | $G_0$ | $G_\infty$ | |
| Brain | 1040 | 35 | 5.28E-04 | 1.68E-04 | 0.5 |

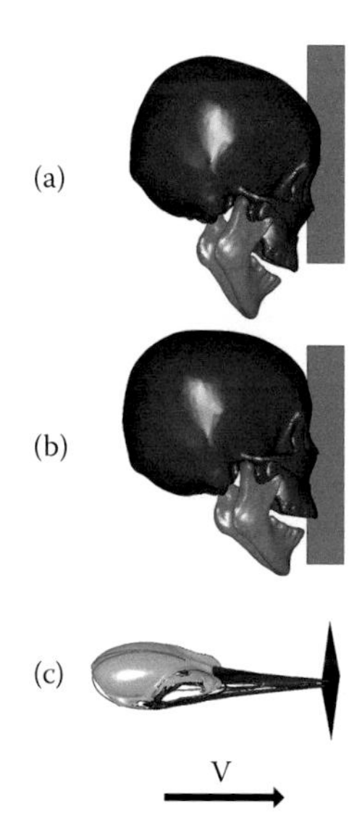


**FIGURE 19.4 (See color insert.)** Boundary conditions of the human and woodpecker head model: (a) human forehead collision, (b) human frontal collision, and (c) woodpecker pecking.

the dynamic FE commercial package LS-Dyna Version 971 (Livermore Software Inc.) (Wollensak and Spoerl, 2004). The results predicted by the FE model of the human head were compared with the published data during impact to validate the FE model (Lowenheilm, 1975; Zhang et al., 2001; Kleiven, 2002). It was closely correlated with the published data (Kleiven, 2002; Motherway et al., 2009). The human head collided with a rigid wall at an initial velocity of 1 m/s with a duration of 10–20 milliseconds (ms) at the location of the forehead and nose, respectively (Figure 19.4a,b). The results predicted by the FE model of the woodpecker head were compared with the corresponding test results during impact in order to validate the FE model (Wang et al., 2011a). The whole head collided with a rigid wall at an initial velocity of 1 m/s with the duration of 10–20 ms based on the kinematics recording (Figure 19.4c). Quantitative studies have been done by analyzing the time histories of effective stress on the skull, brain, and the tip of the upper/lower beak under the initial velocity; the effects of the hyoid bone and the length of the beak on the dynamic response at the selected points of the brain were compared using the FE method. The skull and brain were grouped into a "multi-body part" to permit node-sharing for the human and woodpecker. The boundary relationships between the upper beak and skull and the lower beak and skull were defined with bonded contacts, obligating them to move together at their respective interfaces.

## 19.3 STRESS DISTRIBUTION OF THE SKULL AND THE STRAIN RATE OF THE BRAIN

It is assumed that the woodpecker's head could be protected against acceleration-deceleration impact-related head injury, although no studies have been carried out to prove it comparatively. Simple reasoning would indicate that if woodpeckers got headaches, they would stop pecking. To clarify why woodpeckers have no head injuries, three-dimensional kinematics, mechanical properties, and macro/micro morphological structures were observed in our previous study (Wang et al., 2011a; Wang et al., 2011b; Wang et al., 2013). The dynamic responses of human and woodpecker heads were analyzed quantitatively in view of biomechanics. It was shown that the woodpecker's beak plays an important role in resisting head injuries. In addition, there was some evidence that sudden changes of relevant mechanical parameters in terms of effective stress, shear strain and stress, and relative motion between the brain and skull do indeed cause surface contusions, concussion, and diffuse axonal injury (DAI), as well as acute subdural hematoma. Shear deformation of the brain due to head rotation has long been postulated as a major cause of brain injury, since brain tissue has low shear stiffness (Ruan et al., 1993; Bandak and Eppinger, 1994).

Unfortunately, the measurement of stress on the skull or strain rate on the brain was almost impossible during an impact, particularly in vivo. Alternatively, the FE method can be adopted. Previous studies had developed a two-dimensional FE model of the woodpecker's head using the relevant mechanical parameters of the human head (Oda et al., 2006). The model in this chapter has the exact three-dimensional geometry obtained from micro-CT images, and the measured elastic modulus of the woodpecker's skull and beak may make the results closer to biological reality. The correlation of predicted responses obtained in the FE model and experiment during impact was good (Wang et al., 2011b).

### 19.3.1 Stress Distribution of the Skull at the Selected Points

Parametric analysis was done by changing the impact location, such as the forehead and nose, mainly for the developed human head model, to evaluate the biomechanical effects during pecking. It was expected that variation of the impact location would influence the impact mechanics and load transmission. Two points at the forehead and occiput on the skull and brain were selected to study the time history of the effective stress and strain rate, respectively. As shown in Figure 19.5, the effective stress-time and stress rate-time histories at all of the selected points (forehead and occiput on the skull and brain, respectively) were described when the human head collided with a rigid wall at the location of the forehead and nose respectively, woodpecker's pecking.

### 19.3.2 Strain Rate of the Brain at the Selected Points

Figure 19.6 shows the stress distribution of human and woodpecker heads in the process of colliding with a rigid wall. Maximum effective stress and shear stress concentration of the woodpecker's skull always occurred near the point of collision for all three simulations. It was shown that brain injury is correlated with strain and strain rate (Orsini and Longhi, 2013). Strain rates at the forehead and occiput of the brain were analyzed using the present models. It was found that upper and lower beaks with equal lengths consistently induced higher strains at all three locations on the woodpecker's brain after the comparisons of FE predicted stress on the skull and brain strain rate on the anterior and posterior of the skull and brain during impact, respectively. In addition, the occurrence time of maximum stress was later than that of the beak and skull. The hyroid bone did

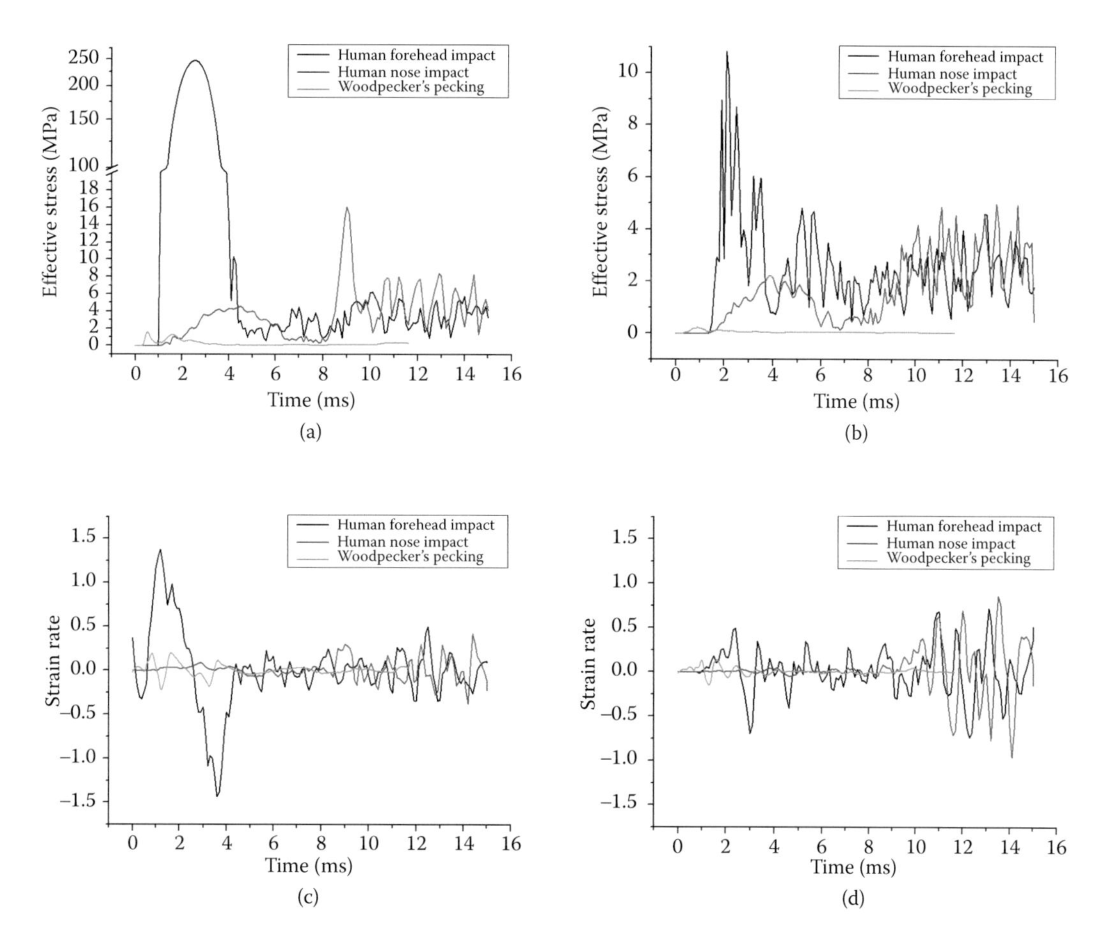


**FIGURE 19.5** **(See color insert.)** Stress/strain rate time histories of the skull and brain for the human and woodpecker: (a) stress history on the forehead on the skull; (b) stress history on the occiput on the skull; (c) strain rate history on the forehead on the brain; and (d) strain rate history on the occiput on the brain.

not work until the end of collision. It seems that the hyoid bone may play a role as a "safety belt" for the woodpecker's head to some extent.

## 19.4 SUMMARY AND APPLICATIONS OF THE MODELS

In this chapter, the finite element models of the human head and woodpecker head were developed to compare the effect of different factors on the load transfer during impact. The modeling method, including the geometric structure of the human skull and woodpecker skull, material properties of the cranial bone and brain, meshing technology, and boundary conditions, are presented in detail. The dynamic FE commercial package LS-Dyna Version 971 (Livermore Software Inc.) (Wollensak and Spoerl, 2004) was used to simulate the dynamic responses during impact. As a robust analysis tool, explicit dynamic solution is ideal for simulating physical events that happen in a short time. It provided a solid platform for parametric analysis.

The conclusions of the present study are summarized as follows: The special beak and hyoid bone were major factors to nonimpact injuries. The long hyoid bone has played a role as a "safety belt" to the woodpecker's head, especially after impact. Beak morphology was found to affect impact force and brain strain rate. The woodpecker's sophisticated shock absorption system is a cooperative phenomenon, with no single factor able to achieve the function. The design of an intelligent helmet or impact-related, injury-resistant device would be lightened greatly by optimization of the

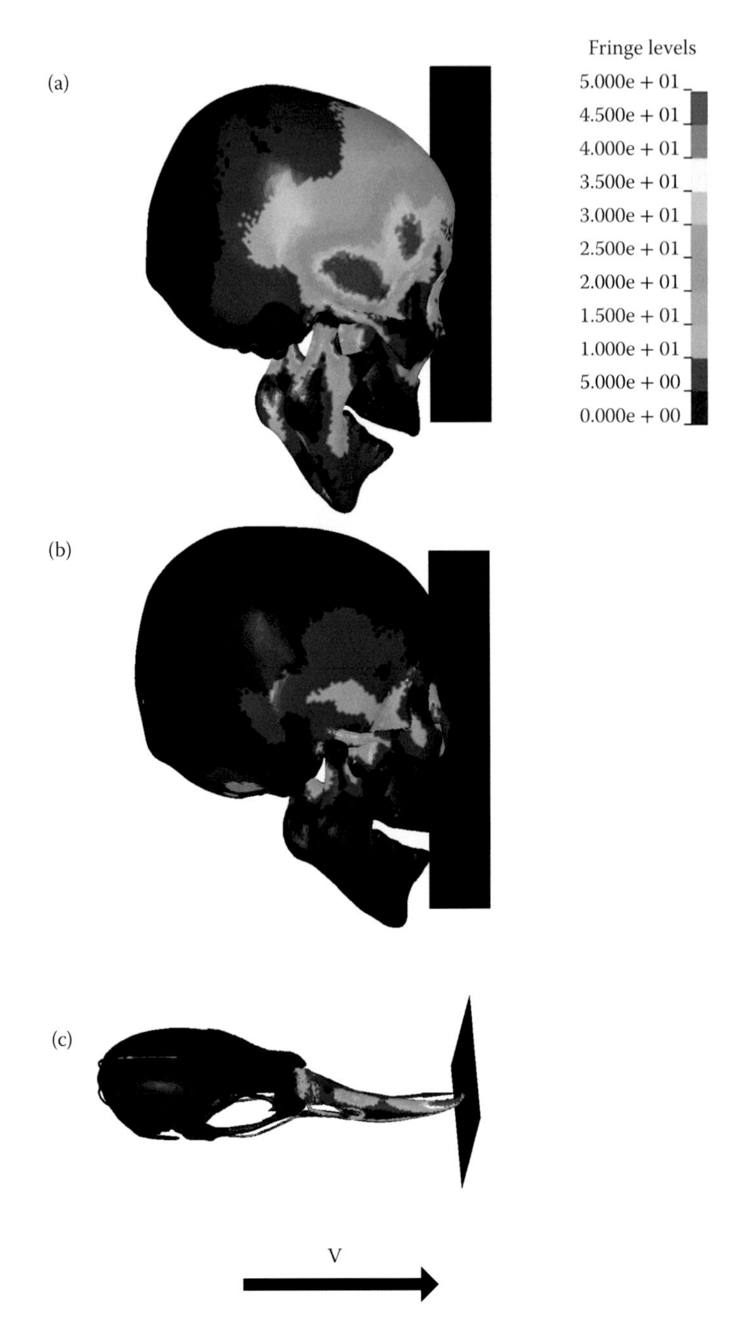


**FIGURE 19.6** **(See color insert.)** Stress distribution of the human and woodpecker heads during impact.

woodpecker's skull morphology, which is helpful in developing new concepts in future work for minimizing head impact injuries.

However, the current head model is far from complete. The most significant limitation is lack of muscle. In addition, the woodpecker's brain dynamic properties remain unknown in order to protect the bird's life. So it is necessary to evaluate the other's birds' dynamic properties in future work for their similar brain with woodpecker. Our works confirm that the FE method is an effective approach to studying the biomechanism of impact injuries.

## ACKNOWLEDGMENTS

This work was supported by the National Natural Science Foundation of China (10925208, 11120101001, 11202017), the Beijing Natural Science Foundation (7133245), Young Scholars for the Doctoral Program of the Ministry of Education of China (20121102120039), the National Science and Technology Pillar Program (No. 2012BAI18B05, 2012BAI18B07), and the 111 Project (No. B13003).

## REFERENCES

Bandak F, Eppinger RA. (1994). Three-dimensional FE analysis of the human brain under combined rotational and translational accelerations. *Proc of 38th Stapp Car Crash Conference*, Society of Automotive Engineers, Florida.

Carter DR, Hayes WC. (1977a). Compact bone fatigue damage—I. Residual strength and stiffness. *Journal of Biomechanics* 10(5–6): 325–337.

Carter DR, Hayes WC. (1977b). The compressive behavior of bone as a two-phase porous structure. *The Journal of Bone and Joint Surgery* 59(7): 954–962.

Carter DR, Fyhrie DP, Whalen RT. (1987). Trabecular bone density and loading history: Regulation of connective tissue biology by mechanical energy. *Journal of Biomechanics* 20(8): 785–794.

Carter DR, Schwab GH, Spengler DM. (1980). Tensile fracture of cancellous bone. *Acta Orthopaedica Scandinavica* 51(5): 733–741.

Ding M, Danielsen CC, Hvid I. (2006). Age-related three-dimensional microarchitectural adaptations of subchondral bone tissues in guinea pig primary osteoarthrosis. *Calcified Tissue International* 78(2): 113–122.

Ding M, Hvid I. (2000). Quantification of age-related changes in the structure model type and trabecular thickness of human tibial cancellous bone. *Bone* 26(3): 291–295.

Ding M, Odgaard A, Linde F, Hvid I. (2002). Age-related variations in the microstructure of human tibial cancellous bone. *Journal of Orthopaedic Research* 20(3): 615–621.

Finfer S, Cohen J. (2001). Severe traumatic brain injury. *Resuscitation* 48: 77–90.

Holbourn AHS. (1943). Mechanics of head injuries. *The Lancet* 2: 438–441.

Hosey R. (1982). Finite elements in biomechanics: A homeomorphic finite element model of the human head and neck, in *Finite Element in Biomechanics*, ed. RH Gallagher, pp. 379–401.

Junge A, Langevoort G, Pipe A, Peytavin A, Wong F, Mountjoy M, Beltrami G, Terrell R, Holzgraefe M, Charles R, Dvorak J. (2006). Injuries in team sport tournaments during the 2004 Olympic Games. *The American Journal of Sports Medicine* 34(4): 565–576.

King AI, Yang KH, Zhang LY, Hardy W. (2003). Is head injury caused by linear or angular acceleration? *IRCOBI Conference*, Lisbon, Portugal.

Kleiven, S. (2002). Finite element modeling of the human head. KTH Royal Institute of Technology. PhD.

Lowenheilm P. (1975). Mathematical simulation of gliding contusions. *Journal of Biomechanics* 8: 351–356.

Martin E, Lu WC, Helmick K, French L, Warden DL. (2008). Traumatic brain injuries sustained in the Afghanistan and Iraq wars. *American Journal of Nursing* 108(4): 40–47.

May PR, Fuster JM, Haber J, Hirschman A. (1979). Woodpecker drilling behavior—An endorsement of the rotational theory of impact brain injury. *Archives of Neurology* 36: 370–373.

May PR, Fuster JM, Newman P, Hirschman A. (1976). Woodpeckers and head injury. *Lancet* 1(7957): 454–455.

McIntosh AS, McCrory P. (2005). Preventing head and neck injury. *British Journal of Sports Medicine*. 39(6): 314–318.

Motherway J, Doorly MC, Curtis M, Gilchrist MD. (2009). Head impact biomechanics simulations: A forensic tool for reconstructing head injury? *Legal Medicine* 11(S1): S220–S222.

Oda J, Sakamoto J, Sakano K. (2006). Mechanical evaluation of the skeletal structure and tissue of the woodpecker and its shock absorbing system. *JSME International Journal Series A* 49(3): 390–396.

Orsini F, Longhi L. (2013). Mannose binding lectin is associated with injured vessels in clinical and experimental traumatic brain injury and its deletion is protective. *Molecular Immunology* 56(3): 256–256.

Ruan JS. (1994). Impact biomechanics of head injury by mathematical modelling, Wayne State University. PhD dissertation.

Ruan JS, Khalil TB, King AI. (1993). Finite element modeling of direct head impact. 37 th Stapp Car Crash Conference Proceedings. Society of Automotive Engineers. Warrendale, PA, 933114.

Shuck LZ, Advani SH. (1972). Rheological response of human brain tissue in shear. ASME Journal Series. *Basic Eng* 94: 905–911.

Spring LW. (1965). Climbing and pecking adaptations in some North American woodpeckers. *Condor* 67: 457–488.

Stalnaker, R. (1969). Mechanical properties of the head. Morgantown, West Virginia University, Ph.D dissertation.

Wang LZ, Cheung JT, Pu F, Li DY, Zhang M, Fan YB. (2011a). Why do woodpeckers resist head impact injury: a biomechanical investigation. *Plos One* 6(10): e26490.

Wang LZ, Lu S, Liu XY, Niu XF, Wang C, Ni YK, Zhao MY, Feng CL, Zhang M, Fan YB. (2013). Biomechanism of impact resistance in the woodpecker's head and its application. *Science China-Life Sciences* 56(8): 715–719.

Wang LZ, Zhang HQ, Fan YB. (2011b). Comparative study of the mechanical properties, micro-structure, and composition of the cranial and beak bones of the great spotted woodpecker and the lark bird. *Science China-Life Sciences* 54(11): 1036–1041.

Wollensak G, Spoerl E. (2004). Biomechanical characteristics of retina. *Retina—the Journal of Retinal and Vitreous Diseases* 24(6): 967–970.

Yang K, Hu J, White NA, King AI, Chou CC, Prasad P. (2006). Development of numerical models for injury biomechanics research: A review of 50 years of publications in the Stapp Car Crash Conference. *Stapp Car Crash Journal* 50: 429–490.

Zhang L, Yang KH, King AI. (2001). Biomechanics of neurotrauma. *Neurological Research* 23: 144–156.

# 20 Tooth Model in Orthodontics and Prosthodontics

*Chao Wang, Yi Zhang, Wei Yao, and Yubo Fan*

## CONTENTS



## SUMMARY

This chapter focuses on two specific fields: orthodontics and prosthodontics. Although finite element analysis (FEA) as a numerical technique has been widely used in the field of dental biomechanics to assess engineering and biomechanical problems, new methods should be developed to offer a more detailed understanding of the alveolar bone response to dental treatment. This chapter reviews the development and application of bone remodeling simulation techniques in the field of dentistry. Three numerical models are presented to illustrate the role of bone remodeling simulations in the dental biomechanical research. In the study of orthodontic tooth movement, both external and internal remodeling mechanisms were incorporated into the tooth movement algorithms to predict the position of the tooth and changes in alveolar bone density. In the study of internal bone architecture around dental implants, the morphology of trabeculae was reconstructed using combined bone remodeling simulation techniques. In the case of alveolar bone remodeling induced by implant-supported restorations, the different bone response induced by different dental restorations was assessed. It is hoped that such simulation methodology can be helpful in improving the effectiveness and reliability of future dental treatments.

## 20.1 INTRODUCTION

This chapter intends to outline the contributions offered by numerical techniques, in particular by finite element (FE)–based bone remodeling simulation for the investigation of biomechanical problems in the dental field. The chapter focuses on two specific fields: orthodontics and prosthodontics.

Orthodontic procedures are usually performed to treat malocclusions (improper bites) resulting from irregular teeth or to control as well as modify facial growth for purely aesthetic reasons (Verna, Zaffe, and Siciliani 1999). Tooth movements resulting from orthodontic treatment are realized through biological responses induced by mechanical stimulus (Henneman, Von den Hoff, and Maltha 2008). In prosthodontics, prosthetics are used to restore the bite to aid in chewing and maintain the position of teeth. A dental implant that resembles a tooth or group of teeth to replace missing teeth is often used to support restoration (Shillingburg 1997). Functional loading induced by the implant can result in changes in the surrounding bone tissue (Palmer et al. 2012).

In the last couple of decades, FEA has been used extensively in the dentistry field (Beaupre, Orr, and Carter 1990a, 1990b; Li et al. 2011; Lin, Lin, and Chang 2010; Ojeda et al. 2011; Kojima and Fukui 2005). A series of computational procedures are employed to calculate the stress and strain resulting from external force, pressure, thermal change, and other factors. FE analysis has evolved as an effective tool to study tooth movement as well as dental implants. However, a more detailed understanding of the alveolar bone's response is necessary for the further development of orthodontics and prosthodontics. As human bone, including the alveolar bone, can rebuild itself in accordance with the mechanical environment, it is imperative that not only biomechanical analysis but also mechanobiological factors are taken into account in further research.

The ability of bone tissue to adapt to external mechanical loading is called bone remodeling (Fyhrie and Carter 1986). Bone remodeling is a biological process during which old bone tissue is removed (resorption) and new bone tissue is formed (Huiskes et al. 1987). The process involves groups of osteoclasts and osteoblasts that function as organized units called basic multicellular units (BMUs). Resorption lacunae created by osteoclasts are subsequently filled by osteoblasts as part of the BMUs. Normally, there is a balance between resorption and formation; however, bone loss will occur as a result of an imbalance in the remodeling process.

At present, when an artificial fixation is implanted or an orthodontic force is applied, bone remodeling can be simulated in FE models to predict the bone response. According to the level of research, these simulation studies can be classified into two main groups:

1. *Macroscale model*: These models focus on the prediction of apparent bone density and elastic modulus distribution as a consequence of the bone remodeling process. In the field of prosthodontics, preliminary studies emphasized the adaptive behavior of alveolar bone under load via FE-based remodeling simulations (Mellal et al. 2004). These simulations led to an important conclusion: peak strains and strain energy densities were consistent with in vivo data, which indicates that both parameters can be employed as a bone remodeling stimulus. Furthermore, based on the traditional model developed for long bone, Li et al. (2007) proposed a new mathematical equation relating the rate of change in bone density with a mechanical stimulus that can simulate both underload and overload resorptions. This is a very useful modification, because the new remodeling algorithm can describe bone loss under excessive loading. Meanwhile, a remodeling simulation by Chou et al. predicted non-homogeneous density/elastic modulus distribution around various dental implant systems (Chou, Jagodnik, and Muftu 2008). Additionally, Daniel Lin and his coworkers reviewed existing numerical studies, analyzed published remodeling data, and assessed different biomechanical remodeling stimuli (Field et al. 2009). Since then, they have developed a series of numerical dental models correlated to clinical computerized tomography (CT) data, and contributed significantly to the development of alveolar bone

remodeling simulations in the field of dentistry (Field et al. 2009; Lin et al. 2009; Lin et al. 2010). A numerical simulation undertaken by Wang et al. (2013) investigated dental bone remodeling induced by an implant-supported fixed partial denture with and without a cantilever.

On the other hand, Beaupre et al. suggested that tooth movement is controlled predominantly by mechanical deformations of the periodontal ligament (PDL) (Beaupre, Orr, and Carter 1990a, 1990b). Schneider et al. revealed that it is possible to integrate a mechanical bone-remodeling algorithm into a realistic three-dimensional tooth and jawbone model (Schneider, Geiger, and Sander 2002). For development of the bone remodeling algorithms, a real model (CT image data) of the individual tooth was developed to increase the accuracy of the model. Subsequently, Marangalou et al. constructed a computational model to calculate the rate of orthodontic tooth bodily movement (Marangalou, Ghalichi, and Mirzakouchaki 2009). The normal strain of the PDL was employed as the key mechanical stimulus for alveolar bone remodeling (Bourauel et al. 1999). Based on the external bone remodeling mechanism, Qian et al. developed a numerical model to reproduce an orthodontic treatment for mandibular canine tipping and predicted tooth bodily movement (Qian et al. 2008; Qian, Liu, and Fan 2010). More recently a set of computational algorithms incorporating both external and internal remodeling mechanisms was implemented into a patient-specific three-dimensional FE model to investigate and analyze orthodontic treatment under four typical modes of orthodontic loading (Wang, Han et al. 2012).

2. *Microscale model*: At the micro level, FE models are more concerned with the morphology of bony trabeculae around the dental implants. Recently, Hasan et al. (2012) applied a set of bone remodeling algorithms to predict the distribution of bone trabeculae around a dental implant. In an idealized bone segment, an FE model of a screw-shaped dental implant was tested. The model succeeded in achieving a trabeculae-like structure around the osseointegrated dental implants. Meanwhile, bone remodeling simulations by Wang and Fan predicted the evolution of the architecture around four implant systems using a novel model that combines both adaptive and microdamage mechanisms (Wang, Wang et al. 2012). The proposed algorithms were shown to be effective in simulating the remodeling process of the trabecular architecture.

The above simulation research can offer a deeper understanding of complex problems in the field of dental biomechanics, such as the bone-implant interaction and the response of natural dentition under different types of loadings. This numerical approach provides bioengineers and dentists with a useful tool for investigating the problems pertaining to the biomechanical response of dental implants and natural teeth. In this chapter, three simulation cases are presented to illustrate the role of bone remodeling numerical technology in the field of dental biomechanics.

## 20.2 MODEL DEVELOPMENT

This section details the process of model development. Generally speaking, there are two main steps. The first step is to build a patient-specific FE model of the jawbone, natural tooth, and dental implant. The second step is to incorporate bone-remodeling algorithms into the application of the FE model.

### 20.2.1 Finite Element Model of the Jawbone, Natural Tooth, and Dental Implant

First, a set of CT images was obtained for bone geometry. Based on these images, a three-dimensional model of maxillary bone and mandibular bone was constructed. The CT images consisted of 652 transversal sections with a slice thickness of 0.5 mm and a pixel width of 0.398 mm. The geometrical shapes of bones and teeth were developed in MIMICS 10.0 (Materialise, Leuven, Belgium). The PDL was generated around the root of the tooth with an average thickness of 0.2 mm. The entire model of maxillary bone is shown in Figure 20.1, and a bone segment including the PDL and central incisor is shown in Figure 20.2. The commercial CAD software SolidWorks 2007 (SolidWork Corp., Dassault Systemes, Concord, MA) was used to build the orthodontic bracket, arch wire, and dental implants. According to the Straumann Standard Plus Implant System, the cylindrical implant used was 10 mm long with a diameter of 4.1 mm (Φ 4.8 mm Regular Neck, RN) and a neck height of 1.8 mm

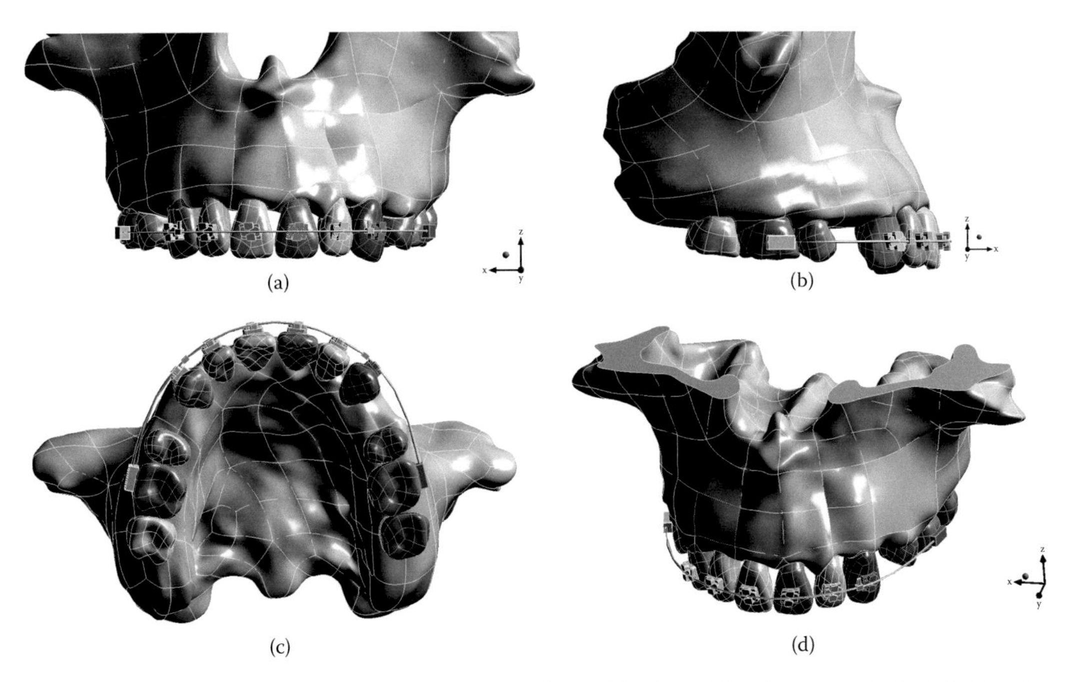


**FIGURE 20.1** For orthodontic treatment, a geometric model of maxillary bone was built including the cortical bone, trabecular bone, teeth, PDL, bracket, and arch wire. (a) Frontal view, (b) side view, (c) bottom view, and (d) isometric view.

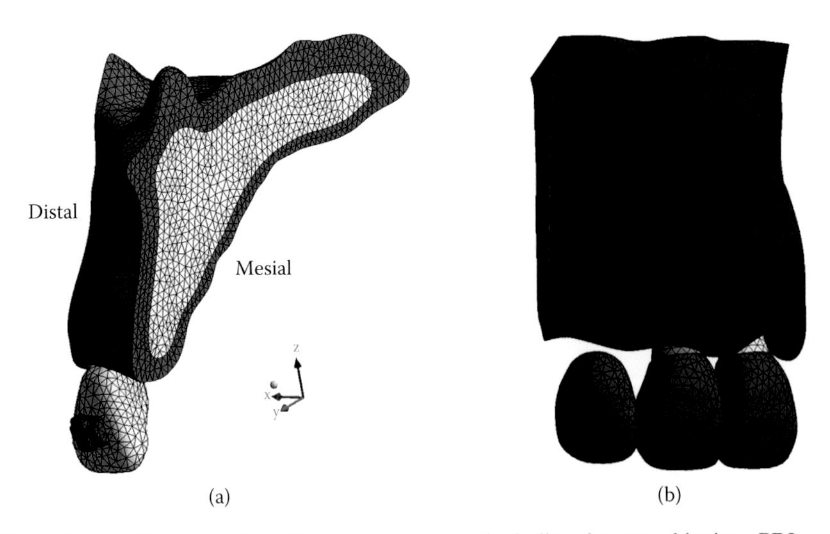


**FIGURE 20.2** (a) An FE model of maxillary bone segment (including the central incisor, PDL, and bracket) used in the simulation of orthodontic tooth movement; (b) an FE model of a portion of the maxilla with cantilever bridgework connecting two implants.

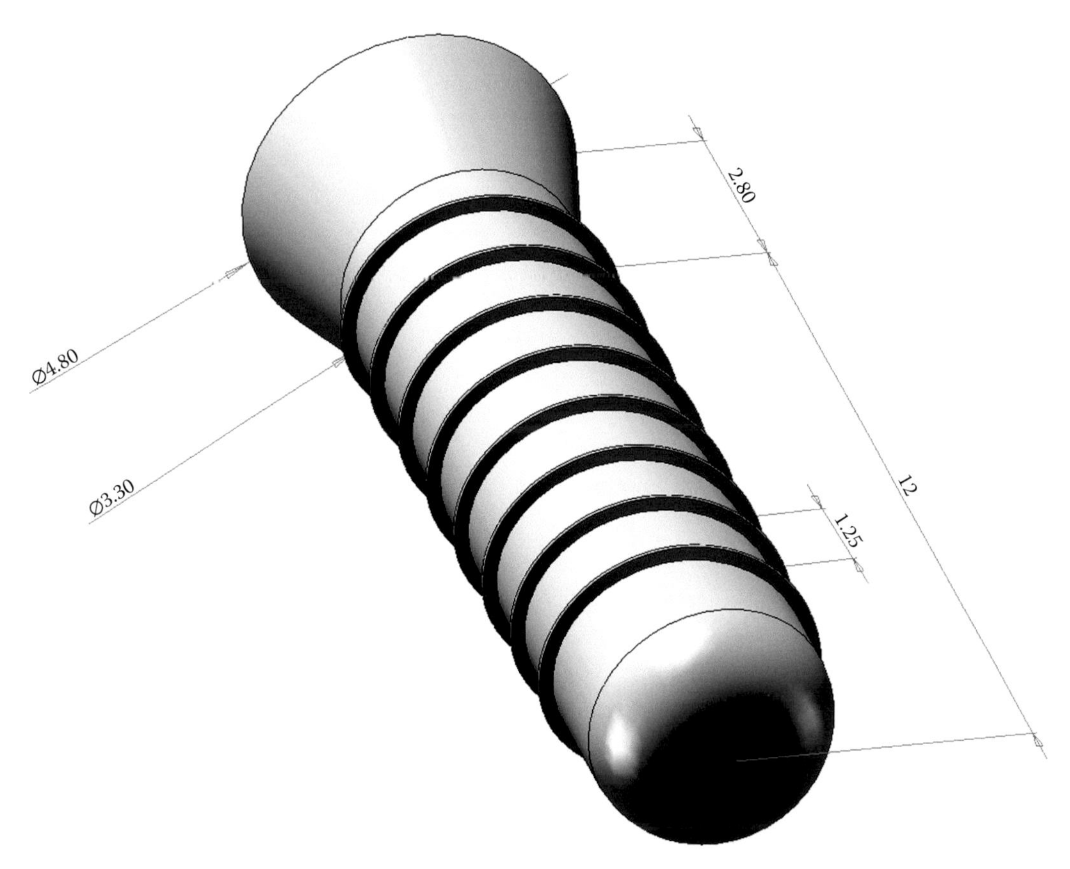


**FIGURE 20.3** CAD model of Straumann Standard Plus Implant used in the simulations.

with an 8° Morse taper abutment-to-implant connection with an internal octagon (Figure 20.3). The dental implant was manufactured in commercially pure titanium. A porcelain-fused-to-metal (PFM) crown with gold alloy was employed. In order to ensure accurate representation, the three-dimensional models of the bracket, arch wire, and dental implants were assessed by a clinical dentist.

The FE models were meshed using 10-node solid tetrahedral elements in ANSYS software (Swanson Analysis System Co., Houston, TX). To check for model convergence, models were created with element sizes of 0.5, 1, 1.5, 2, and 2.5 mm. For each model, the total strain energy was calculated and compared under the same loading and boundary conditions. The total strain energy was reduced by 1.7% with each increase in element size from 0.5 to 1 mm and further reduced by 5.3% with each increase from 1 to 1.5 mm. With elements from 1.5 to 2 mm and 2 to 2.5 mm, strain energy decreased 13.2% and 16.8%, respectively. The computing time for the 0.5 mm model was 20 minutes, but it was substantially reduced to 8 minutes for the 1 mm model. As shown in Figure 20.4, a 1-mm-per-element size was adopted.

### 20.2.2 Mechanoregulation Bone Remodeling Algorithms

According to different clinical problems, different mechanoregulation bone remodeling algorithms were developed and implemented into the FE model.

1. *Tooth movement algorithms incorporating both external and internal remodeling mechanisms.* The algorithms were divided into external and internal remodeling components. An external remodeling principle was employed to predict tooth movement, keeping

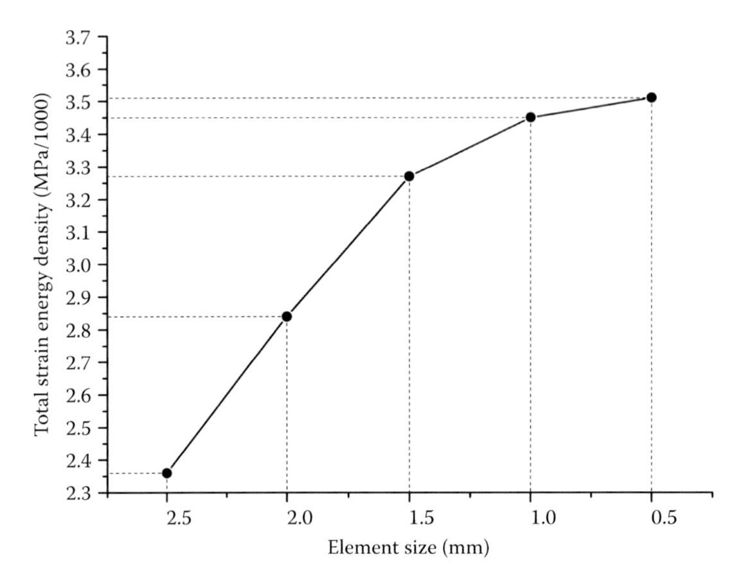


**FIGURE 20.4** The convergence results of the FE model.

mechanical properties unchanged. The normal strains on the PDL surface were regarded as biomechanical stimuli initiating orthodontic tooth movement. A linear relationship was assumed between external remodeling and strain magnitude within the range of 0.03%–0.3% strain and the tooth movement rate was expected to reach a limit after a certain strain (ε). Tooth movement velocity ($\dot{R}$) was determined by Equation 20.1:

$$\dot{R} = \begin{cases} 0.2898 & \varepsilon > 0.3 \\ 1.071\varepsilon - 0.0322 & 0.03 < \varepsilon > 0.3 \\ 0 & \varepsilon > 0.3 \end{cases} \tag{20.1}$$

Internal remodeling was considered the mechanism for the redistribution of cancellous bone density under the action of orthodontic loading. According to Huiskes' approach (Huiskes et al. 2000; Weinans et al. 1993), the strain energy density (SED) $U$ was adopted as the biomechanical signal that controls internal remodeling of the cancellous bone, with

$$U = \frac{1}{2}\varepsilon_{ij}\sigma_{ij} \tag{20.2}$$

where $\varepsilon_{ij}$ is von Mises strain and $\sigma_{ij}$ is von Mises stress. The value of the SED was compared to a homeostatic one ($k$). A "lazy zone," in which no net change in bone density occurs, was incorporated in the formulations. The set of governing equations for internal remodeling was expressed in the following equation:

$$\frac{d\rho}{dt} = \begin{cases} B(U - (1+s)k & \text{for } U > (1+s)k \\ 0 & \text{for } (1-s)k \le U \le (1+s)k \\ B(U - (1-s)k & \text{for } U < (1-s)k \end{cases} \tag{20.3}$$

where $s$ is the threshold level that marks the borders of the lazy zone and $B$ is the remodeling coefficient.

The local cancellous bone density $\rho$ was regulated by algorithms of internal remodeling. The elastic modulus $E$ was related to the local density value $\rho$ using the following equation:

$$E = C\rho^{n} \tag{20.4}$$

where both $C$ and $n$ are experiential constants.

2. *Three-dimensional bone remodeling algorithm incorporating an overloading bone resorption process.* Using Euler's forward integration method, the bone-remodeling rate equations were formulated by relating the bone density increment $\Delta\rho$ to the time increment $\Delta t$ in four different phases (Lin, Lin, and Chang 2010; Li et al. 2007).
Bone disuse resorption:

$$\Delta\rho = B(\Psi - (1-\delta)K_{ref}) \cdot \Delta t \qquad if\, \Psi < (1-\delta)K_{ref} \tag{20.5}$$

Bone equilibrium:

$$\Delta\rho = 0 \qquad if\,(1-\delta)K_{ref} \le \Psi \le (1+\delta)K_{ref} \tag{20.6}$$

Bone formation:

$$\Delta\rho = B(\Psi - (1+\delta)K_{ref}) \cdot \Delta t \qquad if\,(1+\delta)K_{ref} < \Psi < K_{overloading} \tag{20.7}$$

Bone overload resorption:

$$\Delta\rho = B(K_{overloading} - \Psi) \cdot \Delta t \qquad if\, \Psi \ge K_{overloading} \tag{20.8}$$

where $\Psi$ denotes SED per unit bone mass ($U/\rho$), $U$ is the mechanical stimulus (i.e., strain energy density herein), $\rho$ is the bone density, $B$ is the remodeling rate constant, $K_{ref}$ and $K_{overloading}$ are the remodeling reference values, and $\delta$ is the bandwidth of the lazy zone. In this simulation, the time increment $\Delta t$ represents three days and thus 10 time units denote one month. Consequently, based on the work of Lin, Lin, and Chang (2010) the remodeling constant is rescaled to unit days. Because trabecular bone is more actively remodeled than cortical bone, the remodeling constant $B$ in trabecular bone is twice than that in cortical bone.
The dental implants and fixed partial dentures (FPDs) were assumed to be linear elastic, homogeneous, and isotropic in the initial FE models. For trabecular and cortical bone, the elastic moduli were 1.37 GPa and 13.7 GPa, respectively. Then, via the remodeling calculation (Li et al. 2011), the elastic moduli were updated according to the relationship illustrated in Equations 20.9 and 20.10:

$$E_{trabecular} = 2.349\rho^{2.15}\left(0.9 \le \rho_{trabecular} \le 1.2\right) \tag{20.9}$$

$$E_{cortical} = -23.93 + 24\rho\left(1.2 \le \rho_{cortical} \le 2.0\right) \tag{20.10}$$

where $E$ is the elastic modulus of bone (MPa) and $\rho$ denotes the bone density (g/cm$^3$). The initial density of trabecular bone $\rho_{trabecular}$ and cortical bone $\rho_{cortical}$ were assumed to be 0.80 and 1.74 g/cm$^3$, respectively. By utilizing the above remodeling equations, the local bone density values can be calculated. Then the elastic modulus of bone can be updated according to Equations 20.9 and 20.10 for the next remodeling calculation.

3. *Trabeculae remodeling algorithm with both adaptive and microdamage remodeling mechanisms.* Initially, damage rate $\dot{\omega}$ was defined to assess the state of damage. If damage accumulation $\omega = 1$ was a failure, and if the rate of damage accumulation was assumed to be linear, then Miner's rule was employed to define the damage rate (McNamara and Prendergast 2007):

$$\dot{\omega} = \frac{1}{N_f} \tag{20.11}$$

where $N_f$ denotes the number of cycles required for a failure of the material at a given stress. In the case of bones, the empirical equation of Carter et al. (1976) was used to calculate $N_f$:

$$\log N_f = H \log \sigma^i + JT + K\rho^i + M \tag{20.12}$$

where $\sigma^i$ is the stress (MPa), $T$ is the temperature (°C), and $H$, $J$, $K$, and $M$ are all empirical constants. Then, the damage accumulates as

$$\omega = \int_0^t \dot{\omega} dt \tag{20.13}$$

where $\omega$ denotes the state of local damaged bone tissue and $t$ the duration of calculation. When the damage accumulation $\omega$ was below a certain damage threshold, the adaptive remodeling program was used. Based on previous research (Ruimerman et al. 2005; Xinghua, He, and Bingzhao 2005), strain energy density at location $i$ was given by

$$U^i = \frac{\varepsilon^i E^i \varepsilon^i}{2\rho^i} \tag{20.14}$$

where $E^i$ is the elastic modulus (MPa) at location $i$ (assumed isotropic), $\varepsilon^i$ is the strain, and the unit of $\rho^i$ is g/cm$^3$. At the given location, the stimulus for an element is $S^i$, and is defined as

$$S^i = U^i - U_{ref} \tag{20.15}$$

where $U_{ref}$ represents the reference stimulus. The total stimulus received by an osteocyte is a sum of stimuli from other osteocytes surrounding it, hence

$$\frac{d\rho_m(x,t)}{dt} = B\sum_{i=1}^{N} f^i(x)S^i(x) \qquad 0 < \rho \le \rho_{\max} \tag{20.16}$$

where $\rho_m$ is the bone density mediated by adaptive remodeling. Parameter $B$ denotes the proportion coefficient, $N$ is the number of sensor cells that contribute to the total stimuli, and $\rho_{max}$ signifies the maximal bone density. $f^i(x)$ is a spatial influence function, describing the decay in signal intensity relative to distance $d$ and decay parameter $D$, according to

$$f^i(x) = e^{-d(x,x')/D} \tag{20.17}$$

where $d(x, x')$ is the distance from osteocyte $x$ to surface location $x'$ and $D$ is the distance from an osteocyte location where its effect has been reduced to $e^{-1}$. The density of bone tissue at a specified region is estimated using a forward Euler method, and then obtained by Equation 20.18:

$$\rho_{i+1} = \rho_i + \frac{d\rho_m}{dt}\Delta t \tag{20.18}$$

If damage accumulation $\omega$ was above the critical value, a random number generator was used to determine which element becomes a remodeling site according to BMU activation probability, which depends on the local damage state. Once an element is activated, it begins a microdamage-remodeling program; otherwise, it will return back to the adaptive remodeling program. Therefore, when undergoing remodeling, the rate of change of localized bone density is given by

$$\frac{d\rho_r(x,t)}{dt} = \begin{cases} \dot{\rho}_R(x,t) & for\, t \in [t, t+T_R] \\ 0 & for\, t \in (t+T_R, t+T_R+T_O] \\ \dot{\rho}_F(x,t) & for\, t \in (t+T_R+T_O, t+T_R+T_O+T_F] \end{cases} \tag{20.19}$$

where $\rho_r$ is the bone density experiencing remodeling, $T_R$ is resorption period, $T_O$ is reversal period, and $T_F$ is formation period. $\dot{\rho}_R$ represents the rate of osteoclastic resorption and $\dot{\rho}_F$ represents the rate of osteoblastic formation. Therefore, at the different phases of bone remodeling, bone density is determined by

$$\rho_r(x,t) = \begin{cases} \rho_r(x,t+T_R) = \rho_r(x,t) - \int_t^{t+T_R} \dot{\rho}_R(x,t)\,dt \\ \rho_r(x,t+T_R+T_O) = \rho_r(x,t+T_R) \\ \rho_r(x,t+T_R+T_O+T_F) = \rho_r(x,t+T_R+T_O) + \int_{t+T_R+T_O}^{t+T_R+T_O+T_F} \dot{\rho}_F(x,t)\,dt \end{cases} \tag{20.20}$$

The combination of adaptive remodeling and microdamage remodeling algorithms changes the localized bone density. As illustrated in Equations 20.9 and 20.10, the elastic modulus depends on the density value, elastic modulus per element is updated to calculate new stress/strain field using FEA, and then a new loop is started. The process is repeated time after time to simulate the evolution of trabeculae configuration.

## 20.3 APPLICATIONS IN TOOTH MOVEMENT AND DENTAL IMPLANTATION

### 20.3.1 Simulation of Orthodontic Tooth Movement

Figure 20.5 presents the final simulation results of apparent bone density distribution. The model cross-section reveals the effects of four different orthodontic loadings. For case (a), under the action of the tipping torque, a buccal tipping around the center of resistance was achieved at the end of simulation. As indicated in Figure 20.5a, the cancellous bone density increased near the tooth and formed a high-density region connecting the two sides of the cortical bone. For case (b), the surrounding bone density increased dramatically to support the torque loading around the $z$-axis (Figure 20.5b). For case (c), besides a buccal tipping torque, a force along the $y$-axis was applied to achieve bodily tooth movement in the coronal direction (Figure 20.5c). The pattern of bone density distribution was similar to case (a); however, because the tooth was extruded from the alveolar bone,

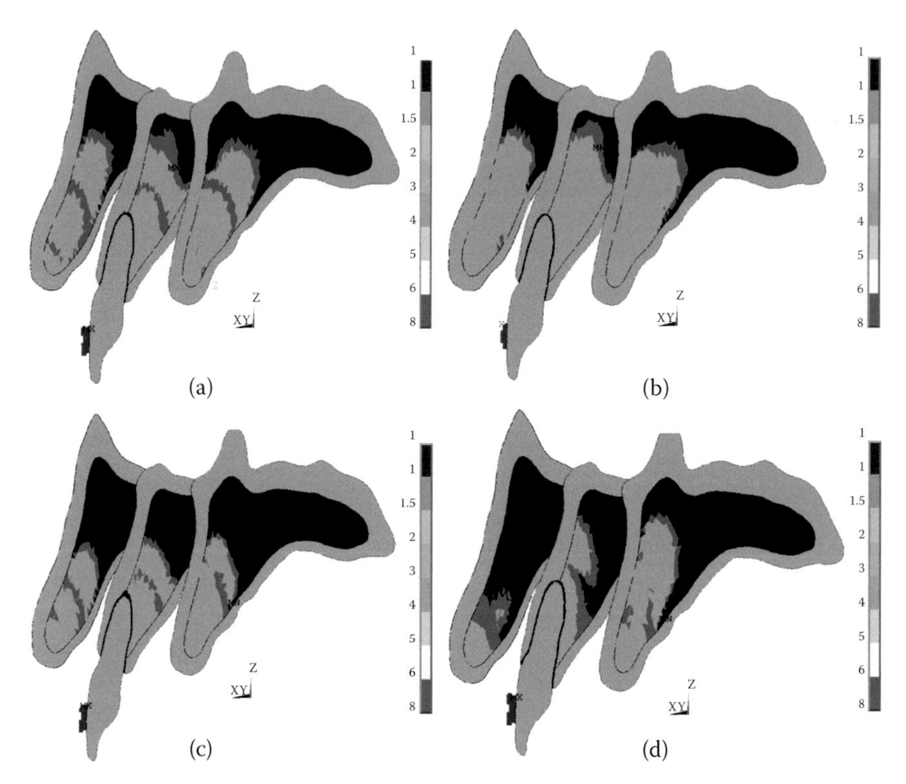


**FIGURE 20.5** **(See color insert.)** Simulated distribution of apparent bone density (g/cm$^3$) under the different actions of orthodontic loading. (a) Tipping; (b) rotation; (c) extrusion; and (d) intrusion. The cross-section of the model is shown in each case.

the affected region was smaller than that of case (a). For case (d), the direction of orthodontic loading was opposite to case (c). When the tooth was intruded into the alveolar bone, the cortical bone undertook most of the mechanical loadings, thus the cancellous bone densification was less than that of other three kinds of orthodontic treatment (Figure 20.5d).

### 20.3.2 Simulation of Alveolar Bone Remodeling Induced by Implant-Supported Fixed Partial Denture with or without Cantilever Extension

Figure 20.6 shows changes in bone density, which varied from 0.20 to 2.0 g/cm$^3$, from the beginning to the end of the 18-month simulation. The relative variation in Model A (a single implant-supported two-unit cantilever FPD) was small and stable under loading, and the highest value was concentrated in the area closer to the alveolar margin. In contrast, the maximum density in Model B (a double implant-supported two-unit non-cantilever FPD) was evidently adjacent to the root of the dental implant and the minimum value was located around the cortical neck area. The blue color around the neck of the implant represents regions with very low density caused by overloading resorption.

### 20.3.3 Simulation of the Bone Remodeling Process of the Internal Bone Architecture around the Dental Implant

Figure 20.7 shows the complete history of the simulated remodeling process of the internal bone architecture around the threaded cylinder implant during iterations. As demonstrated in this figure, the morphology of the trabecular architecture changes gradually from an initial uniform state to a

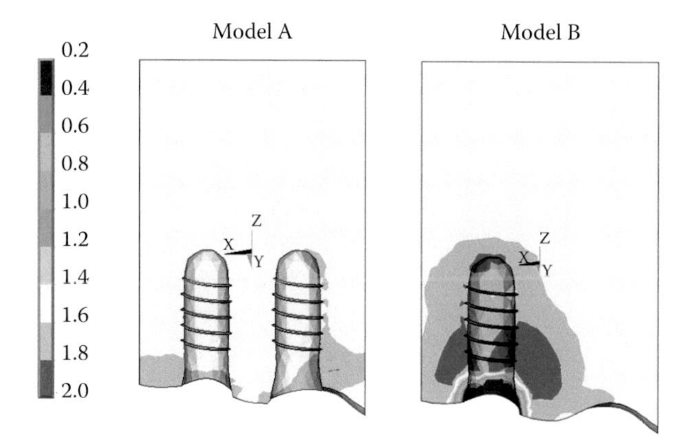


**FIGURE 20.6 (See color insert.)** Comparison of final results showing the variation of alveolar bone density (g/cm³) under mechanical conditions. Model A: non-cantilever model; Model B: cantilever model.

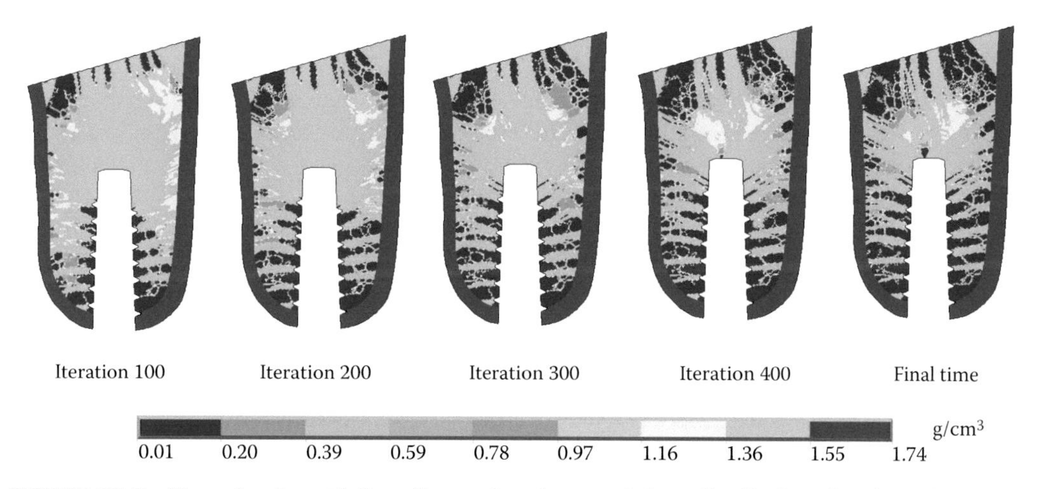


**FIGURE 20.7 (See color insert.)** Cancellous trabeculae morphology distribution after the 100th, 200th, 300th, 400th, and final time steps. Colors represent bone mineral density, shown in g/cm³. Red is the cortical bone. The total number of iteration steps for this case to achieve the converged result is 510.

state more representative of the bone's natural response. The trabecular structure adapts to its surrounding mechanical environment through approximately 510 iterations. The colors from blue to red indicate the range 0.01–1.74 g/cm³. Red represents the density of cortical bone whereas blue characterizes the bone resorption spaces. From this figure, we can see that around the prosthetic system the cancellous trabeculae are oriented outwards from the apex into the cortical bone. Correspondingly, the threads are connected to the cortical bone by newly formed trabeculae. The final results presented are very similar to the actual patterns observed in the existing literature, as shown in Figure 20.7.

## 20.4 MODEL VALIDATION

### 20.4.1 Orthodontic Tooth Movement

Various study design characteristics complicate the comparison with existing data and limit their contribution to the clarification of related outcomes. With respect to the rotation angle of tooth movement, during a 12-week period Yee et al. (2009) reported that the average canine rotations were 22.36°±18.59° and 1.29°±2.67° in the heavy force (3 N) group and light force (0.5 N) group, respectively. Moreover, in a study by Iwasaki et al. (2000), after applying a continuous retraction force averaging 0.18 N, the maxillary canines underwent approximately 0.6° of rotation, and another force averaging 0.6 N produced approximately 5.9° rotation. Overall, the distal crown tip averaged 3.2° over the 84 days (12 weeks). In addition, under an average 8 N•mm torque, Qian et al. (2008) calculated a tipping degree of 4.59° over four weeks. However, in our buccal tipping simulation (case (a)), the rotational change in a tooth loaded by a 10 N•mm torque was only 5.5° during a 16-week period. On the other hand, as far as tooth displacement is concerned, Pilon et al. reported that after 16 weeks the displacement of three experimental sides in different dogs with the same force of 1 N were 0.5, 4.7, and 7.2 mm, respectively, which implies that the rate of tooth movement averaged 0.26 mm/week (Pilon, Kuijpers-Jagtman, and Maltha 1996). Furthermore, experimental values reported by Ren et al. (2004) indicated that the mean maximum velocity of human canine retraction was 0.29 mm/week when the force magnitude was 2.72 N. In our simulation for bodily tooth movement (cases (c) and (d)), the averaged velocity of tooth movement was 0.15 mm/week under a force of 1 N, which is consistent with the abovementioned experimental results.

### 20.4.2 Trabecular Structure around Dental Implant

A systematic review of the literature was undertaken to relate the numerical predictions to existing in vivo data. In a study by Kingsmill and Boyde (1998), the real trabecular distribution in a photograph of 2.0-mm thick bone sections from the mental foramen region of an edentulous mandible of a human cadaver was investigated. They reported that the trabeculae had a strong horizontal component and in some cases possessed a very ladder-like arrangement. Interestingly, a paper by Watzak et al. (2005) presented an undecalcified thin ground section of the upper jaw at the molar region (20 μm) of a baboon with different implant shapes after 18 months of occlusal loading, in which cancellous bone trabeculae tended to be oriented in an apico-coronal direction from the host bone to the peri-implant bone. The loaded dental implants in their experiments included a commercially available pure titanium screw, grit-blasted and acid-etched screws, and a titanium plasma-sprayed cylinder. Watzak et al. also suggested that these trabeculae were well oriented around screw-shaped implants, whereas they were unoriented around cylindrical implants. Moreover, Gross et al. (1990) reported on animal experiments in which they studied the effect of implant-to-bone bonding on bone structure near the implant. When an implant was placed in living bone tissue, the tissue remodeled itself to accommodate the implant and yielded the spoke trabecular architecture around it (Geramy 2000). In addition, a canine mandible was used in experimental studies of dental implant incorporation by Schenk and Buser et al. (1998). After a healing period of 3–5 months, bony anchors formed in the cancellous part of the implant site and the threads were connected to pre-existing trabeculae by newly formed bone bridges. Generally speaking, these experimental and clinical observations can be mirrored by our numerical outcomes.

### 20.4.3 Alveolar Bone Remodeling Induced by Implant-Supported Restorations

This numerical simulation can be validated in order to prove the reliability of the calculations. It should be noted that supportive observations from in vivo or clinical studies are scarce, especially regarding data from patients with bruxism or malocclusion. In this study, the radiographs were provided by a dentist to qualitatively compare the resemblance between the computational remodeling results and the clinical data (Wang et al. 2013). For the cantilever prosthesis, it can be seen that at 18 months after the placement an increase in bone density was observed adjacent to the implant. By comparison, for the non-cantilever prosthesis, the radiographic view at 18 months post-surgery demonstrated that marginal bone density around the neck of the implants was lower than that associated with the cantilevered FPD. This confirms the previously described regions of simulated bone remodeling under mechanical conditions (Figure 20.6). It can be concluded that a relatively good agreement was achieved between the simulation result and the clinical bone density distribution after implant placement.

## 20.5 CONCLUSIONS

The study involving numerical simulations presented in this chapter was intended to demonstrate how bone remodeling computational methods can contribute to a great extent to the field of dental biomechanics, including orthodontics and prosthodontics. To assess the biomechanical response from orthodontic loading and implantation, the existing FE method has proven to be an effective way to capture the geometrical and material complexities involved. However, bone is a metabolically active tissue capable of forming its structure and material properties via the process of bone remodeling. Simulations of bone remodeling can be employed in the investigation of bone biological responses. Remodeling numerical techniques have the potential to create specific tools to help dentists in their preoperational planning to estimate the effectiveness of operational practices as well as assist bioengineers in their implant design process to optimize solutions for the improvement of an implant's longevity.

Lastly, it should be pointed out that although the present mathematical algorithm has the ability to predict alveolar bone response through bone remodeling, more substantial biological modeling algorithms should be introduced into the computer simulation. New methods should be developed that consider further biomechanical and mechanobiological influences. The gap between biological theory and computational biomechanics communities should be minimized in the future.

## ACKNOWLEDGMENTS

This work was supported by the National Natural Science Foundation of China (Nos. 10925208, 11120101001, 11202017), the Beijing Natural Science Foundation (7133245), Young Scholars for the Doctoral Program of the Ministry of Education of China (20121102120039), and the 111 Project (B13003).

## REFERENCES

Beaupre, G. S., T. E. Orr, and D. R. Carter. 1990a. An approach for time-dependent bone modeling and remodeling—theoretical development. *J Orthop Res* 8 (5):651–61.

Beaupre, G. S., T. E. Orr, and D. R. Carter. 1990b. An approach for time-dependent bone modeling and remodeling-application: a preliminary remodeling simulation. *J Orthop Res* 8 (5):662–70.

Bourauel, C., D. Freudenreich, D. Vollmer, D. Kobe, D. Drescher, and A. Jager. 1999. Simulation of orthodontic tooth movements. A comparison of numerical models. *J Orofac Orthop* 60 (2):136-151.

Carter, D. R., W. C. Hayes, and D. J. Schurman. 1976. Fatigue life of compact bone—II. Effects of microstructure and density. *J Biomech* 9 (4):211–218.

Chou, H. Y., J. J. Jagodnik, and S. Muftu. 2008. Predictions of bone remodeling around dental implant systems. *J Biomech* 41 (6):1365–73.

Field, C., I. Ichim, M. V. Swain, E. Chan, M. A. Darendeliler, W. Li, and Q. Li. 2009. Mechanical responses to orthodontic loading: a 3-dimensional finite element multi-tooth model. *Am J Orthod Dentofacial Orthop* 135 (2):174–181.

Fyhrie, D. P., and D. R. Carter. 1986. A unifying principle relating stress to trabecular bone morphology. *J Orthop Res* 4 (3):304–17.

Geramy, A. 2000. Alveolar bone resorption and the center of resistance modification (3-D analysis by means of the finite element method). *Am J Orthod Dentofacial Orthop* 117 (4):399–405.

Gross U., Muller-Mai C., Fritz T. et al. 1990. *Implant surface roughness and mode of load transmission influence peri implant bone structure [M]*. New York: Elsevier.

Hasan, I., A. Rahimi, L. Keilig, K. T. Brinkmann, and C. Bourauel. 2012. Computational simulation of internal bone remodeling around dental implants: a sensitivity analysis. *Comput Methods Biomech Biomed Engin* 15 (8):807–14.

Henneman, S., J. W. Von den Hoff, and J. C. Maltha. 2008. Mechanobiology of tooth movement. *Eur J Orthod* 30 (3):299–306.
Huiskes, R., R. Ruimerman, G. H. van Lenthe, and J. D. Janssen. 2000. Effects of mechanical forces on maintenance and adaptation of form in trabecular bone. *Nature* 405 (6787):704–706.
Huiskes, R., H. Weinans, H. J. Grootenboer, M. Dalstra, B. Fudala, and T. J. Slooff. 1987. Adaptive bone-remodeling theory applied to prosthetic-design analysis. *Journal of Biomechanics* 20 (11–12):1135–50.
Iwasaki, L. R., J. E. Haack, J. C. Nickel, and J. Morton. 2000. Human tooth movement in response to continuous stress of low magnitude. *Am J Orthod Dentofacial Orthop* 117 (2):175–183.
Kingsmill, V. J., and A. Boyde. 1998. Variation in the apparent density of human mandibular bone with age and dental status. *J Anat* 192 (Pt 2):233–244.
Kojima, Y., and H. Fukui. 2005. Numerical simulation of canine retraction by sliding mechanics. *Am J Orthod Dentofacial Orthop* 127 (5):542–551.
Li, J., H. Li, L. Shi, A. S. Fok, C. Ucer, H. Devlin, K. Horner, and N. Silikas. 2007. A mathematical model for simulating the bone remodeling process under mechanical stimulus. *Dent Mater* 23 (9):1073–8.
Li, W., D. Lin, C. Rungsiyakull, S. Zhou, M. Swain, and Q. Li. 2011. Finite element based bone remodeling and resonance frequency analysis for osseointegration assessment of dental implants. *Finite Elements in Analysis and Design* 47:898–905.
Lin, C. L., Y. H. Lin, and S. H. Chang. 2010. Multi-factorial analysis of variables influencing the bone loss of an implant placed in the maxilla: prediction using FEA and SED bone remodeling algorithm. *Journal of Biomechanics* 43 (4):644–51.
Lin, D., Q. Li, W. Li, N. Duckmanton, and M. Swain. 2010. Mandibular bone remodeling induced by dental implant. *Journal of Biomechanics* 43 (2):287–93.
Lin, D., Q. Li, W. Li, P. Rungsiyakull, and M. Swain. 2009. Bone resorption induced by dental implants with ceramics crowns. *Journal of the Australian Ceramic Society* 45 (2):1–7.
Marangalou, J., F. Ghalichi, and B. Mirzakouchaki. 2009. Numerical simulation of orthodontic bone remodeling. *Orthodontic Waves* 68 (2):64–71.
McNamara, L. M., and P. J. Prendergast. 2007. Bone remodeling algorithms incorporating both strain and microdamage stimuli. *J Biomech* 40 (6):1381–1391.
Mellal, A., H. W. Wiskott, J. Botsis, S. S. Scherrer, and U. C. Belser. 2004. Stimulating effect of implant loading on surrounding bone. Comparison of three numerical models and validation by in vivo data. *Clin Oral Implants Res* 15 (2):239–248.
Milne, T. J., I. Ichim, B. Patel, A. McNaughton, and M. C. Meikle. 2009. Induction of osteopenia during experimental tooth movement in the rat: alveolar bone remodeling and the mechanostat theory. *Eur J Orthod* 31 (3):221–231.
Ojeda, J., J. Martinez-Reina, J. M. Garcia-Aznar, J. Dominguez, and M. Doblare. 2011. Numerical simulation of bone remodeling around dental implants. *Proc Inst Mech Eng H* 225 (9):897–906.
Palmer, R. M., L. C. Howe, P. J. Palmer, and R. Wilson. 2012. A prospective clinical trial of single Astra Tech 4.0 or 5.0 diameter implants used to support two-unit cantilever bridges: results after 3 years. *Clin Oral Implants Res* 23 (1):35–40.
Pilon, J. J., A. M. Kuijpers-Jagtman, and J. C. Maltha. 1996. Magnitude of orthodontic forces and rate of bodily tooth movement. An experimental study. *Am J Orthod Dentofacial Orthop* 110 (1):16–23.
Qian, Y., Y. Fan, Z. Liu, and M. Zhang. 2008. Numerical simulation of tooth movement in a therapy period. *Clin Biomech* 23 (Suppl 1):S48–52.
Qian, Y., Z. Liu, and Y. B. Fan. 2010. Numerical simulation of canine bodily movement. *International Journal for Numerical Methods in Biomedical Engineering* 26 (2):157–163.
Ren, Y., J. C. Maltha, M. A. Van't Hof, and A. M. Kuijpers-Jagtman. 2004. Optimum force magnitude for orthodontic tooth movement: a mathematic model. *Am J Orthod Dentofacial Orthop* 125 (1):71–77.
Ruimerman, R., P. Hilbers, B. van Rietbergen, and R. Huiskes. 2005. A theoretical framework for strain-related trabecular bone maintenance and adaptation. *J Biomech* 38 (4):931–941.
Schenk, R. K., Buser D. 1998. Osseointegration: a reality [J]. *Periodontology 2000* 17(1): 22–35.
Schneider, J., M. Geiger, and F. G. Sander. 2002. Numerical experiments on long-time orthodontic tooth movement. *Am J Orthod Dentofacial Orthop* 121 (3):257–265.
Shillingburg, H.T. 1997. *Fundamentals of fixed prosthodontics*, 3rd ed. Hanover Park, IL: Chicago Quintessence Publishing Co.
Verna, C., D. Zaffe, and G. Siciliani. 1999. Histomorphometric study of bone reactions during orthodontic tooth movement in rats. *Bone* 24 (4):371–379.
Wang, C., J. Han, Q. Li, L. Wang, and Y. Fan. 2014. Simulation of bone remodelling in orthodontic treatment. *Computer Methods in Biomechanics and Biomedical Engineering* 17(9):1042–50.

Wang, C., Q. Li, C. Mcclean, and Y. Fan. 2013. Numerical simulation of dental bone remodeling response around implant-supported fixed partial dentures with or without cantilever extension. *International Journal for Numerical Methods in Biomedical Engineering* 29(10):1134–47.

Wang, C., L. Wang, X. Liu, and Y. Fan. 2014. Numerical simulation of the remodelling process of trabecular architecture around dental implants. *Computer Methods in Biomechanics and Biomedical Engineering* 17(3):286–95.

Watzak, G., W. Zechner, C. Ulm, S. Tangl, G. Tepper, and G. Watzek. 2005. Histologic and histomorphometric analysis of three types of dental implants following 18 months of occlusal loading: a preliminary study in baboons. *Clin Oral Implants Res* 16 (4):408–416.

Weinans, H., R. Huiskes, B. van Rietbergen, D. R. Sumner, T. M. Turner, and J. O. Galante. 1993. Adaptive bone remodeling around bonded noncemented total hip arthroplasty: a comparison between animal experiments and computer simulation. *Journal of Orthopaedic Research* 11 (4):500–13.

Xinghua, Z., G. He, and G. Bingzhao. 2005. The application of topology optimization on the quantitative description of the external shape of bone structure. *J Biomech* 38 (8):1612–20.

Yee, J. A., T. Turk, S. Elekdag-Turk, L. L. Cheng, and M. A. Darendeliler. 2009. Rate of tooth movement under heavy and light continuous orthodontic forces. *Am J Orthod Dentofacial Orthop* 136 (2):150 e1-9; discussion 150–1.

# 21 Eye Model and Its Application

*Xiaoyu Liu, Lizhen Wang, Deyu Li, and Yubo Fan*

## CONTENTS



## SUMMARY

Injury to the eye can be a serious threat to vision if not treated immediately or appropriately. Blunt impact is a main cause of eye injuries, which usually occur in young people. Experimental impacts to the eye by a standardized projectile have been considered an accepted method to study blunt ocular trauma. However, in vivo high-speed imaging of the interior ocular structures is difficult to obtain experimentally. As an alternative method, finite element simulation can provide visualization of the dynamic response of the interior structures, as well as quantitative analysis. Previous studies have confirmed that the finite element method is an effective tool to analyze various ocular injuries in different blunt impact situations. In this chapter, a finite element model of the human eye is introduced. Modeling methods, including the geometric structure of a human eye, material properties of ocular tissues, meshing technology, and boundary conditions, are described in detail. Traumatic retinal detachment following blunt impact was simulated. Dynamic responses including stress wave propagation in the eye and resulting detachment are discussed in this chapter.

## 21.1 INTRODUCTION

The eye catches and focuses light and converts it to neural signals that are relayed to the brain, allowing us to interpret images. Epidemiological studies conclude that approximately 500,000 people suffer serious eye injury leading to blindness each year (Kuhn and Piermici, 2002). Trauma can lead to a wide spectrum of ocular tissue damage, including corneal abrasion, angle recession, detachment and segmentation of the iris and ciliary body, lens dislocation, zonules and lens capsule rupture, choroid and retinal detachment, and globe rupture (Sponsel et al., 2011; Gray et al., 2008). The Ocular Trauma Classification Group has developed a system to classify ocular trauma. According to the features of injuries, mechanical trauma (injuries caused by chemical, electrical, and thermal sources are not included) to the eye may be categorized into open and closed globe injuries. Blunt impact is a main cause of eye trauma and usually occurs in high-impact sports, such as boxing (Bianco et al., 2005), tennis (Nadeem et al., 2007), and diving (Chorich et al., 1998; Xu et al., 2006). In addition, the eye is also susceptible to blunt trauma when impacting an airbag in vehicle collisions (Kuhn et al., 1995; Manche et al., 1997).

Experimental testing of eye impact by a standardized projectile (a BB with a diameter of 4.5 mm and a mass of 0.375 g) is an effective approach to study blunt ocular trauma. Using high-speed cameras, previous studies have been conducted on human or animal eyes to investigate various ocular blunt traumas (Delori et al., 1969; Scott et al., 2000; Pahk and Adelman, 2009; Sponsel et al., 2011). Although high-speed cameras can record deformation and displacement of the globe and the projectile object, in vivo imaging of the inner structures of the eye is still unavailable. Compared to experimental methods, numerical simulation not only allows various mechanical conditions and parametric tests to be performed but also provides quantitative analysis. Various ocular injuries have been previously investigated using the finite element method. Uchio et al. (1999) developed the first dynamic eye model to study mechanical conditions for intraocular foreign body (IOFB) injury. Later, a more sophisticated eye model named the Virginia Tech eye model (VTEM) was introduced by Stitzel et al. (2002). The VTEM was empirically validated to predict globe rupture at a stress of 23 MPa in the corneoscleral shell. Other simulation studies related to traumatic ocular injuries investigated optic nerve damage (Cirovic et al., 2006), corneal rupture after photorefractive keratectomy (PRK) and laser-assisted in situ keratomileusis (LASIK) (Mousavi et al., 2012), retinal detachment (Hans et al., 2009; Rangarajan et al., 2009), and retinal damage (Rossi et al., 2011). Previous studies have confirmed that the finite element method is an effective tool to analyze various ocular injuries under different impact conditions.

## 21.2 DEVELOPMENT OF THE FINITE ELEMENT EYE MODEL

The human eye is roughly spherical in shape and about 24 mm in diameter. Several structures compose the human eye. Among the most important anatomical components are the cornea, sclera, crystalline lens, ciliary, zonules, iris, optic nerve, choroid, retina, and the vitreous and aqueous humors (Figure 21.1).

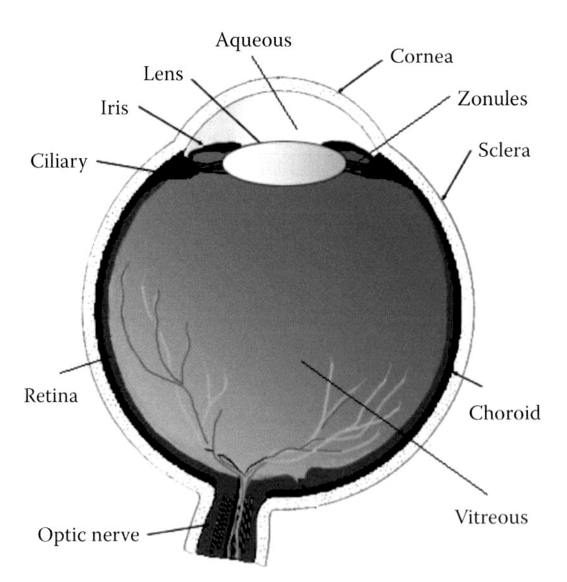


**FIGURE 21.1** **(See color insert.)** Ocular structure: cornea, sclera, iris, ciliary body, zonules, lens, optic nerve, retina, choroid, and vitreous and aqueous humors.

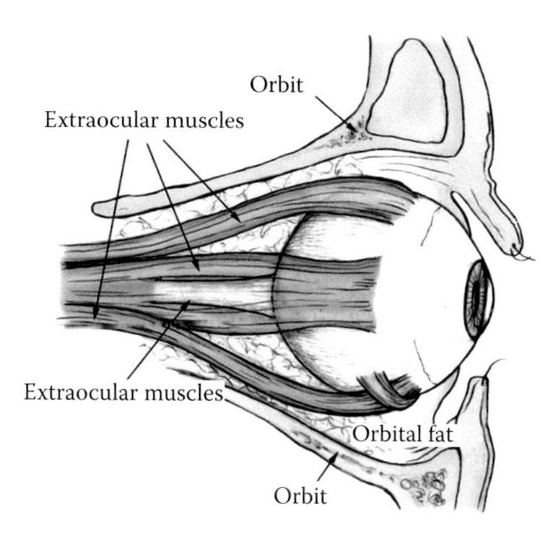


**FIGURE 21.2** Orbit and eye appendages: orbit, orbital fat, and extraocular muscles.

### 21.2.1 Eye Anatomy

The outermost layer of the eyeball is the sclera, which is a thin shell protecting intraocular tissues in the eye. The front part of the sclera is known as the cornea. It is a transparent tissue that provides about two-thirds of the total focusing power of the eye. Behind the cornea is a fluid known as aqueous humor, which helps to maintain the shape of the anterior chamber of the eyeball. The tissue behind the aqueous humor is a colored ring called the iris. It controls the amount of light entering the eye through the pupil, an opening in the iris. The area of the eye bounded by the cornea, iris, pupil, and lens is known as the anterior chamber. The posterior chamber of the eye is the space behind the iris and in front of the lens. It is filled with aqueous humor. Behind the pupil and iris are the crystalline lens and the ciliary muscle. The muscle holds the lens in place and changes its shape. Its function is to focus light rays onto the retina.

Before light enters the retina, it first must pass through the entire inside of the eye, known as the vitreous chamber. This chamber constitutes almost 70% of the total volume of the eyeball and is filled with a transparent gelatinous material called the vitreous humor. The retina is the innermost layer of the eye and converts images into electrical impulses sent along the optic nerve for transmission back to the brain. The extraocular tissues include the orbit, orbital fat, and extraocular muscles (Figure 21.2). The orbital fat functions as a protective cushion that lines the bony orbit supporting the eye. The orbit is the cavity of the skull in which the eye and its appendages are situated. In an adult human, the volume of the orbit is approximately 30 ml, of which the eye occupies 6.5 ml. The extraocular muscles are a set of six muscles that control movement of the eye. Understanding the anatomy of the eye is necessary for constructing a geometrical model.

### 21.2.2 Geometry

An idealized model of a human eye was derived from anatomical observations and a published eye model (Stitzel et al., 2002). The eye model was created in a computer-aided design (CAD) software package, SolidWorks 2010 (Dassault Systems, SolidWorks Corp., SA). The model includes the cornea, sclera, lens, ciliary, zonules, retina, and aqueous and vitreous humors, while the structure of the iris and choroid were ignored. The optic nerve head (ONH) was assumed to be located at the posterior pole of the eye. An opening with a 1.8 mm radius was made in the sclera shell, through which the optic nerve passes (Sigal et al., 2010). The cross-sectional

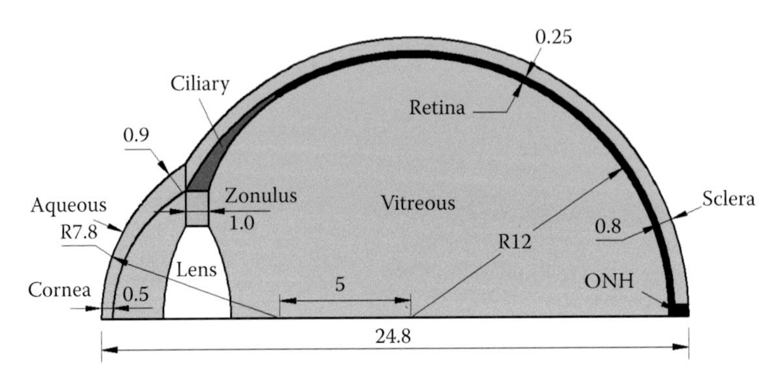


**FIGURE 21.3** Structures and dimensions of the geometric eye model in cross-section.

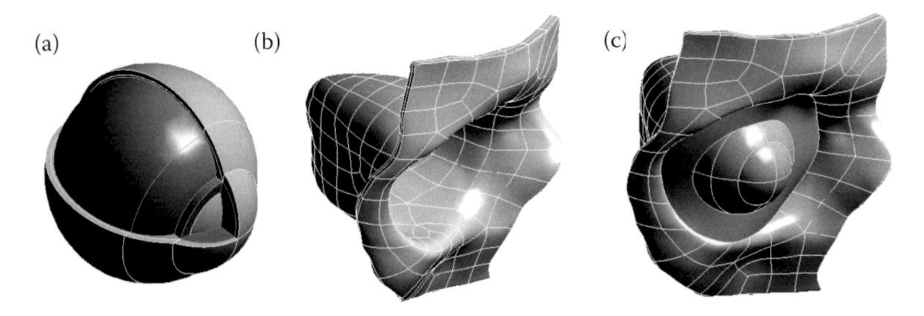


**FIGURE 21.4** Geometric model of the globe (a), the orbit bone (b), and the filling fat (c).

structure of the eye model is shown in Figure 21.3. A three-dimensional orbit model was reconstructed based on computed tomography (CT) image sequences from a 28-year-old male subject. The subject gave for use of his CT images for medical research purposes. Model reconstruction was performed in MIMICS 10.01 (Materialise NV, Belgium). The space between the eyeball and the orbital wall was filled with orbital fat. The geometric model of the orbit bone and fat is shown in Figure 21.4. Extraocular muscles were excluded from the model because previous studies concluded that the muscles have little influence on the structural mechanical response when the eye is subjected to dynamic impact (Kennedy and Duma, 2008).

### 21.2.3 Material Properties

The material properties of the cornea, sclera, lens, ciliary body, zonules, and aqueous humor were derived from the Virginia Tech-Wake Forest University (VT-WFU) eye model (Duck, 1990; Czyan et al., 1996; Uchio et al., 1999; Stitzel et al., 2002; Power et al., 2002). The retina was assumed to be elastic with a Young's modulus of 20 kPa (Jones et al., 1992). Vitreous and fatty tissues were both treated as viscoelastic materials (Lee et al., 1992; Schoemaker et al., 2006). The orbital bone was assigned with a Young's modulus of 14.5 GPa (Robbins and Wood, 1969). It should be noted that a four-parameter model equivalent to the Burgers model was used to describe the viscoelastic properties of the vitreous in Lee's work. However, the Burgers model is incompatible with the related parameters in the current model. In ANSYS, viscoelastic materials perform as a linear solid viscoelastic model with a Maxwell element in parallel with a spring element. The compliance function $J_0(t)$ for Burgers viscoelastic model can be expressed as

$$J_0(t)^B(t)=\left(\frac{1}{G_m^B}+\frac{1}{G_k^B}\right)+\frac{t}{\eta_m^B}-\frac{1}{G_k^B}\exp(-\frac{G_k^B}{\eta_k^B}t)$$

**TABLE 21.1**
**Mechanical Properties and Element Types in the Eye Model**

| Structures | Density (Kg m$^{-3}$) | Material Model | Material Parameters | Source |
|---|---|---|---|---|
| Cornea | 1076 | Elastic | Nonlinear stress-strain | Union et al. (1999) |
| Sclera | 1243 | Elastic | Nonlinear stress-strain | Union et al. (1999) |
| Lens | 1078 | Elastic | E = 6.88 MPa | Czyan et al. (1996) |
| Zonules | 1000 | Elastic | E = 357.78 MPa | Power et al. (2002) |
| Ciliary | 1600 | Elastic | E = 11 MPa | Power et al. (2002) |
| Retina | 1100 | Elastic | E = 20 kPa | Jones et al. (1992) |
| Aqueous | 1000 | Liquid | Shock EOS linear $C_1$ = 1530 m/s, $s_1$ = 2.1057 | Duck (1990) |
| Vitreous | 950 | Viscoelastic | $G_0$ = 10 Pa, $G_\infty$ = 0.3 Pa, B = 14.26 1/s, K = 2.0 GPa | Lee et al. (1992) |
| Fat | 970 | Viscoelastic | $G_0$ = 0.9 kPa, $G_\infty$ = 0.5 kPa, B = 0.2, K = 2.2 GPa | Schoemaker et al. (2006) |
| Orbit | 1610 | Elastic | E = 14.5 GPa | Robbins and Wood (1969) |

*Notes:* E, elastic modulus; K, bulk modulus; $G_0$, initial shear modulus; $G_\infty$, infinite shear modulus; β, viscoelastic decay constant; $C_1$, speed of sound through the material; $s_1$, coefficient related to the speed of the shocked material.

The compliance function $J_0(t)$ for a standard linear solid viscoelastic model can be expressed as

$$J_0^m(t) = \frac{1}{G_e^M} - \frac{G_m^M}{G_e^m(G_e^M + G_m^M)} \exp(-\frac{t}{\tau_s^M})$$

The parameters of the current model were obtained by curve fitting the experimental data for Burgers model. Parameters from the fitting were obtained and applied to our eye model. The material parameters of ocular tissues are listed in Table 21.1.

### 21.2.3 Mesh Creation and Boundary Condition

The mesh was created using special meshing software, ANSYS ICEM CFD. This software can generate high-quality volume or surface meshes for finite element solutions. During simulation, tetrahedron elements are known to exhibit mesh locking for incompressible materials. In the eye model, the vitreous, aqueous, and fat have the properties of incompressibility (Poisson's ratio close to 0.49). A hexahedron mesh was applied to these tissues. A hexahedron mesh also has greater calculation accuracy and efficiency compared to a tetrahedron mesh. Figure 21.5 illustrates the process of mesh creation in ANSYS ICEM CFD (taking the vitreous as an example). ICEM can mesh geometry created using other dedicated CAD packages, such as SolidWorks. According to the geometrical features of the vitreous, the model was separated into several independent surfaces (Figure 21.5a). Once the geometry was created, the next step was to generate a suitable block. Using the blocking tab, a block can be created around the entire geometry and then split up into several subsections (Figure 21.5b). Once the blocks have been created and a meshing density set, the premesh tool can then be used to view the meshing (Figure 21.5c). Finally, after confirming the quality of the mesh, a satisfactory model can be obtained (Figure 21.5d).

All the structures, except the orbit, were meshed using hexahedron elements. The orbit model used a tetrahedron mesh because of its irregular geometry and because a tetrahedral mesh is more suitable for complicated volumes (Figure 21.6). The cornea, sclera, lens, zonules, and ciliary were grouped into a "multibody part" to permit node-sharing. The boundary relationships between orbit

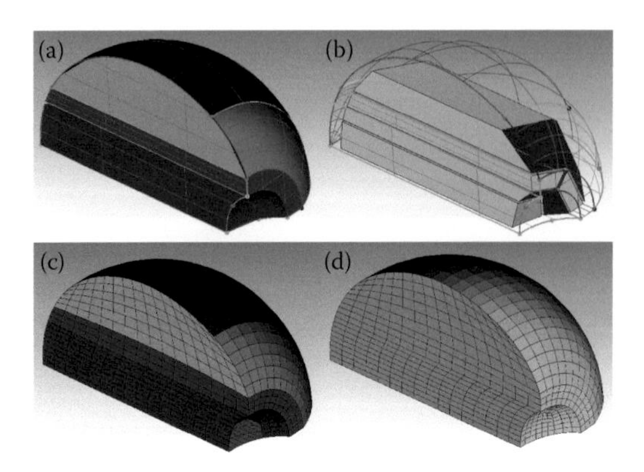


**FIGURE 21.5** **(See color insert.)** Process of mesh generation for the vitreous model. (a) Surface creation. (b) Block division. (c) Mesh preview. (d) Mesh generation.

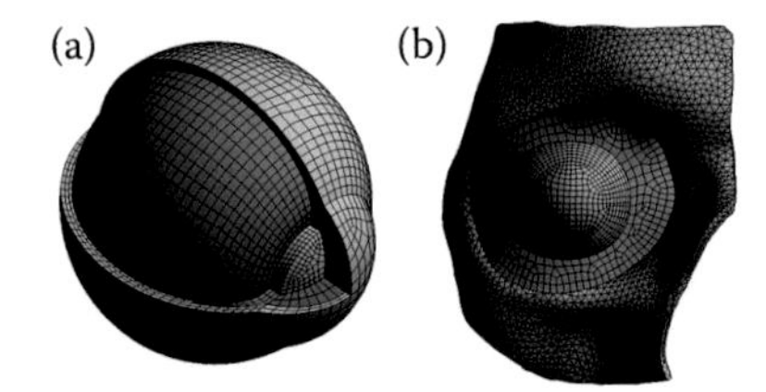


**FIGURE 21.6** Finite element model of the human eye. (a) A hexahedron mesh was used for the cornea, sclera, lens, ciliary body, zonules, retina, and aqueous and vitreous humors. (b) A tetrahedron mesh was used for the complex structure of the orbit.

and fat, optic nerve and retina or sclera, and vitreous or aqueous and its neighboring tissues were defined with bonded contacts, obligating them to move together at their respective interfaces.

### 21.2.4 Model Validation

To validate the current eye model, three types of blunt projectile impacts (BB, foam, and baseball) were undertaken to reproduce the simulations of the VT-WFU eye model by Stitzel et al. (2002). Two different initial speeds were given to each type of projectile. Table 21.2 lists the details on the loading information.

**TABLE 21.2**
**Loading Information for the Matched Simulations**

| Simulation | Objects | Mass (g) | Diameter (mm) | Velocity (m/s) | Modulus (MPa) |
|---|---|---|---|---|---|
| S1 | BB | 0.375 | 4.50 | 56.0 | 200,000 |
| S2 | BB | 0.375 | 4.50 | 92.0 | 200,000 |
| SF1 | Foam | 0.077 | 6.35 | 10.0 | 2.208 |
| SF2 | Foam | 0.077 | 6.35 | 30.0 | 2.208 |
| SB1 | Baseball | 146.5 | 76.10 | 34.4 | 12 |
| SB2 | Baseball | 146.5 | 76.10 | 41.2 | 12 |

The matched simulation results are shown in Figure 21.7. The comparison of the results includes the peak stress, peak strain, and peak deflection of the two models. Significant positive correlation was found with the Pearson correlation test (R = 0.982 for peak stresses, R = 0.983 for peak strains, and R = 0.996 for peak deflections). It indicates that the current eye model has the capability to predict dynamic responses to different impact loads. Table 21.3 lists the comparison of results between the current and VT-WFU eye models.

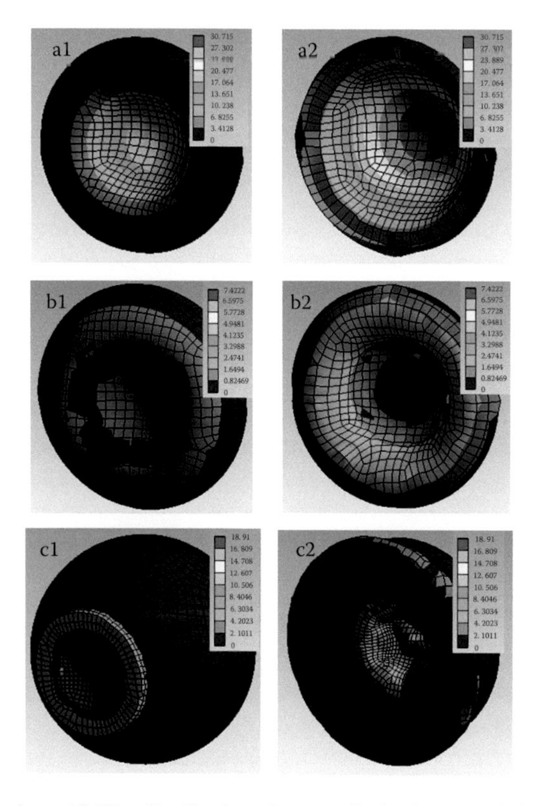


**FIGURE 21.7** **(See color insert.)** The distribution of max principal stress of the globe in the six matched simulations: (a) stress distribution for low (left) and high (right) speed BB simulation; (b) stress distribution for low (left) and high (right) speed foam simulation; and (c) stress distribution for low (left) and high (right) speed baseball simulation.

## TABLE 21.3
## Comparison of Results between the Current and VT-WFU Eye Models in Matched Simulations

| | Peak Stress (MPa) | | Peak Strain | | Peak Deflection (mm) | |
|---|---|---|---|---|---|---|
| **Simulations** | **Current** | **VT-WFU** | **Current** | **VT-WFU** | **Current** | **VT-WFU** |
| S1 | 23.561 | 22.812 | 0.519 | 0.437 | 6.060 | 6.420 |
| S2 | 31.734 | 32.757 | 0.702 | 0.732 | 7.836 | 7.790 |
| SF1 | 3.091 | 3.180 | 0.007 | 0.009 | 1.126 | 1.300 |
| SF2 | 7.109 | 7.830 | 0.182 | 0.250 | 2.891 | 2.670 |
| SB1 | 22.081 | 22.153 | 0.261 | 0.309 | — | — |
| SB2 | 23.340 | 24.167 | 0.331 | 0.360 | — | — |
| **Correlation** | R = 0.981 | | R = 0.977 | | R = 0.997 | |

## 21.3 MODEL APPLICATION: TRAUMATIC RETINAL DETACHMENT

The retina is a thin-layer tissue that covers the back inside wall of the human eye. A detachment occurs when the retina is pulled away from its normal position. Epidemiological studies indicate that the annual incidence of retinal detachment could be about 0.1 per thousand. Among them, traumatic detachments account for 10%–20% of the total detachments (Brinton and Wilkinson, 2009). For retinal detachment to occur, the retina must break and then the liquid vitreous passes through the break into the subretinal space (Brinton and Wilkinson, 2009). Blunt ocular trauma can lead to both retinal break and vitreous liquefaction. A finite element model of the human eye was developed to simulate retinal detachment when subjected to BB impact. Retinal adhesion was incorporated into the eye model to provide more realistic interaction between the retina and its support tissue. The simulation results provide a complete understanding of the mechanism of traumatic retinal detachment.

Retinal adhesion (maintaining the retina attached to the supporting tissue) was incorporated into the eye model using a breakable bonded contact. Initially, the retina and the corneoscleral shell were connected with a bonded contact. When subjected to blunt impact, stress can be induced in the contact (Figure 21.8). The contact may break when the stress exceeds a specified stress limit, which can be expressed as

$$\left(\frac{\sigma_n}{\sigma_n^{limit}}\right)^n + \left(\frac{|\sigma_s|}{\sigma_s^{limit}}\right)^m \geq 1$$

where $\sigma_n$ is the normal stress computed at the breakable contact, $\sigma_n^{limit}$ is the specified normal stress limit, $n$ is the normal stress exponent, $\sigma_s$ is the shear stress computed at the breakable contact, $\sigma_n$ is the specified shear stress limit, and $\sigma_s^{limit}$ is the normal stress exponent.

The retinal adhesive force was defined as the normal stress limit $\sigma_n^{limit}$ while the shear stress limit $\sigma_s^{limit}$ was ignored. The adhesive force can be obtained using methods detailed by Kita et al. (1990). However, different from Kita's study, we aimed to obtain the force per unit area (pressure). By duplicating Kita's experiment, the retinal adhesive force per unit area was obtained in living rabbits. Calculated from ten measurements, the retinal adhesive force was evaluated to be 340 ± 78 Pa.

Simulations of BBs impacting toward the corneal apex were conducted with different projectile speeds in ANSYS 14.0. Previous research indicated that retinal damage would occur when the eye was subjected to BB impact at a speed of 62.3 m/s (Delori et al., 1969). At such speeds, the corneoscleral shell can protect the eye from rupture (Stitzel et al., 2002). Therefore, three different speeds were selected (20 m/s, 40 m/s, and 62.5 m/s) to provide sufficient dynamic loading for retinal detachment without global rupture. The interaction between the BB and the cornea was defined as frictionless. The boundary of the orbit model was fixed to limit its movement.

According to Delori et al. (1969), the process of eye globe deformation is divided into four stages: compression, decompression, overshooting, and longtime oscillation. The stress

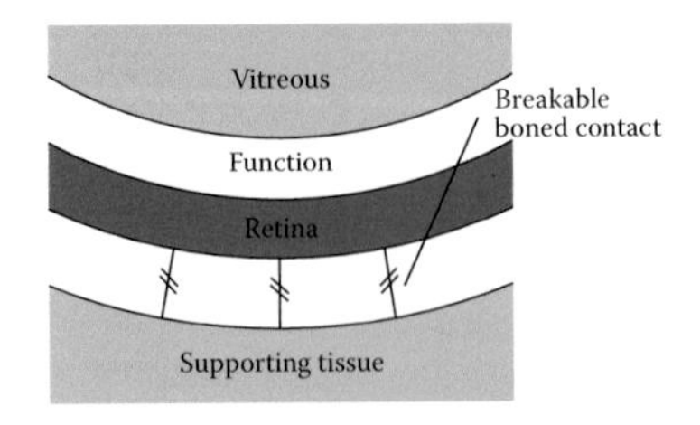


**FIGURE 21.8** Diagram of breakable bonded contact for retinal adhesion.

wave propagation in the retina and the peak stress time history from 0 to 3 ms are shown in Figure 21.9. At six typical time nodes, the stress distributions in the retina are drawn (Figure 21.9a–f). The propagation of the stress wave starts in the ora serrata at 0.14 ms, then extends to the equator and concentrates on the macula at 0.73 ms. High stresses in the retina are found in the region of the ora serrata, equator, and macula. After 1.5 ms, the stress in the retina begins to fluctuate but never exceeds 10 KPa.

After the BB collides with the eye globe, deformation and stress were generated by the dynamic loading from the projectile impact; the location of high stress constantly varies under dynamic loading. The disturbance produced by the collision travels throughout the globe to form stress waves. The propagating stress wave in the intraocular structures causes transient local stress concentration, which could lead to retinal break. Retinal break is related to soft tissue failure, whose mechanism has not been well understood due to its complex material properties. However, stress-based criterion for the material failure of soft tissue has been widely accepted in numerical simulations. Based on this criterion, the soft tissue will damage or break when the stress exceeds its ultimate failure strength. The stress distribution and variation in the retina indicate that a peak stress of 34 KPa in the macula exceeded the failure threshold from experimental measurements (Wollensak and Spoerl, 2004).

Detachment between the retina and the corneoscleral shell is shown in Figure 21.10. The detachments show various extents under different projectile speeds. For lower speeds, the separations are limited to the small areas around the equator (Figure 21.10a). When subjected to

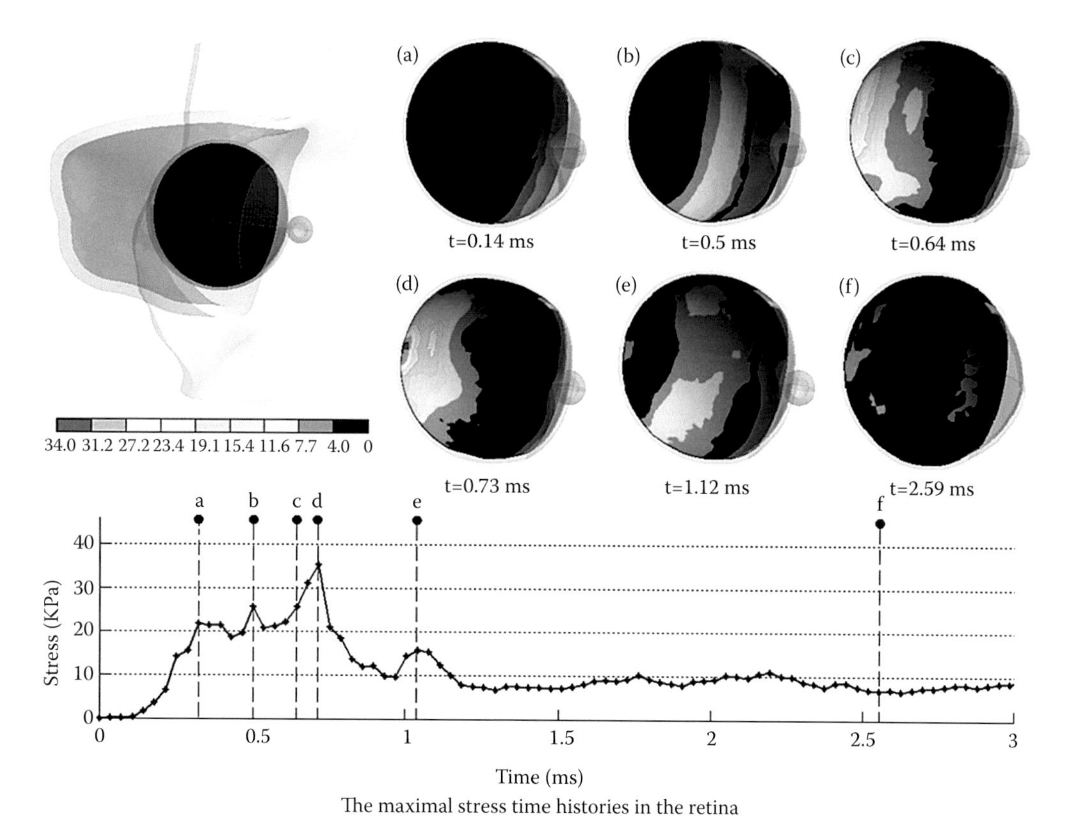


**FIGURE 21.9 (See color insert.)** The stress wave propagating in the retina (projectile speed of 62.5 m/s). The profile of the maximal stress time histories during the simulation from 0 to 3 ms was drawn. The stress distributions in the retina at six time nodes are illustrated.

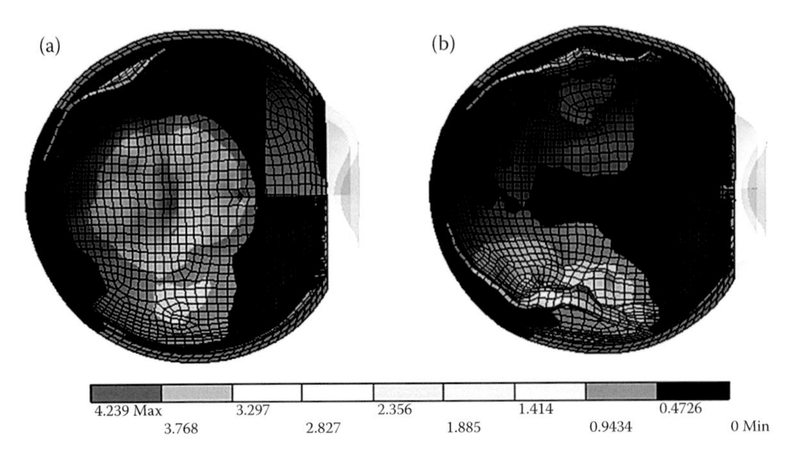


**FIGURE 21.10** **(See color insert.)** Retinal detachments simulated at lower (a) and higher (b) impacting speeds by BB projectile.

greater impact speeds, the separation extends toward the posterior pole and increases in size (Figure 21.10b).

## 21.3 CONCLUSIONS

In this chapter, a finite element model of the human eye was introduced. The modeling method was presented in detail, including the geometric structure of the human eye, material properties of ocular tissues, meshing technology, and boundary conditions. ANSYS Explicit Dynamics was employed to simulate retinal dynamic responses to projectile impact. As a robust analysis tool, the explicit dynamic solution is ideal for simulating physical events that happen over a short time frame. The eye model highlights modeling retinal adhesion using breakable bonded contact between the retina and the supporting tissue. This contact, originally used to model spot welds, was then used to simulate retinal detachment.

However, the current eye model is far from complete. The most significant limitation is lack of experimental validation. At present, high-speed imaging for the intraocular tissues is unavailable, making such experiments extremely difficult, if not impossible, to conduct. Therefore, other methods were used to validate the model. By comparison with a well-known eye model by Stitzel et al. (2002), the current model was verified to simulate the dynamic response of the cornea and sclera effectively. The model can provide details of the retinal dynamic response under blunt impact, which enables us to understand the mechanism of traumatic retinal detachment. This work confirms that the finite element method is an effective approach to studying eye injuries caused by blunt trauma.

## ACKNOWLEDGMENTS

This work was supported by the National Natural Science Foundation of China (Nos. 10925208, 11120101001, 11202017), the Beijing Natural Science Foundation (7133245), Young Scholars for the Doctoral Program of the Ministry of Education of China (20121102120039), and the 111 Project (No. B13003).

## REFERENCES

Bianco M., Vaiano A.S., Colella F., Coccimiglio F., Moscetti M., Palmieri V., Focosi F., Zeppilli P., Vinger P.F., 2005. Ocular complications of boxing. *British Journal of Sports Medicine* 39, 70–74.

Brinton D.A., Wilkinson C.P., 2009. *Retinal Detachment: Principles and Practice*. Oxford University Press, New York, pp. 9–39.

Chorich L.J., Davidorf F.H., Chambers R.B., Weber P.A., 1998. Bungee cord-associated ocular injuries. *American Journal of Ophthalmology* 125, 270–272.

Cirovic S., Bhola R.M., Hose D.R., Howard I.C., Lawford P.V., Marr J.E., Parsons M.A., 2006. Computer modeling study of the mechanism of optic nerve injury in blunt trauma. *British Journal of Ophthalmology* 90, 778–783.

Czygan G., Hartung C., 1996. Mechanical testing of isolated senile human eye lens nuclei. *Medical Engineering & Physics* 18, 345–349.

Delori F., Pomerantzeff O., Cox M., 1969. Deformation of the globe under high speed impact: its relation to contusion injuries. *Investigative Ophthalmology & Visual Science* 8, 290–301.

Duck F.A., 1990. *Physical Properties of Tissue: A Comprehensive Reference Book*. Academic Press, London, pp: 77–78, 138.

Gray W., Weiss C.E., Walker J.D., Sponsel W.E., 2008. Computational and experimental study of paintball impact ocular trauma. *Proceedings of the 24th International Symposium on Ballistics*, New Orleans, LA, 2, 1260–1267

Hans S.A., Bawab S.Y., Woodhouse M.L., 2009. A finite element infant eye model to investigate retinal forces in shaken baby syndrome. *Graefes' Archive for Clinical and Experimental Ophthalmology* 24, 561–571.

Jones I.L., Warner M., Stevens J.D., 1992. Mathematical modeling of the elastic properties of retina: a determination of Young's modulus. *Eye (Lond)* 6, 556–559.

Kennedy E., Duma S., 2008. The effects of the extraocular muscles on eye impact force-deflection and globe rupture response. *Journal of Biomechanics* 41, 3297–3302.

Kita M., Marmor M.F., 1992. Retinal adhesive force in living rabbit, cat, and monkey eyes. *Investigative Ophthalmology & Visual Science* 33(6), 1879–1882.

Kita M., Negi A., Kawano S., Honda Y., Maegawa S., 1990. Measurement of retinal adhesive force in the in vivo rabbit eye. *Investigative Ophthalmology & Visual Science* 31, 624–628.

Kuhn F., Morris R., Witherspoon C.D., 1995. Eye injury and the air bag. *Current Opinion in Ophthalmology* 6, 38–44.

Kuhn F., Piermici D.J., 2002. *Ocular Trauma: Principles and Practice*. Thieme, New York, pp. 206–234.

Lee B., Litt M., Buchsbaum G., 1992. Rheology of the vitreous body. Part I: viscoelasticity of human vitreous. *Biorheology* 29, 521–533.

Manche E.E., Goldberg R.A., Mondino B.J., 1997. Air bag-related ocular injuries. *Ophthalmic Surgery and Lasers* 11, 246–250.

Nadeem Q., Muhanmmad A., Mizan R.M., Nadeem I., Muhanmmad M.C., Wajid A.K., 2007. Traumatic retinal detachment due to tennis ball injury. *Pakistan Journal of Ophthalmology* 23, 151–154.

Pahk P.J., Adelman R.A., 2009. Ocular trauma resulting from paintball injury. *Graefes' Archive for Clinical and Experimental Ophthalmology* 247, 469–475.

Power E.D., Duma S.M., Stitzel J.D., Herring I.P., West R.L., Bass C.R., Crowley J.S., Brozoski F.T., 2002. Computer modeling of airbag-induced ocular injury in pilots wearing night vision goggles. *Aviation, Space and Environment Medicine* 73, 1000–1006

Rangarajan N., Kamalakkannan S.B., Hasija V., Shams T., Jenny C., Serbanescu I., Ho J., Rusinek M., Levin A.V., 2009. Finite element model of ocular injury in abusive head trauma. *Journal of AAPOS* 13, 364–369.

Robbins D.H., Wood J.L., 1969. Determination of mechanical properties of the bones of the skull. *Experimental Mechanics* 9, 236–240.

Rossi T., Boccassini B., Esposito L., Iossa M., Ruggiero A., Tamburrelli C., Bonora N., 2011. The pathogenesis of retinal damage in blunt eye trauma: Finite element modeling. *Investigative Ophthalmology & Visual Science* 52, 3994–4002.

Schoemaker I., Hoefnagel P.P.W., Mastenbroek T.J., Kolff F.C., Picken S.J., Helm F.C.T., Simonsz H.J., 2004. Elasticity, viscosity and deformation of retrobulbar fat in eye rotation. *Investigative Ophthalmology & Visual Science* 2004, 45: E-Abstract 5020.

Schoemaker I., Hoefnagel P.P., Mastenbroek T.J., 2006. Elasticity, viscosity, and deformation of orbit fat. *Investigative Ophthalmology & Visual Science* 47, 4819–4826.

Scott W.R., Lloyd W.C., Benedict J.V., Meredith R., 2000. Ocular injures due to projectile impacts. *Annual Proceedings of the Association for the Advancement Automotive Medicine* 44, 205–217.

Mousavi S.J., Nassiri N., Masoumi N., Nassiri N., Majdi N.M., Farzaneh S., Djalilian A.R., Peyman G.A., 2012. Finite element analysis of blunt foreign body impact on the cornea after PRK and LASIK. *Journal of Refractive Surgery* 28, 59–64.

Sigal I.A., Flanagan J.G., Tertinegg I., Ethier C.R., 2010. 3D morphometry of the human optic nerve head. *Experimental Eye Research* 90, 70–80.
Sponsel W.E., Gray W., Scribbick F.W., Stern A.R., Weiss C.E., Groth S.L., Walker J.D., 2011. Blunt eye trauma: empirical histopathologic paintball impact thresholds in fresh mounted porcine eyes. *Investigative Ophthalmology & Visual Science* 52, 5157–5166.
Stitzel J.D., Duma S.M., Cormier J.M., Herring I.P., 2002. A nonlinear finite element model of the eye with experimental validation of the prediction of globe rupture. *Stapp Car Crash Journal* 46, 81–102.
Uchio E., Ohno S., Kudoh J., Aoki K., Kisielewicz L.T., 1999. Simulation model of an eyeball based on finite element analysis on a supercomputer. *British Journal of Ophthalmology* 83, 1106–1111.
Uchio E., Watanabe Y., Kadonosono K., Matsuoka Y., Goto S., 2003. Simulation of airbag on eyes after photorefractive keratectomy by finite element analysis method. *Graefes' Archive for Clinical and Experimental Ophthalmology* 241, 497–504
Wollensak G., Spoerl E., 2004. Biomechanical characteristics of retina. *Retina* 24, 967–970.
Wu W., Peters W.H. 3rd, Hammer M.E., 1987. Basic mechanical properties of retina in simple elongation. *Journal of Biomechanical Engineering* 109, 65–67.
Xu L., Zhang X.Y., Liu A.Z. et al., 2006. Intraocular pressure monitoring in diving experiment of model eye with tension sensor. *Ophthalmology in China* 15, 271–273.

# 22 Temporomandibular Joint Model for Asymptomatic and Dysfunctional Joints

*Zhan Liu, Yuan-li Zhang, Ying-li Qian, and Yubo Fan*

## CONTENTS



## SUMMARY

The temporomandibular joints (TMJ) are the only bilateral linked joints in the human body, and are necessary for chewing, swallowing, speech, facial expression, and so on. Computational models can be used to understand the biomechanics of the joint and provide some clinical guidance. Three-dimensional finite element (FE) models of the human mandible and TMJs were developed from reconstructed computed tomography (CT) images of volunteers. The interfaces between the disc and cartilages of the condyle and the temporal bone were bonded together and simulated as gap elements and contact elements, respectively. The results showed that the final configuration and the stress distribution of the TMJ were reasonable when contact elements were simulated on the upper and lower interfaces of the disc, and nonlinear cable elements were used to simulate the disc attachments and mandibular ligaments. Based on these findings, this method was used to analyze the stress distribution of the TMJ under different occlusal loads and the biomechanical effects of typical temporomandibular disorders (TMD), that is, relaxation of discal attachment, disc displacement, and disc perforation. The validated models can be also used for studies of various clinical issues, such as the biomechanical effects of orthognathic surgery, oral and maxillofacial surgery, oral implants, orthodontics, prosthodontics, and so on.

## 22.1 INTRODUCTION

The temporomandibular joints (TMJs), located between the condyle of the mandible and the articular fossa-eminence of the temporal bone, are unique joints in the jaw. The TMJs are the only bilateral linked joints in the human body, and are necessary for chewing, swallowing, speech, and facial expressions. Moreover, TMJs have been proven to be load-bearing joints and in remodeling all the life. (Brehnan et al. 1981; Faulkner et al. 1987; Boyd et al. 1990; Throckmorton and Dechow 1994). The joint loads produced by mandibular movement play an important role in the joints' structure and function, and may be related to the etiology and treatment of TMJ disorders (Chen and Xu 1994; Chen et al. 1998; Tanaka et al. 2001a, 2004). Although some studies have measured TMJ loads in vivo (Brehnan et al. 1981; Faulkner et al. 1987) and in vitro (Throckmorton and Dechow 1994), the stress distributions in these joints are still not well known because of the difficulties in experimental measurements. The FE method has been largely used in dental biomechanics to understand the stresses and the deformations in the normal mandibles (Hart et al. 1992; Korioth and Hannam 1994a; Hu et al. 2003), to simulate the motions of the TMJ components (Chen and Xu 1994; Devocht et al. 1996; Chen et al. 1998), to analyze the stress distribution in the two discs and ligaments of each side during nonsymmetrical movements of a healthy joint (del Palomar and Doblare 2006c), to analyze the differences in the stress distribution of the TMJ between subjects with and without internal derangement (Tanaka et al. 2000, 2004; del Palomar and Doblare 2007), and so on. The FE method has been proven to be a useful tool for evaluating the TMJ.

In the past three decades, many FE models of TMJs or mandibles have been constructed for various biomechanical investigations. The articular surfaces of the condyle and the articular fossa-eminence complex are covered with cartilage layers, which are separated by an articular disc. The interaction between the cartilage layers and the disc is key for accurate simulation of the TMJ. The disc and cartilage layers are bonded together in many mandibular models to analyze the stress distributions within the mandibles or the TMJs (Tanaka et al. 2000, 2001a; Hart et al. 1992; Korioth and Hannam 1994a). In these models, the TMJs were treated as inactive joints that could not simulate the motions and the interactions of the joint components. Contact elements were introduced into the two-dimensional (2D) FE models of the TMJ to simulate the interaction between the disc and the articular surfaces (Chen and Xu 1994; Devocht et al. 1996; Chen et al. 1998). Similar research using three-dimensional (3D) FE analyses has been reported (Tanaka et al. 2004; Beek et al. 2000, 2001). However, the single TMJ modeled in these studies failed to portray the linked characteristics of the bilateral joints, and the load conditions were only applied through the condylar displacements. Subsequently, contact elements were also used to simulate the upper and lower interfaces of the discs in the 3D FE models of the mandible (Hu et al. 2003; del Palomar and Doblare 2006b, c, 2007; Koolstra and van Eijden 2005). Gap elements have also been used to simulate the interfaces between the articular disc and the cartilage layers in several FE models developed to analyze the stress distribution of the TMJs (Zhou et al. 1999; Castano et al. 2002). The normal behavior of the disc and the articular surfaces could also be simulated using gap elements, whereas the tangential behavior could not been simulated. Up to now, bond, gap, and contact elements have been used to simulate the interaction between the disc and the articular cartilages. The type of modeling is very important in order to obtain accurate results in stress analyses. The biomechanical environment in the TMJ is key to understanding the origin and progression of TMD (Tanaka et al. 2004; del Palomar and Doblare 2007), so a verified model is also useful for clinical practice. Therefore, it is necessary to compare the interaction in the TMJ with bond, contact, and gap elements (Liu et al. 2008).

TMD is the most common of TMJ diseases. The major symptoms of TMD are joint pain, abnormal joint tone, and movement disorders of the mandible. TMD can be caused by many factors, such as abnormal occlusion, joint trauma, bad habits (bruxism, clenching, unilateral chewing, etc.), loss of posterior teeth, and psychological factors. Except for masticatory muscle disorders, TMD is related to abnormal positioning between the disc and facies articularis of the condyle and articular fossa, such as occurence with disc displacements, relaxation of discal attachment, and disc perforation.

TMD may be caused by abnormal loads in the TMJ. On the other hand, TMD could also lead to abnormal stress distributions in the TMJ. Tanaka et al. evaluated the differences in stress distribution in the TMJ for models with and without anterior disc displacement during maximum occlusion (2000) and jaw opening (2004). They also analyzed the regularity of stress distribution in the TMJ during prolonged clenching and found that the maximum stress was located in the central and lateral zones of the disc (Tanaka et al. 2008). del Palomar and Doblare (2007) simulated anteriorly displaced discs with and without reduction during jaw opening. The results showed that anterior displacement of the disc could lead to greater compressive stress in the posterior band of the disc. They also compared the displacements and stress distributions of an anteriorly displaced disc without reduction and a surgically repositioned one with those of a healthy disc during jaw opening (del Palomar and Doblare 2006a). Roh et al. (2012) used MRI scans to study the relationships between anterior disc displacement, joint effusion, and degenerative changes of the TMJ with TMD. The results showed that anterior disc displacement was significantly related to degenerative changes of the condyles and joint effusions in patients with TMD. Koolstra (2012) analyzed the influence of friction on anterior disc displacement, and suggested that an increase in friction may not be the cause of anterior disc displacement. These studies of TMD were focused on anterior disc displacement. Other types of TMD should also be investigated, for example other directions of disc displacements, relaxation of discal attachment, and disc perforation.

In this chapter, a 3D FE model of the human mandible, disc, and fossa-eminence complex was developed based on CT images. The mandibular ligaments and the attachments of the discs were considered. Bond, gap, and contact elements were used to simulate the interaction between the discs and the cartilages in the TMJs. Differences between the three simulations were compared with the stress analyses of the models under centric occlusion. Then the effects of different occlusal loads (i.e., centric occlusion, anterior occlusion, and right side molar occlusion) on the stress distributions in TMJ were analyzed. FE models of three types of TMD were developed: relaxation of discal attachment (relaxation of anterior and posterior attachments), disc displacement (anterior, posterior, medial, and lateral disc displacement), and disc perforation. The stress distributions in TMJs were compared with a healthy TMJ.

## 22.2 COMPARATIVE RESEARCH OF TMJ SIMULATIONS

### 22.2.1 Development of a Mandible and TMJ Model

#### 22.2.1.1 Finite Element Modeling

The geometry of the model was based on CT images of a volunteer with normal occlusion and asymptomatic joints. The contours of the cortical and cancellous bone were digitized and imported into ANSYS 8.0 (Swanson Analysis System Co., Houston, TX) for constructing the 3D model of the mandible and the articular fossa-eminence complex. The teeth and periodontal tissues have been shown to have little influence on the stress distribution in the mandible, far from the alveolar bone (Hart et al. 1992). Because this study focused on simulations of the TMJs, the teeth were not included in the models. Fibrocartilage layers were simulated on the articular surfaces of the condyle and the temporal bone. Based on the anatomical structure, the thickness of the cartilage layers varied from 0.2 mm at the crests of the condyle and the articular fossa to 0.5 mm at the anterior surfaces of the condyle and the articular eminence (Pullinger et al. 1990). According to the shapes of the articular surfaces and the anatomy of the disc (Hansson and Nordstrom 1977), the models of the two articular discs were constructed with an anterior band of 2 mm, an intermediate zone of 1 mm, and a posterior band of 2.7 mm, as shown in Figure 22.1.

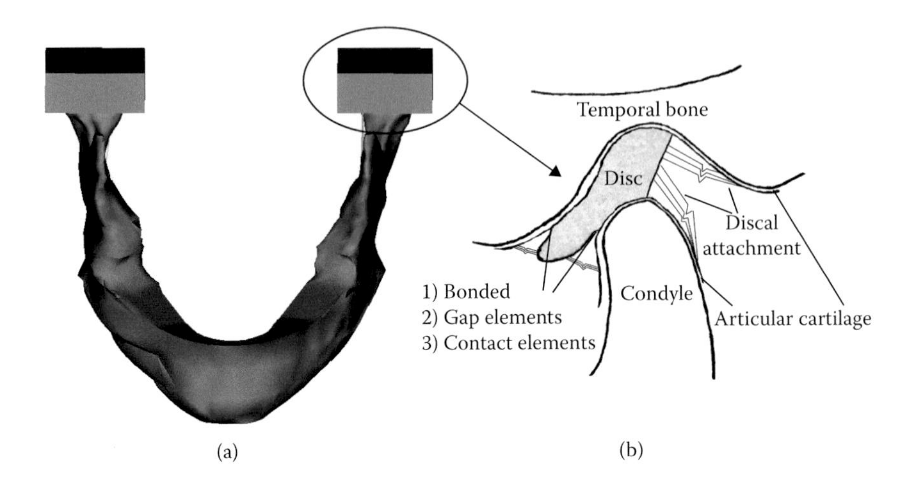


**FIGURE 22.1** (a) 3D model of the mandible and TMJ. (b) Details of the TMJ.

An FE mesh with 10-node quadratic tetrahedral elements was built using ANSYS free meshing techniques because of the inherently irregular geometries. The FE model consisted of 110,781 tetrahedral elements and 171,405 nodes in total after convergence.

The mechanical properties of the cortical bone, cancellous bone, and cartilage were assumed to be homogeneous, isotropic, and linearly elastic, and the articular disc to be non-linearly elastic (Tanaka et al. 2001b; Liu et al. 2008). The Young's modulus and Poisson's ratio of the cortical bone were 13,700 MPa and 0.3, respectively; those of the cancellous bone were 7,930 MPa and 0.3, respectively; and those of the cartilage were 0.79 MPa and 0.49, respectively. When the disc stress was equal to or less than 1.5 MPa, the Young's modulus of the disc was 44.1 MPa, and when the stress was greater than 1.5 MPa, the Young's modulus of disc was 92.4 MPa. The Poisson's ratio of the disc was 0.4.

#### 22.2.1.2 Simulations of TMJs

The temporomandibular ligaments, sphenomandibular ligaments, and stylomandibular ligaments were considered in the models due to their close relation to the functions of the TMJ. Ligaments scarcely bear compressive stress, so the nonlinear cable elements were used to simulate the ligaments. Stiffness values were assigned to the ligaments according to previous experiments (Race and Amis 1994; Muralidhar et al. 2000). The articular disc attachments, including the temporal anterior attachment, mandibular anterior attachment, and bilaminar zones, were modeled as nonlinear cable elements with referenced stiffness (Chen and Xu 1994).

In this study, the interface between the disc and the cartilages of the condyle and the temporal bone were bonded together and simulated as gap elements or contact elements, called the bond model, gap model, and contact model, respectively. In the bond model, the upper interfaces of the discs and the temporal cartilages were tied, and the lower interfaces of the discs and the condylar cartilages were also tied. Therefore, the discs could not move between the cartilage layers. In the gap model, the discs and the cartilage layers were linked by 171 gap elements. The gap elements undergo compressive forces as the two objects approach each other, but could not undergo tensile forces. Moreover, the frictional behavior between the discs and the cartilage layers could not be simulated using the gap elements. In the contact model, 10,542 contact elements between the discs and the cartilage layers were obtained, and friction/slide was considered. The frictional coefficient between the disc and the articular cartilages was important for determining the stress distribution in the TMJ and the disc movement. In the synovial joints, the frictional coefficients ranged from 0.001 to 0.1 (Mabuchi et al. 1999; Forster and Fisher 1996). In this study, a frictional coefficient of 0.001 was chosen according to a related study on asymptomatic TMJs (Tanaka et al. 2004).

#### 22.2.1.3 Loading and Boundary Conditions

The magnitude of masticatory muscle forces reaches its maximum under centric occlusion. Therefore, the force vectors of the bilateral masticatory muscles (i.e., the superficial and deep masseter, anterior and posterior temporalis, medial pterygoid, and superior and inferior lateral pterygoid) corresponding to centric occlusion were applied to the three models. The magnitude of each muscle force was assigned according to its physiological cross section and the scaling factor (Koolstra et al. 1988). In addition, the origin and direction of each muscle force were defined from anatomical measurements (Faulkner et al. 1987). Boundary conditions had the models fixed at the occlusal surface and the external regions of the temporal bone (Tanaka et al. 2000, 2001b; Hu et al. 2003).

### 22.2.2 Analysis of the TMJ Model and Applications

As expected, the movement of the condyles and the discs was observed to be different among the three models. The interaction between the discs and the cartilages and the tensile forces in the disc attachments was related to the movement of the discs. The condyles moved backwards in the three models, and the maximum displacements were 0.07 mm in the bond model and 0.3 mm in the contact and gap models. In the bond model, the discs and the articular cartilages were bonded together, so the movement of the condyles was restricted by the fixed temporal bones and the discs couldn't move between the articular surfaces. The anterior and posterior attachments of the discs were in tension, but the bonded discs and the articular cartilages did not produce sufficient tensile forces in the discal attachments.

In the gap model, the movements of the condyles could be simulated similar to the contact model. The discs slightly rotated around the condyles with the anterior band rotating upwards and the posterior band rotating downwards. The rotation of the discs resulted in an abnormal condyle-disc-fossa position. The anterior and posterior attachments of the discs were also in tension during rotation around the condyles. However, the maximum tensile force occurred in the temporal posterior attachments, which should be slack (Chen et al. 1998). Meanwhile, the maximum compressive stresses of the gap elements were located between the center of the condyle and the posterior band of the disc and between the intermediate zone of the disc and the anterior of the temporal bone, respectively.

In the contact model, the discs were observed to move along with the condyles without rotation, consistent with previous biomechanical research (Chen and Xu 1994; Chen et al. 1998; Devocht et al. 1996; Beek et al. 2000, 2001; del Palomar and Doblare 2006b, c, 2007). Moreover the backward and upward slide of the discs occurred between the articular surfaces of the condyles and the fossa-eminences. The posterior bands of the discs were finally located between the crests of the condyles and the articular fossas, in agreement with the normal disc position (Incesu et al. 2004). Thus, the contact elements between the disc and the articular cartilages could provide more accurate characterization of movement of the condyle and the disc. The anterior attachments of the discs were in tension, with the bilaminar zones slack, produced by the backward and upward slide of the discs. The maximum tensile force in the temporal anterior attachment was 0.41 N, similar to Chen et al. (1998). Moreover, the maximum contact stresses occurred between the intermediate zone of the disc and the anterior of the condyle and the temporal bone. The contact area and the maximum contact stress on the lower interface were higher than those on the upper interface, consistent with previous results (Beek et al. 2001).

The stress distributions in the discs induced by the interaction between the discs and the cartilages varied among the three models. In the bond model, the average von Mises stresses were 0.15 MPa in the anterior bands, 0.23 MPa in the intermediate zone, and 0.42 MPa in the posterior band (Figure 22.2). Likewise, the maximum tensile and compressive stresses of the discs fell over the posterior band (Table 22.1), in combination with a poor ability to resist tensile loading (Kang et al.

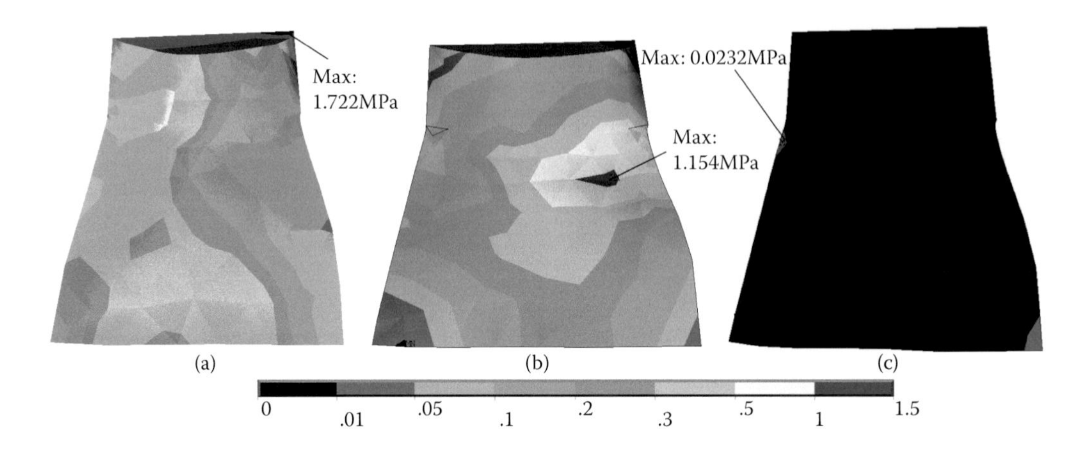


**FIGURE 22.2** **(See color insert.)** The von Mises stresses in the left discs of the three models (the maximum stresses were listed): (a) the bond model, (b) the contact model, and (c) the gap model.

**TABLE 22.1**
**Maximum Stresses of Discs in Three Models**

| | | Bond | Contact | Gap |
|---|---|---|---|---|
| Maximum tensile stresses (MPa) | Anterior bands | 0.716 | 0.430 | 0.013 |
| | Intermediate zone | 0.759 | 0.542 | 0.056 |
| | Posterior band | 2.580 | 0.318 | 0.586 |
| Maximum pressures (MPa) | Anterior bands | −0.539 | −0.178 | −0.009 |
| | Intermediate zone | −1.873 | −0.651 | −0.039 |
| | Posterior band | −2.098 | −0.586 | −0.231 |

2006). The maximum tensile stress (2.58 MPa) was much greater than the tensile strength (1.35 MPa [Kang et al. 1998]). In the gap model, stress concentration was observed in the posterior band, while the stresses in the other bands were very low (Figure 22.2). The tensile stress was dominant in the three bands (Table 22.1), inconsistent with the load-resistant properties of the discs (Kang et al. 2006). In the contact model, the maximum pressures of the discs occurred in the intermediate zones (Table 22.1), in agreement with the physiological function of the discs and related biomechanical research (Koolstra and van Eijden 2005; del Palomar and Doblare 2006c). Moreover, tensile stress was dominant in the anterior band, accordant with the tension-bearing structure; compressive stress was dominant in the posterior band, accordant with the pressure-bearing structure (Kang et al. 2006). Therefore, a more biologically appropriate stress distribution of the discs was obtained in the contact model.

The stress distributions in the articular cartilages were also caused by the interaction between the discs and the cartilages. In the condylar cartilages, the stresses in the anterior and central regions were much greater than the posterior, medial, and lateral regions in all models. However, the magnitudes of stresses were too small in the gap model due to the interaction between the discs and the cartilages, similar to those in the discs. The maximum pressure was located at the central region in the bond and gap models and at the anterior end in the contact model (Figure 22.3). The stress distributions in the temporal cartilages were similar to the stress distributions in the condylar cartilages, and the maximum pressures of the temporal cartilages fell over the anterior end in the three models (Figure 22.3). Anatomical studies have found that injury to the condylar and temporal cartilages usually occurs in the anterior segment (Oberg et al. 1971). Thus, the stress distributions of the condylar and temporal cartilages were reasonable in the contact model.

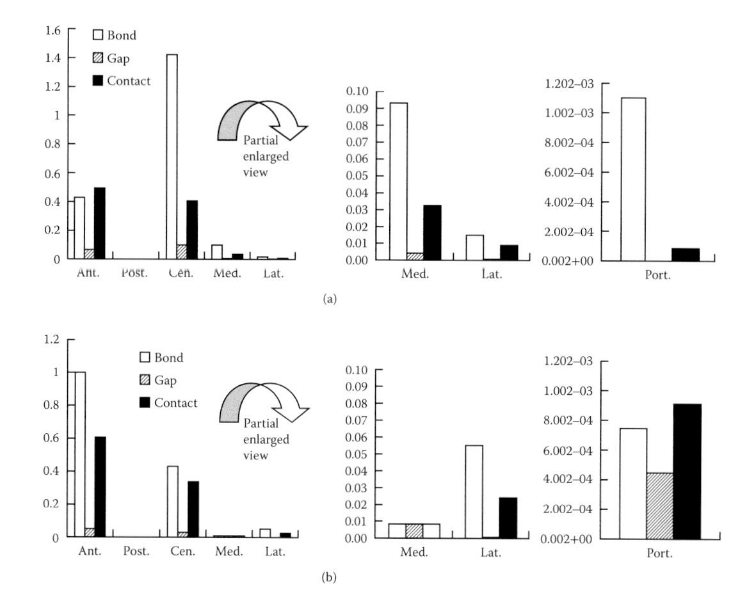


**FIGURE 22.3** The maximum pressures (MPa) of the condylar and temporal cartilages. Bond, gap, and contact indicate the interfaces between the disc and the articular cartilages were bonded together and simulated as gap elements or contact elements. "Ant.," "Post.," "Cen.," "Med.," and "Lat." indicate the anterior, posterior, central, medial, and lateral regions, respectively. (a) The maximum pressures of condylar cartilage. (b) The maximum pressures or temporal cartilage.

Internal derangement of TMJ (TMJ-ID) is one of the most common TMD. Degenerative joint disease and increased friction between the moving parts are considered to play a major role in the etiology of disc displacement. Therefore, the biomechanical environment in the TMJ is a key feature to understanding the inducing and progressive mechanisms of TMJ-ID (Tanaka et al. 2004). Moreover, functional overloading could cause other TMJ diseases. For example, excessive compressive and shear stresses are probably the most essential factors of condylar resorption and disc perforation (Tanaka et al. 2000, 2001b). Accurate modeling of TMJ is helpful for understanding the mechanical characteristics of TMJ diseases and to forecast suitable treatments.

### 22.2.3 Summary and Significance

Three FE models were developed to compare simulations of the interaction between the disc and the cartilages on the condyle and the temporal bone. The results showed that bonding the disc and articular cartilages led to an inactive joint and increased stresses in the TMJ. Gap elements between the disc and the cartilages resulted in abnormal movement and stress distribution in the TMJ and low stresses in the disc and articular cartilages. Otherwise, the final configuration and stress distribution of the TMJ were reasonable when contact elements were simulated on the upper and lower interfaces of the disc. The model could provide a reliable simulation of interaction in the TMJ.

## 22.3 EFFECTS OF OCCLUSAL LOADS ON STRESS DISTRIBUTIONS IN TMJ

### 22.3.1 Development of the Finite Element Model and Parameter Settings

The geometry of the model was based on CT images of a male volunteer with normal occlusion and asymptomatic joints. The teeth, cortical, and cancellous bone of the mandible were constructed according to the gray values using MIMICS 8.1 (Materialise, Leuven, Belgium). Likewise, the

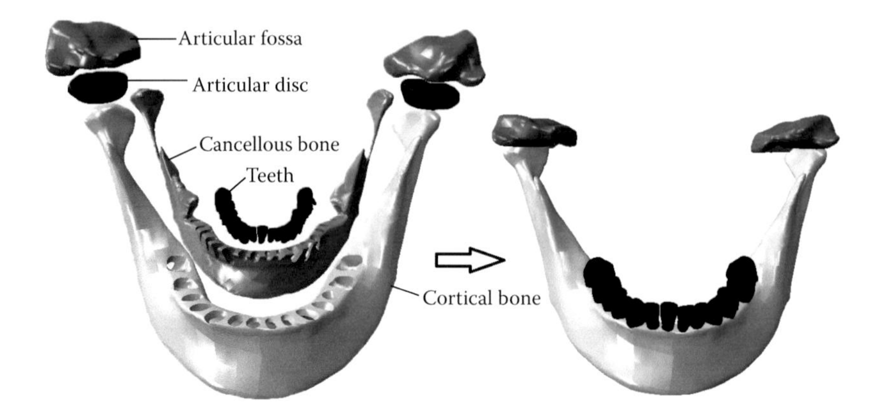


**FIGURE 22.4** The 3D model of the mandible and TMJ.

bilateral articular fossa-eminence complexes were also constructed. The models of the articular discs were also constructed according to the CT images, the articular surfaces, and the anatomy of disc. Then the 3D models of the mandible and TMJ were imported into ABAQUS 6.6 (ABAQUS Inc, USA), as shown in Figure 22.4. The mechanical properties of the mandible and TMJ were the same as described in Section 22.2, and the Young's modulus and Poisson's ratio of the teeth were 30,000 MPa and 0.3 (Tanaka et al. 2004), respectively.

Tetrahedral elements were also used. The mesh of the disc was refined in order to improve the calculation accuracy. Based on the comparative study in Section 22.2, the interfaces between the disc and the articular surfaces of the condyle and temporal bone were simulated as contact elements (Liu et al. 2008). The friction coefficient was also considered 0.001. The ligaments and bilaminar zones were modeled as cable elements with a referenced stiffness (Liu et al. 2008). The force vectors of the bilateral masticatory muscles corresponded to centric occlusion, anterior occlusion, and right molar occlusion. The magnitude of each muscle force was assigned according to its physiological cross section and the scaling factor (Pruim et al. 1980; Weijs and Hillen 1984; Korioth and Hannam 1994b). The models were fixed at the occlusal surface and the external regions of the temporal bone.

### 22.3.2 Calculation of the TMJ Model

Due to the symmetry of the model and loading, the stress distributions of the condyles, discs, and articular fossas were almost the same under the central occlusion and anterior occlusion. However, there were significant differences in stress distribution between the right and left TMJs under the right molar occlusion, so the results of both TMJs were analyzed.

Stress at the anterior condyle was greater than at other regions under the three occlusal conditions. The average von Mises stress of the anterior condyle was 0.6 MPa under the central occlusion, which was twice as large as the stress at the posterior condyle. For the anterior occlusion, the average von Mises stress of the anterior and posterior condyle was 2.6 MPa and 2.1 MPa, respectively. However, for the right molar occlusion, the stresses of the left condyle were significantly greater than the right side. On the left side, the average von Mises stress of the anterior and posterior condyle was 2.3 MPa and 1.8 MPa, respectively. Otherwise, the average von Mises stress was 0.25 MPa in the anterior condyle and one-third of this in the posterior condyle on the right side.

The maximum stresses were located in the intermediate zone of the disc under the three occlusions. For the central occlusion, the average von Mises stress of the disc was 0.48 MPa in the intermediate zone, which was eight and four times that in the anterior posterior band, respectively (Figure 22.5). For the anterior occlusion, the average von Mises stress of the disc was 2.0 MPa in the intermediate zone, 1.2 MPa in the posterior band, and 0.26 MPa in the anterior band (Figure 22.5).

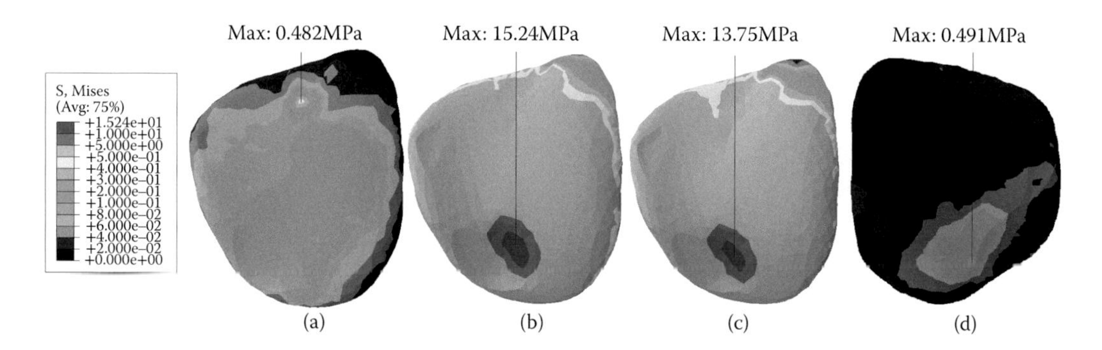


**FIGURE 22.5** **(See color insert.)** The von Mises stress distributions of the disc under the three occlusions (the maximum stresses were listed): (a) central occlusion (left disc), (b) anterior occlusion (left disc), (c) right side molar occlusion (left disc), and (d) right side molar occlusion (right disc).

Similar to the condyle, the stress of the left disc was significantly greater than the right side under the right side molar occlusion. The average von Mises stress of the left disc was 0.5 MPa in the anterior band, 3.2 MPa in the intermediate zone, and 1.4 MPa in the posterior band. The average von Mises stress of the right disc was 0.32 MPa in the intermediate zone and much smaller in the anterior and posterior bands (Figure 22.5). The stresses in the dorsal side (contacting with the articular fossa) were less than those in the ventral side (contacting with the condyle).

The stress of the articular fossa-eminence was much less than the condyle and disc under the three occlusions. The maximum stress of articular fossa-eminence also occurred at the right molar occlusion with a value of 0.12 MPa located at the posterior articular eminence of the left articular fossa-eminence.

### 22.3.3 Analysis and Application

The stress of the anterior condyle was greater than at other regions under the three occlusal conditions, consistent with the related research (Beek et al. 2001; Liu et al. 2008). In the same sagittal plane, most regions of the ventral disc contacted the condyle, but only small regions of the dorsal disc contacted the articular fossa-eminence, similar to results reported by Tanaka et al. (2001b). The fiber network and proteoglycans of the discs dispersed the load between the condyle and articular fossa through their interaction, consistent with the physiological structure and the function of the disc. The discs can buffer shocks imposed on the TMJ and protect the tissues of the brain. The maximum disc stress occurred at the intermediate zone, which is clinically susceptible to perforation and rupture (Tanaka et al. 2004; del Palomar and Doblare 2007, 2008; Liu et al. 2008). This phenomenon showed that the collagen fibers in the intermediate zone of the disc absorbed a large amount of energy and acted as a stress buffer, consistent with the structure and weight-bearing characteristics. The stress of the posterior articular eminence, the functional area of the TMJ, was greater than at other regions. Therefore, the models of this study can be deemed valid.

The stresses of the condyle under the anterior occlusion and the right molar occlusion were significantly greater than the central occlusion. Stress concentrations occurred at some areas. The tensile and compressive stresses of the disc under the anterior occlusion were 4.0 MPa and 17.4 MPa, respectively, greater than the magnitudes under the central occlusion and the right side molar occlusion. This can explain why some patients with TMD feel pain when biting using anterior teeth. The deformation of the left disc was more than the right disc under right molar occlusion, so chewing with both sides of the molars should be recommended in order to avoid facial deformities. Likewise, the stresses of the contralateral TMJ were significantly greater than the chewing side under the unilateral occlusion. If pain is present on one side of the TMJ, the pain will worsen if the contralateral side is used for chewing.

## 22.4 BIOMECHANICAL MODELS OF TYPICAL TEMPOROMANDIBULAR DISORDERS

### 22.4.1 Model Development

The base model of the dentate mandible was developed in MIMICS 8.1 (Materialise, Leuven, Belgium) according to CT images of a person with normal occlusion and without TMD, as shown in Figure 22.6. Likewise, the bilateral articular fossa-eminence complexes were also constructed. The geometries of the cortical bones, cancellous bones, and the teeth were imported into ANSYS 8.0 (Swanson Analysis System Co., Houston, TX). The models of the fibrocartilage layers and the articular discs were also constructed using the method detailed in Section 22.2.1. The minimum thickness of the intermediate zone was about 1 mm, while the maximum thickness of the anterior and posterior bands was 2 and 3 mm, respectively (Liu et al. 2007). The attachments of the articular disc and the mandibular ligaments were also considered nonlinear cable elements, as described in Section 22.2.2.

Seven models of typical TMDs were established based on the base model with normal TMJs: relaxation of discal attachment (relaxation of anterior and posterior attachments), disc displacement (anterior, posterior, medial, and lateral disc displacement), and disc perforation. Only the right TMJs in the models were modified and the left joints were treated as healthy joints. Relaxation of anterior and posterior attachments was simulated by decreasing the stiffness of the cable elements. According to the typical morphology of the displaced discs, four models of disc displacement were developed. Moreover, the bilaminar zones were not modeled in the anterior disc displacement without reduction. Likewise, the temporal anterior attachment and mandibular anterior attachment were not modeled in the posterior disc displacement; the inner layer of the temporomandibular ligament was not considered in the medial disc displacement. Disc perforation typically occurs at the lateral intermediate zone, so this region was perforated in the model of disc perforation.

The mechanical properties of the cortical bone, cancellous bone, articular cartilage, teeth, and disc were identical to those described in Sections 22.2 and 22.3. The interfaces between the disc and the articular surfaces of the condyle and the temporal bone were treated as contact elements, as described in Section 22.2. Loading and boundary conditions of the centric occlusion were applied to the models as described in Section 22.2.

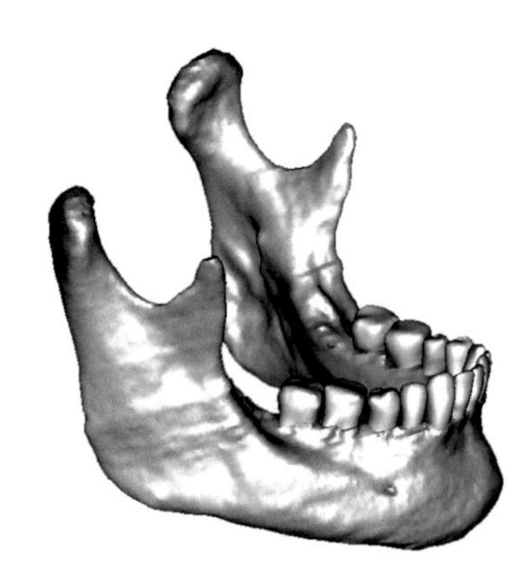

**FIGURE 22.6** The model of the dentate madible.

### 22.4.2 Analysis of TMD Models and Applications

In the base model with normal TMJ, the maximum contact stresses were located between the anterior condyle and the lower surface of the intermediate zone of the disc and between the upper surface of the intermediate zone of the disc and the posterior articular eminence (i.e. anterior temporal bone), respectively. Therefore the maximum stresses occurred at the intermediate zone of the disc, and the anterior of the condyle and temporal bone, accordant with the physiological function and related biomechanical research (del Palomar and Doblare 2006c; Koolstra and van Eijden 2005; Oberg et al. 1971).

#### 22.4.2.1 Relaxation of the Discal Attachment

Abnormal positioning of the pathologic disc with respect to the condyle and articular fossa-eminence was caused by relaxation of disc attachments, which altered the biomechanical environment of the pathologic side of the TMJ, but scarcely affected the asymptomatic side. Thus, the results of the pathologic TMJs were analyzed.

The interaction between the lateral intermediate zone of the disc and the lateral side of the condyle and the articular fossa-eminence was evident in the model with relaxation of the anterior disc attachments. Therefore, the maximum stresses of the pathologic disc were located at the lateral intermediate zone. The stress level in the pathologic disc was significantly greater than the identical disc in the base model, as shown in Figure 22.7. The maximum compressive stress of the pathologic disc was about five times greater than the identical disc in the base model, and the maximum tensile and shear stresses of the pathologic disc were about triple the magnitude of the base model. Thinning or perforation of the lateral intermediate zone of the disc with the relaxed anterior attachment may result from long-term effects of occlusion and mandibular movement. According to the contact status, the maximum stresses of the condyle and the temporal bone were located at the lateral region. However, the maximum stresses of the anterior of the condyle and the temporal bone were much lower than those in the base model.

In the model with relaxation of the bilaminar zones (posterior disc attachments), the maximum contact stress, located at the intermediate zone of the disc and posterior articular eminence, was much greater than that of the base model. Contrarily, the interaction between the disc and the condyle was weaker than that of the base model. The maximum stresses of the pathologic disc were located at the intermediate zone. The stress level in the pathologic disc was slightly greater than the identical disc in the base model, as shown in Figure 22.7. Based on the contact status, the maximum stresses of the condyle and the temporal bone were located at the anterior region. The maximum

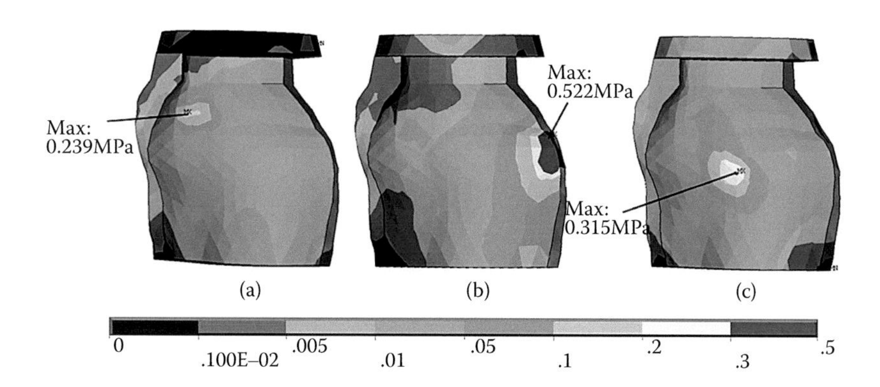


**FIGURE 22.7 (See color insert.)** The von Mises stress distributions of the normal disc and the discs with relaxed attachment (the maximum stresses were listed; units: MPa): (a) normal disc, (b) disc with relaxation of anterior attachments, and (c) disc with relaxation of bilaminar zones.

stresses of the condyle were much lower than the identical condyle in the base model. On the other hand, the maximum compressive and shear stresses of the posterior articular eminence increased by 41.6 and 6.4 times those in the base model, respectively. Damage to the posterior articular eminence may be induced by the significantly high stress.

#### 22.4.2.2 Disc Displacements

Contrary to normal TMJs, the interaction between the disc and the posterior articular eminence was much more intensive than that between the disc and the condyle in various disc displacements. Stress concentrations occurred at the intermediate zone of the anteriorly displaced disc. The maximum stresses of the disc were much greater than the identical disc in the base model (Figure 22.8). The maximum compressive stress of the disc was 14.6 times that of the base model, which is similar to related publications (del Palomar and Doblare 2007; Arnerr et al. 1996), and greater than the other displaced disc. The maximum tensile stress of the disc, located at the intermediate zone, reached 1.43 MPa, close to the tensile failure stress obtained through experimental means (1.53 MPa) (Kang et al. 1998). Thus, long-term occlusion could lead to thinning or perforation of the intermediate zone of the disc.

In models with posterior, medial, and lateral disc displacements, the stress levels in all disc bands were greater than the discs in the base model and the anteriorly displaced model (Figure 22.8). The stress distributions of the discs were similar in the models with posterior and lateral disc displacements. The maximum stresses of the anterior bands and intermediate zones of the discs were much greater than those of the posterior bands of the discs in the two models. The maximum tensile stresses of the discs with posterior and lateral displacements, located at the intermediate zones, were 1.91 MPa and 1.56 MPa, respectively. The magnitudes were greater than the tensile failure stress of the intermediate zones (1.53 MPa) (Kang et al. 1998), so the significantly high tensile stresses could cause some pathological changes in the discs, such as perforation. The compressive stresses were greater than the tensile stresses in the anterior band with posterior and lateral displacements, opposite from the tension-bearing structure. The maximum tensile stress of the anterior band of the posteriorly displaced disc was 1.47 MPa, close to its tensile failure stress (1.85 MPa) (Kang et al. 1998).

In the model with medial disc displacement, stress concentrations occurred at each band of the disc. The stress levels of the anterior and posterior bands were significantly greater than those in the base model and other displaced discs (Figure 22.8). The maximum tensile stress of the anterior band of the disc was significantly greater than that of the intermediate zone and reached 2.23 MPa, exceeding the tensile failure stress (1.85 MPa) (Kang et al. 1998). The tensile stresses of the posterior band were greater than the compressive stresses, inconsistent with the pressure-bearing structure. The maximum tensile stress of the posterior band of the disc was 2.13 MPa, which is much greater than the tensile failure stress (1.35 MPa) (Kang et al. 1998). Disc perforation and

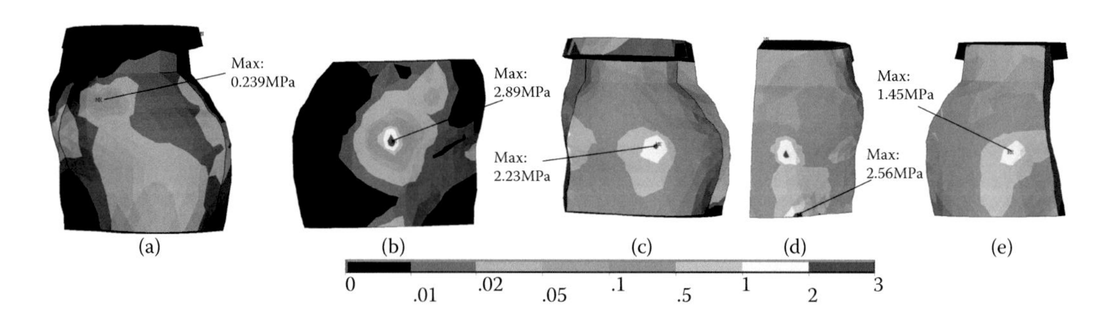


**FIGURE 22.8 (See color insert.)** The von Mises stress distributions of the discs in models with disc displacements (the maximum stresses were listed; units: MPa): (a) normal disc, (b) anteriorly displaced disc, (c) posteriorly displaced disc, (d) medially displaced disc, and (e) laterally displaced disc.

other pathological changes in the anterior and posterior bands could probably be induced by such significantly high tensile stresses.

The maximum von Mises stresses of the condyle cartilages with anterior and posterior disc displacements were approximately half as much as those in the base model, and the magnitudes were close to the base model in models with medial and lateral disc displacements. However, the stresses of the posterior articular eminences with various disc displacements were much greater than those in the base model. The maximum von Mises stresses of the posterior articular eminences with posterior and medial disc displacements were 59 and 46 times greater than the base model, respectively. Moreover, the maximum tensile stress of the posterior articular eminences occurred after posterior disc displacement and reached 5.08 MPa, significantly exceeding the tensile failure stress (2.15 MPa) (Yi and Kang 2001). The significantly increased stresses in the posterior articular eminence could cause flattening and perforation of the cartilage and possible bone resorption.

In summary, various disc displacements could lead to an increase in stress in the discs and posterior articular eminences, and the excessive stresses may lead to worse disc displacement. Intensive disc displacements will also increase the magnitude of stresses and alter the stress distributions in the TMJs. Therefore, an effective method to restore the normal position and function of the disc is to treat TMD when initial symptoms are noticed, such as joint clicking.

#### 22.4.2.3 Disc Perforation

Similar to disc displacement, the interaction between the disc and the articular fossa-eminence was much more intensive than that between the disc and the condyle in the model with disc perforation. The maximum contact stresses occurred between the anterior condyle and the lower surface of the intermediate zone of the disc and between the upper surface of the intermediate zone of the disc and the posterior articular eminence, respectively. Therefore, the stress concentration was located at the intermediate zone of the perforated disc (Figure 22.9). The stresses of all the bands of the perforated disc were greater than the identical disc in the base model. The compressive stresses in the anterior band of the perforated disc were much greater than the tensile stresses, inconsistent with the tension-bearing structure. On the contrary, the posterior band of the perforated disc undertook higher tensile stresses than compressive stresses, inconsistent with the pressure-bearing structure. The maximum stress of the three bands of the perforated disc may not lead to continued perforation, accordant with Stratmann's study (Stratmann and Schaarschmidt 1996). However, disc perforation could lead to increased stress and unreasonable stress distribution, possibly inducing other pathological changes to the perforated disc.

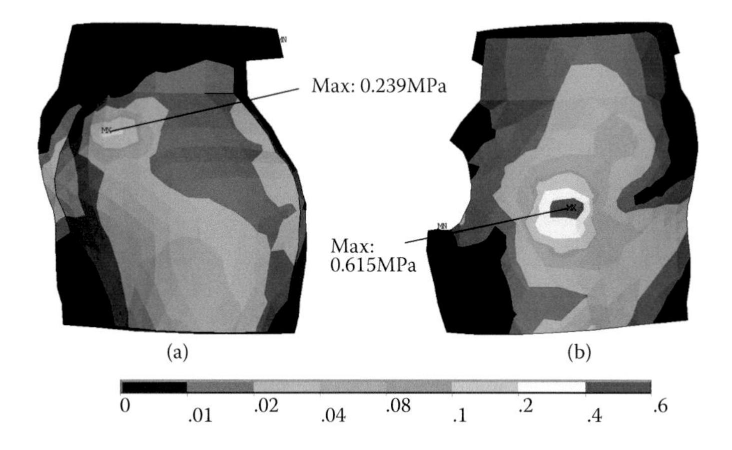


**FIGURE 22.9** **(See color insert.)** The von Mises stress distributions of the normal and disc perforation models (the maximum stresses were listed; units: MPa): (a) normal disc and (b) perforated disc.

The stress level of the condylar cartilage with the perforated disc was lower than that in the base model. On the other hand, disc perforation significantly increased the stress of the temporal cartilage, especially at the posterior articular eminence. The excessive stresses may cause some pathological changes, such as flattening and perforation of the cartilage. Obviously, treatment should be carried out prior to the perforation of the articular disc.

### 22.4.3 Summary and Significance

Relaxation of the anterior disc attachment could lead to increased stress in the lateral intermediate zone of the disc, which is often susceptible to perforation. Relaxation of bilaminar zones could significantly increase the stress of the posterior articular eminence and may cause some pathological changes.

Anterior disc displacement could produce stress concentrations in the intermediate zone of the disc. Posterior and lateral displacements of the disc could induce significantly increased stresses in the intermediate zones and the anterior bands of the discs. The maximum tensile stresses of the displaced discs, located at the intermediate zones, were greater than the tensile failure stress. Moreover, the maximum tensile stress of the anterior band of the posteriorly displaced disc was close to the tensile failure stress. Medial disc displacement could lead to stress concentration in each band of the disc. The maximum tensile stresses of the anterior and posterior bands of the disc significantly exceeded the tensile failure stress. These significantly high stresses in the displaced discs could cause thinning or perforation of the discs. On the other hand, disc displacements could lead to stress concentration in the posterior articular eminence, especially with posterior and medial disc displacements. The significantly increased stresses in the posterior articular eminence could cause flattening and perforation of the cartilage and bone resorption.

Disc perforation could lead to stress concentration at the intermediate zone of the disc and the posterior articular eminence. The excessive stresses may cause some pathological changes, such as thinning of the disc and flattening and perforation of the temporal cartilage.

In summary, the increased stresses and the abnormal stress distribution in the discs and articular cartilages could be induced by various TMDs, or could be the source of TMD development. Therefore, effective methods to maintain and restore the normal structure and function of the TMJ should aim to treat the disorder as soon as initial symptoms are noticed, such as clicking of the joint.

## ACKNOWLEDGMENTS

This work was supported by the National Natural Science Foundation of China (11202143).

## REFERENCES

Arnerr, G.W., Milam, B., Gottesman, L. 1996. Progressive mandibular retrusion-idiopathic condylar resorption. *American Journal of Orthodontics and Dentofacial Orthopedics* 110: 117–127.

Beek, M., Koolstra, J.H., van Ruijven, L.J., van Eijden, T.M.G.J. 2000. Three-dimensional finite element analysis of the human temporomandibular joint disc. *Journal of Biomechanics* 33: 307–316.

Beek, M., Koolstra, J.H., van Ruijven, L.J., van Eijden, T.M.G.J. 2001. Three-dimensional finite element analysis of the cartilaginous structures in the human temporomadibular joint. *Journal of Dental Research* 80: 1913–1918.

Boyd, R.L., Gibbs, C.H., Mahan, P.E., Richmond, A.F., Laskin, J.L. 1990. Temporomandibular joint forces measured at the condyle of macaca arctoides. *American Journal of Orthodontics and Dentofacial Orthopedics* 97: 472–479.

Brehnan, K., Boyd, R.L., Laskin, J., Gibbs, C.H., Mahan, P. 1981. Direct measurement of loads at temporomandibular joint in macaca arctoides. *Journal of Dental Research* 60: 1820–1824.

Castano, M.C., Zapata, U., Pedroza, A., Jaramillo, J.D., Roldan, S. 2002. Creation of a three-dimensional model of the mandible and the TMJ in vivo by means of the finite element method. *International Journal of Computerized Dentistry* 5: 87–99.

Chen, J., Akyuz, U., Xu, L.F., Pidaparti, R.M.V. 1998. Stress analysis of the human temporomandibular joint. *Medical Engineering & Physics* 20: 565–572.
Chen, J., Xu, L.F. 1994. A finite element analysis of the human temporomandibular joint. *Journal of Biomechanical Engineering* 116: 401–407.
del Palomar, A.P., Doblare, M. 2006a. Anterior displacement of the TMJ disk: repositioning of the disk using a mitek system. A 3D finite element study. *Transactions of the ASME, Journal of Biomechanical Engineering* 128: 663–673.
del Palomar, A.P., Doblare, M. 2006b. Finite element analysis of the temporomandibular joint during lateral excursions of the mandible. *Journal of Biomechanics* 39: 2153–2163.
del Palomar, A.P., Doblare, M. 2006c. The effect of collagen reinforcement in the behavior of the temporomandibular joint disc. *Journal of Biomechanics* 39: 1075–1085.
del Palomar, A.P., Doblare, M. 2007. An accurate simulation model of anteriorly displaced TMJ discs with and without reduction. *Medical Engineering & Physics* 29: 216–226.
del Palomar, A.P., Doblare, M. 2008. Dynamic 3D FE modelling of the human temporomandibular joint during whiplash. *Medical Engineering & Physics* 30: 700–709.
Devocht, J.W., Goel, V.K., Zeitler, D.L., Lew, D. 1996. A study of the control of disc movement within the temporomadibular joint using the finite element technique. *Journal of Oral and Maxillofacial Surgery* 54: 1431–1437.
Faulkner, M.G., Hatcher, D.C., Hay, A., 1987. A three-dimensional investigation of temporomandibular joint loading. *Journal of Biomechanics* 20: 997–1002.
Forster, H., Fisher, J. 1996. The influence of loading time and lubricant on the friction of the articular cartilage. Proceedings of the Institution of Mechanical Engineers, Part H. *Journal of Engineering in Medicine* 210: 109–119.
Hansson, T., Nordstrom, B. 1977. Thickness of the soft tissue layers and articular disc in temporomandibular joints with deviations in form. *Acta Odontologica Scandinavica* 35: 281–288.
Hart, R.T., Hennebel, V.V., Thongpreda, N., van Buskirk, W.C., Anderson R.C. 1992. Modeling the biomechanics of the mandible: a three-dimensional finite element study. *Journal of Biomechanics* 25: 261–286.
Hu, K., Rong, Q.G., Fang, J., Mao, J.J. 2003. Effects of condylar fibrocartilage on the biomechanical loading of the human temporomandibular joint in a three-dimensional nonlinear finite element model. *Medical Engineering & Physics* 25: 107–113.
Incesu, L., Takaya-Yilmaz, N., Ogutcen-Tollerb, M., Uzun, E. 2004. Relationship of condylar position to disc position and morphology. *European Journal of Radiology* 51: 269–273.
Kang, H., Bao, G.J., Qi, S.N. 2006. Biomechanical responses of human temporomandibular joint disc under tension and compression. *International Journal of Oral and Maxillofacial Surgery* 35: 817–821.
Kang, H., Yi, X.Z., Chen, M.S. 1998. A study of tensile mechanical property of human temporomandibular joint disc. *West China Journal of Stomatol* 16: 253–255.
Koolstra, J.H. 2012. Biomechanical analysis of the influence of friction in jaw joint disorders. *Osteoarthritis and Cartilage* 20: 43–48.
Koolstra, J.H., van Eijden, T.M.G.J. 2005. Combined finite-element and rigidbody analysis of human jaw joint dynamics. *Journal of Biomechanics* 38: 2431–2439.
Koolstra, J.H., van Eijden, T.M.G.J., Weijs, W.A., Naeije, M. 1988. A three-dimensional mathematical model of the human masticatory system predicting maximum possible bite forces. *Journal of Biomechanics* 21: 563–576.
Korioth, T.W.P., Hannam, A.G. 1994a. Deformation of the human mandible during simulated tooth clenching. *Journal of Dental Research* 73: 56–66.
Korioth, T.W.P., Hannam, A.G. 1994b. Mandibular forces during simulated tooth clenching. *Journal of Orofacial Pain* 8: 178–189.
Liu, Z., Fan, Y.B., Qian, Y.L. 2007. Biomechanical simulation of the interaction in the temporomandibular joint within dentate mandible: a finite element analysis. *Proceedings of 2007 International Conference on Complex Medical Engineering* 1873–1877.
Liu, Z., Fan, Y.B., Qian, Y.L. 2008. Comparative evaluation on three-dimensional finite element models of the temporomandibular joint. *Clinical Biomechanics* 23: 53–58.
Mabuchi, K., Obara, T., Ikegami, K., Yamaguchi, T., Kanayama, T. 1999. Molecular weight independence of the effect of additive hyaluronic acid on the lubricating characteristics in synovial joints with experimental deterioration. *Clinical Biomechanics* 14: 352–356.
Muralidhar, S., Jagota, A., Bennison, S.J. 2000. Mechanical behaviour in tension of cracked glass bridged by an elastomeric ligament. *Acta Materialia* 48: 4577–4588.

Oberg, T., Carlsson, G.E., Fajers, C.M. 1971. The temporomandibular joint. A morphologic study on a human autopsy material. *Acta Odontologica Scandinavica* 29: 349–384.

Pruim, G.J., De Jongh, H.J., Ten Bosch, J.J. 1980. Forces acting on the mandible during bilateral static bite at different bite force levels. *Journal of Biomechanics* 13: 755–763.

Pullinger, A.G., Baldioceda, F., Bibb, C.A. 1990. Relationship of TMJ articular soft tissue to underlying bone in young adult condyles. *Journal of Dental Research* 69: 1512–1518.

Race, A., Amis, A. 1994. The mechanical properties of the two bundles of the human posterior cruciate ligament. *Journal of Biomechanics* 27: 13–24.

Roh, H.S., Kim, W., Kim, Y.K., Lee, J.Y. 2012. Relationships between disk displacement, joint effusion, and degenerative changes of the TMJ in TMD patients based on MRI findings. *Journal of Cranio-Maxillofacial Surgery* 40: 283–286.

Stratmann, U., Schaarschmidt, K., Santamaria, P. 1996. Morphometric investigation of condylar cartilage and disc thickness in the human temporomandibular joint: Significance for the definition of osteatthrotic changes. *Journal of Oral Pathology & Medicine* 25: 200–205.

Tanaka, E., del Pozo, R., Tanaka, M. et al. 2004. Three-dimensional finite element analysis of human temporomandibular joint with and without disc displacement during jaw opening. *Medical Engineering & Physics* 26: 503–511.

Tanaka, E., Hirose, M., Koolstra, J.H. et al. 2008. Modeling of the effect of friction in the temporomandibular joint on displacement of its disc during prolonged clenching. *Journal of Oral and Maxillofacial Surgery* 66: 462–468.

Tanaka, E., Rodrigo, D.P., Miyawukiy, Y., Lee, K., Yamaguchi, K., Tanne, K. 2000. Stress distribution in the temporomandibular joint affected by anterior disc displacement: a three-dimensional analytic approach with the finite-element method. *Journal of Oral Rehabilitation* 27: 754–759.

Tanaka, E., Rodrigo, D.P., Tanaka, M., Kawaguchi, A., Shibazaki, T., Tanne, K. 2001a. Stress analysis in the TMJ during jaw opening by use of a three-dimensional finite element model based on magnetic resonance images. *International Journal of Oral and Maxillofacial Surgery* 30: 421–430.

Tanaka, E., Tanaka, M., Watanabe, M., del Pozo, R., Tanne, K. 2001b. Influences of occlusal and skeletal discrepancies on biomechanical environment in the TMJ during maximum clenching: an analytic approach with the finite element method. *Journal of Oral Rehabilitation* 28: 888–894.

Throckmorton, G.S., Dechow, P.C. 1994. In vitro measurements in the condylar process of the human mandible. *Archives of Oral Biology* 39: 853–867.

Weijs, W.A, Hillen, B. 1984. Relationship between the physiological cross-section of the human jaw muscles and their cross-sectional area in computer tomograms. *Acta Anatomica* 118: 129–138.

Yi, X.Z., Kang, H. 2001. Biomechanics of the cartilage in temporomandibular joint. In *Chinese Stomatology* 769–772.

Zhou, X.J., Zhao, Z.H., Zhao, M.Y., Fan, Y.B. 1999. Analysis of the condyle in the state on the mandibular protraction by means of the three-dimensional finite element method. *Chinese Journal of Stomatology* 34: 85–87.

# 23 Fingertip Model for Blood Flow and Temperature

*Ying He, Hongwei Shao, Yuanliang Tang, Irina Mizeva, and Hengdi Zhang*

## CONTENTS



## 23.1 INTRODUCTION

It has been shown that chronic illnesses such as systemic sclerosis and diabetes mellitus are related to some disorders at the capillary level, such as increased vessel permeability, the presence of avascular areas, enlarged loops, poor circulation, and increased tortuosity (Daly and Leahy 2013). There has been an increasing interest in using microcirculation as a marker for cardiovascular health and metabolic functions, as it may be related to the development of instruments for detecting a variety of pathological processes in the circulatory system.

Since the skin is readily accessible, it provides an appropriate site to assess peripheral microvascular reactivity. Vascular reactivity is a primary feature of the circulatory system that enables the vasculature to respond to physiological and physical stimuli that require adjustments in blood flow, vessel tone, and vessel diameter. For more than two decades, methods that focus on the noninvasive exploration of cutaneous microcirculation have been mainly based on optical microscopy and laser Doppler techniques (Roustit and Cracowski 2012).

In recent years, laser Doppler flowmetry (LDF), in combination with wavelet analysis of blood flow oscillations, has been increasingly used to detect alterations in vessel tone. Blood flow modulations in microvessels form five non-overlapping frequency bands within the wave range 0.0095–3 Hz (Stefanovska et al. 1999) and the lowest frequency range (0.0095–0.021 Hz) is related to the functional activity of the microvessel endothelium. Fedorovich (2012) investigated the correlation between metabolic and microhemodynamic processes in the skin and showed that improved oxygen uptake and glucose disposal by tissues is accompanied by a significant increase in endothelial rhythm amplitude. Bernjak et al. (2008) showed that congestive heart failure exhibits abnormally attenuated blood flow oscillations in the frequency interval 0.005–0.021 Hz, and that treatment with $\beta_1$-blockers (Bisoprolol) can move the spectral amplitude 0.005–0.0095 Hz to that of the healthy control subjects.

Laser speckle contrast imaging is a recently developed optical noncontact technique that allows for the continuous assessment of skin perfusion over wide areas. In comparison to laser Doppler flowmetry, laser speckle contrast imaging can provide real-time monitoring of the microcirculation with high resolution (Basak et al. 2012). An extensive overview of these optical techniques may be found in a review of blood flow imaging by Daly and Leahy (2013). A major drawback of these methods is that the instruments they employ are complicated and may not be suitable for daily use.

The fingertip temperature response to a thermal stimulus or to pressure loading depends on the amount of blood perfusion, which implies that monitoring alterations in fingertip temperature could be employed to study microvascular reactivity. Based on a bioheat transfer equation, Haga et al. (2012) employed an inverse analysis method for examining blood perfusion. The instrument that was developed from this method can be used to estimate blood perfusion according to the observed fingertip temperature response under a certain thermal stimulus and shows good measurement repeatability and sensitivity. Yue et al. (2008) presented a three-point method to measure blood perfusion, whereby a heater is wrapped around a cylindrical section of living tissue, and three-point skin temperatures are measured. The characteristic points are located at the center of the heater, 1 cm away from the edge of the heater, and 2 cm away from the heater, respectively. By constructing an objective function between the measured and calculated temperatures and minimizing the function, the optimal value for blood perfusion and thermophysical properties can be obtained.

Apart from methods that are needed to load heat sources in living tissues, Nagata et al. (2009) established a passive method to evaluate blood perfusion only by using thermal information for the human body, whereby blood perfusion can be expressed in terms of the rate of temperature change at the contact sensor point and the initial skin temperature. Furthermore, a novel thermal method has been presented that uses the fingertip temperature, forearm temperature minus rectal temperature, and their changes across time to predict finger blood flow (Carrillo et al. 2011). Recently, a new thermal peripheral blood flowmeter has been developed that is integrated with a force sensor for force-compensated blood flow measurement (Sim et al. 2012). The important feature of this device is that, apart from the conventional metal resistance of the temperature detector, there is a membrane fabricated by surface and bulk micromachining techniques that is embedded with a piezoresistive force sensor. The compensated blood flow can be determined by detecting the rate of temperature change and the contact force.

Due to simple structures and low costs, many thermal methods for peripheral blood flow measurement have been developed. Not only can the peripheral blood flow rate be determined from the skin temperature, but vasodilated function may also be detected. Fingertip digital thermal monitoring (DTM) during cuff-occlusive reactive hyperemia (RH) is a new, noninvasive method of vascular function assessment that is based on the premise that changes in fingertip temperature during and after an ischemic stimulus reflect changes in blood flow and endothelial function. DTM technology usually requires 2 min of cuff inflation to cause brachial artery occlusion and 5 min of deflation later to bring about blood reperfusion. During the occlusion stage, a vasodilatory response occurs in the peripheral arteries and capillaries due to the absence of blood flow. After the brachial occlusion is finished, blood rushes into the forearm and hand, and thus causes a transient temperature rebound in the fingertip.

Ley et al. (2008, 2009) presented two mathematical models of heat transfer based on Pennes equation to estimate the influence of different factors on the dynamic temperature response in the fingertip during vascular occlusion and reperfusion. The models are based on different dimensions and anatomical details. One is a lumped parameter model that neglects tissue composition and the other is a two-dimensional thermal model of a simplified finger. A comparison of the models indicates that the lumped parameter model is effective as long as the proper variation between the initial condition and environmental parameters that affect the response is adopted. For both models, the blood perfusion rate was considered to be an input and distributed uniform in any places of the finger.

In subsequent work, Ley et al. (2011) measured the fingertip temperature, heat flux, and blood perfusion of 12 healthy volunteers. According to their heat transfer analysis, the magnitude of the laser Doppler signal was correlated with the local tissue temperature in an exponential manner. Moreover, they found that the initial and minimum temperatures had a significant effect on the thermal response during post-occlusion.

Akhtar et al. (2010) proposed a three-dimensional thermal model of a simplified finger to estimate the sensitivity of DTM parameters currently in clinical use to assess RH. According to their simulation, temperature rebound was shown to have the best correlation with the level of RH with good sensitivity for the range of flow rates that were studied. Their results indicated that temperature rebound and the equilibrium initial temperature are necessary to identify the amount of RH and to establish criteria for predicting the state of a patient's cardiovascular health.

McQuilkin et al. (2009) developed a signal-processing model to establish the basis for the relationship between finger temperature and blood flow reactivity following a brachial artery occlusion and reperfusion procedure. They expressed the fingertip temperature as the convolution of the flow-domain signal and the exponential impulse response. The normalized flow-domain signals that were computed from Doppler and DTM sensors showed favorable agreement. Their work demonstrates that vascular reactivity indices measured by DTM can be evaluated accurately.

Many researchers have also attempted to determine the correlation between skin temperature and blood flow oscillations. Liu et al. (2011) employed functional infrared imaging to identify low-frequency temperature fluctuations on the surface of the skin. They found that large veins have stronger contractility in the range of 0.005–0.006 Hz, compared to microvasculature and skin areas without infrared(IR)-detectable vessels. Sagaidachnyi et al. (2012) calculated the rate of blood flow from measured skin temperature signals with the use of Pennes bioheat equation. By adding a time delay of 10–20 s caused by temperature waves, the calculated blood flow waves corresponded with measured blood flow signals by using photoplethysmography.

Compared to detecting blood flow oscillations under normal thermal conditions, investigation of blood flow oscillations after a cooling test may reveal endothelial functions more accurately. Cold-induced vasodilation is an acyclic oscillation of blood flow that occurs upon exposure to the cold and commonly occurs in the extremities. The initial response to cold exposure is sympathetically mediated peripheral vasoconstriction, which results in reduced local tissue temperature. With continued cold exposure, this vasoconstriction may be interrupted, which results in periods of vasodilation. The vasodilation shows a characteristic cycle of increasing and decreasing blood flow.

Smirnova et al. (2013) employed wavelet analysis to investigate skin temperature oscillations during a contralateral cooling test on type 2 diabetic patients. Fifty-five healthy subjects and 35 type 2 diabetic patients with retinopathy and nephropathy participated in the study. Their skin temperatures were measured on the palm surface of their index fingers of the right hand while their left hands were immersed in cold water at 5°C for 3 min. The results of a wavelet analysis show that during the cold stimulus test the amplitudes of the skin temperature oscillations on the index finger of the contralateral hand significantly decreased and increased after unloading the cold stimulus for the healthy subjects. However, for the type 2 diabetic patients, after the decrease in temperature fluctuation during the cold stimulus, there was no further reliable increase in the amplitudes of the temperature fluctuations. These results demonstrate that cold-induced vasodilation for type 2 diabetic patients is much weaker than that for healthy subjects.

So far, we have seen that there are significant benefits in obtaining information on blood flow by employing thermal methods. In particular, temperature variation after brachial artery occlusion and a contralateral cooling test is closely related to significant variation in microvascular resistance. However, the mechanisms with regard to how vasomotion and peripheral temperature interact with each other remain unclear. Quantitatively predicting the relationship between vascular reactivity, vessel diameter, and blood flow may give insights into the assessment of endothelial functions. In this review, the methods for modeling temperature and blood flow at the fingertips during cuff-occlusive RH and a contralateral cooling test are discussed. The discussion centers on the developments of the authors' modeling work in this aspect.

This review is arranged into five parts. The first part deals with a two-dimensional model for assessing fingertip temperature variation during cold-water stimulation. Next, approaches for the simulation of peripheral thermoregulation are discussed. The third part focuses on the development of an image-based human model, a Darcy's model, and a heat transfer model of living tissue. Finally, the applications of this model for assessing vascular reactivity after brachial artery occlusion and a contralateral cooling test are discussed.

## 23.2 TWO-DIMENSIONAL THERMAL MODEL FOR ASSESSING COLD-STRESSED EFFECTS ON THE HUMAN FINGER

Figure 23.1 shows the anatomical structure of a finger. It can be seen that the arteries, veins, and nerves pass through the tissue and are located very close to each other. According to the anatomical structure of the finger cross-section, the finger is modeled as an elliptical cross-sectional area and consists of four parts—bone, tendon, dermis, and epidermis—as shown in Figure 23.2a. Figure 23.2b depicts a grid generation of this computation (He et al. 2001).

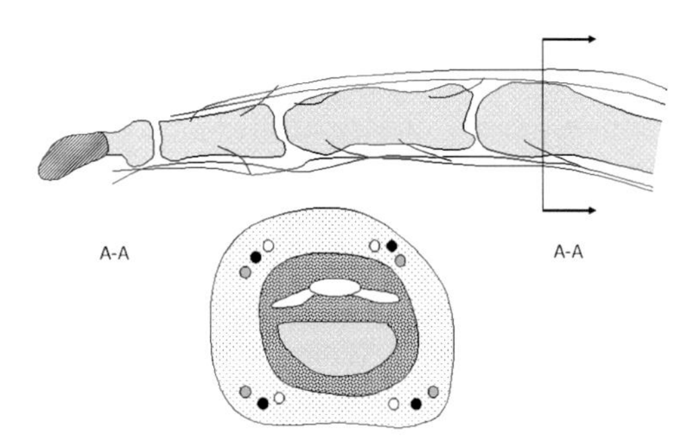


**FIGURE 23.1** Anatomical structure of a human finger. (From Kahle W et al., *Tachen atlas Der Anatomie*, Bunkoudo, 1990.)

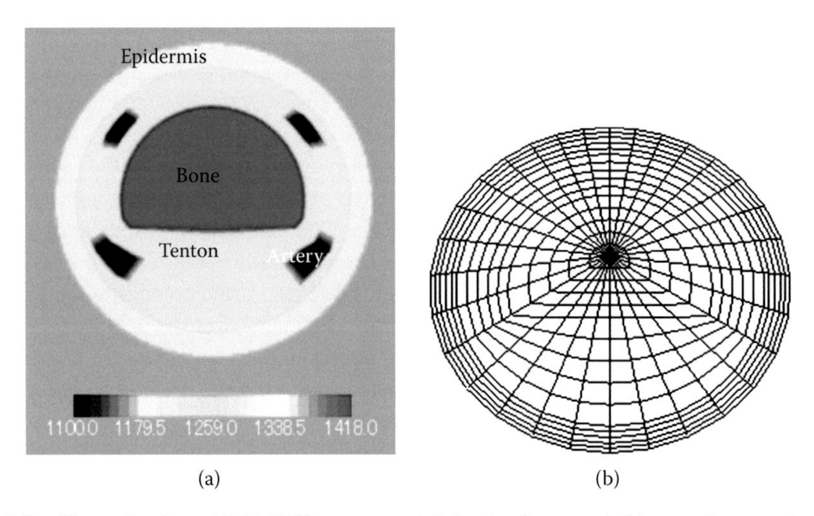


**FIGURE 23.2** **(See color insert.)** (a) Different materials in the finger and (b) a mesh network of the cross section of the finger.

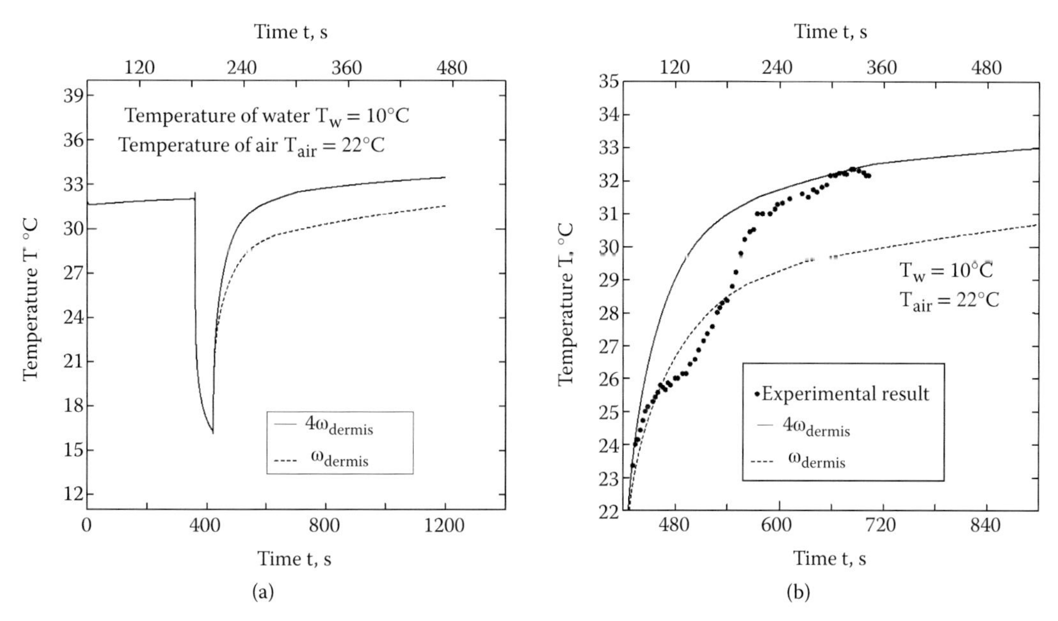


**FIGURE 23.3** (a) Predicted average skin temperature under cold water stimulation. (b) Comparison between predicted average skin temperature and experimental data. (From Clark RP and Edholm OG, *Man and his thermal environment*, Edward Arnold (Publishers) Ltd., London, 1985.)

In the simulation, the following computational conditions were set up. The environmental temperature was 22°C and the wind velocity was 0.1 m/s. After immersion in 10°C cold water for 1 min, the finger was then exposed to indoor air at 22°C. The arterial temperature was considered to be constant at 37°C. The venous temperature was set to be equal to the tissue temperature. The predicted average skin temperature is plotted in Figure 23.3a. It can be seen that the skin temperature became the same as that at the resting stage within 5 min, when the blood perfusion of the dermis was four times as large as that at the resting stage. If the blood perfusion in the recovery stage remained the same as that of the resting perfusion, the skin temperature could not recover even up to 10 min later. Figure 23.3b clearly shows that the predicted temperature with larger blood perfusion after cold-water stimulation was in a good agreement with the experimental values.

We can also see that there was a fluctuation in the measured skin temperature that was not revealed in the prediction, and that the fluctuation was not different from the heart rate. Hence, the fluctuation may have been related to regulation of the autonomic nervous system (ANS).

## 23.3 THERMAL REGULATION MODELING

In order to simulate the temperature oscillation, we coupled a model of regulation of the ANS (Zhang et al. 2010). We developed an ANS model on the basis of earlier studies by Liang (2007) and Xu et al. (2008). The peripheral thermoregulation pathway comprises sensory nerves, receptors, the central nervous system (CNS), efferent nerves, and effectors. Receptors are located in the hypothalamus, efferent nerves are sympathetic vasoconstrictors, and effectors adhere to arteriole smooth muscle. A flow chart for signal transport is shown in Figure 23.4.

The thermoregulation signal is considered to start from the current through the opening of ion channels in the sensory nerves, and the intensity of the current is related to the environmental temperature and the subject's comfort threshold. The receptors receive the frequency of impulses from

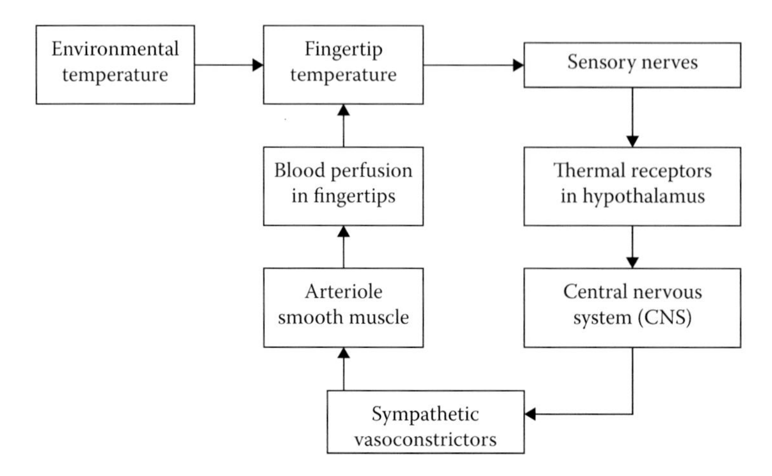


**FIGURE 23.4** Flow chart of the thermoregulation pathway. (From Zhang HD et al., *Computers in Biology and Medicine*, 40, 2010. With permission.)

external stimulation, and this frequency is a function of the current. Therefore, the frequency in the afferent fibers can be expressed as a function of temperature as follows:

$$f_{af} = \frac{f_{af,\min} + f_{af,\max} \exp[(T_{ftip,t} - T_{cri}) / K]}{1 + \exp[(T_{ftip,t} - T_{cri}) / K]} \tag{23.1}$$

Here, $f_{af}$ is the frequency of spikes in the afferent fibers, and $f_{af,min}$ and $f_{af,max}$ are the minimum and maximum values, respectively, of the frequency. $K$ is the parameter that controls the slope between temperature $T_{ftip,t}$ and afferent frequency $f_{af}$. If $K$ is equal to 3, $f_{af}$ will reach the maximum value (15 Hz) when the fingertip temperature is 43°C. The threshold value of the tissue is 43°C (Xu et al. 2008). Thus, it is reasonable to choose $K$ as being 3. $T_{cri}$ was considered to be 30°C.

After the thermal receptor receives the frequency in the hypothalamus, it delivers these signals to the CNS. The CNS is instrumental in gathering input information from the receptors and sending the lowest signal to the effectors. The expression for this process is as follows:

$$f_{ef,in} = \min(f_{ef,in,\max}, T_{af}) \tag{23.2}$$

Here, $f_{ef,in,max}$ is the maximum constant frequency of discharge from the sympathetic nerves. The control signal from the CNS is subsequently transferred to the effectors. For thermoregulation during cold-water stimulation, a cutaneous vasoconstrictor nerve may be activated as the effector to allow the arteriole smooth muscle to constrict, thereby causing changes in blood perfusion. This variation is expressed as follows:

$$\sigma(t) = G_A \log[f_{ef,in}(t - D_A) - f_{efin,\min} + 1] \tag{23.3}$$

Here, $f_{ef,in,\min}$ is the threshold of the sympathetic nerve, $G_A$ is the gain factor, and $D_A$ is the latency time of the nerves. Therefore, blood perfusion after cold-water stimulation can be expressed as follows:

$$\frac{d\Delta\omega(t)}{dt} = \frac{1}{\tau_A}[-\Delta\omega(t) + \sigma(t)]$$

$$\omega(t) = \omega_0 + \Delta\omega(t) \tag{23.4}$$

Here, $\omega(t)$ is the blood perfusion at time $t$, $\omega_0$ is the initial blood perfusion before stimulation, and $\Delta\omega(t)$ is the variation in blood perfusion at time $t$. The equation for skin temperature variation in the fingertip is adopted from Ley et al. (2008) as follows:

$$\rho V C_p \frac{dT}{dt} = h_{air} A(T_{air} - T_{ftip,t}) + \rho_b C_{pb} \omega(t)(T_A - T_{ftip,t}) \tag{23.5}$$

Here, $\rho$ is the tissue density of the fingertip, $C_p$ is the specific heat of tissue, $\rho_b$ is the density of blood, $C_{pb}$ is the specific heat of blood, $T_A$ is the temperature of arterial blood in the fingertip (assumed to be 34°C), $T_{air}$ is the environmental temperature, $T_{ftip,t}$ is the fingertip temperature at time $t$, and $h_{air}$ is the heat transfer coefficient. $V$ is defined as being the volume of the fingertip, which is assumed to be a hemisphere, and is expressed as follows:

$$V = \pi D^3 / 12 \tag{23.6}$$

Here, $D$ is the average diameter of the fingertip. $A$ is the skin area of the fingertip and is defined as follows:

$$A = \pi D^2 / 2 \tag{23.7}$$

Equations 23.1 to 23.5 were solved numerically by using the fourth-order Runge-Kutta method.

Some important parameters related to the ANS model are listed in Table 23.1. The thermo-physical properties that are used in Equation 23.5 are listed in Table 23.2. Figures 23.5a and b show simulated recovery temperatures with different nervous control signals after 1 min of immersion

**TABLE 23.1**
**Some Important Parameters Used in the Thermal Regulation Model**

| | | |
|---|---|---|
| Minimum frequency | $f_{ef,min}, f_{af,min}$ | 1 Hz |
| Maximum frequency | $f_{ef,max}, f_{af,max}$ | 15 Hz |
| Sensitivity of receptor | $K$ | 3 |
| Time constant for vessels | $\tau_A$ | 6 s |
| Latency time of nerves | $D_A$ | 3 s |

**TABLE 23.2**
**Values of Thermophysical Parameters**

| Parameters | Bone | Muscle | Blood | Skin |
|---|---|---|---|---|
| Density $\rho$ (kg/m$^3$) | 1418 | 1270 | 1100 | 1200 |
| Heat capacity $C_p$(J/kg K) | 2049 | 3768 | 3300 | 3391 |
| Heat conductivity (W/mK) | 2.21 | 0.35 | 0.50 | 0.37 |
| Heat transfer coefficient at skin level (W/m$^2$K) | | | | 6 |
| Initial blood perfusion (m$^3$/s) | | $2.3 \times 10^{-10}$ | | $2.3 \times 10^{-10}$ |

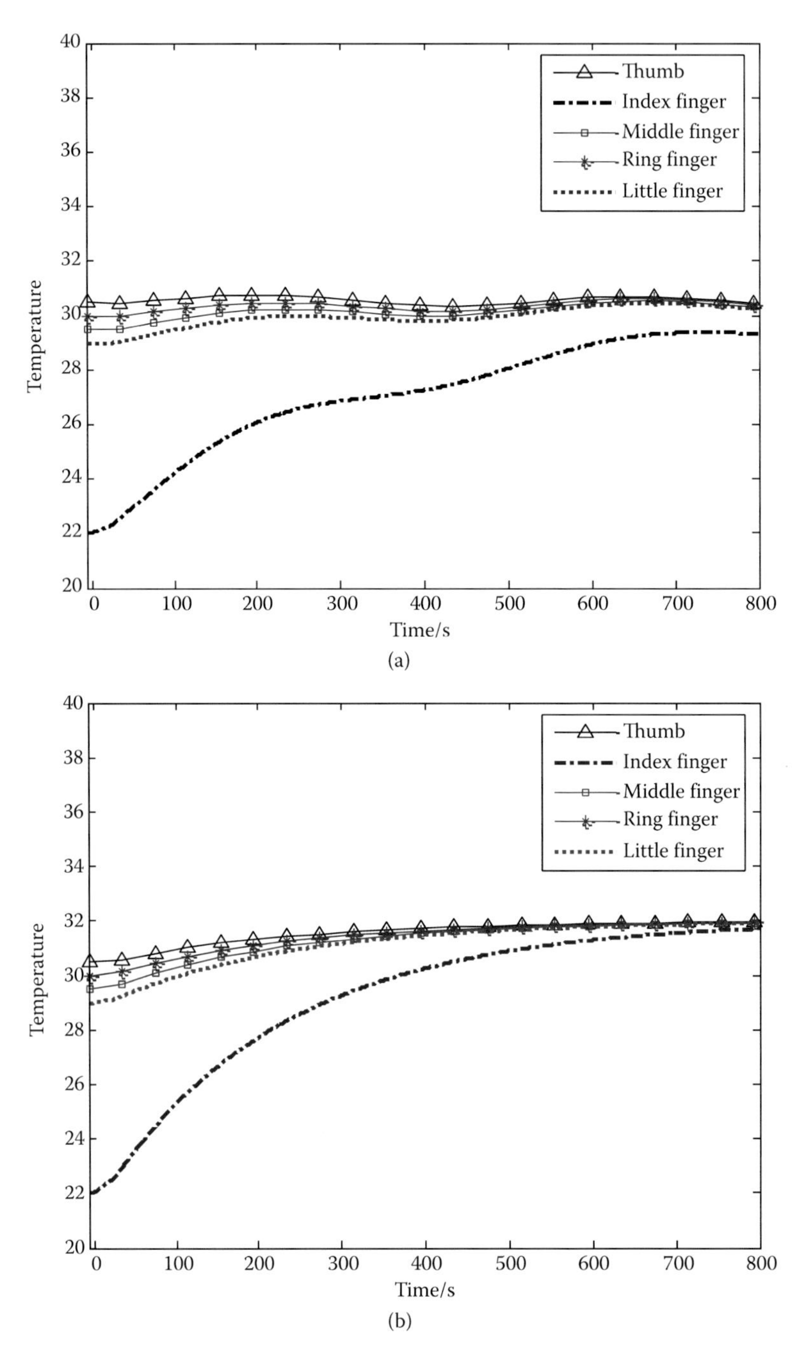


**FIGURE 23.5** Temperature recovery profiles with (a) periodic and (b) constant nervous control signals. (From Zhang HD et al., *Computers in Biology and Medicine*, 40, 2010. With permission.)

in 10°C cold water. When the nervous control signal was periodic and the efferent frequency was expressed as Equation 23.8, the temperature recovery profile was as shown in Figure 23.5a.

$$f_{ef} = \left[1 + \frac{1}{2}\sin(\pi t)\right] f_{af,pre} \tag{23.8}$$

Here, $f_{af,pre}$ is the afferent frequency in the previous time step. We can see that simulated variation profiles follow the same trend as those obtained from the measurements. When the nervous control signal was constant, the temperature recovery profiles were smooth and oscillation-free, similar to those plotted in Figure 23.5b. From these results, we can speculate that temperature oscillation in the measurements may have been due to periodic variation of the nervous control signal.

## 23.4 THREE-DIMENSIONAL ANALYSIS OF BLOOD FLOW AND TEMPERATURE DISTRIBUTION

### 23.4.1 Image-Based Modeling of a Human Hand

Original magnetic resonance (MR) images were acquired from a volunteer's hand. All of the procedures were handled according to an ethical protocol that was approved by RIKEN in Japan. A hand-fitted supporter made of sponge was designed in order to fix the volunteer's hand before taking the images. A 1.5-T scanner (Excelart, Toshiba Medical Systems) was used with different sequences for taking images of blood flow and different tissues. The resolution of the MR images was 320 × 320 and the distance between two adjacent slices was set to be 1 mm.

During image processing, a MATLAB program was developed to extract information on the edges and bones from the original sequential MR images semi-automatically. First, several image processing operators were applied in turn, such as blurring, sharpening in order to reduce noise, and enhancing the contrast. Usually, the gradient of the grayscale near the edge is largest so that the object can be extracted from the background. Due to the limited image quality of the bone regions, reparation was implemented manually to ensure the integrity of the bone regions. The edges and the bone were set in different pixel values separately, such as 255 for the edges and 100 for the bone. Finally, arteries and veins were identified for every slice with specific pixel values (200 for arteries and 180 for veins) according to their anatomical structures.

### 23.4.2 Mesh Generation

Based on the pre-processed cross-section images of the hand, the transfinite interpolation (TFI) method was applied to generate a mesh for every cross-section. Subsequently, the mesh for every cross-section was connected successively by stacking them in turn.

The TFI method is a kind of algebraic mapping method based on mesh generation. The concept of this method is that the physical coordinates of the nodes, which are treated as a function of the computational coordinates, are interpolated based on their values along the edges of the computational domain (Farrashkhalvat and Miles 2003). The one-dimensional interpolations are expressed as follows:

$$P_\xi(\xi,\eta) = (1-\xi)r(0,\eta) + \xi r(1,\eta) \tag{23.9}$$

$$P_\eta(\xi,\eta) = (1-\eta)r(\xi,0) + \eta r(\xi,1) \tag{23.10}$$

where $\xi$, $\eta$ are the computational coordinates, $r(\xi,\eta)$ are the physical coordinates on the edges of the computational domain, and $P_\xi$, $P_\eta$ are the transformations that map the nodes in the computational coordinates to the nodes in the physical coordinates.

A two-dimensional interpolation can be constructed as a linear combination of two one-dimensional interpolations and their product as follows:

$$P_\xi \oplus P_\eta = P_\xi + P_\eta - P_\xi P_\eta \tag{23.11}$$

This formula is the basis of the TFI method in two dimensions. Based on this formula, a mesh is generated by taking discrete values $\xi_i$, $\eta_j$ of $\xi$ and $\eta$ with

$$0 \le \xi_i = \frac{i-1}{i_M - 1} \le 1 \text{ and } 0 \le \eta_j = \frac{j-1}{j_M - 1} \le 1$$

$$i = 1,2,\ldots,i_M,\ j = 1,2,\ldots,j_M \tag{23.12}$$

where $i_M$ and $j_M$ are the maximum numbers of nodes in the two directions.

The TFI method, which is the most common approach to algebraic mesh generation, can produce excellent meshes effectively when other methods are difficult to apply, and it is convenient for adjusting the locations of the mesh nodes.

In the present work, the TFI method was automatically applied to the processed image for quadrilateral mesh generation. Figure 23.6 shows the mesh generation for different layers. It can be seen that the numbers of domains are different for different layers. Since the mesh on each slice was generated, hexahedral mesh generation could be executed by connecting the corresponding quadrilateral elements in two adjacent slices. However, due to the existence of branches in the human hand, a special rule should be made to connect two elements in these areas to ensure the correct geometric shape. For example, there may be one region in the current slice, but three regions in the next slice. Therefore, a rule should be prescribed to determine which two elements will be connected. In this work, the distance between two corresponding regions that should be connected is the minimum. Therefore, by calculating the distances between the corresponding subregions in the adjacent

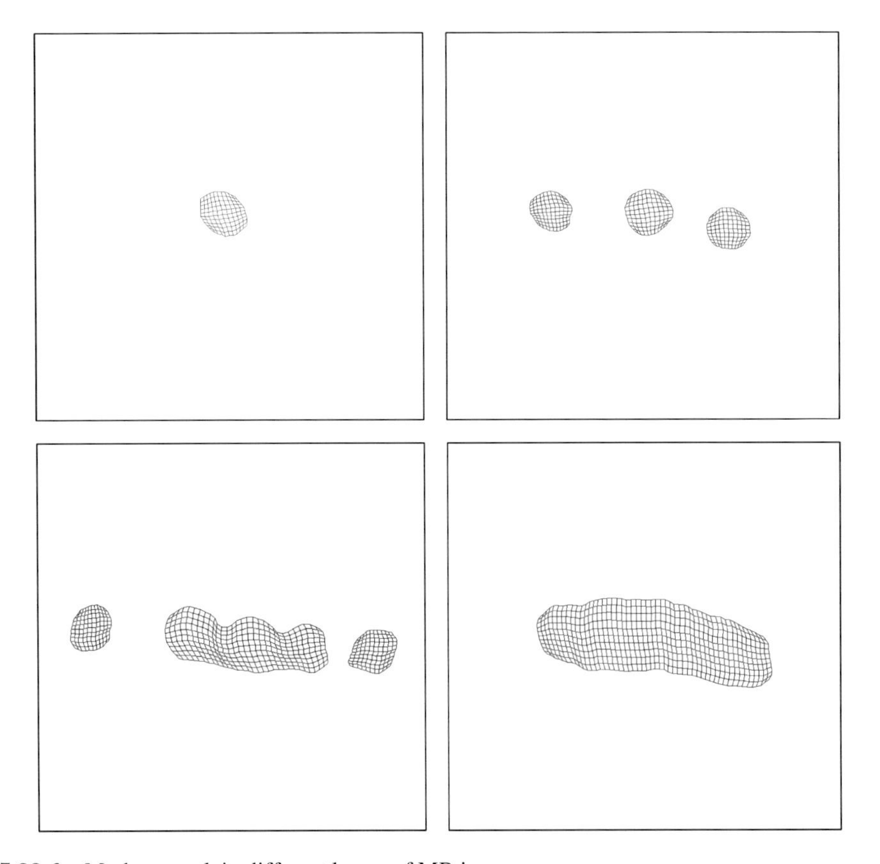

**FIGURE 23.6** Mesh network in different layers of MR images

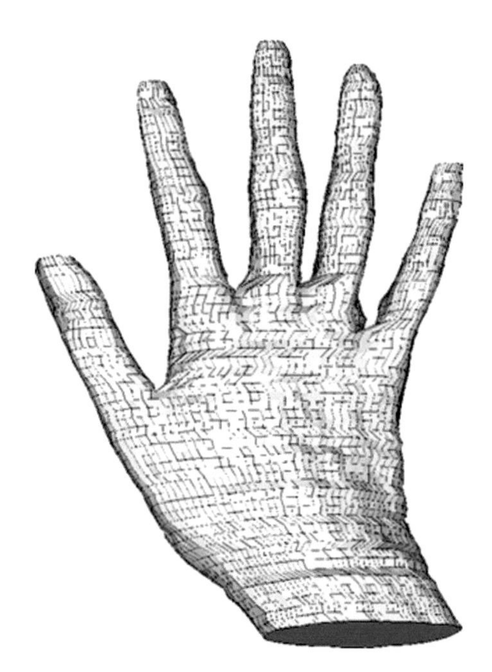

**FIGURE 23.7** Mesh network of the real geometrical human hand. (From Shao HW et al., *Computer Methods in Biomechanics and Biomedical Engineering*, 2012. With permission.)

slices, the regions with the minimum distance were connected to generate the hexahedral mesh. We believe that the hexahedral mesh is beneficial for finite element method (FEM) to generate acceptable results, and it requires less solver time than the other kinds of mesh. Hence, this method for generating hexahedral mesh is more suitable for the finite element method solver and can be applied to the mesh generation of other tissues. A mesh network of the hand that was generated for further analysis is shown in Figure 23.7.

### 23.4.3 Modeling Blood Perfusion and Heat Transfer in Biological Tissues

#### 23.4.3.1 Blood Perfusion Modeling Based on Darcy's Equation

Blood in the hand flows through arteries, capillaries, and veins in turn. In this work, blood flow was divided into two parts: blood flow in large vessels (diameters > 1 mm) and blood perfusion in microvessels. Blood flow in the large vessels was set as a time-dependent input in the numerical model. Blood perfusion in microvessels was considered to be a fluid phase in porous media. Pennes equation was numerically solved to describe the dynamic temperature distribution in the hand.

Because of the huge number of microvessels, modeling all of them would have been difficult and unnecessary. An effective method to deal with this problem is to simplify the tissue with microvessels as porous media. The earliest model for fluid transport in porous media is considered to be Darcy's law, which can be expressed as

$$\nabla P = -\frac{\mu}{k} \boldsymbol{V}_{Darcy} \tag{23.13}$$

where $\mu$ is the viscosity of blood, $k$ is the permeability of the porous media, and $\boldsymbol{V}_{Darcy}$ is the Darcy's velocity. In the present study, blood perfusion in the tissue through the microvessels was considered to occur through seepage in the porous media. For different parts of the human hand, the permeability $k$ varies with the density of the microvessels. In the analysis that is presented in Section 23.5, the value of $k$ for the fingertips, whereby abundant microvessels exist, is set to be $5.0\times10^{-13}\ m^2$, while it is set to be $1.0\times10^{-13}\ m^2$ for the other parts of the hand (Nield and Bejan 1998).

Considering the continuity equation and the momentum equation, an equation for pressure in porous media can be obtained that is expressed as

$$\nabla^2 P = \nabla\cdot\left(-\frac{\mu}{k}\boldsymbol{V}_{Darcy}\right) = -\frac{\mu}{k}\nabla\cdot\boldsymbol{V}_{Darcy} = 0 \tag{23.14}$$

The dimensionless form of Equations 23.13 and 23.14 can be written as

$$\nabla^2 P^* = 0 \tag{23.15}$$

$$\boldsymbol{V}^*_{Darcy} = -Da\cdot Re\cdot\nabla P^* \tag{23.16}$$

where $Da$ is the Darcy number such that $Da = k/D^2$ and $Re$ is the Reynolds number such that $Re = \rho UD/\mu$.

#### 23.4.3.2 Heat Transfer Modeling Based on Pennes Equation

In order to obtain the temperature of living tissue, energy equations for the blood and tissue phase are needed, and this may cause the simulation to become complex. As an alternative, Pennes equation is used to investigate heat transfer in living tissues, which is expressed as

$$\rho c\frac{\partial T}{\partial t} = \lambda\nabla^2 T + Q_m + \omega_b\rho_b c_b(T_b - T) \tag{23.17}$$

where $\rho$ indicates the tissue density, $c$ is the specific heat of the tissue, $\rho_b$ and $c_b$ represent the density and specific heat of blood, respectively, $T_b$ is the temperature of blood that perfuses the tissue, which is assumed to be constant, $\lambda$ is the thermal conductivity of the tissue, $Q_m$ is the heat production per unit volume, and $\omega_b$ indicates the blood perfusion rate.

If the relationship between Darcy's velocity and the blood perfusion rate is known, the local blood perfusion can be obtained. Thus, coupling Darcy's equation and Pennes equation is an elegant method to describe the non-uniformity of blood perfusion. It is assumed that the diameter and length of the microvessels are the same in different parts of the tissue, thus, the ratio of the microvessel blood flow and microvessel volume can be written as

$$Q_b / V_b = v_b / L \tag{23.18}$$

where $v_b$ is the blood velocity and $L$ is the length of the microvessel. Since the microvascular volume and Darcy's velocity are written as $V_b = \varphi V_t$ and $V_{Darcy} = \varphi v_b$, where $\varphi$ is the porosity of the tissue, the blood perfusion can be expressed as

$$\omega_b = Q_b / V_t = \varphi Q_b / V_b = \varphi v_b / L = V_{Darcy} / L \tag{23.19}$$

Substituting Darcy's velocity for the blood perfusion rate, the dimensionless form of the energy equation can be expressed as

$$\frac{\partial T^*}{\partial t^*} = \frac{1}{Pe}\nabla^2 T^* + \frac{1}{Pe}Q^*_m + \beta\gamma V^*_{Darcy}(T^*_b - T^*) \tag{23.20}$$

where $Pe$ is the Peclet number such that $Pe = UD/\alpha$, $\beta$ is the ratio of the heat capacity of blood and tissue, and $\gamma$ is the ratio of the diameter to the length of the microvessel.

On the surface of the skin, convection, radiation, and evaporation are the main modes of heat transport between the tissue and the surrounding environment. The boundary condition on the surface of the skin is thus expressed as

$$-\lambda_s \frac{\partial T}{\partial n} = h_c\left(T - T_{ambi}\right) + h_{ra}\left(T - T_{ambi}\right) + E_{sk} \tag{23.21}$$

where $T_{ambi}$ is the ambient temperature, $\lambda_s$ is the thermal conductivity of skin, $h_c$ is the convective heat transfer coefficient, $h_{ra}$ is the radiative heat transfer coefficient, and $E_{sk}$ is the heat loss due to evaporation. The heat loss due to evaporation can be written as

$$E_{sk} = h_e(P_{skin} - P_{ambi}) \tag{23.22}$$

where $P_{skin}$ is the vapor pressure at the surface of the skin (kPa) and $P_{ambi}$ is the vapor pressure of air (kPa). The evaporation coefficient $h_e$ is related to the air velocity and can be expressed as

$$h_e = 124\sqrt{V_{air}}\ \mathrm{W / m^2kPa} \tag{23.23}$$

The dimensionless form of the boundary condition equation can be written as

$$-\frac{\partial T^*}{\partial n^*} = Bi \cdot (T^* \text{-} T^*_{ambi}) + E^*_{sk} \tag{23.24}$$

where $Bi$ is the Biot number such that $Bi = \frac{h_c D}{\lambda_s}$.

When the hand is in moving air whose velocity is in the range of $0 < V_{air} < 0.2\ m/s$, the heat transfer coefficient on the surface of the hand is 4.2 W/m$^2$K. With respect to the radiative heat transfer coefficient, it is set to be a constant value of 4.8 W/m$^2$K as a representative for typical indoor temperatures (ASHRAE 1997).

Equations 23.15, 23.16, and 23.20 have been discretized by using the Galerkin weighted residual method and a FORTRAN program has been developed to solve these equations. The specific procedures for finite element approximation can be referred to by consulting literature references (Reddy and Gartling 2001; Hanspal et al. 2006; He and Himeno 2012). The thermal physical properties and some related parameters are listed in Tables 23.2 and 23.3, respectively.

**TABLE 23.3**
**Parameters Used in Three-Dimensional Finite Element Element Analysis**

| Parameters | Values | Notes |
|---|---|---|
| $E_{sk}$ (W / m$^2$) | 1030 | Evaporative heat |
| $L_c$(m) | 4E-3 | Capillary length |
| $D_c$(m) | 8E-6 | Capillary diameter |
| $U_c$(m / s) | 2E-3 | Capillary flow |
| $\mu$(kg / m · s) | 1.07E-3 | Blood viscosity |
| $Q_m$(W / m$^3$) | 300 | Metabolic heat production rate |
| $T_a$(°C) | 37 | Arterial blood temperature |
| $T_v$(°C) | 35.5 | Venous blood temperature |
| $\bar{P}_a$(KPa) | 7.0 | Arterial pressure |
| $\bar{P}_v$(KPa) | 1.3 | Venous pressure |

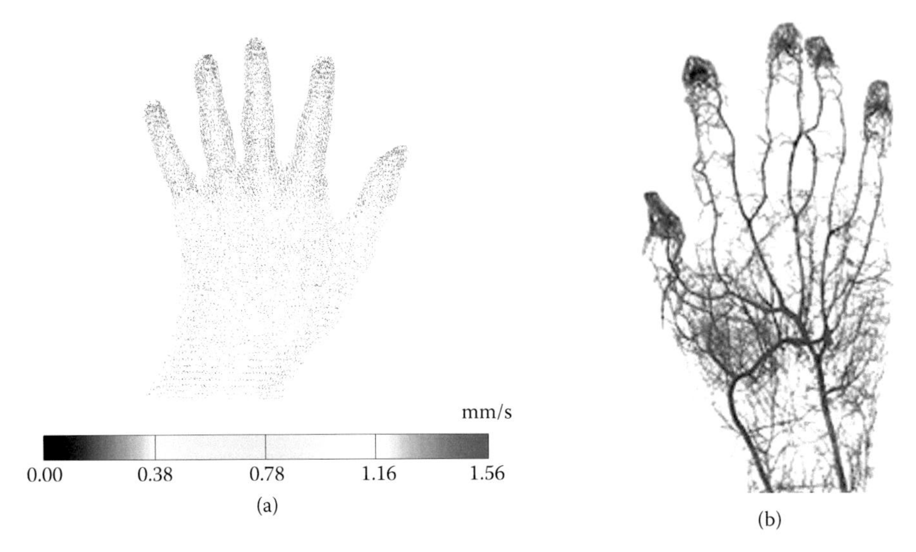


**FIGURE 23.8 (See color insert.)** (a) Simulated blood perfusion of the human hand. (Reproduced from Shao HW et al., *Computer Methods in Biomechanics and Biomedical Engineering*, 2012. With permission.) (b) Vasculature in the human hand. (From http://www.zcool.com.cn/ZMjEzNjg4.html.)

Figure 23.8a represents the distribution of Darcy's velocity at the steady state. The distribution of Darcy's velocity indicates that (a) blood flow in the tissue has a high velocity near the arteries, and (b) in the fingertip the perfusion rate, which is directly proportional to Darcy's velocity, is higher than in other parts of the hand due to the large number of microvessels in the fingertip. Figure 23.8b shows a casting specimen of the human hand, whereby it can be seen that the predicted blood perfusion is in favorable agreement with the microvasculature.

## 23.5 FINITE ELEMENT ANALYSIS OF BLOOD FLOW AND TEMPERATURE DURING DIGITAL THERMAL MONITORING

In the DTM test, hyperemic blood flow for healthy people can reach a peak value that is higher than that for people who are at risk of cardiovascular events, since higher hyperemic blood perfusion values correspond to arteries that are capable of experiencing larger levels of vasodilatation. In order to simulate the fingertip temperature in the DTM test in this work, three sets of inputs for the pressures in the larger arteries and veins of the hand during vessel occlusion and reperfusion were used as shown in Figure 23.9. For all three sets of inputs, an occlusion time period of 180 s was chosen to be consistent with the occlusion time that is used in a majority of clinical trials after a period of steady state lasting 80 s. After occlusion, three kinds of recovery conditions during reperfusion were assumed to simulate different levels of RH: (1) an instantaneous pressure overshoot that was three times higher than that at the steady state and exponential decay back to the steady level, (2) an instantaneous pressure overshoot that was two times higher than that at the steady state and exponential decay back to the steady level, and (3) a direct exponential recovery back to the steady level without overshoot.

The dynamic variation curves for fingertip temperatures are plotted in Figure 23.10, corresponding to the three levels of RH. It can be seen that fingertip temperature variation is closely related to the pressure changes in larger arteries. In comparison with the predicted temperature variation, the measured fingertip temperatures in the DTM tests, which represent the average values of five healthy subjects, are plotted (Wang and He 2010). The simulation results showed good agreement with the experimental results at the rest and occlusion stages. At the recovery stage,

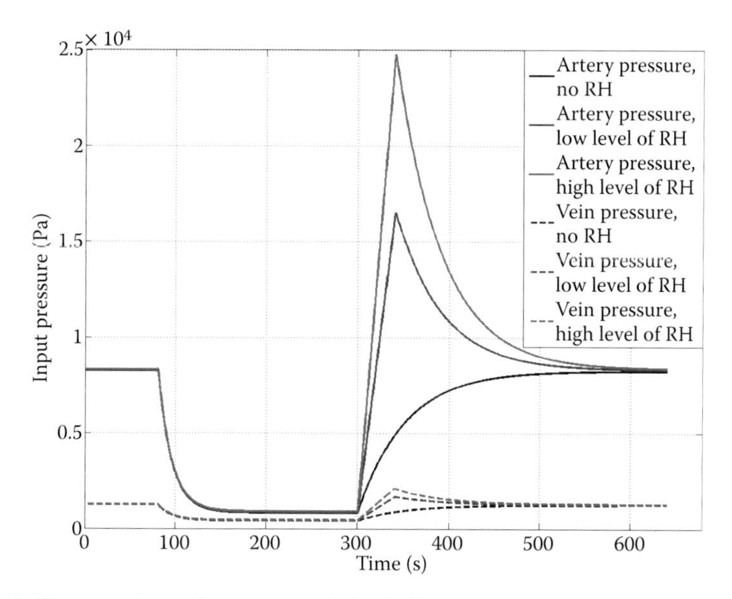


**FIGURE 23.9** Different settings of pressure variation in larger arteries and veins. (From Shao HW et al., *Computer Methods in Biomechanics and Biomedical Engineering*, 2012. With permission.)

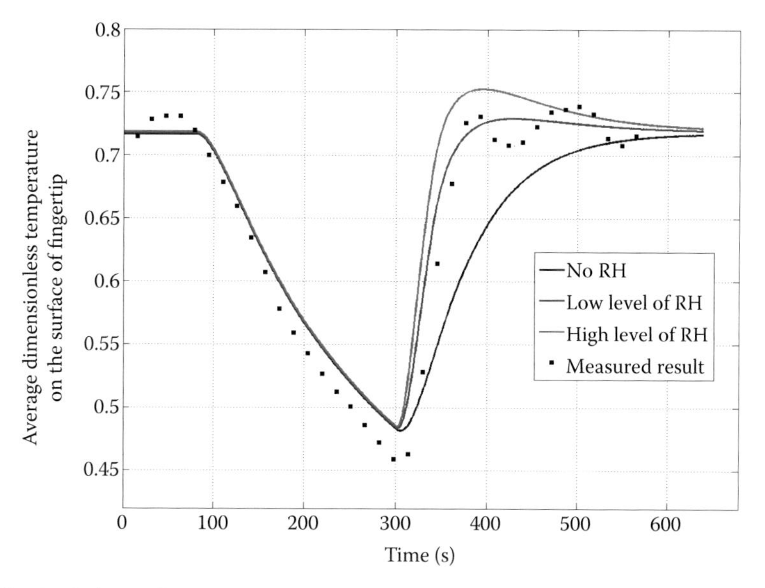


**FIGURE 23.10** Variations of fingertip temperature for different levels of RH. (From Shao HW et al., *Computer Methods in Biomechanics and Biomedical Engineering*, 2012. With permission.)

the predicted values for low-level RH were generally in agreement with the measured data. It is notable that oscillations of the fingertip temperatures at the recovery stage were not revealed in the simulation. Blood flow resulting from oscillations in the vasomotor smooth muscle tone may cause the temperature fluctuations (Podtaev et al 2008). Since the smooth muscle in charge of vasomotion exists in the arterioles and the density of microvessels was modeled as porosity in the present model, it was reasonable and feasible to model periodic vasomotion as the periodic change in porosity.

## 23.6 PREDICTION OF TEMPERATURE OSCILLATIONS AFTER THE CONTRALATERAL COOLING TEST

For the computation of Darcy's velocity, we adopted the following empirical equation to compute the permeability, which is determined by the porosity $\varphi$:

$$k = \frac{D^2 \varphi^3}{180(1-\varphi)^2} \tag{23.25}$$

where $D$ is the diameter of the microvessels. The temperatures of larger arteries and veins were set to be constant, and were the same as those in the DTM simulation.

The porosity $\varphi$ in the model is defined as the volume ratio of microvessels to the tissue volume, which varies with the density of the capillary network. For this, values of 0.2, 0.5, and 0.6 were adopted as the porosities of bone, tissue, and the fingertip when the human hand is in a thermally comfortable state. Vasodilation and vasoconstriction in the tissue are two important control processes of the human thermoregulatory system, and they may be modeled as sinusoidal variation in porosity from the thermal neutral value. If the mean skin temperature is assumed to be the sensory signal for vasomotion, the periodic change in porosity can be expressed as follows (Tang et al. 2013):

$$\varphi = \varphi_0 \left( 1 + \sum_{i=1}^{n} A_i \sin(2\pi f_i \cdot t) \right) \cdot e^{(T-32)/50} \tag{23.26}$$

where $\varphi_0$ is the value in the thermal neutral state, $f_i$ is a frequency of internal activity such as endothelial regulation that leads to vasomotion, $A_i$ is the amplitude of the corresponded frequency, and $n$ is the number of internal activities including neurogenic and endothelial oscillations. 32°C is adopted as the thermal neutral temperature of skin. Without consideration of internal activities, $\varphi$ will become an exponential function of the single variable $T$, as shown in Figure 23.11. It can be seen that if the initial porosity is set to be 0.5, the porosity at 25°C will become 0.4. If the initial porosity is set to be 0.6, the porosity at 25°C will become 0.5, which gives reasonable variation ranges.

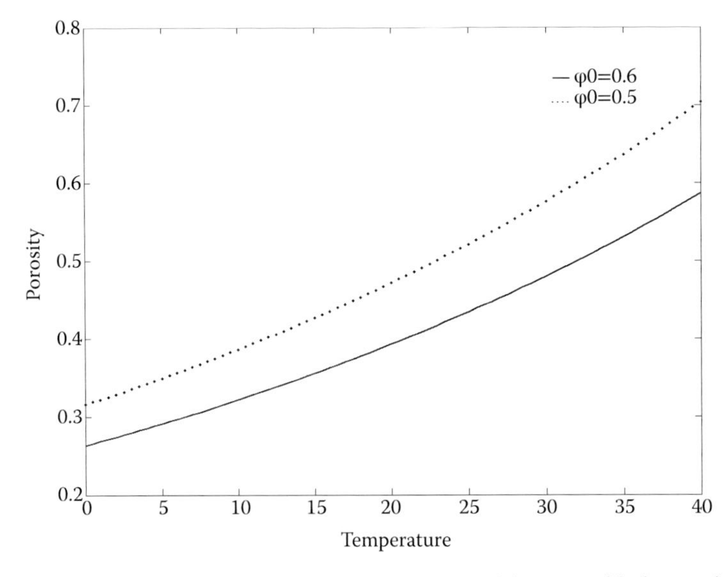


**FIGURE 23.11** Variation in porosity with tissue temperature without considering periodic vasomotion according to Equation 23.26.

We implemented the simulation cooling test by using the following protocol: The hand of a volunteer was first exposed to air at 22°C for 10 min to reach thermal balance. The hand was then immersed in 0°C cold water for 3 min, and after that the hand was returned back to the 22°C air for recovery.

In order to save on computational time, the model of the hand was divided into four parts from the fingertip to the wrist and adiabatic conditions were prescribed between adjacent parts. The simulation results of the average skin temperatures of the fingertip and the palm for the hand in the cooling test are shown in Figure 23.12a. It can be seen that the skin temperatures of the two parts have similar variations in their features: before the cooling test, the skin temperatures stabilized around 30–33°C in the 22°C air. During the cooling stage the temperature of the skin decreased considerably. After cooling, the skin temperature gradually recovered. Fluctuations were observed in the variation of the temperature curves, which indicates the occurrence of periodic impacts on internal activities like endothelial regulation. Furthermore, the fingertip temperature indicated a

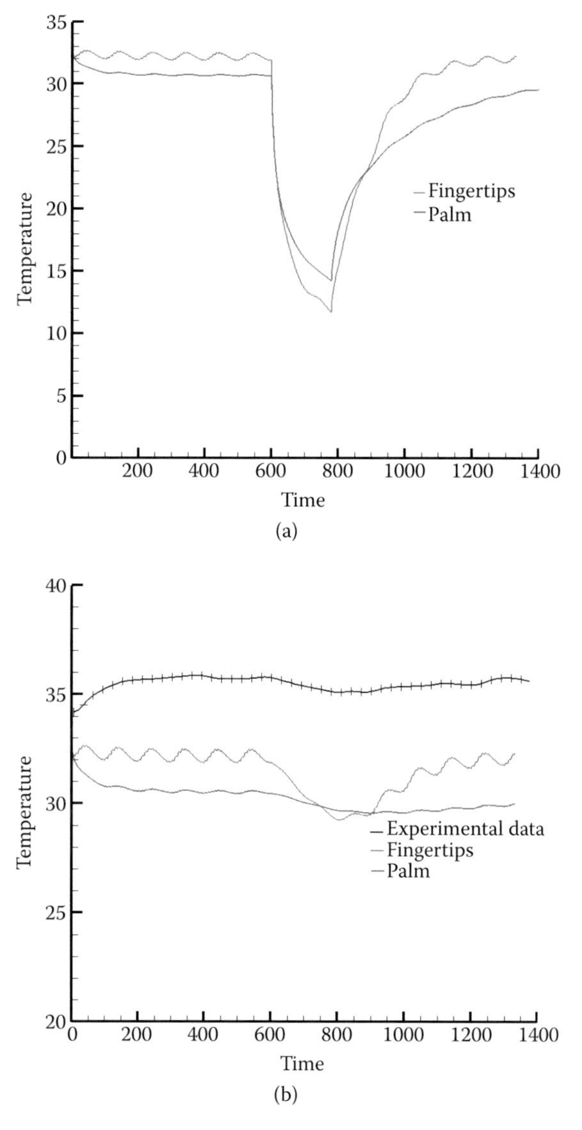


**FIGURE 23.12** (a) Variation of the fingertip and palm temperatures of the hand during the cooling test. (b) Variation in the fingertip and palm temperatures for the contralateral hand during the cooling test.

faster rate of decline and recovery in the 3 min cooling and recovery stage, which implies greater changes in blood perfusion in the fingertips due to more abundant microvessels.

The contralateral skin temperatures of the fingertip and palm are plotted in Figure 23.12b. It can be seen that the fingertip temperature decreases during cooling of the contralateral hand and has fluctuations that are similar to those of the fingertip temperature for the cooling hand. Compared to the measured fingertip temperature during the cooling test of the contralateral hand as plotted in Figure 23.12b, the simulated fingertip temperature underwent larger oscillations. This was because the same changes of porosities that are related to vasoconstriction were set in both hands during the simulation, which may have caused larger variations in the non-cooling hand. The variations in skin temperature can be seen more directly from the temperature contours of the hand in Figure 23.13,

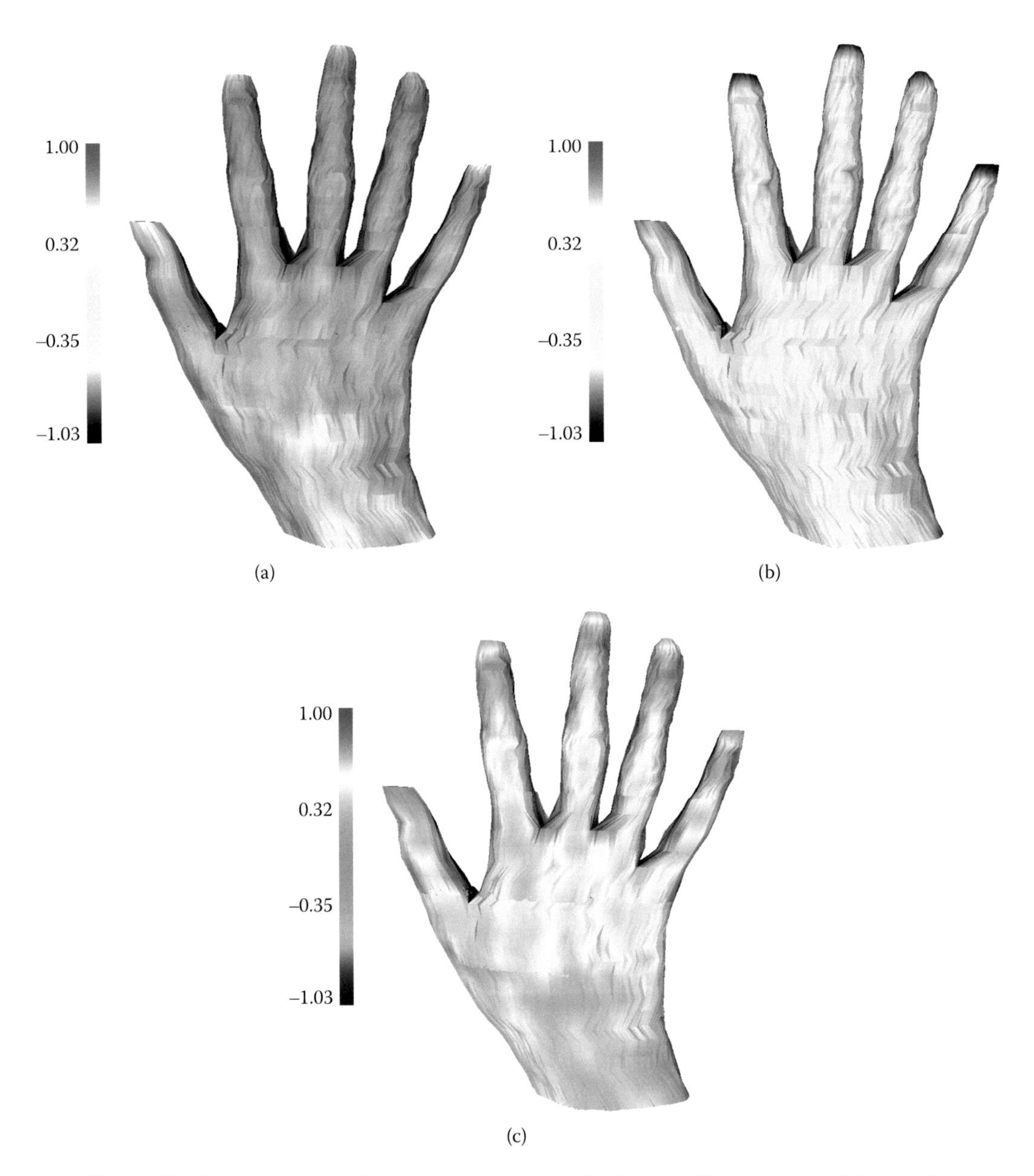


**FIGURE 23.13 (See color insert.)** Skin temperature distribution in different stages of the cooling test. (a) At 600 s. (b) At 780 s. (c) At 1350 s.

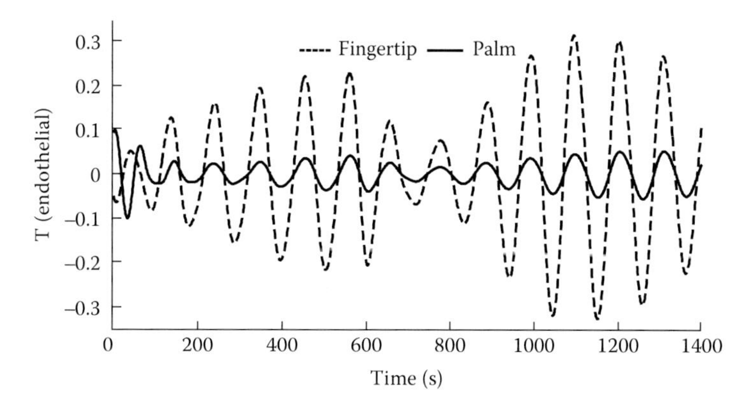


**FIGURE 23.14** Decomposed wavelets of the simulated fingertip and palm temperatures in the endothelial frequency band.

which shows the temperature distributions during the resting (600 s), cooling (780 s), and recovery (1350 s) stages. The finger skin temperatures were distinctly higher than the palm temperatures in the recovery stage, which is in agreement with the physiological data of Clark et al. (1985).

For comparison, wavelet analyses of the simulated data in Figure 23.12b were performed. Figure 23.14 provides the amplitudes of skin temperature fluctuations in the endothelial frequency range. It can be observed that the amplitudes of temperature oscillations both in the fingertip and palm during the cooling test (601–680 s) significantly decreased, which is in good agreement with the measured data of Smirnova et al. (2012). The changes in porosities of the porous media model may reflect the vasomotion during cold water stimulation.

## 23.7 CONCLUDING REMARKS

This review describes several sets of modeling for assessing blood flow variation in the tissue via finger temperatures. In the two-dimensional model, several layers of tissue and arteries were considered and Pennes bioheat equation was employed. The results indicated that the finger skin temperature after cold-water stimulation corresponds to larger blood perfusion. However, the model cannot illustrate how blood perfusion varies after external stimulation. Based on the thermoregulation pathway, a one-dimensional bioheat equation was coupled with a regulation model of the ANS, and it illustrates very well the periodic regulation in blood perfusion after cold stimulus. The problem is that blood perfusion should still be known from the start.

The theory of transport in porous media was found to be quite useful for describing blood flow in biological tissues. We employed Darcy's model to describe blood flow through biological tissues and constructed the relationship between Darcy's velocity and blood perfusion. By coupling Darcy's equation and Pennes equation, the variation in blood perfusion due to vasomotion can be reflected in the tissue temperature.

Two processes that are employed in clinical applications were investigated: digital thermal monitoring and a contralateral cooling test. We proved that porosity that represents the volume ratio of microcirculation in the whole tissue volume has a great influence on skin temperature variation.

Our own numerical developments for finite element modeling demonstrate the advantage in considering various problems regarding peripheral blood flow and temperature.

Some diseases may lead to damage in the microcirculation. Since the detection of microcirculation abnormalities is difficult to implement, this work revealed that skin temperature information, which can be much more easily measured, is a good alternative for detecting microcirculation abnormalities.

## ACKNOWLEDGMENTS

This work was supported by KAKENHI of JSPS, Grant No. 18560218, the National Science Foundation of China (NSFC), Grant No. 10772176. Experimental data for the simulation verification were obtained by the financial support of the Project RSF 14-15-00809.

## REFERENCES

Akhtar MW, Kleis SJ, Metcalfe RW, and Naghavi M. (2010). Sensitivity of digital thermal monitoring parameters to reactive hyperemia. *J Biomech Eng* 132(5): 051005-1.

ASHRAE. (1997). *ASHRAE Fundamentals Handbook*. Atlanta (GA): ASHRAE Press.

Basak K, Manjunatha M, and Dutta, PK. (2012). Review of laser speckle–based analysis in medical imaging. *Med. Bio. Eng. Comput.* 50: 547–558.

Bernjak A, Clarkson PBM, McClintock PVE, and Stefanovska A. (2008). Low-frequency blood oscillations in congestive heart failure and after $\beta_1$-blockade treatment. *Microvascular Research* 76: 224–232.

Carrillo AE, Cheung SS, and Flouris AD. (2011). A novel method to predict cutaneous finger blood flow via finger and rectal temperature. *Microcirculation.* 18(8): 670–676.

Clark RP and Edholm OG. (1985). *Man and his thermal environment*. London: Edward Arnold (Publishers) Ltd.

Daly SM, and Leahy MJ. (2013). "Go with the flow": A review of methods and advancements in blood flow imaging. *J. Biophotonics* 6(3): 217–255.

Farrashkhalvat M and Miles JP. (2003). *Basic structured grid generation—with an introduction to unstructured grid generation*, Butterworth-Heinemann Pub. Co. Ltd. 76–96.

Fedorovich, AA. (2012). Non-invasive evaluation of vasomotor and metabolic functions of microvascular endothelium in human skin. *Microvascular Research* 84: 86–93.

Haga T, Ibe A, Aso Y, Ishiozawa M, Miyajima M, and Takeda K. (2012). Development of methodology for the estimation of skin blood perfusion by applying inverse analysis of skin model. *Trans. Jap. Soc. Med. Bio. Eng.* 50(4): 317–328.

Hanspal NS, Waghode AN, Nassehi V, and Wakeman RJ. (2006). Numerical analysis of coupled stokes/Darcy flows in industrial filtrations. *Transport in Porous Media* 64: 73–101.

He Y and Himeno R. (2012). Finite element analysis on fluid filtration in system of permeable curved capillary and tissue. *J. Mech. Med. Bio.* 12(4): 1250077-1-20.

He Y, Kawamura T, and Himeno, R. (2001). A two-dimensional thermal model for determining cold-stressed effects on a human finger. *Computational Biomechanics-RIKEN Sympo*, June 4–5. 129–132.

Kahle W, Leonhardt H, and Platzer W. (1990). *Atlas of Human Anatomy.* (Japanese edition, translated by Ochi J.) Bunkoudo. 246. http://www.zcool.com.cn/ZMjEzNjg4.html

Ley O and Deshpande CV. (2009). Comparison of two mathematical models for the study of vascular reactivity. *Comput. Bio. Med.* 39: 579–589.

Ley O, Deshpande CV, Prapamcham B, and Naghavi M. (2008). Lumped parameter thermal model for the study of vascular reactivity in the fingertip. *J. Biomech Eng.* 130: 031012.

Ley O, Dhindsa M, Sommerlad SM, Barnes JN, DeVan AE, Naghavi M, and Tanaka H. (2011). Use of temperature alterations to characterize vascular reactivity. *Clin. Physiol. Funct. Imaging* 31: 66–72.

Liang FY. (2007). An integrated computational study of multiscale hemodynamics and multi-mechanism physiology in human cardiovascular system. PhD thesis, Chiba University.

Liu WM, Meyer J, Scully CG, Elster E, and Gorbach AM. (2011). Observing temperature fluctuations in human using infrared imaging. *QIRT Journal* 8(1): 21–36

McQuilkin GL, Panthagani D, Metcalfe RW, Hassan H, Yen AA, Naghavi M, and Hartley CJ. (2009). Digital thermal monitoring (DTM) of vascular reactivity closely correlates with doppler flow velocity. *31st Annual International Conference of the IEEE EMBS*. Minneapolis, Minnesota, September 2–6: 1100–1103.

Nagata K, Hattori H, Sato N, Ichige Y, and Kiguchi M. (2009). Heat transfer analysis for peripheral blood flow measurement system. *Review of Scientific Instruments* 80: 064902-1-6.

Nield DA and Bejan A. (1998). *Convection in Porous Media*, 2nd ed. New York (NY): Springer Publishing Company.

Podtaev S, Morozov M, and Frick P. (2008). Wavelet-based correlations of skin temperature and blood flow oscillations. *Cardiovasc Eng* 8: 185–189.

Reddy JN and Gartling DK. (2001). *The Finite Element Method in Heat Transfer and Fluid Dynamics*, 2nd ed. Boca Raton, FL: CRC Press, 79–147.

Roustit M, and Cracowski J. (2012). Non-invasive assessment of skin microvascular function in humans: an insight into methods. *Microcirculation* 19(1): 47–64.

Sagaidachnyi AA, Usanov DA, Skripal AV, and Fomin AV. (2012). Correlation of skin temperature and blood flow oscillations. *Saratov Fall Meeting 2011: Optical Technologies in Biophysics and Medicine XIII, Proc. of SPIE* 8337: 83370A-1-8.

Shao HW, He Y, and Mu LZ. (2012). Numerical analysis of dynamic temperature in response to different levels of reactive hyperemia in a three-dimensional image-based hand model. *Computer Methods in Biomechanics and Biomedical Engineering*. Doi: 10.1080/10255842.2012.723698.

Sim JK, Youn S, and Cho YH. (2012). A thermal peripheral blood flowmeter with contact force compensation. *J. Micromech. Microeng.* 22: 125014-1-7.

Smirnova E, Podtaev S, Mizeva I, and Loran E. (2013). Assessment of endothelial dysfunction in patients with impaired glucose tolerance during a cold pressor test. *Diabetes and Vascular Disease Research.* 10(6): 489–497.

Stefanovska A, Bracic M, and Kvernmo HD. (1999). Wavelet analysis of oscillations in peripheral blood circulation measure by Doppler technique. *IEEE Trans. Biomed. Eng.* 46: 1230–1239.

Tang YL, He Y, Shao HW, and Mizeva I. (2013). A porous media model to study the relationship between endothelial function and fingertip temperature sscillation. *10th IEEE International Conference on Control & Automation*, June 12–14, Hangzhou.

Wang X and He Y. (2010). Experimental study of vascular reactivity in the fingertip: an infrared thermography method. *2010 3rd International Conference on Biomedical Engineering and Informatics (BMEI2010)*, Vol. 1–7: 1180–1184.

Xu F, Wen T, Lu TJ, and Seffen KA. (2008). Modeling of nociceptor transduction in skin thermal pain sensation. *J. Biomech Eng.* 130: 041013-1–041013-13.

Yue K, Zhang XX, and Zuo YY. (2008). Noninvasive method for simultaneously measuring the thermophysical properties and blood perfusion in cylindrically shaped living tissues. *Cell Biocem., Biophys.* 50: 41–51.

Zhang HD, He Y, Wang X, Shao HW, Mu LZ, and Zhang J. (2010). Dynamic infrared imaging for analysis of fingertip temperature after cold water stimulation and neurothermal modeling study. *Computers in Biology and Medicine*, 40: 650–656.

# Section VII

## Bone

# 24 Micro-Finite Element Model for Bone Strength Prediction

*He Gong, Ming Zhang, and Ling Qin*

## CONTENTS



## SUMMARY

A quantitative assessment of trabecular bone strength at the tissue level is essential for understanding bone failure mechanisms associated with osteoporosis, osteoarthritis, loosening of implants, and cell-mediated adaptive bone remodeling. The material properties of the trabeculae, in combination with their microarchitectures at the tissue level, determine the strength of trabecular bone. Micro-CT (computed tomography) image-based micro-finite element analysis takes most determinants of trabecular bone strength into consideration, such as the material properties of the trabeculae and their micro-architectures. Under large loading conditions, bone tissue material nonlinearity and large deformations within the trabecular network, such as bending and buckling, are important for the characterization of the apparent and tissue-level failure behaviors. In this chapter, two trabecular specimens with different microstructures are used as examples to illustrate a numerical technique that offers a better understanding of trabecular yield behaviors under different loading directions.

## 24.1 INTRODUCTION

There are two types of bones in the skeleton: dense, compact cortical bone and porous, spongy trabecular bone. Trabecular bone is located inside cortical bone shells and responds faster to biophysical stimuli, and thus is most susceptible to osteoporosis, particularly in the elderly population (Melton et al., 1992). Tissue level trabecular bone strength is essential for understanding bone failure mechanisms, which is determined by the material properties of trabeculae in combination with their microarchitectures at the tissue level. Direct mechanical tests on trabecular specimens have been used to obtain the mechanical properties of trabecular bone (Morgan et al., 2003). However, mechanical tests can be performed on each specimen only once due to the tests' destructive nature. Alternatively, with enhanced computational power and advanced imaging techniques, mechanical tests can be simulated by micro-finite element analysis (also written as μFEA) based on high-resolution micro-CT computer reconstruction models for different loading conditions (Judex et al., 2003; Gong et al., 2007, 2011). It is the most efficient way to determine the tissue-level stress/strain distributions of trabecular bone.

Micro-CT image-based micro-finite element analysis takes most of the determinants of trabecular bone strength into consideration, such as the material properties of the trabeculae and their microarchitectures. There is little experimental data available in the literature about the nonlinear mechanical properties of bone material at the tissue level. Some investigations have used a bilinear asymmetric elastic-plastic constitutive relationship with hardening and different yield strengths in tension and compression (Bayraktar et al., 2004; Niebur et al., 2000). Although this tissue-level constitutive model cannot be extended to predict the apparent strain softening or trabecular fracture (Verhulp et al., 2008), it is capable of describing the apparent yield behavior, and thus provides a conservative estimation of bone strength (Niebur et al., 2002). Hence, the bilinear asymmetric elastic-plastic constitutive relationship of bone material is important for a better understanding of the overall nonlinear mechanical properties of trabecular bone.

In this chapter, two trabecular specimens with different microstructures are used as examples to illustrate the numerical technique for simulating trabecular yield behaviors under different loading directions.

## 24.2 DEVELOPMENT OF A MICRO-FINITE ELEMENT MODEL FOR BONE STRENGTH PREDICTION

### 24.2.1 Micro-CT Scanning Procedure

Two trabecular specimens, 4×4×4 mm$^3$ in size with different microarchitectures, were selected from a cohort of 90 specimens investigated previously (Gong et al., 2005, 2007). The two trabecular specimens in this study were from two L4 vertebral bodies of two male cadavers. Specimen A was from the anterior part of a 70-year-old vertebral body and specimen B was from the central part of a 69-year-old vertebral body. The images were obtained from micro-CT scanning (CT40, Scanco Medical AG, Bassersdorf, Switzerland). The voxel size was 20 μm × 20 μm with a 20 μm slice thickness. Table 24.1 lists all details of the two specimens, that is, bone volume fraction (BV/TV), structure model index (SMI), connectivity density (Conn.D), trabecular number (Tb.N), trabecular thickness (Tb.Th), and the number of finite elements, from which we can see that the two trabecular specimens had distinct microarchitectures.

### 24.2.2 Three-Dimensional Modeling of Trabecular Specimens

Three-dimensional modeling of trabecular specimens was performed in MIMICS software (Materialise, Inc.) (Figure 24.1).

### 24.2.3 Nonlinear Micro-Finite Element Analysis

The nonlinear micro-finite element analyses were performed by ABAQUS software to simulate axial compression and tension in the longitudinal and transverse directions. A fixed displacement boundary condition was chosen so that all nodes at the bone-platen interface were constrained in the

**TABLE 24.1**
**Detailed Information of the Two Trabecular Specimens**

| | Age (years) | BV/TV (%) | SMI (-) | Conn.D (1/mm$^3$) | Tb.N (1/mm) | Tb.Th (μm) | Number of Elements (-) |
|---|---|---|---|---|---|---|---|
| **Specimen A** | 70 | 8.0855 | 1.3656 | 2.5354 | 0.9447 | 98.1 | 1444218 |
| **Specimen B** | 69 | 6.5007 | 2.5024 | 2.7326 | 1.3366 | 100.3 | 1151332 |

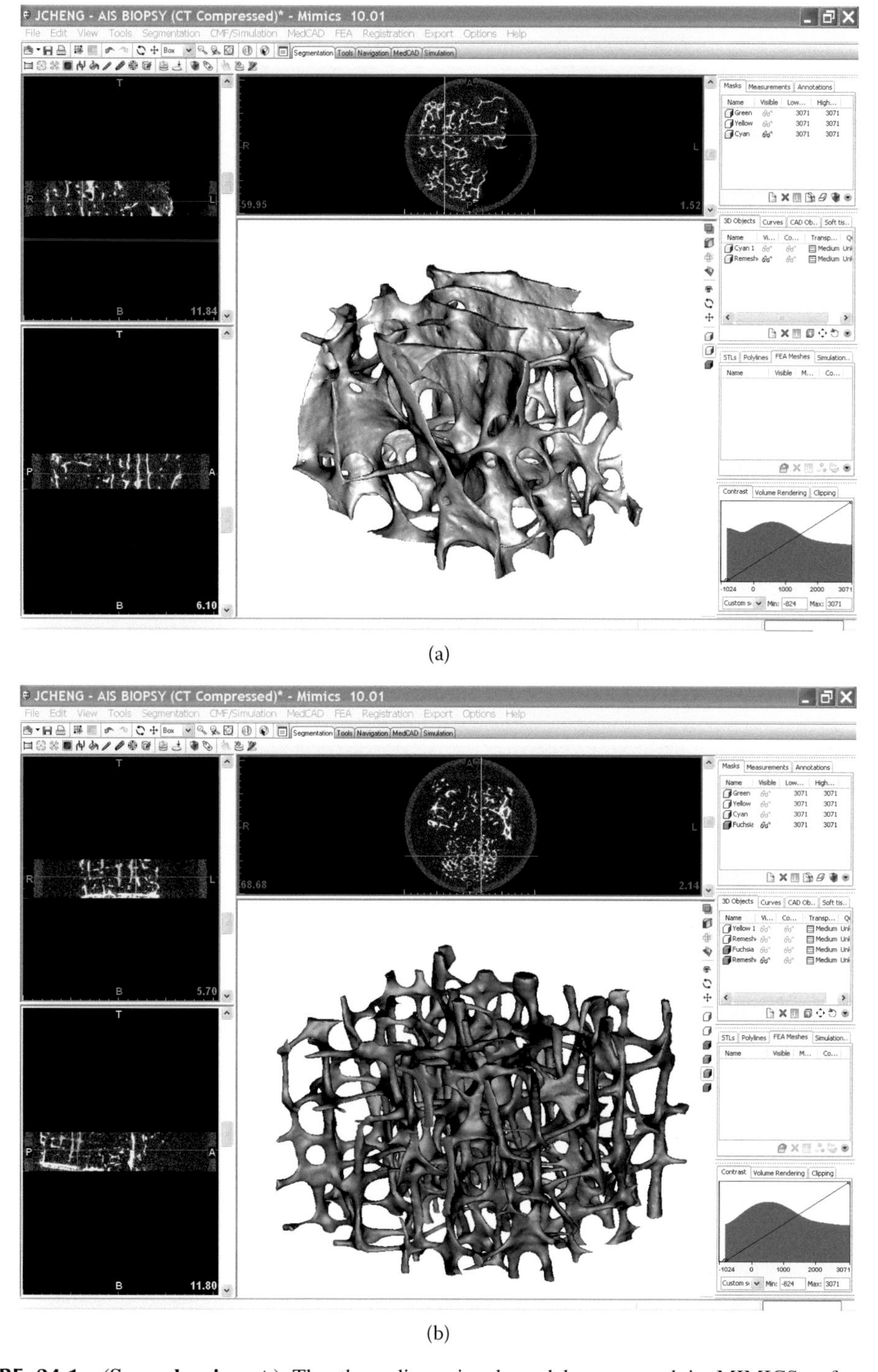


**FIGURE 24.1 (See color insert.)** The three-dimensional models generated in MIMICS software: (a) trabecular specimen A and (b) trabecular specimen B.

plane of the platen, with all other surfaces unconstrained (Gong et al., 2007; Morgan et al., 2003). Figure 24.2 shows the finite element models of the trabecular specimens, with loading and boundary conditions, in compressive loading in the longitudinal direction.

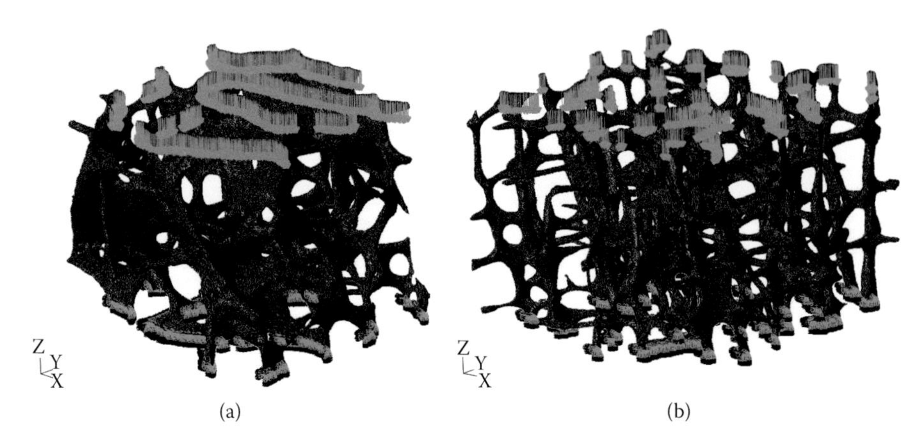


**FIGURE 24.2** **(See color insert.)** Finite element models of the trabecular specimens with loading and boundary conditions in compressive loading conditions in the longitudinal direction, respectively: (a) trabecular specimen A and (b) trabecular specimen B.

Both geometrical nonlinearity and bone tissue material nonlinearity were considered in each analysis. The bilinear tissue level constitutive model in Chapter 9 (Figure 9.1) was used to describe bone tissue material nonlinearity. The four parameters in the model were $E$ = 18 GPa, $E_u$ = 5% $E$, $\varepsilon_t^T = 0.48\%$, and $\varepsilon_c^T = 0.8\%$ (Bayraktar et al., 2004; Rho et al., 1997).

The cast iron plasticity material constitution in ABAQUS was used to describe the elastic-plastic behavior of trabecular bone material with asymmetric yield strength and hardening in tension and compression (Bayraktar et al., 2004; Gong et al., 2011). Linear analyses were done first to determine the loading mode (tension or compression) of each element in each finite element mesh. The maximum principal stress criterion was used to determine the loading mode of each element. Suppose the three principal stresses of an element are $\sigma_1$, $\sigma_2$, and $\sigma_3$. If $|\sigma_1| > |\sigma_3|$, the element is predominantly tensile; if $|\sigma_1| < |\sigma_3|$, the element is predominantly compressive (Schileo et al., 2008). In each nonlinear analysis, the initial apparent yield point was determined using the 0.2% offset method.

## 24.3 RESULTS FROM THE MODEL ANALYSIS

Figure 24.3 shows the apparent stress-strain curves for the two trabecular cubes in compression and tension in the longitudinal and transverse directions. The linear portion of each curve shows the elastic response and the slope gives the apparent Young's modulus in that direction. The 0.2% offset lines were drawn on the figure to determine the initial apparent yield point in each loading case.

Table 24.2 lists the distribution of tissue von Mises stresses of the trabecular specimens and the number of elements that yielded in compression and tension at the apparent yield point in each loading direction. It was found that for both longitudinal and transverse loading, when loaded in tension, very few elements yielded in compression and the majority of the yielding elements were in tension; but when loaded in compression, there were still a considerable amount of elements yielding in tension. Figure 24.4 shows the von Mises stress distributions of the trabecular specimens at the yield point in apparent compressive and tensile loading conditions in the longitudinal and transverse directions, respectively.

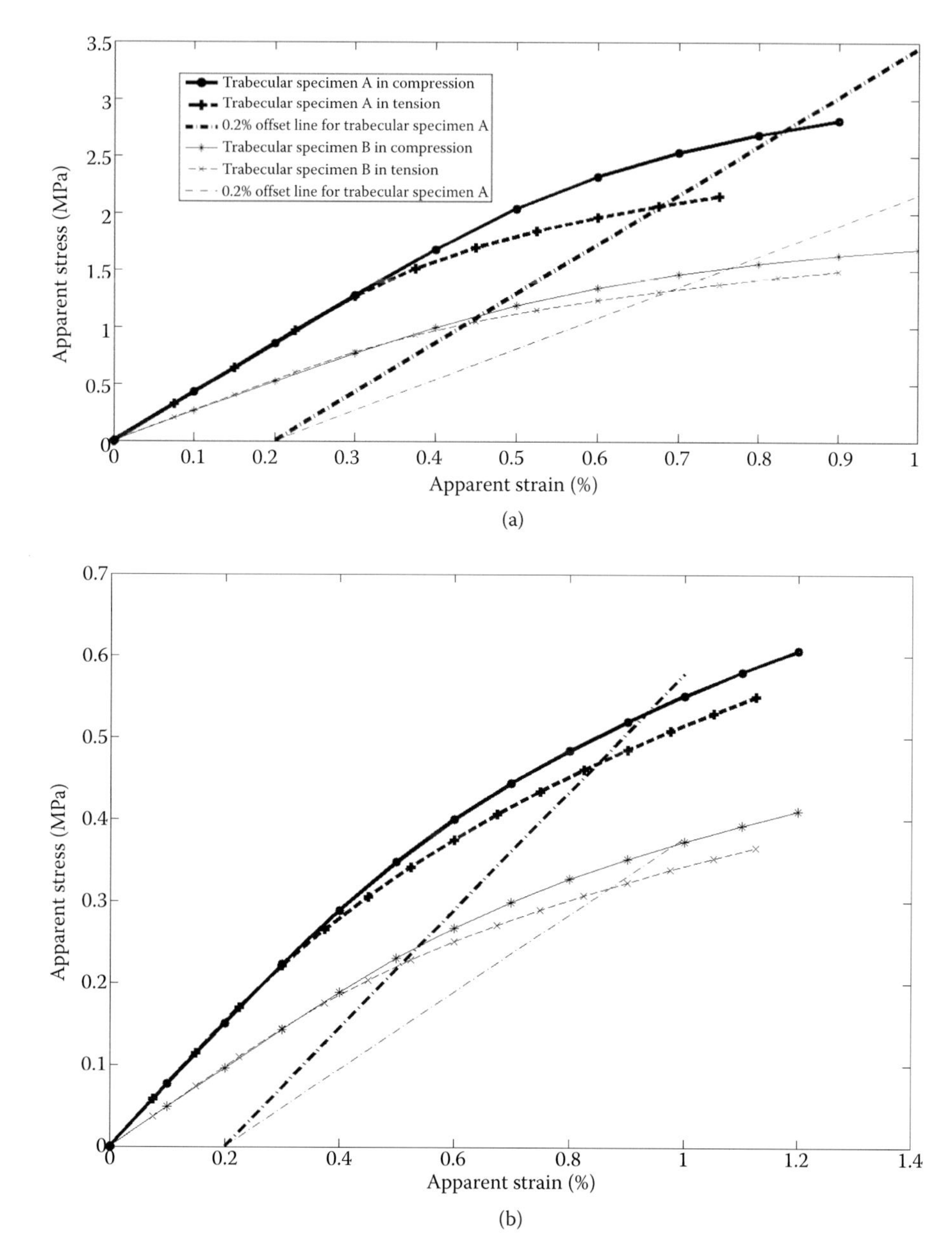


**FIGURE 24.3** The apparent stress-strain curves for the two trabecular cubes in compression and tension in the longitudinal direction, as well as the transverse direction, respectively: (a) in the longitudinal direction and (b) in the transverse direction.

## 24.4 APPLICATIONS OF THE MODEL

Trabecular bone yield behaviors are determined by the tissue-level mechanical parameters and the microstructures, and are direction-related. The two trabecular specimens had distinct micro-architectures, especially regarding the structure model index (SMI). SMI can be used

**TABLE 24.2**
**Distribution of Tissue von Mises Stress, and the Amounts of Tissue Elements Yielded in Compression and Tension at the Apparent Yield Point in Each Loading Condition**

| | | Lontitudinal Direction | | Transverse Direction | |
|---|---|---|---|---|---|
| | | Compression | Tension | Compression | Tension |
| Distribution of tissue von Mises stresses (MPa) (mean±SD) | Trabecular specimen A | 55.41±38.10 | 40.02±26.68 | 24.04±23.52 | 20.76±19.06 |
| | Trabecular specimen B | 44.74±33.95 | 35.29±25.20 | 23.74±20.58 | 19.47±17.53 |
| Amount of tissue elements yielded in compression (%) | Trabecular specimen A | 7.4730 | 0.1928 | 1.0072 | 0.0640 |
| | Trabecular specimen B | 4.1022 | 0.1581 | 0.9128 | 0.0941 |
| Amount of tissue elements yielded in tension (%) | Trabecular specimen A | 2.7220 | 10.6903 | 0.6426 | 2.3876 |
| | Trabecular specimen B | 2.3514 | 2.9208 | 0.8224 | 2.0527 |

to evaluate the structure model type, which relates to the convexity of the structure to a model type (Hilderbrand and Rüegsegger, 1997). Our previous study found that vertebral trabecular bone with different three-dimensional microstructural properties based on their SMI values had different fracture risks (Gong et al., 2006). Trabecular specimen 1 in this study belonged to a low-SMI group, and specimen 2 belonged to a high-SMI group. They underwent different yield behaviors, which can be quantified by apparent yield stress and apparent yield strain at the apparent level, and von Mises stress distribution, the amount of tissue elements yielded in compression as well as in tension at the tissue level. The vertebral body is a weight-bearing bone. The internal trabecular bone represented more superior mechanical properties in the longitudinal direction in comparison with the transverse direction, that is, a higher apparent Young's modulus, yield stress, and tissue von Mises stresses, and more elements yielded in compression and tension at the apparent yield point when compressed in the longitudinal direction, showing more elements performing at the maximum load-bearing capacity. More trabecular specimens need to be investigated for a more detailed relationship between trabecular bone yield behaviors and their microstructures.

In the current study, the hard tissue was highly homogeneous, which can be seen in Figure 24.1. Hence, all the bone tissue material was assumed to have the same bilinear, tissue-level, constitutive behavior. The role of bone material heterogeneity on the apparent and tissue-level yield behaviors cannot be derived accordingly. Trabecular specimens from other locations with hard tissue heterogeneity need to be further investigated.

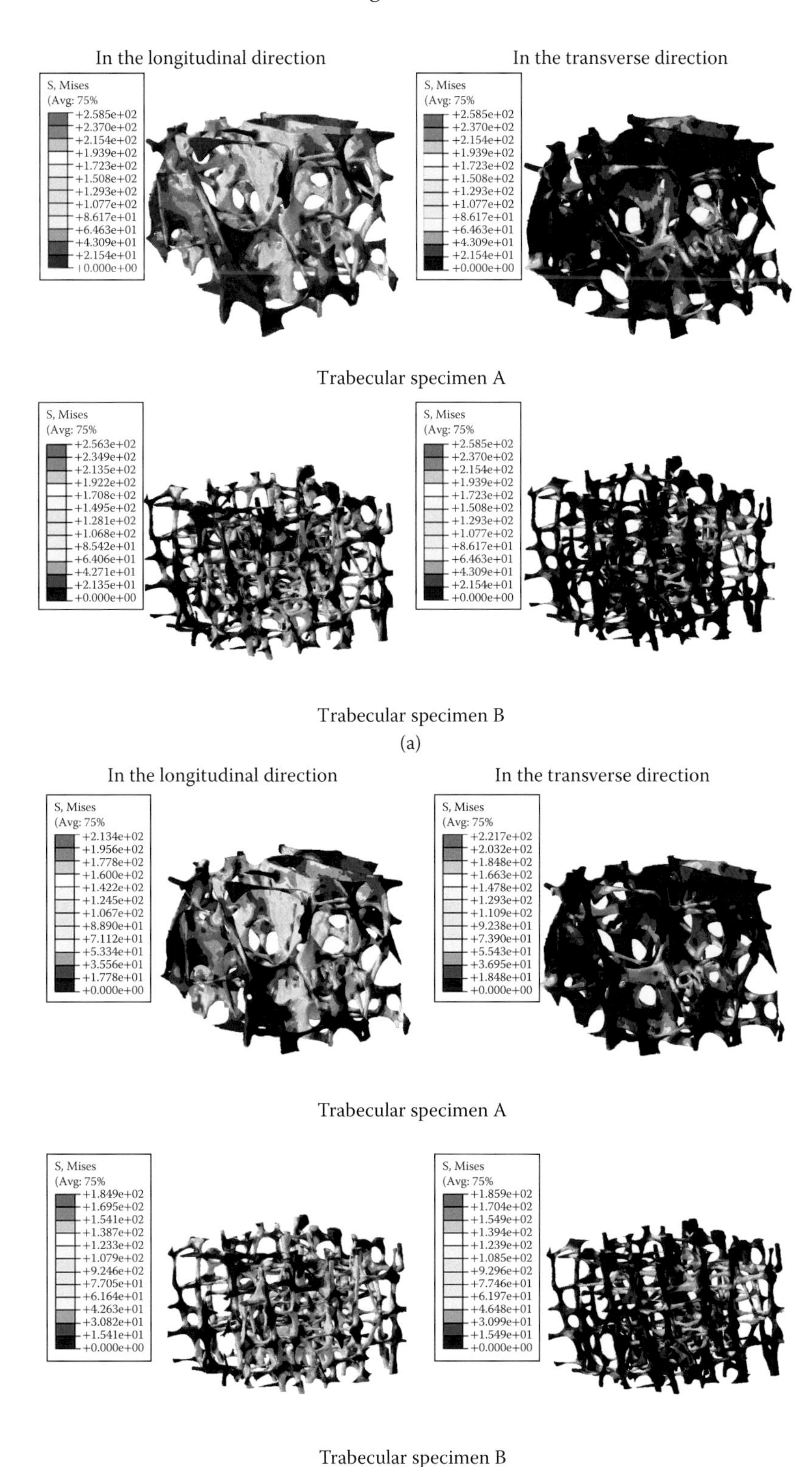


**FIGURE 24.4** **(See color insert.)** von Mises stress distributions of the trabecular specimens at the yield point in apparent compressive and tensile loading conditions in the longitudinal and transverse directions, respectively: (a) compression and (b) tension.

## ACKNOWLEDGMENTS

This work is supported by a grant from the National Natural Science Foundation of China (No. 11120101001, 11322223, 11272273) and the Program for New Century Excellent Talents in University (NCET-12-0024).

## REFERENCES

Bayraktar, H. H., Morgan, E. F., Niebur, G. L., Morris, G. E., Wong, E. K., and T. M. Keaveny. 2004. Comparison of the elastic and yield properties of human femoral trabecular and cortical bone tissue. *J Biomech* 37:27–35.

Gong, H., Zhang, M., and Y. Fan. 2011. Micro-finite element analysis of trabecular bone yield behavior—effects of tissue non-linear material properties. *J Mech Med Biol* 11(3):563–580.

Gong, H., Zhang, M., Qin, L., and Y. J. Hou. 2007. Regional variations in the apparent and tissue level mechanical parameters of vertebral trabecular bone with ageing using micro-finite element analysis. *Ann Biomed Eng* 35(9):1622–1631.

Gong, H., Zhang, M., Qin, L., Lee, K. K. H., Guo, X., and S. Q. Shi. 2006. Regional variations in microstructural properties of vertebral trabeculae with structural groups. *Spine* 31(1):24–32.

Gong, H., Zhang, M., Yeung, H. Y., and L. Qin. 2005. Regional variations in microstructural properties of vertebral trabeculae with ageing. *J Bone Miner Metab* 23(2):174–180.

Hilderbrand, T., and T. Rüegsegger. 1997. Quantification of bone microarchitecture with the structure model index. *Comp Meth Biomech Biomed Eng* 1(1):15–23.

Judex, S., Boyd, S., Qin, Y. X., Turner, S., Ye, K., Muller, R., and C. Rubin. 2003. Adaptations of trabecular bone to low magnitude vibrations result in more uniform stress and strain under load. *Ann Biomed Eng* 31:12–20.

Melton, L. J., Chrischilles, E. A., Cooper, C., Lane, A. W., and B. L. Riggs. 1992. Perspective: How many women have osteoporosis? *J Bone Miner Res* 7:1005–1010.

Morgan, E. F., Bayraktar, H. H., and T. M. Keaveny. 2003. Trabecular bone modulus-density relationships depend on anatomic site. *J Biomech* 36:897–904.

Niebur, G. L., Feldstein, M. J., and T. M. Keaveny. 2002. Biaxial failure behavior of bovine tibial trabecular bone. *J Biomech Eng* 124:699–705.

Niebur, G. L., Feldstein, M. J., Yuen, J. C., Chen, T. J., and T. M. Keaveny. 2000. High-resolution finite element models with tissue strength asymmetry accurately predict failure of trabecular bone. *J Biomech* 33:1575–1583.

Rho, J. Y., Tsui, T. Y., and G. M. Pharr. 1997. Elastic properties of human cortical and trabecular lamellar bone measured by nanoindentation. *Biomaterials* 18:1325–1330.

Schileo, E., Taddei, F., Cristofolini, L., and M. Viceconti. 2008. Subject-specific finite element models implementing a maximum strain criterion are able to estimate failure risk and fracture location on human femurs tested in vitro. *J Biomech* 41:356–367.

Verhulp, E., van Rietbergen, B., Müller, R., and R. Huiskes. 2008. Indirect determination of trabecular bone effective tissue failure properties using micro-finite element simulations. *J Biomech* 41:1479–1485.

# 25 Simulation of Osteoporotic Bone Remodeling

*He Gong, Yubo Fan, and Ming Zhang*

## CONTENTS



## SUMMARY

As a living biological material, bone has an amazing ability to adapt to mechanical load or other biophysical stimuli in terms of bone mass and architecture—an ability that is known as bone functional adaptation. There are two fundamental physiological processes of bone functional adaptation: modeling and remodeling. Remodeling occurs in all in vivo bone tissues as an important bone renewal mechanism. Bone remodeling is performed by groups of osteoclasts and osteoblasts organized into basic multicellular units (BMUs). Osteoporosis is a systematic skeletal disease with a consequent increase in bone fragility and susceptibility to fracture. With osteoporosis, the structural integrity of trabecular bone is impaired and cortical bone becomes more porous and the cortex becomes thinner. Mechanical usage and biological factors are two major determinants of osteoporosis. Computing power has made it possible to quantitatively simulate the bone remodeling process to predict the bone mass and architecture during adaptation to mechanical load and/or biophysical stimuli. In this chapter, the development of a computational simulation of the bone remodeling process at the BMU level in trabecular bone as well as cortical bone will be introduced, and the related osteoporotic processes will be predicted as examples.

## 25.1 INTRODUCTION

The skeleton is made up of cortical bone and trabecular bone. Bone has a functional adaptation ability, that is, it can adapt to mechanical load or other biophysical stimuli in terms of its mass and architecture. There are two fundamental physiological processes of bone functional adaptation: modeling and remodeling. (Frost, 2001). Modeling results in longitudinal bone growth. In this process, continuous bone resorption and formation occur separately at different locations so that the whole bone morphology changes. Remodeling is an important bone renewal mechanism, which is performed by groups of osteoclasts and osteoblasts forming BMUs. Remodeling by BMUs comprises BMU activation, bone resorption by osteoclasts, and bone formation by osteoblasts.

It is possible to quantitatively simulate bone remodeling process to predict the adaptation of bone mass and architecture to mechanical load and/or biophysical stimuli. (Slade et al., 2005; Ritzel et al., 1997). Such simulations can also predict the outcomes of some experimental protocols without the need to physically carry out the experiments. Computational simulation of bone remodeling is a quick and convenient way to quantify the osteoporotic process, which can shorten the research period and significantly reduce costs related to experimental work.

In this chapter, computer simulations of trabecular bone and cortical bone remodeling processes at the BMU level will be developed, and the related osteoporotic processes will be predicted as examples.

## 25.2 DEVELOPMENT OF BONE REMODELING PROCESS SIMULATION MODELS FOR OSTEOPOROSIS

Computational simulation of bone remodeling includes quantitative descriptions of the BMUs, taking into account the birthrate of BMUs, the progression of each BMU through the bone, the resorption and formation of bone performed by each BMU, and variations in the morphological and material properties of the bone tissue, that is, porosity in trabecular bone and the thickness of cortical bone (Hernandez et al., 2001, 2003; Gong et al., 2010).

### 25.2.1 Model Development

#### 25.2.1.1 Computational Simulation of Trabecular Bone Remodeling

The computational simulation of trabecular bone remodeling at the BMU level comprised 8 state variables and 12 constants (see Tables 25.1 and 25.2, respectively), taking into consideration a quantitative description of the contributions of biological factors and their coupling relationships with mechanical load (Gong et al., 2006, 2010).

**TABLE 25.1**
**State Variables in the Computational Simulation of Trabecular Bone Remodeling**

| | State Variable |
|---|---|
| $E$ | Elastic modulus (MPa) |
| $p$ | Porosity |
| $N_R$ | Number of resorbing BMUs (BMUs/mm$^2$) |
| $N_F$ | Number of refilling BMUs (BMUs/mm$^2$) |
| $fa$ | BMU activation frequency (BMUs/mm$^2$/day) |
| $S_A$ | Normalized specific surface area |
| $\varepsilon$ | Strain ($\mu\varepsilon$) |
| $\Phi$ | Mechanical stimulus (Pa) |

**TABLE 25.2**
**Constants in the Computational Simulation of Trabecular Bone Remodeling**

| | Constant | Nominal Values |
|---|---|---|
| $T_r$ | Resorption period (days) | 1 [a] |
| $T_i$ | Reversal period (days) | 4 [a] |
| $T_f$ | Refilling period (days) | 33 [a] |
| $f_{a0}$ | Initial BMU activation frequency (BMUs/mm$^2$/day) | 0.00640 [b] |
| $\Phi_0$ | Initial mechanical stimulus (Pa) | 28.12 |
| $f_{a(\max)}$ | Maximum BMU activation frequency (BMUs/mm$^2$/day) | 0.5 [b] |
| *kbob* | Dose-response coefficient (Pa$^{-1}$) in refilling rate function | 0.31 [c] |
| *kcob* | Dose-response coefficient (Pa) in refilling rate function | 14.06 [c] |
| *kbfa* | Dose-response coefficient (Pa $^{-1}$) in activation frequency function | 0.31 [c] |
| *kcfa* | Dose-response coefficient (Pa) in activation frequency function | 14.06 [c] |
| *Qrmax* | Maximum resorption rate (mm$^2$/day) | 0.031 [d] |
| *Qfmax* | Maximum formation rate (mm$^2$/day) | 9.341×10-4 [e] |

*Note:* The nominal values are for the example in Section 25.2.2.1 about bone loss in the trabecular bone of the rat femur.

[a] Based on Baron et al., *Anatomical Record*, 208, 237–45, 1984.

[b] Based on Hazelwood et al., *Journal of Biomechanics*, 34, 299–308, 2001.

[c] Parametrical sensitivity analyses done for the coefficients to fit the curves within known experimental data ranges. (From Lecoq et al., *Joint Bone Spine*, 73, 189–95, 2006.)

[d] Calculated local resorption rate in Hernandez et al., *Bone*, 32, 357–63, 2003.

[e] Calculated local formation rate in Hernandez et al., *Bone*, 32, 357–63, 2003.

The rate of change of porosity $dP(t) / dt$ was assumed to be a function of the bone resorption rate $Q_r(t)$ and the bone refilling rate $Q_f(t)$ for each BMU and the density of resorbing $N_R(t)$ and refilling $N_F(t)$ of the BMUs/area, as proposed by Hazelwood et al. (2001):

$$\frac{dP(t)}{dt} = Q_r(t)N_R(t) - Q_f(t)N_F(t) \tag{25.1}$$

where the resorption rate $Q_r(t)$ was assumed to be constant, that is, $Q_r(t)=Q_{r\,max}$. The refilling rate $Q_f(t)$ was determined by

$$Q_f(t) = Q_{f\max} / (1 + e^{kbob(kcob - \varphi(t))}) \quad \text{for } \varphi(t) < \varphi_0 \tag{25.2}$$

to account for the reduced refilling on bone surfaces during disuse (Frost, 1998). *kbob* is defined as the slope and *kcob* is the inflection point of the curve. $\phi(t)$ is the mechanical stimulus described by strain energy density (Mullender and Huiskes, 1995):

$$\varphi(t) = \frac{1}{2}E(t)\varepsilon^2(t) \tag{25.3}$$

where $E(t)$ is elastic modulus, $\varepsilon(t)$ is mechanical strain, and $\phi_0$ is the mechanical stimulus at equilibrium:

$$Q_f(t) = Q_{f\max} \quad \text{for } \varphi(t) \geq \varphi_0 \tag{25.4}$$

$N_R(t)$ and $N_F(t)$ are the populations of resorbing BMUs and refilling BMUs, which are calculated as

$$N_R(t) = \int_{t-T_r}^{t} f_a(t')\,dt' \tag{25.5}$$

$$N_F(t) = \int_{t-(T_r+T_i+T_f)}^{t-(T_r+T_i)} f_a(t')\,dt' \tag{25.6}$$

where $f_a(t)$ is BMU activation frequency.

A BMU activation frequency threshold *akfa*(*t*) was introduced by Gong et al. (2006) to obtain a mechanical-biological factor coupled BMU activation frequency function:

$$f_a(t) = \frac{f_{a(\max)}}{1 + e^{kbfa\left(\frac{\varphi(t)}{akfa(t)} - kcfa\right)}} S_A \tag{25.7}$$

where $\phi(t)$ is the mechanical stimulus, $f_{a(\max)} = 0.5$ BMUs/mm$^2$/day (Fyhrie and Carter, 1986; Hazelwood et al., 2001), and *kbfa* and *kcfa* are coefficients defining the slope and inflection point of the curve. Sensitivity analyses were done for the coefficients in these functions to fit the curves within known experimental data ranges (Lecoq et al., 2006). *akfa*(*t*) is the BMU activation threshold and $S_A$ is the specific surface area, as mentioned by Hazelwood et al. (2001).

In this chapter, bone loss in trabecular bone of a rat femur due to mechanical disuse and/or estrogen deficiency was taken as an example to illustrate the implementation of this method. The elastic modulus was found to be 7 GPa for cortical bone in the rat femur and 0.9 GPa for cancellous bone (Westerlind et al., 1997; Ferretti et al., 1993). Assuming a typical bone volume fraction of 0.28 for cancellous bone and 1.0 for cortical bone (Westerlind et al., 1997), and a linear relationship between porosity $P(t)$ and elastic modulus $E(t)$ (Keaveny et al., 2001), the elastic modulus can be expressed as

$$E(t) = (8472 \times (1 - P(t)) - 1472) \text{ MPa} \tag{25.8}$$

In this analysis, the equilibrium strain level was set to 250 $\mu\varepsilon$ for cancellous bone. The system was initiated without any active BMUs in an equilibrium state at $t = 0$ before beginning the experiment. The model was given an initial porosity $P(0) = 72\%$ (Westerlind et al., 1997). The cross-sectional area for the load was 6 mm$^2$. Hence, a compressive force of $F(0) = \sigma \cdot A = E \cdot \varepsilon \cdot A = 1.35\,N$ can provide the mechanical stimulus for the bone resorption and formation to be in equilibrium.

#### 25.2.1.2 Computational Simulation of Cortical Bone Remodeling

A computational simulation of cortical endosteal surface remodeling was developed at the BMU level (Gong and Zhang, 2010). Six state variables and nine constants included in the model are listed in Tables 25.3 and 25.4, respectively. The remodeling analysis was performed on a representative rectangular slice of the cross-section of the cortical bone volume, as shown schematically in Figure 25.1.

An imbalance between bone resorption and refilling leads to change in cortical volume. The rate of change in cortical volume was assumed to be a function of the bone resorption rate ($Q_r(t)$) and bone refilling rate ($Q_f(t)$) for each BMU, and the density of resorbing and refilling BMUs/area ($N_R(x, t)$ and $N_F(x, t)$, respectively):

$$\frac{d\dfrac{h(t)\cdot l}{h_0 \cdot l}}{dt} = Q_r(t)N_R(t) - Q_f(t)N_F(t) \tag{25.9}$$

where the resorption rate $Q_r(t)$ and the refilling rate $Q_f(t)$ were assumed to be linear in time: $Q_r(t) = \dfrac{A}{T_r}$ and $Q_f(t) = \dfrac{A}{T_f}$ with $A$ representing the area of bone resorbed by each BMU. $h(t)$ was the cortical thickness at time $t$, $l$ was the length of the representative rectangular slice, and $h_0$ was the initial cortical thickness, as shown in Figure 25.1.

**TABLE 25.3**
**State Variables in the Computational Simulation of Cortical Bone Remodeling**

| | State Variable |
|---|---|
| $h$ | Cortical thickness (mm) |
| $N_R$ | Number of resorbing BMUs (BMUs/mm$^2$) |
| $N_F$ | Number of refilling BMUs (BMUs/mm$^2$) |
| $fa$ | BMU activation frequency (BMUs/mm$^2$/day) |
| $\varepsilon$ | Strain ($\mu\varepsilon$) |
| $\Phi$ | Mechanical stimulus (MPa) |

**TABLE 25.4**
**Constants in the Computational Simulation of Cortical Bone Remodeling**

| | Constant | Nominal Values |
|---|---|---|
| $A$ | Cross-sectional area of each BMU (mm$^2$) | $2.84\times10^{-2}$ [a] |
| $T_r$ | Resorption period (days) | 24 [b] |
| $T_i$ | Reversal period (days) | 8 [b] |
| $T_f$ | Refilling period (days) | 64 [b] |
| $f_{a(max)}$ | Maximum BMU activation frequency (BMUs/mm$^2$/day) | 0.299 [c] |
| $kb$ | Dose-response coefficient (Pa$^{-1}$) | 0.005759 [c] |
| $kc$ | Dose-response coefficient (Pa) | 1129.493 [c] |
| $f_{a0}$ | Maximum BMU activation frequency (BMUs/mm$^2$/day) | $4.8768\times10^{-4}$ [d] |
| $\Phi_0$ | Initial mechanical stimulus (Pa) | 2258.986 |

*Note:* The nominal values are for the example in Section 25.2.2.2 about bone loss simulations of the cortical bone of the femur and tibia in paralysis state measured in Eser et al., Bone, 34, 869–80, 2004.

[a] Based on Parfitt, *Bone Histomorphometry: Techniques and Interpretation*, 143–223, CRC Press, Boca Raton, FL, 1983.

[b] From Hazelwood et al., *Journal of Biomechanics*, 34, 299–308, 2001, based on several histomorphometric studies.

[c] Parametrical sensitivity analyses were done for the coefficients to fit the curves within known experimental data ranges. (Eser et al., *Bone*, 34, 869–80, 2004.)

[d] Based on Equation 25.3.

Hence

$$\frac{dh(t)}{dt} = h_0(Q_r(t)N_R(t) - Q_f(t)N_F(t)) \tag{25.10}$$

$N_R(t)$ and $N_F(t)$ are the populations of resorbing BMUs and refilling BMUs, respectively, and were calculated according to Equations 25.5 and 25.6, respectively, and $f_a(t)$ is the BMU activation frequency. The relationship between BMU activation frequency and mechanical load was assumed to be sigmoidal, similar to the response found in pharmacological applications (Hazelwood et al., 2001):

$$f_a(t) = \frac{f_{a(\max)}}{1 + e^{kb(\varphi(t)-kc)}} \tag{25.11}$$

Mechanical stimulus $\phi(t)$ was described by strain energy density, having the same form as Equation 25.3.

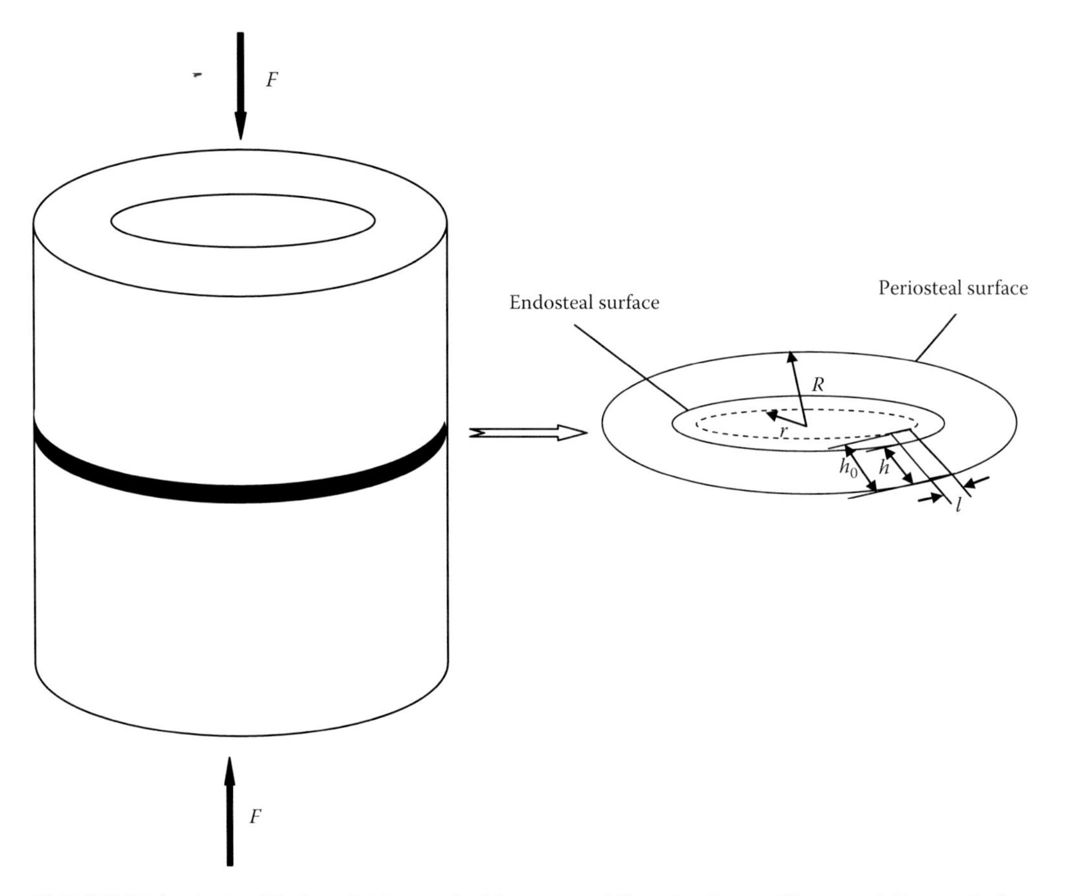


**FIGURE 25.1** A simplified model for cortical bone remodeling simulation. The remodeling analysis was performed on a representative rectangular slice of the cross section of the cortical bone volume, as shown schematically here. Bone material was assigned as linearly elastic and isotropic. Axial compressive loads $F$ were placed on the cortical bone volume. For simplification, the cortical bone volume was assumed to be cylindrical. $r$: Initial endosteal radius; $R$: periosteal radius; $l$: length of the representative rectangular slice; $h$: cortical thickness; and $h_0$: initial cortical thickness with $h_0=R-r$.

### 25.2.2 Model Validation

#### 25.2.2.1 Model Validation for the Computational Simulation of Trabecular Bone Remodeling

The computational simulation algorithm of trabecular bone remodeling (Equations 25.1–25.8) was validated through an animal experiment performed by Lecoq et al. (2006). Thirty-six 12-week old female Wistar rats were randomized into groups with (1) bilateral surgical ovariectomy without tail suspension for comparison with the estrogen deficiency simulation model; (2) bilateral surgical ovariectomy with tail suspension for comparison with the dual-factor simulation model; (3) sham surgery with tail suspension for comparison with the mechanical unloading simulation model; or (4) sham surgery without tail suspension as the control group.

Time-dependent computer simulations using Equations 25.1–25.8 of the bone remodeling process were performed on a representative cross section of 6 $mm^2$ trabecular bone in the distal femoral metaphysis of the rats from $t = 0$ day to $t = 30$ day to model the experimental time.

The nominal values of the constants in the model are listed in Table 25.2. Gong et al. (2006) suggested that the BMU activation threshold increased due to estrogen deficiency and the mechanical loading decreased in the disuse state. In this analysis, the changes of the BMU activation threshold and mechanical loading were assumed as exponential functions. The coefficients were chosen to make the simulated changes of BMD correspond with the experimental data from ovariectomy and hind-limb suspension. The combined effects of changes of the BMU activation threshold and mechanical loading were simulated by assuming that the values were the same when both factors changed as when each factor changed alone.

#### 25.2.2.2 Model Validation for the Computational Simulation of Cortical Bone Remodeling

The computational simulation algorithm of cortical bone remodeling (Equations 25.3, 25.5, 25.6, 25.10, and 25.11) was validated by the pQCT data showing the relationship between the duration of paralysis and bone structure of spinal-cord-injured patients by Eser et al. (2004) as an example of mechanical disuse. Eighty-nine men with injury to the motor complex of the spinal cord (24 tetraplegics and 65 paraplegics) with a duration of paralysis between 2 months and 50 years were included. The age range was 41.5 ± 14.2 for all subjects. The reference group comprised 21 healthy able-bodied men of the same age range.

The remodeling behaviors of cortical bone in the femur and tibia were simulated. The representative cross section of the femur cortical bone was located at 25% of the total bone length from the distal end for the femur, and at 38% for the tibia, mirroring the locations of the diaphyseal scans in Eser et al. (2004). For simplification, the cortical bone volume was assumed to be cylindrical and the endosteal surface and periosteal surface were assumed to be concentric circles (Figure 25.1). The geometrical and mechanical conditions of the two cortical samples are shown in Table 25.5.

**TABLE 25.5**
**Geometrical and Mechanical Properties of the Representative Rectangular Slices for Femur and Tibia Cortical Models**

| | Tibia | Femur | Notes |
|---|---|---|---|
| Initial cortical thickness $h_0$ (mm) | 5.84 | 3.33 | From Eser et al. (2004). |
| Length $l$ (mm) | 1 | 1.64 | The length of the tibia model was chosen to guarantee a small slice and that of the femur model was chosen such that both models occupied the same percentage of total cortical area, i.e., approximately 1.69%. |
| Initial area $A_0$ of the model (mm$^2$) | 5.84 | 5.46 | From equation $A_0 = h_0 \cdot l$. |
| Equilibrium strain $\varepsilon_0$ ($\mu$ ε) | 500 | | From Beaupré et al. (1990) and Turner et al. (1997). |
| Porosity $p_0$ (%) | 4.43 | | From Hazelwood et al. (2001). Eser et al. (2004), found no decrease in cortical BMD of the diaphyses with time after injury. Hence, the cortical porosity in this analysis was held constant. |
| Elastic modulus $E$ (MPa) | 18000 | | From the relationship between elastic modulus and porosity of cortical bone proposed by Currey (1988): $E = (23440 \times (1 - p_0)^{5.74}) MPa$. |
| Equilibrium load $F_0$ (N) | 52.770 | 49.336 | From equation $F_0 = \varepsilon_0 \cdot A_0 \cdot E$. |

The mechanical load in paralytic disuse was assumed to decrease exponentially with time:

$$F(t) = F_0 \cdot e^{-t \cdot a} \tag{25.12}$$

with $t$ = time (year), and $a$ as a decreasing coefficient, which was chosen such that the simulation results could be in accordance with experimental data ranges (Eser et al., 2004). $F_0$ was the equilibrium load.

Using a time-dependent approach (Equations 25.3, 25.5, 25.6, 25.10, 25.11, and 25.12), computer simulations of cortical endosteal remodeling processes in the femur, as well as in the tibia, were performed from $t$ = 0 days to $t$ = 50 years to cover the course of the clinical investigation.

## 25.3 RESULTS FROM THE MODEL ANALYSIS

### 25.3.1 Results for Trabecular Bone Loss Associated with Mechanical Disuse and Estrogen Deficiency

In the animal experiment by Lecoq et al. (2006), the bone mineral density of the distal femoral metaphysis was measured by dual x-ray absorptiometry in all animals on days 0, 7, 14, and 30. The differences between the experimental groups and the control groups were compared with the simulation results.

Figure 25.2 shows the predicted bone losses due to mechanical unloading, estrogen deficiency, or both over the 30-day experimental period. The data points were calculated from a comparison of the experimental groups and control groups in Lecoq et al. (2006). The simulated bone loss patterns due to mechanical disuse, estrogen deficiency, or both all corresponded well with the experimental observations.

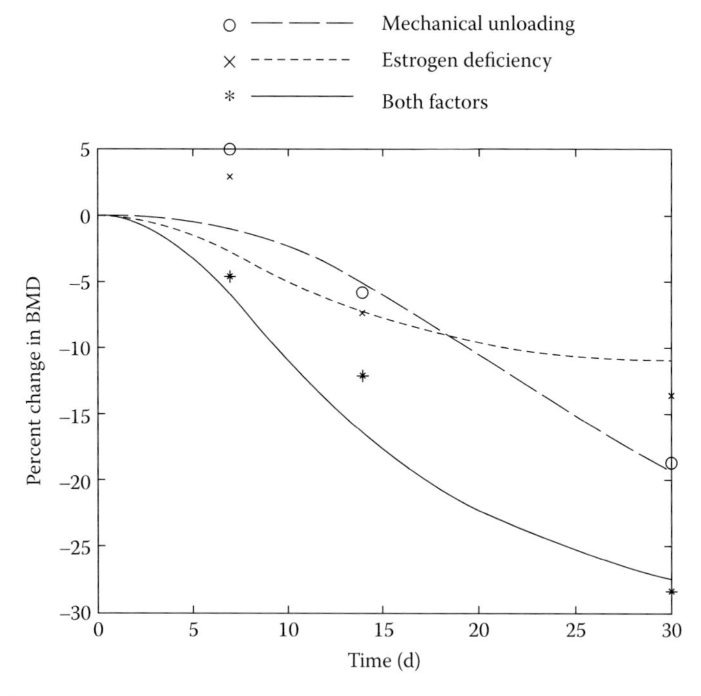


**FIGURE 25.2** Simulation results for the percent change in BMD due to mechanical disuse, estrogen deficiency, or both. (Data from Lecoq et al., *Joint Bone Spine*, 198–95, 2006.)

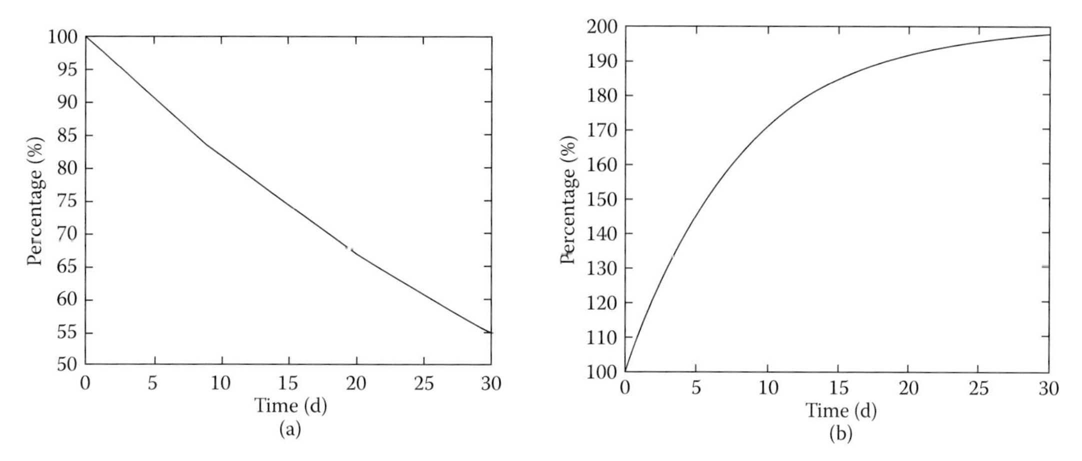


**FIGURE 25.3** Percentage changes in the mechanical loading and BMU activation threshold in the simulations relative to their magnitudes at equilibrium. (a) Mechanical loading and (b) BMU activation threshold.

Figure 25.3 shows the relative percentage changes of the mechanical loading and BMU activation threshold in the simulations relative to their magnitudes at equilibrium. The mechanical loading decreased to 54.59% at the end of the experiment and followed the equation $F(t) = 2.42^{-t/44.0}F(0) \quad 0 \le t \le 30$ days. The BMU activation threshold increased to 197.55% and followed the equation $akfa(t) = 2.0 - 1.6^{-t/3.8} \quad 0 \le t \le 30$ days.

### 25.3.2 Results for Cortical Bone Loss Associated with Mechanical Disuse

Figure 25.4 shows the numerical outcomes of the cortical thickness of the femur and tibia models. The simulated cortical bone thicknesses for both models were consistent with the clinical data measured by Eser et al. (2004). The correlation coefficients between the clinical data points and the simulated values at the same time points were $R^2 = 0.43$ ($p < .001$) for the femur model, and $R^2 = 0.33$ ($p < .001$) for the tibia model. The simulated steady state cortical thicknesses in both models were close to the clinical measurements. For the femur model, the simulated steady state cortical thickness was 2.038 mm and 2.17 mm according to the clinical data. For the tibia model, the simulated steady state value was 4.134 mm and 3.94 mm from the clinical data.

Figure 25.5 shows the changes of the mechanical load for both models. The mechanical load in the tibia model followed the equation $F(t) = F_{0\,tibia} \cdot e^{-0.2t}$ with $t$ = time (year) and that in the femur model followed the equation $F(t) = F_{0\,femur} \cdot e^{-0.3t}$ with $t$ = time (year). The values of $F_{0\,tibia}$ and $F_{0\,femur}$ are listed in Table 25.5.

## 25.4 APPLICATIONS OF THE MODEL

Mechanical disuse and estrogen deficiency are two main causes of osteoporosis. The human musculoskeletal system has evolved under the continuous influence of Earth's gravity. Removal of gravity during long-duration space flight results in a loss of homeostasis in the musculoskeletal system. In the general population, skeletal muscle forces decrease significantly by 20%–40% with aging (Frost 1999; Polla et al., 2004). Loads on bones generate strain-dependent signals (Martin, 2000; Jones et al., 1991). Frost (1983) used the minimum effective strain (MES) as a threshold or mechanical set point to describe bone's adaptation to its mechanical environment. When dynamic strains stay below a lower remodeling threshold range (MESr), "disuse-mode" remodeling removes bone next to

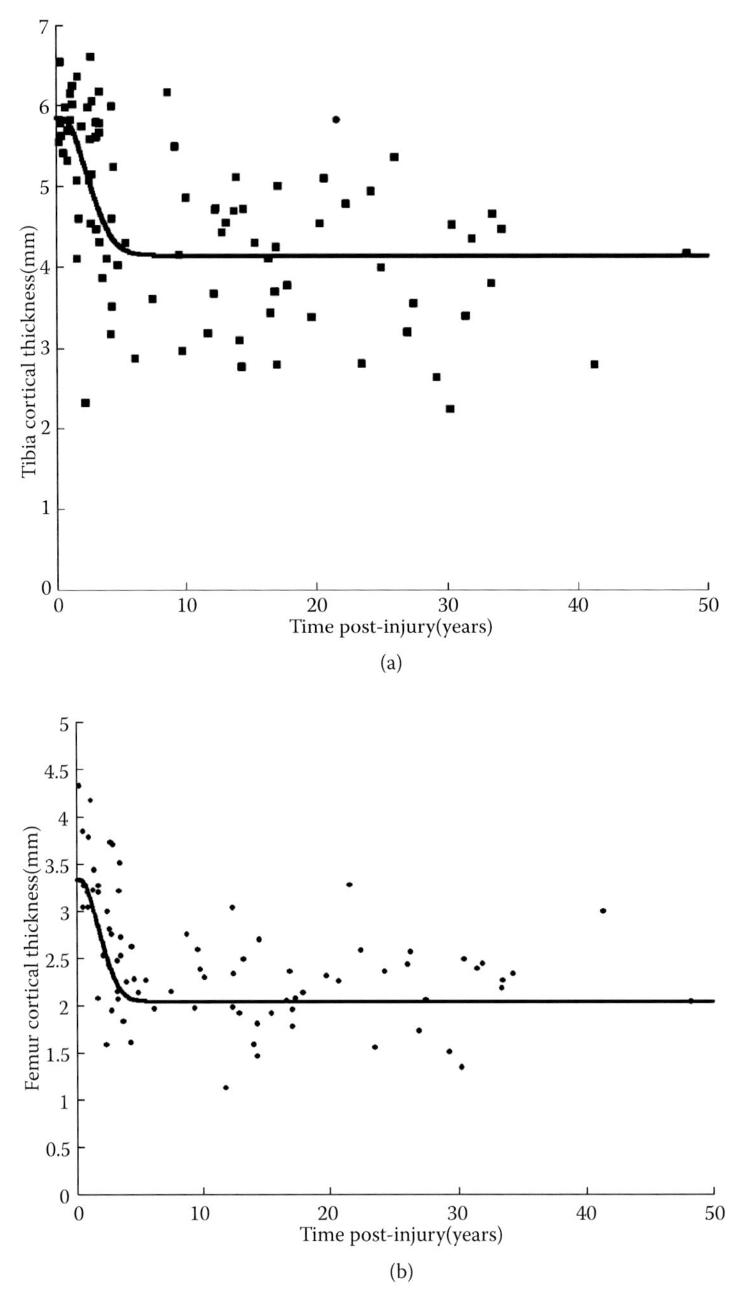


**FIGURE 25.4** Simulation results of cortical thickness of tibia and femur models. The curves representing cortical thickness are the simulation outcomes. (a) Tibia model and (b) femur model. (Data from Eser et al., *Bone*, 34, 869-80, 2004.

marrow. This removal causes a "disuse-pattern" osteopenia characterized by widening of the marrow cavities and thinning of cortical bone, as well as enlarged pores and thinning and fracture of the trabeculae of cancellous bone. This is the physiological basis for our computational simulations of trabecular and cortical osteoporosis induced by mechanical disuse.

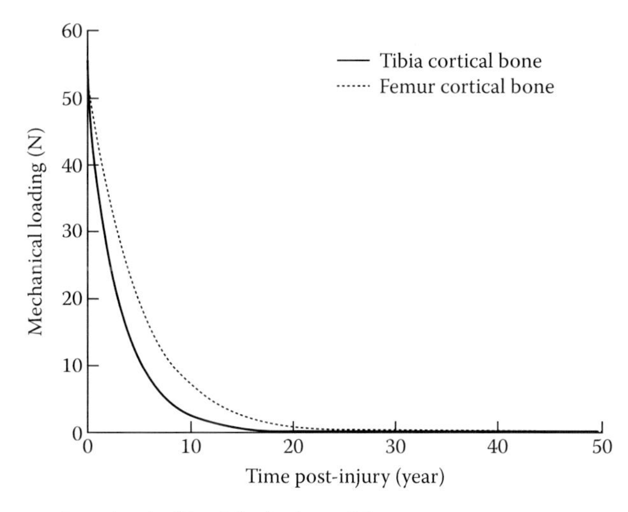


**FIGURE 25.5** Changes of mechanical load for both models.

Biological factors such as aging, menopause, and drug treatment can modulate the system by changing the thresholds or changing the intensities of certain signals (Li et al., 2001). In the bone remodeling system, the contributions of mechanical factors and biological factors are coupled.

Consideration for changes in the mechanical environment, in biological factors, and in particular their coupling relationships may offer a deeper understanding of osteoporosis. Because of the discrepancies in different individuals, this simulation method can be used to predict bone loss patterns for each individual according to his or her personal characteristics, for example mechanical environment and biological conditions, such as the level of physical activity, gender, age, hormone levels, and so on. In addition, if the change of bone mass could be obtained, and changes in one of the biological or mechanical factors were known, the simulation method developed may also be used to predict the changing patterns of the other. For example, age-related bone loss can be simulated when the age-related muscle strength is available (e.g., Brooks and Faulkner, 1994). The effects of mechanical loading can also be simulated given the bone loss patterns (Gong et al., 2006). These parameters are important for evaluating the biological and mechanical status of bone, and may help to provide more insight into the mechanism of osteoporosis and identify improved osteoporosis treatment and prevention.

## ACKNOWLEDGMENTS

This work is supported by a grant from the National Natural Science Foundation of China (No. 11120101001, 11322223, 11272273) and the Program for New Century Excellent Talents in University (NCET-12-0024).

## REFERENCES

Baron, R., Tross, R., and A. Vignery. 1984. Evidence of sequential remodeling in rat trabecular bone: morphology dynamic histomorphometry and changes during skeletal maturation. *Anatomical Record* 208:137–45.

Beaupré, G. S., Orr, T. E., and D. R. Carter. 1990. An approach for time-dependent bone modeling and remodeling—application: a preliminary remodeling simulation. *J Orthop Res* 8:662–70.

Brooks, S. V., and J. A. Faulkner. 1994. Skeletal muscle weakness in old age: underlying mechanisms. *Medical Science of Sports and Exercise* 26:432–39.

Currey, J. D. 1988. The effect of porosity and mineral content on the Young's modulus of elasticity of compact bone. *J Biomech* 21:131–39.

Eser, P., Frotzler, A., Zehnder, Y., Wick, L., Knecht, H., Denoth, J., and H. Schiessl. 2004. Relationship between the duration of paralysis and bone structure: a pQCT study of spinal cord injured individuals. *Bone* 34:869–80.

Ferretti, J. L., Capozza, R. F., Mondelo, N., Montuori, E., and J. R. Zanchetta 1993. Determination of femur structural properties by geometric and material variables as a function of body weight in rats. Evidence of a sexual dimorphism. *Bone* 14:265–70.

Frost, H. M. 1983. A determinant of bone architecture: the minimum effective strain. *Clinical Orthopaedics & Related Research* 175:286–92.

Frost, H. M. 1998. On rho, a marrow mediator, and estrogen: their roles in bone strength and "mass" in human females, osteopenias, and osteoporoses—insights from a new paradigm. *Journal of Bone and Mineral Metabolism* 16:113–23.

Frost, H. M. 1999. Why do bone strength and "mass" in aging adults become unresponsive to vigorous exercise? Insights of the Utah paradigm. *Journal of Bone and Mineral Metabolism* 17:90–7.

Frost, H. M. 2001. Seeking genetic causes of "osteoporosis": insights of the Utah paradigm of skeletal physiology. *Bone* 29:407–12

Fyhrie, D. P., and D. R. Carter. 1986. A unifying principle relating stress to trabecular bone morphology. *Journal of Orthopedic Research* 4:304–17.

Gong, H., and M. Zhang. 2010. A computational model for cortical endosteal surface remodeling induced by mechanical disuse. *Molecular and Cellular Biomechanics* 7(1):1–11.

Gong, H., Zhang, M., Zhang, H., Zhu, D., and L. Yang. 2006. Theoretical analysis of contributions of disuse, basic multicellular unit activation threshold, and osteoblastic formation threshold to changes in bone mineral density at menopause. *Journal of Bone and Mineral Metabolism* 24:386–94.

Gong, H., Zhu, D., Zhang, M., and X. Zhang. 2010. Computational model for the underlying mechanisms regulating bone loss by mechanical unloading and estrogen deficiency. *Tsinghua Science and Technology* 15(5):540–46.

Hazelwood, S. J., Martin, R. B., Rashid M. M., and J. J. Rodrigo. 2001. A mechanistic model for internal bone remodeling exhibits different dynamic responses in disuse and overload. *Journal of Biomechanics* 34:299–308.

Hernandez, C. J., Beaupre, G. S., and D. R. Carter. 2003. A theoretical analysis of the changes in basic multicellular unit activity at menopause. *Bone* 32:357–63.

Hernandez, C. J., Beaupre, G. S., R. Marcus et al. 2001. A theoretical analysis of the contributions of remodeling space, mineralization, and bone balance to changes in bone mineral density during alendronate treatment. *Bone* 29(6):511–16.

Jones, D. B., Nolte, H., Scholubbers, J. G., Turner, E., and D. Veltel. 1991. Biochemical signal transduction of mechanical strain in osteoblast-like cells. *Biomaterials* 12:101–4.

Keaveny, T. M., Morgan, E. F., Niebur, G. L., and O. C. Yeh. 2001. Biomechanics of trabecular bone. *Annual Review of Biomedical Engineering* 3:307–33.

Lecoq, B., Potrel-Burgot, C., Granier, P., Sabatier, J. P., and C. Marcelli. 2006. Comparison of bone loss induced in female rats by hindlimb unloading, ovariectomy, or both. *Joint Bone Spine* 73:189–95.

Li, X. J., Frost, H. M., Webster, S. S. J., and H. Z. Ke. 2001. The update review of basic bone biology (II). *Chinese Journal of Osteoporosis* 7(3):253–61.

Martin, R. B. 2000. Towards a unifying theory of bone remodeling. *Bone* 26:1–6.

Mullender, M. G., and R. Huiskes. 1995. Proposal for the regulatory mechanism of Wolff's Law. *Journal of Orthopedic Research* 13:503–12.

Parfitt, A. M. 1983. The physiologic and clinical significance of bone histomorphometric data. In *Bone Histomorphometry: Techniques and Interpretation*, ed. R. R. Recker, 143–223. Boca Raton, FL: CRC Press.

Polla, B., D'Antona, G., Bottinelli, R. and C. Reggiani. 2004. Respiratory muscle fibers: specialization and plasticity. *Thorax*, 59(9):808–17.

Ritzel, H., Amling, M., Pösl, M., Hahn, M., and G. Delling. 1997. The thickness of human vertebral cortical bone and its changes in aging and osteoporosis: a histomorphometric analysis of the complete spinal column from thirty-seven autopsy specimens. *J Bone Miner Res* 12:89–95.

Slade, J. M., Bickel, S., Modlesky, C. M., Majumdar, S., and G. A. Dudley. 2005. Trabecular bone is more deteriorated in spinal cord injured versus estrogen-free postmenopausal women. *Osteoporosis Int* 16:263–72.

Turner, C. H., Anne, V., and R. M. V. Pidaparti. 1997. A uniform strain criterion for trabecular bone adaptation: do continuum-level strain gradients drive adaptation? *J Biomech* 30:555–63.

Westerlind, K. C., Wronski, T. J., Ritman. E. L., Luo, Z. P., An, K. N., Bell, N. H., and R. T. Turner 1997. Estrogen regulates the rate of bone turnover but bone balance in ovariectomized rats is modulated by prevailing mechanical strain. *Proceedings of National Academy of Science USA* 94:4199–204.

# Index

## D



## E

## G



## H



## I

## J



## K



## L

## M

## N



## O

## P



## Q



## R

## T



## U



## V

## W



## Y